Photoprocesses in Transition Metal Complexes, Biosystems and Other Molecules. Experiment and Theory

NATO ASI Series

Advanced Science Institutes Series

A Series presenting the results of activities sponsored by the NATO Science Committee, which aims at the dissemination of advanced scientific and technological knowledge, with a view to strengthening links between scientific communities.

The Series is published by an international board of publishers in conjunction with the NATO Scientific Affairs Division

A Life Sciences **B Physics**	Plenum Publishing Corporation London and New York
C Mathematical and Physical Sciences **D Behavioural and Social Sciences** **E Applied Sciences**	Kluwer Academic Publishers Dordrecht, Boston and London
F Computer and Systems Sciences **G Ecological Sciences** **H Cell Biology** **I Global Environmental Change**	Springer-Verlag Berlin, Heidelberg, New York, London, Paris and Tokyo

NATO-PCO-DATA BASE

The electronic index to the NATO ASI Series provides full bibliographical references (with keywords and/or abstracts) to more than 30000 contributions from international scientists published in all sections of the NATO ASI Series.
Access to the NATO-PCO-DATA BASE is possible in two ways:

– via online FILE 128 (NATO-PCO-DATA BASE) hosted by ESRIN, Via Galileo Galilei, I-00044 Frascati, Italy.

– via CD-ROM "NATO-PCO-DATA BASE" with user-friendly retrieval software in English, French and German (© WTV GmbH and DATAWARE Technologies Inc. 1989).

The CD-ROM can be ordered through any member of the Board of Publishers or through NATO-PCO, Overijse, Belgium.

Series C: Mathematical and Physical Sciences - Vol. 376

Photoprocesses in Transition Metal Complexes, Biosystems and Other Molecules. Experiment and Theory

edited by

Elise Kochanski
UPR 139 du CNRS,
Université Louis Pasteur,
Strasbourg, France

SPRINGER SCIENCE+BUSINESS MEDIA, B.V.

Proceedings of the NATO Advanced Study Institute on
Photoprocesses in Transition Metal Complexes, Biosystems and Other Molecules.
Experiment and Theory
Aussois, France
September 1–13, 1991

ISBN 978-0-7923-1936-8 ISBN 978-94-011-2698-4 (eBook)
DOI 10.1007/978-94-011-2698-4

Originally published by Kluwer Academic Publishers in 1992

TABLE OF CONTENTS

INTERNATIONAL ORGANIZING COMMITTEE

E. Kochanski (Director)
Laboratoire de Chimie Théorique
Institut de Chimie, Université Louis Pasteur
1, rue Blaise Pascal, BP 296
67008 Strasbourg, France

V. Balzani
Universita degli studi di Bologna
Dipartimento di Chimica "G. Ciamician"
Via Selmi2
40126 Bologna, Italy

A. Riera
Departamento de Quimica C XIV
Universidad Autonoma de Madrid
Ciudad Universitaria de Canto Blanco
28049 Madrid, Spain

J.P. Sauvage
Laboratoire de Chimie Organo-Minérale
Institut de Chimie, Université Louis Pasteur
1, rue Blaise Pascal, BP 296
67008 Strasbourg, France

R. Voltz
Centre de Recherches Nucléaires
67037 Strasbourg, France

LOCAL COMMITTEE

Chambron Jean-Claude
Laboratoire de Chimie Organo-Minérale
Institut de Chimie, Université Louis Pasteur
1, rue Blaise Pascal, BP 296
67008 Strasbourg, France

Coudret Christophe
Laboratoire de Chimie Organo-Minérale
Institut de Chimie, Université Louis Pasteur
1, rue Blaise Pascal, BP 296
67008 Strasbourg, France

Daniel Chantal
Laboratoire de Chimie Quantique
Institut Le Bel, Université Louis Pasteur
4, rue Blaise Pascal
67000 Strasbourg, France

Heitz Valérie
Laboratoire de Chimie Organo-Minérale
Institut de Chimie, Université Louis Pasteur
1, rue Blaise Pascal, BP 296
67008 Strasbourg, France

Rohmer Marie-Madeleine
Laboratoire de Chimie Quantique
Institut Le Bel, Université Louis Pasteur
4, rue Blaise Pascal
67000 Strasbourg, France

PREFACE AND ACKNOWLEDGEMENTS

Exchange of ideas among scientists of different fields may be unefficient, even when these fields overlap, if they do not share a common language. This point was noted in discussions among the members of the board of the Chemical Physics Section of the European Physical Society. This type of difficulty is frequently encountered when theoreticians and experimentalits meet. If they work in a field with a tradition of common meetings, they develop a common language. In new fields, this language has to be built. It appears that this is the case for the different areas of photochemistry and photophysics which have their own dynamics independantly and could have most benefit by sharing their experience. This is the origin of this school which was initially planned as the sixth Europhysics Summer School on Chemical Physics. Since it has been sponsored by NATO, it turned into a NATO Advanced Institute which has been held in Aussois (France), September 1-13, 1991.

The field of photoprocesses is very promising. Special emphasis has been put on transition metal complexes and biosystems. It is interesting to show what similarities can exist between these two very active areas. However, other areas of photoprocesses have also been treated, because the basics may be common to them, but also with the purpose to build bridges between these fields. This insures that this book may be useful to a wide range of researchers and give, in fact, a broad view of photoprocesses. The ultimate goal of this ASI was to initiate collaborations between attendants. This has been the case and was made possible by the excellent communication which has been established between the participants.

The first week of the Institute has been devoted to the acquisition of the bases absolutely necessary to develop a common language : basics on photophysics have been given in the lectures of Professors Tramer (elementary notions) and Marcus (electron transfer processes); the quantum chemistry approach has been treated by Professors Salem (orbitals, surfaces, diagrams, diradical intermediates, sudden polarization), Roos (CI techniques) and Peyerimhoff (lifetimes and relativistic effects). A very successful tutorial work (A. Strich) dealt with this part of the program. Spectroscopic techniques have been described by Professor Turner for the study of organometallic intermediates and Professor Schneider for the study of biliproteins.

Emphasis on the photochemistry of organometallics has been found in the lectures of Professors Turner, Grevels, Veillard, Stufkens, Balzani and Scandola. The photochemistry of metal-metal bonded carbonyls (J.J. Turner,

F.W. Grevels, A. Veillard, D.J. Stufkens) and polynuclear systems in supramolecular photochemistry (V. Balzani, F. Scandola) deserved special attention. A transition toward large systems has been provided by the lectures of Professors Verhoeven (alkane-bridged donor-acceptor systems) and Marcus (theory of electron transfer). The photophysics of biosystems has been treated by Professors Schneider (energy transfer in biological antenna systems), Mathis (photoinduced electron transfer in biological reaction centers), Ogrodnik and Michel-Beyerle (charge separation in reaction centers). The final lecture, by Professor Sessler, was devoted to artificial systems (new approaches to photosynthetic model systems).

The articles in this book are presented in the same order as the lectures during the school. We also include the list of communications at the two very successful poster sessions which have allowed the participants to present their own results.

The program of the ASI has been set up with the help of the members of the organizing committee and the active participation of the lecturers. Many thanks are due to all of them for their efficient assistance and for the very stimulating lectures which they have given.

We also thank the members of the local committee: Drs. Jean-Claude Chambron, Christophe Coudret, Chantal Daniel, Valérie Heitz and Marie-Madeleine Rohmer who devoted most of their free time during the school to make everything going smooth. Their help was crucial and has been very much appreciated by the participants.

We thank Mrs Jahiel for secretarial services. The facilities of the Centre de Calcul CNRS de Strasbourg-Cronenbourg have been of great help and special thanks should go to Mrs Grange. The help of Daniel Ott in the preparation of the school and of the book was very precious.

Finally, and most importantly, we thank NATO for the financial support which made possible this Summer School and has provided funds for many of the participants.

E. Kochanski
UPR 139 du CNRS, Laboratoire de Chimie Théorique,
Institut de Chimie, Université Louis Pasteur,
BP 296, 67008 Strasbourg-Cedex, France

ELEMENTARY NOTIONS OF PHOTOPHYSICS.

by A.Tramer
(Laboratoire de Photophysique Moléculaire CNRS, Université de Paris-Sud, 91405-Orsay, France)

E. Kochanski (ed.), Photoprocesses in Transition Metal Complexes, Biosystems and Other Molecules. Experiment and Theory, 1–47.

1. Introduction.

The scope of this paper is to recall fundamental notions of the molecular spectroscopy and dynamics, necessary for discussion of photophysical and photochemical processes *in condensed phases*. We will thus treat in a more detailed way the specific features which are important for molecular systems strongly interacting with their environment. Other aspects such as the time evolution of isolated molecules, single-level excitation and state-to-state chemistry, important for the gas-phase photophysics are omitted.

We start (Sec.2) with a brief description of radiative processes (light absorption and emission) in molecules. In the quantum-mechanical treatment of this problem, the appropriate basis is that of so-called *zero-order* states, corresponding to the traditional scheme of electronic states (singlets, doublets, triplets etc.) and vibrational levels belonging to each state. The most important point will be deduction of *selection rules* for radiative transitions. At this stage all molecular states are considered as *stationary states*.

In order to treat the breakdown of simple selection rules and *non-radiative transitions* between individual molecular states, it is necessary to take into account the mechanisms coupling the zero-order states (Sec.3). We will first focus on intramolecular coupling effects and then discuss the solvent effects on intramolecular relaxation processes. The problem of the non-radiative transfer of the electronic energy between different molecules - closely related to that of the energy dissipation within a single molecule - will be treated in Sec.4.

In the last section, we will enumerate briefly some specific relaxation paths of electronically excited molecules involving the solvent-solute interaction (photo-dissociation, photo-ionization and photoinduced electron transfer, formation of donor-acceptor and hydrogen-bonded complexes).

2. Electronic States of Molecules in the Zero-Order Approximation.

2.1 GENERAL REMARKS.

The aim of the spectroscopic experiment is to get maximum of data concerning the internal energy and essential properties (geometry, dipole moment, time evolution, chemical reactivity etc.) of a molecule in its ground and excited states. The interpretation of this data will be then carried out on the basis of quantum-mechanical calculations of *eigenfunctions* and *eigenergies* of the molecule obtained by solution of the Schrödinger equation with a *molecular Hamiltonian* describing the molecule.

Since the Schrödinger equation cannot be exactly solved for a many-body system such as a molecule (N nuclei + n electrons i.e. 3(N+n) degrees of freedom), the first step will consist in replacement of the *exact* molecular hamiltonian - H_{mol} by a truncated hamiltonian H_0 (hamiltonian of the *zero-order approximation* or simply *zero-order hamiltonian*) :

$$H_0 = H_{mol} - H'$$

allowing separation of nuclear - $\vec{R}_N$ and electronic - $\vec{r}_n$ coordinates. This necessitates the neglect in H_0 of cross terms between different types of coordinates. These terms are "relegated" to H' and considered as first- and higher-order perturbations. Essential simplifications consist in :

(i) separation of translation and rotation from internal coordinates (neglect of the coupling of rotation with vibration and electron movement),

(ii) neglect of the interaction between spin and orbital movement of electrons (of the *spin-orbit coupling*),

(iii) the Born-Oppenheimer approximation assuming separation of the slow motion of nuclei (moving in the *average potential field* due to electrons) and of a rapid motion of electrons in the field of *fixed nuclei* (for each nuclear configuration). This approximation corresponds to the neglect in the Schrödinger equation of terms containing the derivatives of the electronic wavefunction with respect to nuclear coordinates of the type of $\partial\psi^{el}/\partial X_N$ and $\partial^2\psi^{el}/\partial X_N{}^2$.

Small terms due to spins and quadrupole moments of nuclei are equally neglected.

The *zero-order states* of the molecule (*eigenstates of the zero-order Hamiltonian* H_0) may be then described as products of electronic, vibrational, rotational and spin functions :

$$|\Psi_{nJKS}(\vec{R},\vec{r},s)\rangle = |\psi_n^{el}(\vec{r},\vec{R}_0)\rangle|\chi_v(Q)\rangle|Y_{JK}(\theta,\phi)\rangle|\sigma_S(s)\rangle \ldots (2.1)$$

where :

- ψ_n^{el} is the electronic function depending only on the 3n electronic coordinates - $\vec{r}$ (for a given configuration of nuclei - $\vec{R}_0$). In the so-called *adiabatic Born-Oppenheimer (ABO) approximation* the electronic wavefunctions depend parametrically on nuclear coordinates R_0, while in the *crude Born-Oppenheimer (CBO) approximation* this dependence is neglected : it is assumed that in the whole range of nuclear configurations corresponding to low-amplitude vibrations, the electronic wavefunctions are the same as for the equilibrium configuration - R_{eq} :

$$|\psi_n^{el}(\vec{r},\vec{R}_0)\rangle = |\psi_n^{el}(\vec{r},\vec{R}_{eq})\rangle$$

- χ_v is the vibrational function depending on 3N-6 *normal coordinates* of nuclei - Q_n (linear combinations of cartesian coordinates : X_N, Y_N, Z_N of nuclei) and characterized by an ensemble of 3N-6 vibrational quantum numbers v_n,
- Y_{JK} - the rotational function of angular coordinates and
- σ_S - the function of electronic spins $\vec{s}$; the most important parameter being the total spin $\vec{S} = \sum_i \vec{s}_i$.

The energy of each state is the sum of its electronic, vibrational and rotational energies (translational energy is treated apart and the small terms related to the electron spin effects may be neglected) :

$$E_{nvJK} = E_n^{el} + E_v^{vib} + E_{JK}^{rot} \quad \ldots\ldots\ldots\ldots\ldots (2.2)$$

It is important to note that H_0 is diagonal in the basis of zero-order states : this means that *no transition* between these states will occur *in absence* of external perturbations. Such perturbations may be induced by collisions with other molecules or electrons and by electromagnetic fields. Only the last type of perturbations (*radiative transitions*) will be discussed in this chapter.

The leading term for induction of a radiative transition is the interaction between electric field $\vec{\varepsilon}(t) = \vec{\varepsilon}_0 \cos \omega t$ and the electric dipole moment of the molecule - $\vec{\mu}$. The probability of the radiative transition (absorption or stimulated emission of a photon) is proportional to the square of the *matrix element* of the perturbation :

$$P_{n\to n'} \sim |\langle\Psi_n|\vec{\mu}\vec{\varepsilon}_0|\Psi_{n'}\rangle|^2 = |\mu_{nn'}|^2 \, I_\mu \quad \ldots\ldots\ldots (2.3)$$

where $\vec{\mu}_{nn'} = \langle\Psi_n|\vec{\mu}|\Psi_{n'}\rangle$ is the $n \to n'$ *transition moment* and $I_\mu = \varepsilon_{0\mu}^2$ - the intensity of the exciting radiation polarized along $\vec{\mu}_{nn'}$.
It may be easily shown that for transitions between different electronic states only the electronic part of the molecular dipole moment $\vec{\mu}_{el} = e\sum_i \vec{r}_i$ is efficient. Since $\vec{\mu}_{el}$ commutes with vibrational and spin functions which do not involve electronic coordinates, the transition moment $\vec{\mu}_{nvS\to n'v'S'}$ may be written as :

$$\vec{\mu}_{nvS\to n'v'S'} = \langle\Psi_{nvS}(\vec{R},\vec{r},s)|\vec{\mu}_{el}|\Psi_{n'v'S'}(\vec{R},\vec{r},s)\rangle =$$

$$=(\vec{\mu}_{el})_{n\to n'}\langle\chi_n(Q)|\chi_{n'}(Q)\rangle\langle\sigma_S(s)|\sigma_{S'}(s)\rangle \quad \ldots\ldots\ldots\ldots (2.4)$$

where the electronic transition moment :

$$(\vec{\mu}_{el})_{n\to n'} = -e\langle\psi_n^{el}(\vec{r},\vec{R}_0)|\sum_i \vec{r}_i|\psi_{n'}^{el}(\vec{r},\vec{R}_0)\rangle$$

depends parametrically on nuclear coordinates in the adiabatic but is independent of the nuclear configuration in the crude

Born-Oppenheimer approximation. The rotational part of the wavefunction is omitted : we will not discuss here the rotation effects important only for freely rotating molecules in the gas phase.

The transition moment is thus represented as product of three factors : electronic transition moment - $(\vec{\mu}_{el})_{n\to n'}$, the *overlap integral* of vibrational wavefunctions - $\langle\chi_n(Q)|\chi_{n'}(Q)\rangle$ and a similar integral of spin functions - $\langle\sigma_S(s)|\sigma_{S'}(s)\rangle$. In the following section, we will discuss all of them in order to deduce the most important features of the electronic spectra of molecules.

2.2. SYSTEMATICS OF ELECTRONIC STATES. SELECTION RULES FOR ELECTRONIC TRANSITIONS.

2.2.1. Spin-Allowed and Spin-Forbidden Transitions. The last factor in Eq.(2.4) is different from zero only for S = S', the spin functions being orthogonal. This means that - in the zero-order approximation - the electronic states of different multiplicities (singlet - S=0, triplet - S=1, ... for molecules with even number of electrons, doublet - S=1/2, quartet - S=3/2, ... for ions and radicals) form completely isolated manifolds, the *intercombinations* (radiative transitions with $\Delta S \neq 0$, as e.g. singlet $\leftrightarrow$ triplet) being *strictly forbidden*. This selection rule holds rigorously as long, as the spin-orbit coupling is neglected.

2.2.2. Electronic Transition Moments and Orbital Selection Rules. The first factor in Eq.(2.4) may be rewritten for each component of the $\vec{\mu}_{nn'}$ vector in a developed form as :

$$(\mu_\rho)_{nn'} = e\int\psi_n^{*el}(x_i,y_i,z_i)(\sum_i\rho_i)\psi_{n'}^{el}(x_i,y_i,z_i)d\tau = e\int F d\tau \ldots (2.5)$$

where ρ = x,y,z and $\int d\tau$ indicates triple integration over the whole $(-\infty,+\infty)$ space. The value of $(\mu_\rho)_{nn'}$ depends on the *symmetry* and on the *spatial* properties of the electronic wavefunctions of the initial and final electronic state : ψ_n^{el} and $\psi_{n'}^{el}$ [1].

The symmetry properties may impose for some components of the transition moment the identity $(\mu_\rho)_{nn'} \equiv 0$. Let us take as example a planar molecule (such as an aromatic hydrocarbon) with the symmetry plane σ_z identical with the molecular plane (xy). Because of the symmetry, the electronic wavefunctions ψ^{el} remain unchanged or only change sign on reflection in the σ_z plane i.e. for $z \rightarrow -z$:

$$\psi^{el}(x,y,-z) = \pm \psi^{el}(x,y,z)$$

what corresponds - in the language of the *theory of molecular point groups* to the functions (states) *symmetric (A')* and *anti-symmetric (A")* with respect to σ_z.

The necessary condition for $(\mu_\rho)_{nn'} \neq 0$ is the even character of the integrand F in Eq.(2.5) with respect to the z-coordinate :

$$F(x,y,-z) = F(x,y,z)$$

since for the integral of the odd function we obtain :

$$\int_{-\infty}^{\infty} F(x,y,z)dz = \int_{-\infty}^{0} F(x,y,z)dz + \int_{0}^{\infty} F(x,y,z)dz = \int_{0}^{\infty} F(x,y,z)dz - \int_{0}^{\infty} F(x,y,z)dz \equiv 0$$

$(\mu_z)_{nn'}$ will be different from zero only when the triple product $\psi_n^{el} z \psi_{n'}^{el}$ is even, i.e. when $\psi_n^{el}\psi_{n'}^{el}$ is odd with respect to the z variable, in other words for the A' ↔ A" transitions. On the other hand, in the case of A' ↔ A' and A" ↔ A" transitions $\psi_n^{el}\psi_{n'}^{el}$ is even, so that $(\mu_z)_{nn'} \equiv 0$, while $(\mu_x)_{nn'}$ and $(\mu_y)_{nn'}$ may be different from zero. The A' ↔ A" transitions are thus out-of-plane and A' ↔ A' and A" ↔ A" ones - in-plane polarized.

In the one-electron approximation of the electronic wavefunction :

$$|\psi^{el}(\vec{r}_1,\vec{r}_2,\ldots\vec{r}_n)\rangle = |\varphi_\alpha(\vec{r}_1)\rangle|\varphi_\beta(\vec{r}_2)\rangle\ldots|\varphi_\lambda(\vec{r}_n)\rangle \qquad (2.6)$$

the electronic transition corresponds to a transfer of one electron between two *molecular orbitals* φ_λ and $\varphi_{\lambda'}$, while the remaining part of the wavefunction is unchanged. In view of the normalization of φ-functions, Eq.(2.5) becomes :

$$(\mu_\rho)_{nn'} = e\int\varphi^*_\lambda(x_n,y_n,z_n)\ \rho_n\ \varphi_{\lambda'}(x_n,y_n,z_n)d\tau \quad .....(2.7)$$

In the previously proposed example of a planar aromatic or hetero-aromatic molecules, we differentiate - in the zero-order approximation - the π and π^* (A") orbitals from the σ, σ^* and n (A') ones. In view of selection rules the $n \rightarrow \pi^*$ and $\pi^* \rightarrow \sigma^*$ or $\sigma \rightarrow \pi$ transitions will be polarized perpendicular to the molecular plane :

$(\mu_x)_{nn'} = (\mu_y)_{nn'} = 0, (\mu_z)_{nn'} \neq 0$

while the $\sigma \rightarrow \sigma^*$ and $\pi \rightarrow \pi^*$ ones are in-plane polarized :

$(\mu_z)_{nn'} = 0,\ (\mu_x)_{nn'}$ and $(\mu_y)_{nn'} \neq 0$

as shown in Fig.1.

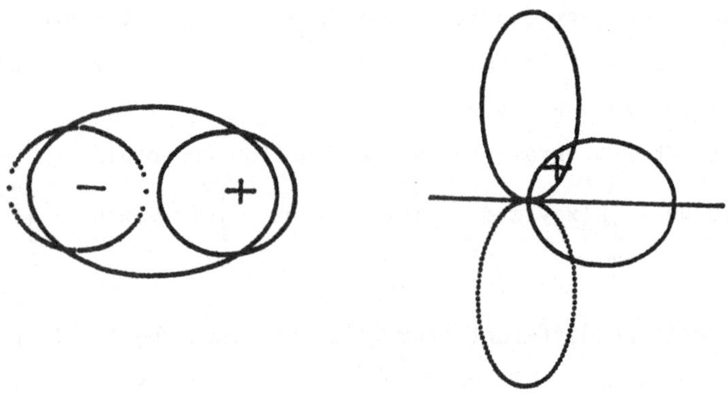

1. Schematic representation of molecular orbitals in the case of $\sigma \rightarrow \sigma^*$(left) and $n \rightarrow p(\pi)$ transitions. Areas corresponding to positive (negative) values of wavefunctions are limited by solid (broken) lines. + and - indicate the sign of the $\varphi_\lambda\varphi_{\lambda'}$ products.

In the molecules of higher symmetry some electronic transitions are entirely *symmetry-forbidden* :

$(\mu_x)_{nn'} = (\mu_y)_{nn'} = (\mu_z)_{nn'} \equiv 0$

It is the case for the first singlet-singlet transitions in formaldehyde (the $^1A_2 \leftarrow {}^1A_1$ transition forbidden in the C_{2v} symmetry) and benzene ($^1B_{2u} \leftarrow {}^1A_{1g}$ in the D_{6h} symmetry).

This selection rule is imposed by the symmetry of the equilibrium configuration (R_{eq}) and valid for $R_0 \neq R_{eq}$ only in the crude Born-Oppenheimer approximation, i.e. when the dependence of ψ_{el} on nuclear configuration (R_0) is neglected. In reality, the transition becomes *weakly allowed* when the symmetry of the molecule is lowered by the non-totally-symmetric vibrations. The transition between vibrationless levels of the ground and excited electronic states (the 0_0^0 band) remains forbidden in absorption and emission spectra, but the vibronic bands corresponding to transitions to the levels containing one quantum (or an odd number of quanta) of a non-totally symmetric mode become allowed. This effect (the *Herzberg-Teller effect)* is usually treated in terms of the *intensity borrowing (or stealing)* from allowed electronic transitions (see below, Sec.3.2.).

In spite of the partial breakdown of the selection rule, the integrated intensity of a symmetry-forbidden transition is lower by 2÷3 orders of magnitude as compared to an allowed one.

Selection rules define conditions for individual $(\mu_\rho)_{nn'}$ components to be different from zero but do not allow us to estimate $|\mu_{nn'}|$ values i.e. intensities of electronic transitions. Such an estimate may be deduced - in the one-electron approximation - from the type of initial and final molecular orbitals. The value of the integral in Eq.(2.5) depends on the overlap of electronic wavefunctions of the initial and final state :

$$\varphi_\lambda^*(x_n,y_n,z_n)\varphi_{\lambda'}(x_n,y_n,z_n)d\tau \quad\quad (2.8)$$

Obviously, when this product is equal to zero in the whole space, the transition moment tends also to zero.The value of $(\mu_\rho)_{\lambda\lambda}$ is thus closely related to that of the *overlap integral* :

$$S_{\lambda\lambda'} = \int\varphi_\lambda^*(x_n,y_n,z_n)\varphi_{\lambda'}(x_n,y_n,z_n)d\tau \quad\quad (2.8a)$$

The overlap is usually larger (in absence of specific symmetry effects) between orbitals of the same type. For this reason, the transition moment (intensity of transition) will be larger for $\pi \rightarrow \pi^*$ and $\sigma \rightarrow \sigma^*$ than for $n \rightarrow \pi^*$ or $\pi \rightarrow \sigma^*$ transitions. In the case of *charge-transfer* transitions (transfer of electron from the orbital of the electron-donor molecule - φ_D to an empty orbital of the electron acceptor - φ_A), the value of the overlap integral $\int \varphi_D{}^* \varphi_A d\tau$ depends strongly on the intermolecular distance R_{AD} and goes to zero for R_{AD} exceeding the extension of the orbitals.

2.2.3. Overlap of vibrational wavefunctions - the Franck-Condon Rule. In the zero-order approximation, the vibrational wavefunctions $\chi(Q)$ are products of $\chi_{v_n}(Q_n)$ functions representing 3N-6 *normal modes* of the molecule. Each of them depends on one *normal coordinate* - Q_n and is characterized by the corresponding vibrational quantum number v_n:

$$|\chi(Q_1,Q_2..Q_n)\rangle = |\chi_{v_1}(Q_1)\rangle|\chi_{v_2}(Q_2)\rangle..|\chi_{v_n}(Q_n)\rangle \quad \text{.........(2.9)}$$

so that the overlap integral of vibrational wavefunctions in Eq.(2.4) takes the form of a product of overlap integrals for each normal mode in the initial (χ'') and final (χ') state :

$$\langle\chi''|\chi'\rangle=\langle\chi''_{v_1}(Q_1)|\chi'_{v_1}(Q_1)\rangle\langle\chi''_{v_2}(Q_2)|\chi'_{v_2}(Q_2)\rangle \ldots$$
$$\ldots\langle\chi_{v_n}{}''(Q_n)|\chi'_{v_n}(Q_n)\rangle$$

The functions $|\chi_{v_n}(Q_n)\rangle$ are orthonormalized. This implies that $|\chi'_{v_n}(Q_n)\rangle$ and $|\chi_{v_n}''(Q_n)\rangle$ are orthogonal when the potential for the n-th normal mode is identical in both states i.e. when they are eigenstates of the same hamiltonian :

$$\langle\chi_{v_n}(Q_n)|\chi'_{v'_n}(Q_n)\rangle = \delta_{v_n v'_n}$$

If it is so, the transition may take place only between the vibrational levels with the same quantum number ($\Delta v_n = 0$ what implies that from $v_n'' = 0$ only the transition to $v'_n = 0$ is allowed in a "vibrationally cold" molecule.The transitions with a change of the vibrational quantum number (vibrational *progressions* with Δv_n = 1, 2, ...,n) will occur only for the normal modes, the potential (and frequency) of which are strongly changed by the electronic excitation of the molecule. The intensity distribution within a given progression depends on this difference. In most polyatomic molecules only a limited number of valence bonds and angles are modified by the electronic excitation so that the number of *optically active* modes giving long progressions is limited. The vibrational structure of the electronic transitions is thus relatively simple even in large molecules (see below).

2.2.4. Electronic Absorption and Emission Spectra. Electronic spectra of polyatomic molecules are composed of a number of *vibronic bands* well resolved for medium-size molecules in the gas phase and for "cold" large molecules in supersonic jets and low-temperature matrices. In the room temperature fluid solutions their spectra are diffuse or even structureless. As shown in Fig.2., the broad-band emission spectrum of a large molecule in a room-temperature fluid solutionsplits into a number of narrow band upon the laser excitation of the same molecule in a low-temperature matrix.

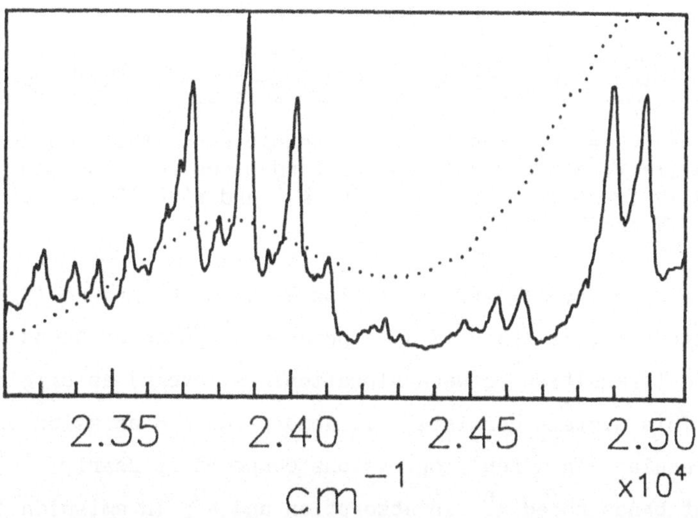

Fig.2. A part of the fluorescence spectrum of 9,10-dichloroanthracene in room-temperature cyclohexane solution (points) and in the argon matrix at T = 10K upon the narrow-band laser excitation (solid line).

The main reasons for the diffuseness of the room-temperature spectra are :

(i) the sequence congestion : overlap of transitions from a large number of thermally populated low-energy vibrational levels involving all (also optically inactive) modes $v_n" = 1,2,...$giving origin to the sequences of "hot bands" $v_n' = v_n"$ accompanying each vibronic transition (Fig.3) and

(ii) the line broadening due to a rapid exchange of vibrational energy between the molecule and the solvent (see below).

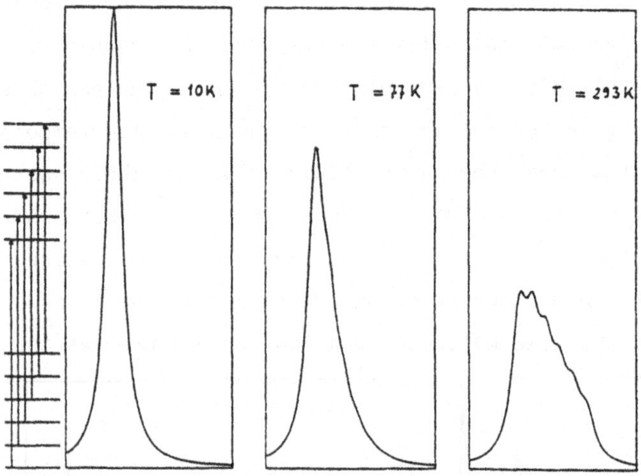

3. Scheme of (v" = v') transitions responsible for the sequence congestion (left) and a simulation of bandshapes for a molecule with one optically inactive mode : $\nu" = 50\ cm^{-1}$ and $\nu' = 45\ cm^{-1}$ at T = 4K, T = 77K and T = 300K.

For a symmetry-allowed transition between the ground (S_0) and first excited singlet state (S_1) we observe a system of bands with the 0_0^0 band (transition between vibrationless levels) as *origin* (Fig.4a). This systems consists - in absorption and emission of a "cold" molecule - in a few progressions composed of nearly equidistant bands noted $A_0^{n'}$ in absorption and $A_{n"}^0$ in emission (where A is the "name" (number) of the normal mode, the superscripts and subscripts indicate the number of quanta of this mode in the excited and ground state, respectively). The frequencies of these transitions are :

$$\nu_{abs} \cong \nu(0_0^0) + \sum_A n_A'' \nu_A''; \quad \nu_{em} \cong \nu(0_0^0) - \sum_A n_A' \nu_A'$$

where ν_A' and ν_A'' are excited and ground-state frequency of the A-mode.

In the spectrum of a symmetry-forbidden transition (Fig.4), the 0_0^0 band is missing. The first band in the absorption spectrum is the X_0^1 vibronic transition involving one quantum of the

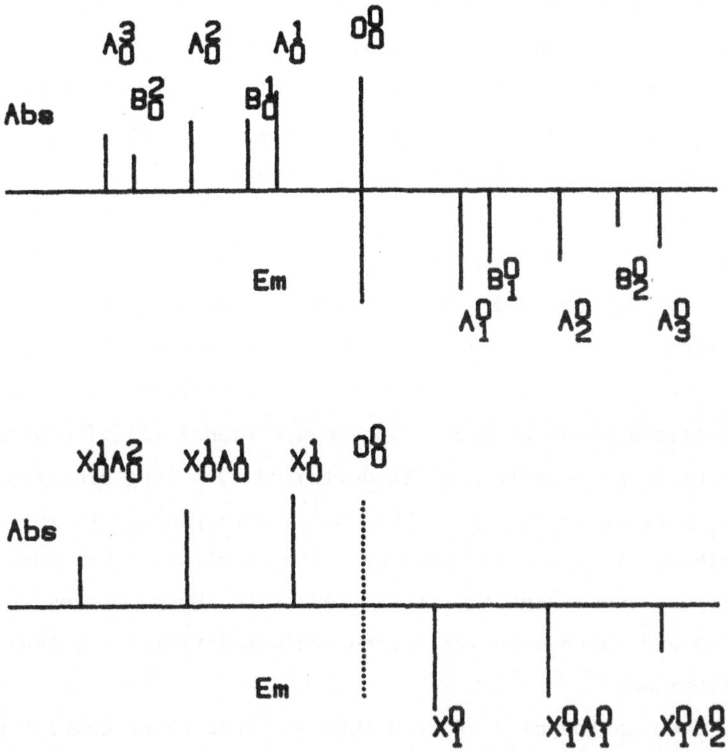

4. Schematic representation of the vibrational structure of an allowed (top) and forbidden (bottom) transition

non-totally symmetric (*promoting*) mode X displaced by ν_X' to higher frequencies with respect to the 0_0^0 transition. The corresponding band in emission is the X^0_1 one, red shifted by ν''_X. X^0_1 and X^1_0 bands are *false origins* of vibronic progressions in emission and absorption spectra involving optically active modes ($X^0_1 A^0_{n''}$ and $X^1_0 A^{n'}_0$)

$$\nu_{abs} = \nu(0^0_0) + \nu'_X + \sum_A n_A \nu'_A \quad ; \quad \nu_{em} = \nu(0^0_0) - \nu''_X - \sum_A n_A \nu''_A$$

The absence of the 0^0_0 transition (the only band common for the absorption and emission spectrum) is an indication of the forbidden character of the electronic transition.

The intensity distribution in each *vibronic progression* (v_n = 0, 1, 2 ,... for the n-th optically active mode) is a source of information about the difference of nuclear configurations in the ground and excited state. As mentioned above, only $\Delta v = 0$ transitions are allowed for the modes whose potential field is unchanged by electronic excitation (e.g. C-H stretching modes of aromatic molecules). Long progressions correspond to a drastic change of the bond order (e.g. the C=O stretching mode of formaldehyde : R_{eq} = 1.22 and 1.32 A, ν = 1746 and 1177 cm^{-1} in the ground and in the excited $n\pi^*$ state, respectively) or of the valence angle (linear-to-bent transition in acetylene) [2]. The absorption spectrum of benzene (Fig.5) is dominated by a long progression in the ν_1 mode (ν' = 923, ν'' = 992 cm^{-1}) corresponding to the ring "breathing" i.e. to the symmetric stretch of the C-C bonds, which are longer and weaker in the excited state. Such changes of bond lengths and valence angles may be evaluated from the intensity distributions.

Even in the case of structureless spectra of complex molecules in room-temperature solutions, it is still possible to estimate roughly the difference between equilibrium configurations in both electronic states from the extension *(half-width)* of absorption and emission bands and from the frequency difference between the maxima in absorption and emission *(absorption-to-fluorescence Stokes shift*: $\delta\nu_{St} = \nu^{abs}_{max} - \nu^{fl}_{max}$). Very large Stokes shifts in non-polar solvents indicate a drastic change of the geometry which often corresponds often to the photo-isomerization of the excited molecule, fluorescence being emitted by an isomeric species different from the absorbing one.

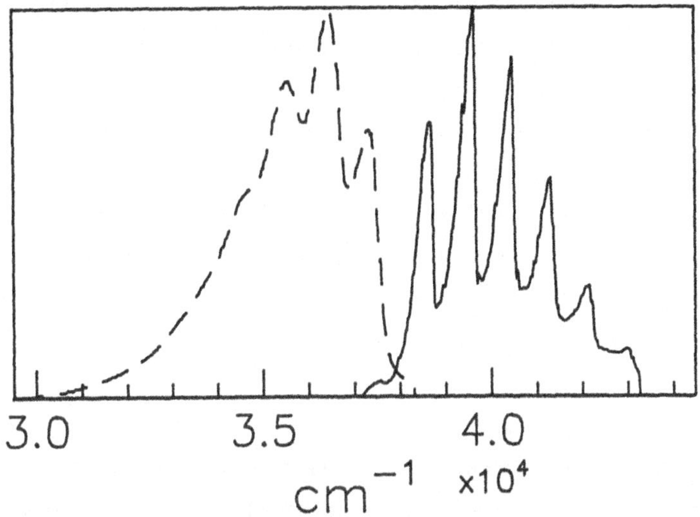

5. Absorption and fluorescence spectrum of benzene-d_6 in the cyclohexane solution (reproduced from I.B.Berlman, Handbook of Fluorescence Spectra of Aromatic Molecules, Acad.Press, New York, 1965).

3. Coupling between Zero-Order States : Breakdown of Selection Rules. Radiationless Transitions.

3.1. GENERAL REMARKS.

As previously indicated (cf.Sec.2.1), the exact molecular Hamiltonian H_{mol} is not diagonal in the basis of zero-order states. The presence of off-diagonal terms $(H_{mol})_{mn}$ $(m \neq n)$ means that the energies E^0_{nvJK} calculated in the zero-order approximation are not the exact ones and that the $|\Psi_{nvJKS}\rangle$ states are not the *stationary states* of the molecule.

In order to account for the presence of these terms, it is possible to proceed in two different ways : either

- rediagonalize the hamiltonian and determine in this way the energies and wavefunctions of *exact molecular eigenstates* or

- conserve the initial basis but take into account the *non-stationary* character of $|n\rangle$ states i.e. the possibility of the population transfer between these states in absence of external perturbations (*radiationless transitions).*

The first way is usually applied in the high-resolution spectroscopy of small molecules, in order to explain an irregular level spacing and intensity anomalies. They are then described in terms of *perturbations* displacing molecular eigenstates with respect to the zero-order states.

In view of our interest in the *time evolution* of electronically excited molecules we will choose essentially the second way (excepted for the problem of the breakdown of selection rules). This choice is particularly well adapted to the case of condensed phases, since it allows us a uniform description of effects due to off-diagonal terms in the molecular hamiltonian and to additional off-diagonal terms induced by the environment.

3.2. BREAKDOWN OF SELECTION RULES.

Selection rules may be relaxed because of intra-molecular terms, neglected in the zero-order approximation, as well as because of the environment effects.

(i) Intercombinations (e.g. singlet $\leftrightarrow$ triplet radiative transitions) rigorously forbidden in the zero-order approximation become weakly allowed by the singlet-triplet mixing induced by the spin-orbit coupling. Its efficiency depends on the presence of heavy atoms in the molecule (spin-orbit coupling is roughly proportional to Z^3) and on the type of excited states : it may be easily shown that the ${}^1n\pi^* \leftrightarrow {}^3\pi\pi^*$ and ${}^1\pi\pi^* \leftrightarrow {}^3n\pi^*$ interaction is much stronger than the ${}^1n\pi^* \leftrightarrow {}^3n\pi^*$ or ${}^1\pi\pi^* \leftrightarrow {}^3\pi\pi^*$ one. For these reasons, the *oscillator strengths* (relative intensities) of $S_0 \leftrightarrow T_1$ radiative transition vary between 10^{-9} in benzene , 10^{-7} in bromobenzene (*internal heavy atom effect*) and 10^{-6} in pyrazine ($n\pi^* \leftrightarrow \pi\pi^*$ interaction) [3]. The effect of environment on singlet $\leftrightarrow$ triplet transitions is weak excepted for solvents containing heavy atoms (*external heavy atom effect*) as e.g. in CH_2I_2 solutions or in matrices of heavy rare gases such as xenon [4].

(ii) The breakdown of the orbital selection rules is related to intramolecular or intermolecular (environment) effects lowering the symmetry of the molecule. In both cases, the mechanisms are described in terms of the parametric dependence of the electronic wavefunction on the ensemble of nuclear coordinates - R contained in the *adiabatic* (but not in the *crude* !) Born-Oppenheimer approximation. If the geometry of the molecule (described by the ensemble R_0) is slightly different from its equilibrium configuration R_{eq}, the perturbation method may be applied for small $R_0 - R_{eq}$ deviations : the electronic wavefunction of the n-th electronic state is developed over the unperturbed functions corresponding to R_{eq} :

$$\Psi_n(r,R_0) = N\{\psi_n(r,R_{eq}) + \sum_{n'} a_{n'}(R_0-R_{eq})\psi_{n'}(r,R_{eq})\} \quad \ldots\ldots(3.1)$$

so that the transition moment from the ground (supposed to be unperturbed) to the n-th state becomes :

$$\vec{\mu}_{0n}(R_0) = N\{\vec{\mu}_{0n}(R_{eq}) + \sum_{n'} a_{n'}(R_0-R_{eq})\vec{\mu}_{0n'}(R_{eq})\} \quad \ldots(3.1a)$$

If the first term in Eq.3.1a is equal to zero because of orbital selection rules, some higher terms will be different from zero so that the transition moment will be *borrowed* from the allowed transitions $|0\rangle \rightarrow |n'\rangle$.

In the theory of the *Herzberg-Teller effect* [5], the deviation from the equilibrium configuration ($R_0 - R_{eq}$) in the vibrating molecule is expressed in terms of normal coordinates Q_s and (3.1a) becomes :

$$\vec{\mu}_{0n}(R_0) = N\{\vec{\mu}_{0n}(R_{eq}) + \sum_s \sum_{n'} a_{n's} Q_s \vec{\mu}_{0n'}(R_{eq})\} \ldots (3.1b)$$

The efficiency of the Herzberg-Teller mechanism (i.e. the intensity and polarization of the transition forbidden in the zero-order crude Born-Oppenheimer approximation) depends on the the excited vibrational mode - Q_s and on the $a_{n's}$ coefficients.

The equilibrium configuration of the molecule and its symmetry may be also modified by its environment : the exact symmetry of the system will be the *site symmetry* : the symmetry of the system consisting of the molecule *and* of its solvation shell. This symmetry is usually lower than that of the isolated molecule.

In fluid solutions, the intensity induced by the solute-solvent interaction may attain the same order of magnitude as that due to the Herzberg-Teller effect. For instance, the 0^0_0 band, missing in the spectrum of benzene vapors, appears in ethanol solution with intensity lower only by a factor of five as compared to that of the 6^1_0 band allowed by the Herzberg-Teller mechanism.

3.3. ELECTRONIC RELAXATION OF EXCITED MOLECULAR STATES.

3.3.1. Interstate Coupling. The time evolution of excited molecular states will be developed in terms of the coupling between non-stationary states. It is useful to differentiate the *doorway states* which are "prepared" by optical excitation and a dense manifold of *dark states*.

Let us suppose for simplicity sake that the initially prepared doorway state is the vibrationless level of an electronically excited second singlet state $S_2(0)$. This level is quasi-isoenergetic with dense manifolds of high vibronic levels of lower electronic states ($S^\#_1$, $T^\#_2$, $T^\#_1$ and $S^\#_0$) as shown in Fig.6. We have thus a number of radiationless relaxation channels, usually divided into two groups : *internal conversion* within the same spin manifold such as $S_2(0) \rightarrow S^{\#}_1$, $S_2(0) \rightarrow S^{\#}_0$, $T_2 \rightarrow T^{\#}_1$ and *intersystem crossing* between different spin manifolds as e.g. $S_2(0) \rightarrow T^{\#}_1$ and $T_1(0) \rightarrow S^{\#}_0$. In usual conditions, the excess of vibrational energy in $S^{\#}_1$ or $T^{\#}_1$ states is rapidly evacuated to the molecular environment *(heat bath)*, so that molecules leaving the $S_2(0)$ levels are rapidly transfered to thermally equilibrated, low vibrational levels of the S_1 or T_1 states : $S_1(0)$ and $T_1(0)$.

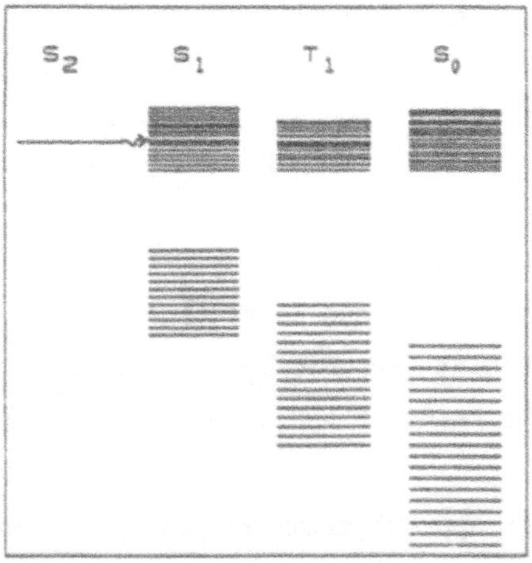

6. Schematic representation of the coupling between the S_2 vibrationless levels and iso-energetic dense manifolds of $S_1^{\#}$, $T_1^{\#}$ and $S_0^{\#}$ vibronic levels.

Each electronic state may decay radiatively by the $S_2(0) \rightarrow S_0$, $S_1(0) \rightarrow S_0$ or $T_1(0) \rightarrow S_0$ emission. The branching ratios between different channels may be estimated by measurements of quantum yields and rates of radiative processes competing with non-radiative channels. The intensities and decay times of the $S_2 \rightarrow S_0$ and $S_1 \rightarrow S_0$ fluorescence components monitor the time-dependent populations of S_2 and S_1 states, while that of the T_1 state may be deduced from the $T_1 \rightarrow S_0$ phosphorescence and/or the $T_1 \rightarrow T_i$ transient absorption. Many channels are, however, "dark" and cannot be directly monitored.

The theoretical treatment of relaxation is based on the picture of a discrete level - *doorway state* $|s\rangle$ (vibrationless S_2 level) coupled to one or a few continua of *dark states* [6]. In the simple case of a single continuum, the time evolution of populations is given by the *Fermi Golden Rule* : the population of a discrete level $|s\rangle$ is irreversibly transfered to the continuum of $|\ell\rangle$ levels. This process is exponential with the rate constant :

$$k_{s\ell} = (1/\hbar)\langle v_{s\ell}^2\rangle\langle\rho_\ell\rangle \quad (3.2)$$

where $v_{s\ell}$ is the coupling constant (off-diagonal term in H') between the $|s\rangle$ level and one of the $|\ell\rangle$ levels :

$$v_{s\ell} = \langle\Psi_s|H'|\Psi_\ell\rangle \quad (3.3)$$

and $\langle\rho_\ell\rangle = 1/\langle|E_\ell - E_{\ell-1}|\rangle$ - inverse of an average of ℓ-level spacing - is the *level density* of the $\{\ell\}$ manifold.

For an $|s\rangle$ level coupled to independent $\{\ell_1\}$, $\{\ell_2\}$, ...$\{\ell_n\}$ continua, the non-radiative lifetime of the $|s\rangle$ level population will be :

$$1/\tau_s^{nr} = k_s^{nr} = \sum_i k_{s\ell_i}$$

The coupling between the vibrationless level of the S_2 electronic state and a quasi-resonant vibronic level $|v_1, v_2, \ldots, v_n\rangle$ of the S_1 state is induced by the terms of the type :

$$\partial\psi^{el}/\partial Q_n \text{ and } \partial^2\psi^{el}/\partial Q_n^2$$

neglected in the Born-Oppenheimer approximation and involving only the electronic wavefunction. By representing Ψ_s and Ψ_ℓ as products of electronic, vibrational and spin functions (cf. Eq.(2.1)), we obtain :

$$v_{s\ell} = \langle\psi_s^{el}|H'|\psi_\ell^{el}\rangle\langle\chi_s(0,0,..0)|\chi_\ell(v_1,v_2,\ldots,v_n)\rangle \quad (3.4)$$

Since both states belong to the same spin manifold, $\langle\sigma_s|\sigma_\ell\rangle = 1$ in view of normalization of spin wavefunctions.

For the states belonging to different spin manifolds (coupling between $S_2 - T_i$, $S_1 - T_j$ and $T_1 - S_0$ states), H' corresponds to the spin-orbit interaction and $v_{s\ell}$ becomes :

$$v_{s\ell} = \langle\psi_s^{el}\sigma_{S=0}|H'_{so}|\psi_\ell^{el}\sigma_{S=1}\rangle \langle\chi_s(0,0,..0)|\chi_\ell(v_1,v_2,\ldots,v_n)\rangle \quad (3.5)$$

In both cases, $V_{s\ell}$ is represented by a product of an electronic coupling constant and of the overlap integral of vibrational wavefunctions. The analogy with radiative transitions is obvious.

3.3.2. Electronic Coupling Constants. The coupling constants are, in general, larger within a given spin manifold that between two electronic states with $\Delta S \neq 0$. If the energy gap between the $S_2(0)$ and $S_1(0)$ state is of the same order of magnitude as between $S_2(0)$ and $T_i(0)$ states (for the discussion of the energy gap dependence, cf. the next section), the $S_2(0) \rightarrow S^{\#}_1$ relaxation *(internal conversion)* will be privileged with repect to the $S_2(0) \rightarrow T^{\#}_1$ direct path; the lowest triplet level is usually populated by a sequential process : $S_2(0) \rightarrow S^{\#}_1 \rightarrow S_1(0) \rightarrow T^{\#}_i \rightarrow T^{\#}_1$ involving the $S_1 \rightarrow T^{\#}_i$ *intersystem crossing* as the slowest (rate determining) step.

The internal conversion of the S_i ($i \geq 2$) states is usually so rapid that the $S_2 \rightarrow S_0$ fluorescence is extremely weak with decay time too short to be directly measured. The rate of the $S_2(0) \rightarrow S^{\#}_1$ internal conversion may be only estimated using the Heisenberg relation :

$$\Delta\tau . \Delta E \geq \hbar$$

from homogeneous widths of bands in the $S_2 \leftarrow S_0$ absorption (or fluorescence-excitation) spectrum. For large aromatic molecules, the rates are of the order of $k_{IC} \cong 10^{13}\ s^{-1}$, i.e. more rapid by 4 ÷ 5 orders of magnitude than those of strong radiative transitions. The fluorescence yield of the $S_2 \rightarrow S_0$ emission is thus typically of the order of $10^{-5} \div 10^{-6}$.

In the case of the intersystem crossing, the relaxation rate strongly depends on the nature of interacting S_1 and T_i states. As previously discussed, the $^1n\pi^* \leftrightarrow {}^3\pi\pi^*$ and $^1\pi\pi^* \leftrightarrow {}^3n\pi^*$ coupling is stronger than the $^1n\pi^* \leftrightarrow {}^3n\pi^*$ and $^1\pi\pi^* \leftrightarrow {}^3\pi\pi^*$ one. The rates of the intersystem crossing vary in wide limits from $3\text{x}10^6\ s^{-1}$ for naphthalene (an aromatic hydrocarbon, for which all low-energy singlet and triplet states of the $\pi\pi^*$ type) to ca. $5\text{x}10^8\ s^{-1}$ for most aza-aromatics and ca. $10^{10}\ s^{-1}$ for aromatic ketones (for which the coupling between $n\pi^*$ and $\pi\pi^*$ states is efficient). Even in the last case, the intersystem crossing occurs at the time scale longer than that of the internal conversion between excited singlet states.

3.3.3. Vibrational Overlap Integrals and the Energy-Gap Law. In the radiationless electronic relaxation such as $S_2(0) \rightarrow S^{\#}_1$ process, a part of the electronic energy is transformed into vibrational energy :

$$E'_{el} + E'_{vib} = E''_{el} + E''_{vib}$$

The molecule is transfered from *low vibrational levels* (v_1', v_2', .., v_n') of the initially excited higher (S_2) electronic state to a set of quasi-isoenergetic *high vibrational levels* (v_1'',v_2'',...,v_n'') of the lower (S_1) state. If, as previously admitted, the (v_1'=0, v_2'=0, ..,v_n'=0) vibrationless level of the S_2 state - $S_2(0)$ is initially excited (E'_{vib} = 0) $\Delta E''_{vib}$ will be equal to the *energy gap* between the initial and final states :

$$E''_{vib} = \Delta E = E'(0) - E''(0).$$

In a polyatomic molecule, we will find in a narrow energy band around E''_{vib} a large number of levels corresponding to different combinations of modes with frequencies ν''_n and quantum numbers v_n'' such that :

$$\sum_{n=1}^{3N-6} v''_n \nu''_n = E''_{vib}$$

each of them being characterized by Δv_{tot} - the overall change of the vibrational quantum number :

$$\Delta v_{tot} = \sum_{n=1}^{3N-6} |v'_n - v''_n|$$

or (when the vibrationless level is initially excited)

$$\Delta v_{tot} = \sum_{n=1}^{3N-6} v''_n$$

As already mentioned (cf.Sec.2.2.3), if the potential for the n-th mode is identical in both electronic states, the overlap integral $\langle \chi_{v_n}'(Q_n)|\chi_{v_n}''(Q_n)\rangle \equiv 0$ for v' ≠ v". In real molecules, the potential is always more or less modified so that

$<\chi_{v_n}'(Q_n)|\chi_{v_n}''(Q_n)>$ integrals are different from zero for $v_n' \neq v_n''$ but their values decrease rapidly with Δv_n. This implies a rapid (quasi-exponential) decrease of the overall integral : $<\chi'(0,0,..0)|\chi''(v_1,v_2,..,v_n)>$ (product of $<\chi_{v_n}'(Q_n)|\chi_{v_n}''(Q_n)>$ factors) with Δv_{tot}. In a molecule with a narrow distribution of vibrational frequencies ν_n'' around their average value $<\nu''>$ one can assume that $\Delta E_{s\ell} = E''_{vib} = \Delta v_{tot}<\nu''>$, what implies also a quasi-exponential dependence of the overlap integral $<\chi'(0,0,..0)|\chi''(v_1,v_2,..,v_n)>$ on the energy gap. By combining this relation with Eq.(3.2a) and (3.4) or (3.5) we obtain - for a series of systems with identical electronic coupling constants - an exponential decrease of non-radiative constants $k_{s\ell}$ with the energy gap $\Delta E_{s\ell}$ [9] :

$$k_{s\ell} \cong Ae^{-\alpha\Delta E_{s\ell}} \quad \text{or} \quad \ln(k_{s\ell}) \cong A' - \alpha\Delta E_{s\ell} \qquad (3.6)$$

This relation may be considered as a particular case of a very general dependence of relaxation processes on the energy difference called the *energy gap law*.

In real molecules the frequency distribution is, however, very broad : its spectrum contains stretching modes of the X-H (C-H, O-H etc.) bonds with much higher frequencies (2800 ÷ 3600 cm^{-1}) than those of the skeleton modes (400 ÷ 1600 cm^{-1}). Δv_{tot} is small (and $k_{s\ell}$ relatively large) when the energy of the final level corresponds to a few quanta of high-frequency modes, while for the levels involving mainly low-frequency modes Δv_{tot} is large and $k_{s\ell}$ - small. The high-frequency modes are thus efficient *energy-accepting modes.* Their role is evidenced by a drastic decrease of the relaxation rate in deuterated compounds. Since the frequency of the X-D stretching vibration is lower by a factor close to $\sqrt{2}$ than that of the corresponding X-H mode, Δv_{tot} will be significantly increased on deuteration for all levels involving high-frequency (energy-accepting) modes.

The exponential dependence of $k_{s\ell}$ on $\Delta E_{s\ell}$ and a pronounced *deuterium effect* have been observed in many series of molecules, the classical case being that of the $T_1 \rightarrow S_0$ intersystem crossing in aromatic hydrocarbons (Fig.7).

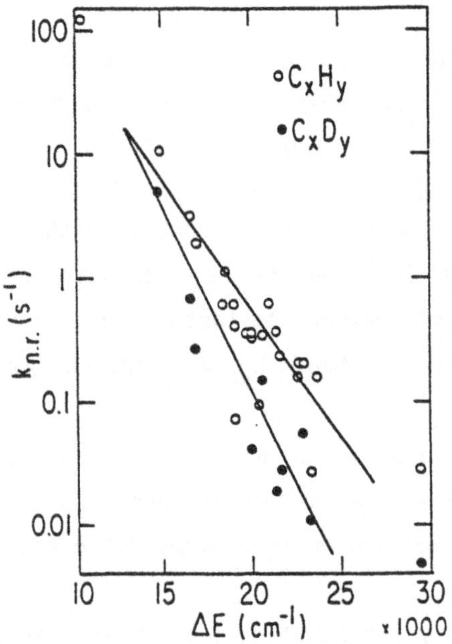

7. Dependence of of non-radiative rates of the $T_1 \rightarrow S_0^{\#}$ intersystem crossing on the $T_1 - S_0$ energy gap for a series of C_xH_y (circles) and C_xD_y (full circles) aromatic hydrocarbons (reproduced from W.Siebrand, J.Chem.Phys.47,2411,(1967))

A direct consequence of the energy gap law is the *Kasha rule* [10] : in almost all polyatomic molecules the excitation of higher electronic states is followed by a rapid $S_i \rightarrow S_1$ ($T_j \rightarrow T_1$) non-radiative relaxation.The fluorescence (or phosphorescence) is emitted only from the *lowest* singlet (or triplet) state. This behavior may be easily explained : the energy gap between the ground (S_0) and the first excited singlet (S_1) state is usually much larger than the $S_{i+1} - S_i$ ($i \geq 1$) energy gaps. The internal conversion $S_i \rightarrow S_2 \rightarrow S_1$ is thus so rapid that the $S_i \rightarrow S_0$ emission cannot be detected. For the same reason, the internal coversion within the triplet manifold is much more rapid than the $T_1 \rightarrow S_0$ intersystem crossing, so that the phosphorescence is always emitted from the lowest triplet state. The exceptions from the Kasha rule are molecules with the $S_1 - S_0$ gap smaller than the $S_2 - S_1$ one (e.g. azulene, thioketones, as well as a number of aromatic radical-cations).

3.4. SOLUTE-TO-SOLVENT ENERGY TRANSFER AND VIBRATIONAL RELAXATION.

The thermal movements in a liquid may be described - by analogy with crystal lattices - in terms of *external modes* (displacements and librations) of molecules with characteristic frequencies ν_{ext} in the 1 ÷ 100 cm^{-1} range. They may be also expressed as chaotic, aperiodic motion of individual molecules with a characteristic time necessary for their re-orientation (*relaxation time* τ_D) which corresponds roughly to the classical period ($T = 1/2\pi\nu_{ext}c$) of the external vibration. τ_D varies for typical, room temperature solvents in the $10^{-10} \div 10^{-12}$ s limits and strongly depends on the viscosity (temperature) of the liquid.

The solvent plays in the condensed-phase molecular spectroscopy and dynamics the role of the *heat bath* allowing dissipation of the excess of energy (thermalization) of the molecule. The efficiency of the solvent-to-solute energy transfer is very different for vibrational and electronic energy of the molecule. The vibrational-energy excess is rapidly transfered to the solvent, while the electronic relaxation is an essentially intramolecular process, only slightly influenced by solvent.

This difference may be easily understood on the basis of the energy gap law. The frequencies of "external" vibrations are low and practically independent of the electronic state of a solute molecule. The *direct* transfer of a large amount of the electronic energy of the solute to the solvent (external) modes necessitates Δv_{tot} so large (and the corresponding overlap integrals so small) that the $v_{s\ell}$ matrix element and the k_{sl} rate constant are practically equal to zero. In contrast to it, the low-frequency vibrational modes of the solute molecule are strongly coupled to the heat bath, the frequency difference between intra-molecular and external modes being relatively small.

The *vibrational relaxation* of polyatomic molecules is a sequential process. When a polyatomic molecule is optically excited, its excess of vibrational energy is usually contained in one or a few optically active normal modes. The first step of the relaxation process is the *intramolecular redistribution of the vibrational energy (IVR)* among a large number of modes (with an important amount of energy going to the low-frequency modes) [11]. This process takes place in isolated molecules at a few picosecond time scale and is accelerated by environment effects (collisions, solute-solvent interactions). The energy transfer from low-frequency internal modes to those of the solvation shell and its dissipation in the bulk may be considered as its further steps. However, the amount of experimental data allowing to follow the relaxation process in real time is still very limited.

The vibrational relaxation rates vary in very wide limits [12] ($k_{rel} \cong 10^{-2}\ s^{-1}$ for N_2 molecules in liquid nitrogen) but for large molecules in room-temperature solvents k_{rel} is of the order of $10^{11} \div 10^{13}\ s^{-1}$ (very close to the solvent relaxation rates). $\tau_{rel} = 1/k_{rel} \sim 10^{-12}$ s is thus very short as compared to typical lifetimes of electronically excited states ($\tau_f = 10^{-7} \div 10^{-9}$ s) : we may, therefore, consider that the fluorescence of the solute is emitted by the entirely relaxed molecule so that its spectrum and lifetime characterize the thermally equilibrated system and do not depend on the wavelength of the exciting radiation i.e. on the the initially excited vibrational level.

4. Intermolecular Energy Transfer.

4.1. GENERAL REMARKS.

The problem of the *intermolecular* electronic energy transfer is closely related to that of the *intramolcular* electronic relaxation. We are interested by the time evolution of a system composed of the *energy donor* A and *energy acceptor* B from its initial state : A* + B to the resonant or quasi-resonant A + B* state. A and B may be two different parts (chromophores) of the same molecule or two different molecules. The rate of the transfer process depends on the strength of the A*B ⟷ AB* coupling dependent in turn on the nature of excited states A* and B* and on the A - B distance R_{AB}

In a realistic case of fluid or rigid solutions, we have to do with a wide statistical distribution of R_{AB} dependent on A and B concentrations : c_A and c_B. Moreover, the diffusion phenomena (A - B distance varying with time) cannot be neglected for low viscosity media. Nevertheless, for simplicity sake, we will limit our present discussion to the case of an A + B pair with a well defined, constant R_{AB} value.

The essential feature for a good description of the A*-B system is the fact that both molecules interact with their environment : the thermal movements of the solvent induce fluctuations of the energies of A and B (A* and B*). It may be easily shown that the kinetic model described in the next sections cannot be applied to an isolated A + B pair in absence of the heat bath [13].

4.2. COUPLING MECHANISMS.

Initial $|\Psi_i\rangle$ and final $|\Psi_f\rangle$ states of the A + B system are - in the zero-order approximation :

$$|\Psi_i\rangle = |\psi_A{}^*\psi_B\rangle|\sigma_A{}^*\sigma_B\rangle|\chi_A{}^*\rangle|\chi_B\rangle \quad \ldots\ldots.(4.1a)$$

$$|\Psi_f\rangle = |\psi_A\psi_B{}^*\rangle|\sigma_A\sigma_B{}^*\rangle|\chi_A\rangle|\chi_B{}^*\rangle \quad \ldots\ldots.(4.1b)$$

where ψ,σ and χ indicate electronic (orbital and spin) and vibrational wavefunctions of A and B molecules in their ground and excited states. Note that the coupling between A and B vibrations is neglected : the vibrational wavefunction of the A.B system is a simple product of χ_A and χ_B, while such an approximation is not applied to the electronic wavefunctions. The probability of the energy transfer $k_{i\to f}$ between a pair of states is proportional to the square of the matrix element $|\langle\Psi_i|V|\Psi_f\rangle|^2$. If the coupling involves only the orbital part of the electronic wavefunction and the spin-orbit coupling may be neglected $k_{i\to f}$ will be (in a close analogy with the formula for the radiative transitions cf.Eq.1.4) :

$$k_{i\to f} \sim |\langle\Psi_i|V|\Psi_f\rangle|^2 = |\langle\psi_A{}^*\psi_B|V_{AB}|\psi_A\psi_B{}^*\rangle|^2 \; |\langle\sigma_A{}^*\sigma_B|\sigma_A\sigma_B{}^*\rangle|^2 \times$$
$$\times|\langle\chi_A{}^*(v_A{}')|\chi_A(v_A{}'')\rangle\langle\chi_B(v_B{}'')|\chi_B{}^*(v_B{}')\rangle|^2 \qquad (4.2)$$

Eq.(4.2) describes a process involving a transition from the initial vibronic state $v_A{}'v_B{}''$ to the final one $v_A{}''v_B{}'$ and its last factor is the product of Franck-Condon factors for $v_A{}' \to v_A{}''$ and $v_B{}'' \to v_B{}'$ transitions. In order to obtain the overall rate of the energy transfer we must sum over all possible final states $v''_A v_B{}'$. Since the total energy of the system : $E_A + E_B$ must be conserved, only the transitions with $\Delta E_B = -\Delta E_A$ ($\nu_B = \nu_A$) are to be taken into account :

$$k_{i\to f} \sim \sum_f |\langle\Psi_i|V|\Psi_f\rangle|^2 = |\langle\psi_A{}^*\psi_B|V_{AB}|\psi_A\psi_B{}^*\rangle|^2 \; |\langle\sigma_A{}^*\sigma_B|\sigma_A\sigma_B{}^*\rangle|^2 \times$$
$$\times\sum_{\nu_A=\nu_B}|\langle\chi_A{}^*(v_A{}')|\chi_A(v_A{}'')\rangle\langle\chi_B(v_B{}'')|\chi_B{}^*(v_B{}')\rangle|^2 \qquad (4.2a)$$

It is practically impossible to evaluate Franck-Condon factors for all vibronic transitions but the value of the sum :

$$\sum_{\nu_A=\nu_B}|\langle\chi_A{}^*(v_A{}')|\chi_A(v_A{}'')\rangle\langle\chi_B(v_B{}'')|\chi_B{}^*(v_B{}')\rangle|^2$$

may be approximated by the overlap integral of the normalized spectra of donor emission - $f_A(\nu)$ and the acceptor absorption B - $a_B(\nu)$:

$$\int_0^\infty f_A(\nu)a_B(\nu)d\nu$$

the so called *Förster overlap integral*, the value of which is deduced from experimental data (Fig.8).

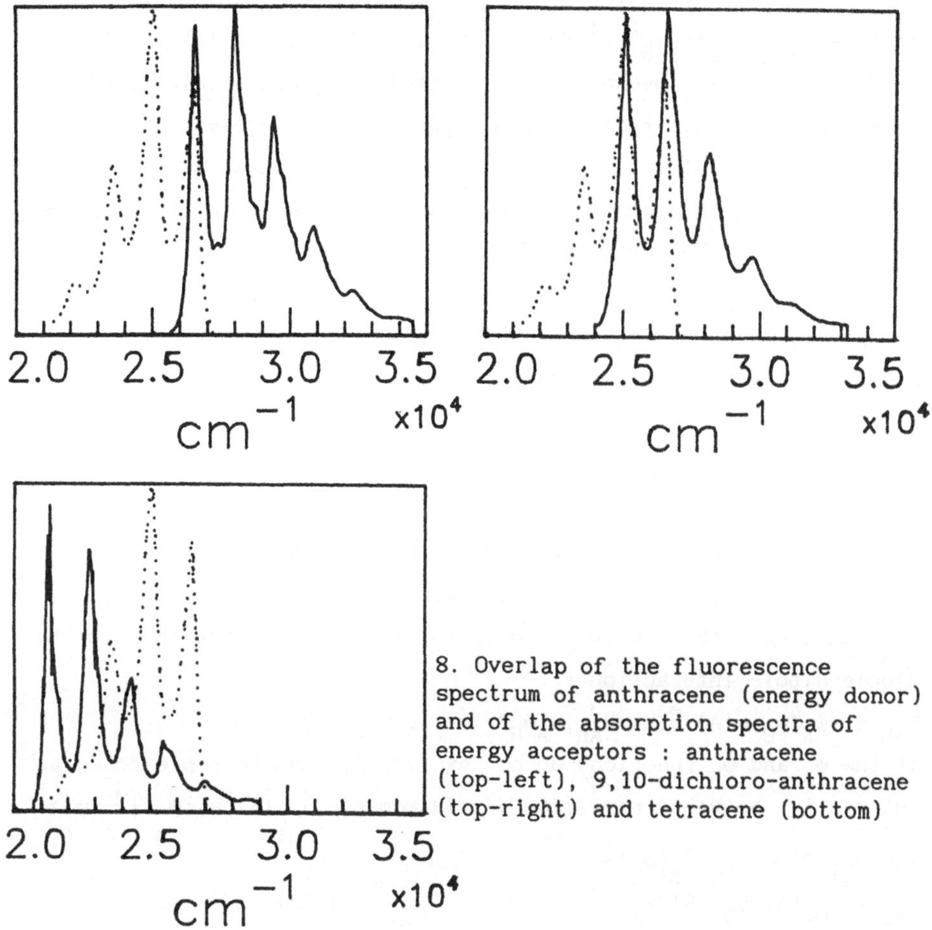

8. Overlap of the fluorescence spectrum of anthracene (energy donor) and of the absorption spectra of energy acceptors : anthracene (top-left), 9,10-dichloro-anthracene (top-right) and tetracene (bottom)

Obviously, the value of this integral is zero for $E_{B*} > E_{A*}$. It will be different from zero when $E_{A*} > E_{B*}$ and its value reaches a maximum for rather small values of $\Delta E = (E_{A*} - E_{B*})$ energy difference. For large ΔE values the overlap will decrease as can be seen in Fig.8. The $A^*B \rightarrow AB^*$ energy transfer is thus efficient in a restrained ΔE range limited by the energy conservation principle for $\Delta E < 0$ and by the energy gap law for large ΔE values.

4.3. SINGLET AND TRIPLET ENERGY TRANSFER.

The selection rule concerning the electron spin may be deduced from the $|\langle\sigma_A{}^*\sigma_B|\sigma_A\sigma_B{}^*\rangle|^2$ factor in Eq.(4.2) : for a large A-B distance (i.e. in absence of the overlap of ψ_A and ψ_B electronic wavefunctions) this factor may be rewritten as $|\langle\sigma_A{}^*|\sigma_A\rangle\langle\sigma_B|\sigma_B{}^*\rangle|^2$. Its physical meaning is simple : the transition is allowed only when the spins of both A and B molecules are the same in their ground and excited states. We have then :

$$|\langle\sigma_A{}^*|\sigma_A\rangle\langle\sigma_B|\sigma_B{}^*\rangle|^2 = 1$$

The energy transfer from the excited singlet state:

$$S_1{}^*(A) + S_0(B) \rightarrow S_0(A) + S_1{}^*(B)$$

is thus allowed, while the triplet energy transfer :

$$T_1{}^*(A) + S_0(B) \rightarrow S_0(A) + T_1{}^*(B)$$

is, in this approximation, forbidden (see below).

For intermolecular distances large with respect to the molecule dimensions (i.e. to the extension of electronic wavefunctions) the leading term in the V_{AB} coupling constant is the dipole-dipole interaction :

$$V_{AB} = \langle\psi_A{}^*\psi_B|\vec{\mu}(A).\vec{\mu}(B)/R_{AB}^3|\psi_A\psi_B{}^*\rangle \text{..........(4.3)}$$

If the ψ_A and ψ_B functions do not overlap V_{AB} may be represented as interaction between *transition dipole moments* $\vec{\mu}_{01}(A) = \langle\psi_A{}^*|\vec{\mu}_A|\psi_A\rangle$ and $\vec{\mu}_{01}(B) = \langle\psi_B|\vec{\mu}_B|\psi_B{}^*\rangle$:

$$V_{AB} = \vec{\mu}_{01}(A).\vec{\mu}_{01}(B)/R_{AB}^3 \text{....................(4.3a)}$$

As a matter of fact, the terms involving higher multipole moments and the overlap of ψ_A and ψ_B functions decrease much more rapidly with the A-B distance than the dipole-dipole term.

We obtain finally :

$$k_{i\rightarrow f} = \text{const.}(|\langle\vec{\mu}_{01}(A).\vec{\mu}_{01}(B)\rangle|^2/R_{AB}^6)\int f_A(\nu)a_B(\nu)d\nu\text{...(4.4)}$$

where $\langle\vec{\mu}_{01}(A).\vec{\mu}_{01}(B)\rangle$ indicates an average value of the dipole-dipole term for a random orientation distribution. The rate of the singlet-singlet energy transfer is thus proportional to the product of intensities I_{01} (i.e. to $|\mu_{01}(A)|^2|\mu_{01}(B)|^2$) of electronic transitions in A and B molecules.

$$k_{i\to f} = \text{const}'.I_{01}(A)I_{01}(B)\int f_A(\nu)a_B(\nu)d\nu/R_{AB}^6 \quad \ldots(4.4a)$$

and decreases as R^{-6} with the intermolecular distance [13].

The singlet-singlet energy transfer is thus described as a *virtual absorption-emission process* (Fig.9), different from the real reabsorption processes (cf.below Sec.4.4.b).

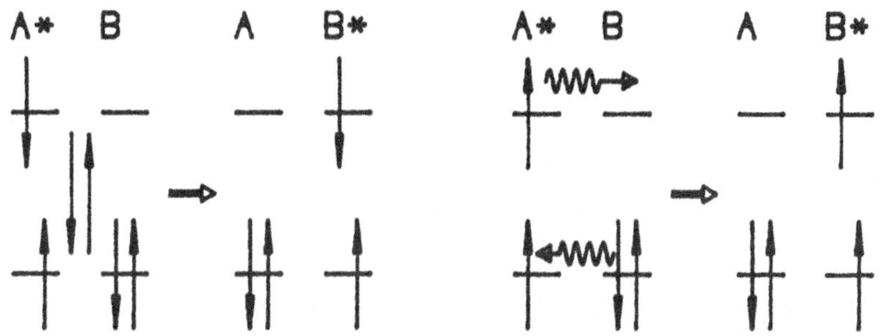

9. Schematic representation of the Förster type energy transfer considered as virtual absorption-emission process (left) and of the Dexter mechanism considered as due to the electron exchange.

From the point of view of experiment, the important parameter is R_{eff} which may be defined as such a value of R_{AB} that the rate of the $A^* \to B$ energy transfer $k_{i\to f}$ is equal to the intrinsic decay rate of the A* molecule - k_i. Note that the radiative rate k_i^{rad} increases with the emission frequency : $k_i^{rad} \sim \nu^3$, while $k_{i\to f}$ is frequency independent. The singlet-singlet energy transfer is thus relatively more efficient in the visible than in the UV region and for molecules with large transition moments (dyes) R_{eff} may attain relatively large values of the order of 10 nm (10÷20 times the molecular radius).

When the intermolecular distance is so strongly reduced that the electronic wavefunctions of the donor and acceptor overlap, the second-order terms in V_{AB} become non-negligible. Note that the electronic wavefunctions $|\psi_A{}^*\psi_B\rangle$ and $|\psi_A\psi_B{}^*\rangle$, represented as products of mono-electronic functions (molecular orbitals) differ only by functions of two electrons, so that V_{AB} contains the term :

$$V_{AB} = \langle \ldots \varphi_a{}^*(1)\varphi_b(2)|\vec{\mu}(A).\vec{\mu}(B)/R_{AB}^3|\ldots\varphi_b{}^*(1)\varphi_a(2)\rangle \quad (4.5)$$

where φ_a, φ_b, $\varphi_a{}^*$, $\varphi_b{}^*$ are, respectively, HOMO and LUMO of A and B molecules, the wavefunctions of other electrons being omitted. This term corresponds to the *electron exchange* between A and B molecules and is, in general, different from zero, as long as these orbitals overlap, i.e. when :

$$\langle\varphi_a{}^*|\varphi_b{}^*\rangle \neq 0 \text{ and } \langle\varphi_a|\varphi_b\rangle \neq 0$$

In this process, the spin of the electron (1) or (2) is not inverted (Fig.9), so that the exchange mechanism (called often the *Dexter mechanism* [14]) enables the *triplet-energy transfer* forbidden in the dipole-dipole (Förster) mechanism.

Since at a large distance from the molecule center, the electronic wavefunctions decay exponentially with R_{AB}, the rate of the transfer decreases as :

$$k_{i\to f} \sim V_{AB}^2(R_{Ab}) \sim e^{-\alpha R_{AB}} = e^{-R_{AB}/R_{eff}} \quad (4.6)$$

Such a dependence is confirmed by experiment showing R_{eff} of the order of 1 nm for a number of organic molecules. In view of the exponential dependence of $k_{i\to f}$ on R_{AB} the rate of the triplet-energy transfer goes rapidly to zero when the intermolecular distance is increased.

4.4. FINAL REMARKS.

a) As previously noted, in a fluid or rigid solution of A and B molecules, we have a wide distribution of intermolecular distances R_{AB}. Since $k_{i\to f}$ varies rapidly with R_{AB}, the energy transfer from a set of excited A* molecules cannot be described by a simple exponential law but shows a more complex time dependence [13].

b) The energy transfer is a *non-radiative process* different from that of reabsorption of the emitted photon by an other, identical or different, molecule (*radiation trapping*). This difference may be visualized in the simple case of identical donor and acceptor molecule. The number of excited centers N^* remains unchanged in the case of the energy transfer :

$$dN^*/dt = -k_i N^* - k_{i\to f}N^* + k_{i\to f}N^* = -k_i N^*$$

(where $-k_{i\to f}N^*$ and $+k_{i\to f}N^*$ indicate identical numbers of de-excited and excited centers). The decay time is not modified. In contrast to it, the reabsorption of the fraction β of $k_i N^*$ emitted photons :

$$dN^*/dt = -k_i N^* + \beta k_i N^* = -(1-\beta)k_i N^* = - k_i' N^*$$

implies an apparent increase of the decay time from τ to $\tau' = \tau/(1-\beta)$.

c) A consequence of an efficient excitation transfer from molecule to molecule is the possibility of a double excitation of one molecule resulting from an "encounter" of two elementary excitations. The important case of such an *excitation fusion* is the transfer of the triplet energy to a molecule already excited to the triplet state :

$$T^* + T^* \to ....$$

The further step depends on the total spin of the interacting pair $\vec{S} = \vec{S}_1 + \vec{S}_2 = 0$, 1 or 2. For $\vec{S} = 0$ ($S_{1z} = +1$, $S_{2z} = -1$), we obtain the *trip)let-triplet annihilation* :

$$T_{+1}^* + T_{-1}^* = S_n^* + S_0$$

which may be monitored by decay of the triplet population and appearance of the *delayed fluorescence* resulting from the $S_n^* \to S_1^*$ relaxation followed by the $S_1^* \to S_0$ emission. For S = 1, the only the quenching of the triplet may take place :

$$T_{+1}^* + T_0^* = T_{+1}^* + S_0$$

while the $T_1^* + T_1^*$ (S=2) "collisions" are unefficient.

5. Solvent-solute Interactions.

5.1. GENERAL REMARKS.

The aim of this section is to treat - from the point of view of the spectroscopic experiment - a few elementary processes which may be situated half-a-way between physics and chemistry of excited molecular states. In the previous discussion, it was implicitly admitted that the interaction between an electronically excited molecule and its environment is limited to the solute-to solvent energy transfer. As a matter of fact, even in absence of "true" photochemical reactions, the excited molecule interacts strongly with its environment and may be involved in such processes, as :

(i) dissociation (predissociation) in the excited state :

$AB + h\nu \rightarrow (AB)^* \rightarrow A + B$ ....(5.1)

(ii) ionization followed by formation of the solvated cation and of the *solvated electron* - $\bar{e}_s$:

$AB + h\nu \rightarrow AB^+ + \bar{e} \rightarrow (AB)_s^+ + \bar{e}_s$ ....(5.2a)

(iii) the electron transfer between specific pairs of molecules AB and CD

$AB + CD + h\nu \rightarrow AB^+ + CD^-$................(5.2b)

with further formation of a strongly polar $(AB^+.CD^-)$ complex or of a *solvated ion pair* $(AB)_s^+ + (CD)_s^-$.

(iv) formation or breakdown of intermolecular bonds.

The electronic excitation may induce an important redistribution of the electronic charge *within* the molecule. In benzene and naphthalene derivatives, the electronic excitation corresponds to the intramolecular charge transfer from electron-donor substituents ($-NH_2$, -OH) to the ring and from the ring to the electron-acceptor substituents (-COOH, $-NO_2$). This transfer may be evidenced by a large difference between permanent dipole moments in the ground and excited electronic states : $\vec{\mu}_g$ and $\vec{\mu}_e$.

All processes involving molecular ions or large electric dipole moments are sensitive to *dielectric properties* of the solvent. The effects of the polar environment will be, therefore discussed in this chapter.

Specific molecular interactions (formation of van der Waals, donor-acceptor or hydrogen-bonded complexes) are deeply modified by the electronic excitation. The redistribution of the electronic density in excited molecules implies often a drastic change of their *acid-base properties.* Its direct consequence is a shift of the acid-base equilibria, i.e. the proton transfer processes :

$AH + B + h\nu \rightarrow AH^* + B \rightarrow A^{-*} + HB^+$ ....(5.3)

which may be considered as the simplest photochemical processes involving the electronically excited solute molecule interacting with the solvent (or with another solute).

We will very briefly discuss the problems related to the dissociation and electron-transfer processes (developed in the further chapters of this book [15,16]) and treat in a more detailed way dielectric solute-solvent interactions and properties of molecular complexes.

5.2 PHOTODISSOCIATION.

Photodissociation of isolated molecules has been extensively studied since a long time. The rate of this process varies in very wide limits between $\sim 10^{-14}$s (direct dissociation) to 10^{-8}s (slow predissociation). On the other hand, the dissociation yield is always high, close to one : since the internal energy of the isolated molecule is conserved, each molecule with energy exceeding its dissociation threshold will fragmentate unless its energy excess is lost by fluorescence emission. The dissociation takes often place after the radiationless relaxation of electronically excited molecule to its "hot" ground state with a large amount of vibrational energy. The dissociation is irreversible : as long as the initial energy of AB and CD fragments (exceeding that of the AB-CD bond) is conserved, AB and CD cannot recombine in a two-body collision.

All these properties are deeply modified by the solute-solvent interaction :

If the initial energy of the molecule is higher than that of the thermally equilibrated S_1 state, this energy excess will be lost by the electronic relaxation and/or by the vibrational energy transfer to the heat bath (cf.Sec.3.3b and 3c). The branching ratio between photodissociation and relaxation depends on the rates of both processes. Since the $S_2 \rightarrow S_1^{\#}$ and $S_1^{\#} \rightarrow S_1$ relaxation rates are typically of the order of 10^{12} s^{-1}, the necessary condition for efficient dissociation is $k_{diss} \geq 10^{12}$ s^{-1}. Otherwise, the molecule is relaxed to the thermally equilibrated S_1 or T_1 states and dissociation will occur only when one of them is dissociative or predissociated.

Dissociation of the AB molecule into A and B fragments takes place within a "cage" (solvation shell) formed by the solvent molecules. The fragments will either escape from the cage and separate or remain in the cage, lose their energy in collisions with the solvent molecules and finally recombine *(geminal recombination)*. The effective photodissociation yield is thus strongly reduced by the *cage effect*. Detailed studies of the escape and recombination probabilities have been recently developed for molecules in rare gas matrices [17].

5.3. ELECTRON TRANSFER PROCESSES.

5.3.1. Dielectric effects and spectral solvent shifts. As already mentioned, the molecular ions and strongly polar zwitterions are stabilized by polarization of their environment, as evidenced by spectral shifts depending on dielectric properties of the solvent.

The leading term in the solute-solvent interactions is that of the electric dipole moment of the molecule - $\vec{\mu}$ with the *reaction field* -$\vec{\mu}$F resulting from the polarization of the dielectric. The strength of the field depends on the solvent polarizibility which may be divided into two parts and expressed in terms of the high-frequency ($\varepsilon_\infty = n^2$) and static (ε_0) parts of its dielectric constant. The first one corresponds to the *electronic polarizibility*; this part of the reaction field follows without delay all changes of μ. The second one - *orientational polarizibility* - is due to the reorientation of the solvent molecules around the solute occuring in a finite time : related to the solvent relaxation time - τ_D (see above, Sec.3.4).

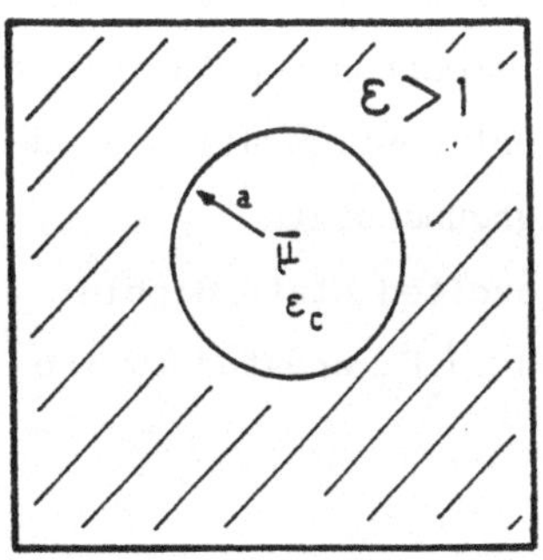

10. Schematic representation of the molecule in the Onsager model.

In the Onsager model [18], the solute-solvent interaction is approximated by that of a point dipole placed in the center of a spherical cavity of radius a (corresponding to the van der Waals radius of the molecule) with an infinite, continuous dielectric (Fig.10).

The energy of this interaction, in thermal equilibrium conditions, is represented by a sum of two terms :

$\Delta E_{sol} = -(\mu^2 F_1 + \mu^2 F_2)/a^3$ where :

$$F_1 = (n^2-1)/(2n^2+1)$$
$$F_2 = (\varepsilon_0 -1)/(\varepsilon_0+2) - (n^2-1)/(n^2+2)$$
corresponding to the electronic and orientational polarization.

We are interested in the solvent effect on thefrequency of an electronic transition i.e. in the difference between solvation energies of both electronic states :

$$\Delta\nu = [(\Delta E_{sol})_e - (\Delta E_{sol})_g]/hc$$

Since the electronic transition takes place at the time scale much shorter than the solvent relaxation time, solvation energies must be evaluated in different ways for initial and final states of the electronic transition. In the case of absorption, the solvation of the ground state corresponds to the *equilibrium configuration* so that :

$$\Delta E_g = - \mu_g^2(F_1 + F_2)$$

but the optical excitation prepares the excited state *out of equilibrium*. The reaction field due to the electronic polarizibility is now equal to $-\vec{\mu}_e F_1$ but the initial solvent configuration around the excited molecule is still that of the ground state (*Franck-Condon configuration*) so that the excited-state dipole moment $\vec{\mu}_e$ interacts with the reaction field $-\vec{\mu}_g F_2$ created by the ground state molecule :

$$\Delta E_e' = - (\mu_e^2 F_1 + \vec{\mu}_e \vec{\mu}_g F_2)$$

If the lifetime of the electronic excited state τ_{fl} is longer than τ_L, the emission will take place from the equilibrium configuration of the solvent around the excited molecule with stabilization energy

$$\Delta E_e = - \mu_e^2(F_1 + F_2)$$

to the Franck-Condon ground state :

$$\Delta E_g' = - (\mu_g^2 F_1 + \vec{\mu}_e \vec{\mu}_g F_2)$$

The solvent-induced frequency shift is thus different in absorption

$$h\Delta\nu_{abs} = \Delta E_e' - \Delta E_g = (\mu_g^2 - \mu_e^2)F_1 + \vec{\mu}_g(\vec{\mu}_g - \vec{\mu}_e)F_2$$

and in emission :

$$h\Delta\nu_{em} = \Delta E_e - \Delta E_g' = (\mu_g^2 - \mu_e^2)F_1 + \vec{\mu}_e(\vec{\mu}_g - \vec{\mu}_e)F_2$$

what implies a *solvent-induced Stokes shift* :

$$h\delta\nu_{St} = h(\Delta\nu_{abs} - \Delta\nu_{em}) = (\vec{\mu}_g - \vec{\mu}_e)^2 F_2 \qquad (5.7)$$

From the variation of the Stokes shift with the solvent polarity one can estimate the dipole moment of the molecule in its electronically excited state (Fig.11). On the other hand, the relaxation time of the solvent may be estimated from the time dependence of the Stokes shift in the time-resolved experiments at the picoseond time scale.

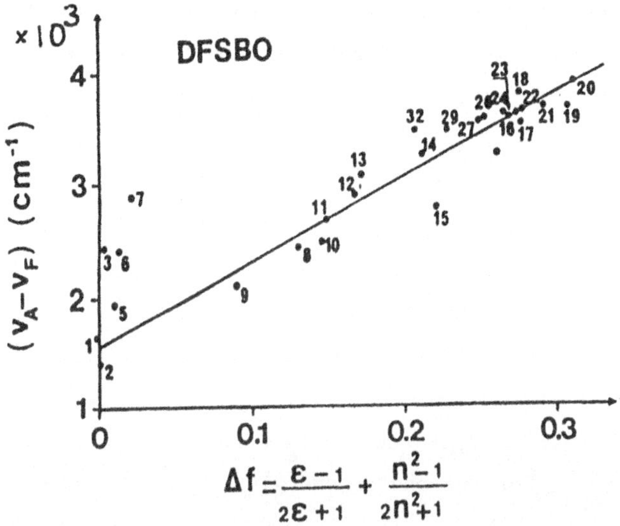

11. Dependence of the Stokes shift $\delta\nu_{St} = \nu_{abs} - \nu_{fl}$ for the DFSBO dye on the dielectric properties (n, ε) of the solvent (reproduced from J.C.Mialocq and M.Meyer, Laser Chem. 10,277,(1990))

5.3.2. Photoionization. Photoionization of an isolated molecule AB is a relatively well understood process. Its threshold corresponds roughly to its ionization potential I_{AB} (difference between the energy between the AB molecule and that of the free electron and AB^+ ion with internal energy of the ion and kinetic energy equal to zero).

The time scale of the *direct ionization* :

$$AB + h\nu \rightarrow AB^+ + \bar{e}$$

and of the *autoionization*

$$AB + h\nu \rightarrow AB^{**} \rightarrow AB^{+} + \bar{e}$$

processes (where AB** is a super-excited state with energy exceeding the ionization threshold) is of the order of 10^{-15} s.

In condensed phases, a close analog of this process is ejection of an electron by a "super-excited" molecule taking place at the time scale shorter than its electronic and vibrational relaxation. The kinetic energy excess of the photo-electron is rapidly dissipated and the electron is localized in some specific site of the solvent "lattice". The overall energy of a sytem : solvated AB^{+}_{s} ion + solvated electron $\bar{e}_{s}$ will be :

$$E_{eq} = I_{AB} - (E_{sol})_{AB^{+}} - (E_{sol})_{\bar{e}} \ldots\ldots\ldots\ldots (5.5)$$

significantly lower than the gas-phase ionization energy. Such a thermal equilibrium is, however, attained with a delay corresponding to the solvent relaxation time, while in the *initial state* (the Franck-Condon configuration of the solvent) the nascent ion + electron pair is only slightly stabilized by the electronic polarizibility of the medium :

$$E_{in} = I_{AB} - (E'_{sol})_{AB^{+}} - (E'_{sol})_{\bar{e}} \ldots\ldots\ldots\ldots (5.5a)$$

where $E'_{sol} \ll E_{sol}$. For the photon energies :

$$E_{in} \leq h\nu \leq E_{eq}$$

the photoionization of the solute may thus be described as a tunneling through a $E_{in} - h\nu)$ energy barrier, as described by the Marcus theory of the electron transfer [15].

5.3.3. Exciplexes and Charge-Transfer Complexes. We will treat here the special case of electron transfer between a pair of molecules : *electron-donor D* and *electron acceptor A* which may induce formation of strongly polar and strongly bonded excited states of the pair: $A^{-}D^{+}$ *exciplexes*. Their properties may be described in terms of the complete set of electronic states of the "super-molecule" A.D. This set may be - in the zero-order approximation - divided into two groups [19] :

(i) *van der Waals* states : the ground state and the excited states resulting from a local excitation of one of the molecules : A*..D and A..D*. The complex is bonded only by a weak van der Waals interaction; the energies of excited states - ε_{vdW} are thus close to the excitation energies of the free A* or D* molecule - ε_{A*} or ε_{B*},

(ii) *ionic* states formed by the electron transfer from the electron *donor* - D to the electron *acceptor* - A with formation of an ion pair $A^{-}..D^{+}$ strongly bonded by the coulombic attraction. The energy of the lowest ionic state - ε_{ion} may be estimated as :

$$\varepsilon_{ion}(R^{\pm}_{eq}) = I_D + E_A + C(R^{\pm}_{eq}) \qquad \text{(IV.6)}$$

where I_D is the ionization potential of the donor, E_A - the electron affinity of the acceptor and $C(R^{\pm}_{eq}) \cong -e^2/4\pi\varepsilon R^{\pm}_{eq}$ - the Coulomb energy at the equilibrium distance of the ionic state - $R^{\pm}_{eq}$ usually much shorter than R_{eq} in the van der Waals states.

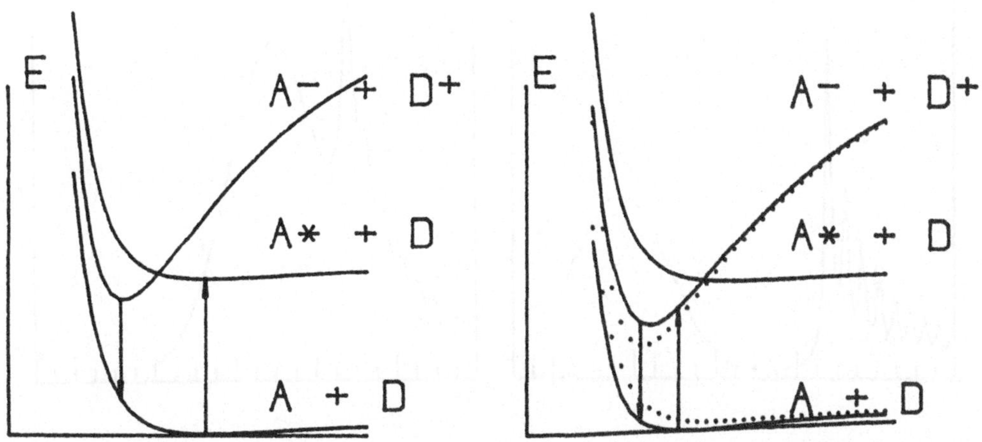

12. Schematic representation of the potential energy curves and electronic transition for an exciplex (left) and a charge-transfer complex (right). Note that the direct transition to the charge-transfer excited state is possible only in the latter case.

The lowest ionic state of the A.D system plays an important role when :

$$\varepsilon_{ion}(R^{\pm}_{eq}) < \varepsilon_{vdw}(R_{eq})$$

One can roughly differentiate two types of A+D systems :

(i) In most cases (Fig.12), we have :

$$\varepsilon_{ion}(R_{eq}) > \varepsilon_{vdw}(R_{eq})$$

so that the optical excitation ("vertical transition") prepares initially the A*D (or AD*) state. In any way, the bonding energy of the ground state of AD is so weak that the complex is unstable in solutions at the room temperature, as evidenced by the absorption spectrum the A + D solution - a simple sum of unmodified absorption bands of the A and D molecules (Fig.13). Upon the excitation of the D (or A), the behavior of the system depends on the (diffusion controlled) probability of the A + D* (or A* + D) encounter with the R_{DA} distance corresponding to the intersection of the (A..D*) and

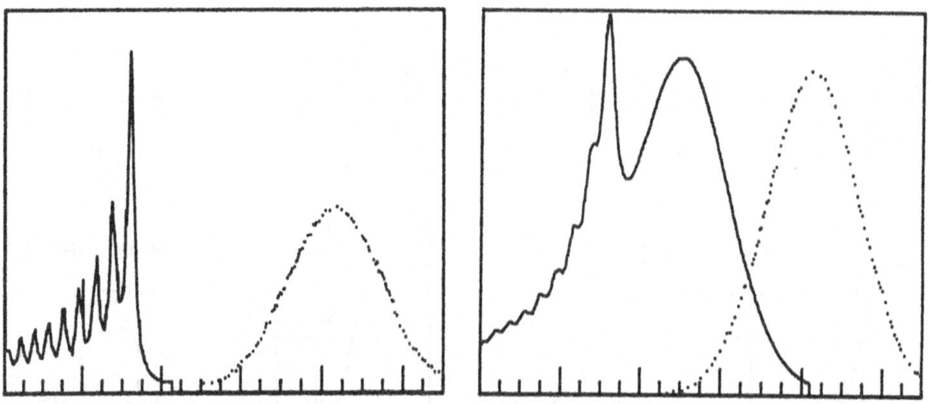

13. Schematic representation of absorption and emission spectra for an exciplex (left) and a charge transfer complex (right). Note the presence of the charge-transfer band in the latter case only. The narrow-band spectrum in (a) corresponding to the transition to the lowest van der Waals state becomes diffuse in (b) because of a rapid relaxation from the van der Waals to the charge-transfer state.

$A^{-}..D^{+}$ potential surfaces. The further evolution may be described in terms of the "harpooning" process [20] : the electron jump from the lowest excited orbital of the donor to the lowest unoccupied orbital (LUMO) of the acceptor creating an $A^{-} + D^{+}$ ion pair. In non-polar solvents the attraction between A^{-} and D^{+} ions results in formation of the strongly bonded $(A^{-}..D^{+})_{s}$ complex *(exciplex)*. This last stage may be monitored by detection of the *exciplex emission* : a broad, structureless band strongly red shifted with respect to the A and D absorption bands. This band corresponds to the *charge-transfer* transition between the ionic $A^{-}D^{+}$ state and the ground state : the polar character of the excited state may be evidenced by a very strong solvent shift in weakly polar solvents.[21-22].

In strongly polar solvents, the electron transfer is not followed by formation of a complex, each radical-ion being efficiently stabilized by the solvent molecules. The presence of the pair of solvated radical-ions : $A^{-}_{s} + D^{+}_{s}$ may be evidenced by their transient absorption spectra. The encounter distance corresponding to the surface crossing was established from the kinetic studies of the D* fluorescence quenching and $A^{-}D^{+}$ fluorescence induction. Its value is of the order of 7÷8 A for aromatic hydrocarbons and their derivatives [21].

(ii) In the case of very low-lying ionic states, the interaction with the ground state is strong enough to induce a non-negligible level shift and mixing of both states (Fig.12). The admixture of the ionic wavefunction ψ_{ion} to that of the ground state is strong enough for formation of the stable, polar complex (*charge-transfer complex)* in the ground electronic state [23] :

$$\psi_{0} = \alpha\psi_{g}(A.D) + \beta\psi_{i}(A^{-}.D^{+}) \text{ with } \alpha^{2} \gg \beta^{2}$$

The evidence for this process is $\mu_{g} \neq 0$ in the ground states of complexes formed by non-polar A and D molecules and appearance in the electronic spectrum of a characteristic *charge-transfer* absorption band - broad, structureless and red shifted with respect to the absorption bands of the complex components (Fig.13). It corresponds to the direct excitation of the A^-D^+ state i.e. to the promotion of an electron from the highest occupied orbital (HOMO) of the donor to the lowest unoccupied orbital (LUMO) of the acceptor. The emission of the complex : a similar broad band with a very large Stokes shift is identical with the previously described exciplex emission. Such a Stokes shift, as well as the width and shape of absorption and emission bands indicate a pronounced difference between equilibrium geometries in the weakly bound ground state and strongly bound excited state ($R^{\pm}_{eq} \ll R_{eq}$).

5.4. PROTOLYTIC EQUILIBRIA AND PROTON TRANSFER.

As previously mentioned, the charge distribution in the molecules of aromatic and hetero-aromatic compounds is deeply modified by electronic excitation. A direct consequence of the increased effective positive charge of the -OH and $-NH_2$ groups will be enhancement of the acidity of hydroxy-aromatics (naphthol,phenol) and aromatic amines (aniline and its conjugated acid : anilinium ion - $C_6H_5-NH_3^+$) in their excited states. This enhancement is expressed usually in terms of ground- and excited-state acidity constants pK_a and pK_a* ($pK^*_a \ll pK_a$ in this case). On the other hand, the acidity of aromatic R-COOH acids is reduced in the excited state ($pK_a^* \gg pK_a$), the carboxylic group showing pronounced basic (proton-acceptor) properties [24].

A direct consequence of the modified acidities of electronically excited molecules is :

(i) the shift of the protolytic equilibrium in a polar solvent

$$AH_s + B_s \rightleftarrows A_s^- + HB_s^+$$

upon the electronic excitation of AH or B molecule and

(ii) the proton transfer in the electronically excited state of the A-H..B. hydrogen bonded complex :

$$AH..B \rightarrow A^-..HB^+$$

(i) A classical example is that of 2-naphthol $C_{10}H_7OH$ (pK_a = 10.0). In a neutral (pH ≅ 7) buffer solution in water the protolytic equilibrium is strongly shifted towards the neutral AH form ($[A^-]/[AH] = 10^{-3}$). On the other hand, for the excited molecule (pK^*_a ≅ 3) the equilbrium would correspond to the predomination of anionic form ($[A^-]/[AH] \cong 10^4$). The absorption spectrum of the solution is thus that of the neutral naphthol. The fluorescence spectrum recorded in the same conditions will be identical with that of the strongly basic naphthol solutions (pH ≥ 12), i.e. with that of the naptholate anion. The fluorescence of neutral naphthol is observed only in stronly acid (pH ≤ 2) solutions.

(ii) similar effects are observed in non-polar solvents for the systems involving an inter-molecular or intra-molecular hydrogen bond between acidic and basic groups. The absorption spectrum of naphthol forming a hydrogen-bonded complex with a strong basis like triethylamine NEt_3 is only slightly modified as compared to that of the free naphthol : the structure of the complex in its ground electronic state corresponds to the interaction between two neutral molecules :

$$Ar\text{-}O\text{-}H..NEt_3$$

The fluorescence spectrum of the complex is intermediary between that of the neutral molecule and that of naphtholate anion [18]; the emitting form may be described as a hydrogen-bonded ion pair :

$$Ar\text{-}O^{-*}..HNEt_3^{\ +}$$

with a large dipole moment, as evidenced by a strong dependence of the fluorescence spectrum on the solvent dielectric constant. The proton transfer is a rapid taking place at the ca.10 ps time scale [25].

Similar effects are observed for intramolecular hydrogen bonds : the well known example being that of methyl salicylate, where the nature of the bond between the -OH and -COOR groups is drastically changed because of the enhanced acidity of -OH and basicity of -COOR groups.

REFERENCES.

1. G.Herzberg, Molecular Spectra and Molecular Structure, vol. I. Spectra of Diatomic Molecules, van Nostrand,Toronto,1950; p.381-386; vol.III.
2. cf. e.g. D.A.Ramsay : Eletronic Spectra of Polyatomic Molecules in : Determination of Organic Structures by Physical Methods, vol.II, p.246-340, (F.C.Nachod and W.D.Phillips eds.) Acad.Press, New York, 1962.
3. S.P.McGlynn, T.Azumi and M.Kinoshita, Molecular Spectroscopy of the Triplet State, Englewood Cliffs, Prentice Hall, 1969
4. a) D.Evans, J.Chem.Soc. 1987,(1961)
   b) A.Grabowska, Spectrochim. Acta 19,307,(1963)
5. H.C.Longuet-Higgins : Some Recent Developments in the Theory of Molecular Energy Levels, in : Advances in Spectroscopy, vol.II. p.429-473, (H.W.Thompson ed.), Interscience, New York,1961
6. a) S.Mukamel and J.Jortner : Time Evolution of Excited Molecular States, in : Excited States, vol.3, p.57, (E.C.Lim ed.), Acad.Press, New York, 1978
   b) A.Tramer and R.Voltz : Time-Resolved Studies of Excited Molecules in : Excited States, vol.4, p.329, (E.C.Lim ed.) Acad.Press, New York, 1979
7. J.R.Lakowicz, Principles of Fluorescence Spectroscopy, Plenum, New York, 1983
8. N.J.Turro, Modern Molecular Spectroscopy, Benjamin, Menlo Park, (1978)
9. B.R.Henry and W.Siebrand : Radiationless Transitions in : Organic Molecular Photophysics (J.B.Birks ed.) Wiley-Interscience, London, 1973
10. M.Kasha, Disc. Faraday Soc. 9,14,(1950)
11. R.D.Levine and R.B.Bernstein, Molecular Reaction Dynamics and Chemical Reactivity, Pergamon, Oxford, 1987
12. a) V.E.Bondybey : Time-Resolved Fluorescence Studies ... in : Chemistry and Physics of Matrix Isolated Species,p.107, (L.Andrews and M.Moskowitz eds.), Elsevier, Amsterdam, 1989
   b) H.Dubost and F.Legay : Energy Transfer and Lifetime Studies... in : Chemistry and Physics of Matrix Isolated Species,p.303, (L.Andrews and M.Moskowitz eds.), Elsevier, Amsterdam, 1989
13. Th.Förster in: Modern Quantum Chemistry, Part III. p.93, (O.Sinanoglu ed.) Acad.Press, New York, 1965
14. D.L.Dexter, J.Chem.Phys. 21,836,(1953)
15. R.A.Marcus, this book p. 49.

16. J.W.Verhoeven, M.N.Paddam-Row and J.M.Warman, this book p.271.
17. for recent works concerning the cage effect in low-temperature matrices see e.g. :
a) R.Alimi, R.B.Gerber and V.A.Apkarian, J.Chem.Phys. 92,3551,(1990)
b) R.Schiever, M.Chergui, Ö.Ünai, N.Schwentner and V.Stepanenko, J.Chem.Phys. 93,3245,(1990)
18. N.Mataga and T.Kubota, Molecular Interactions and Electronic Spectra, Marcel Dekker, New York, 1970
19. M.Castella, P.Millié, F.Piuzzi, J.Caillet, J.Langlet, P.Claverie and A.Tramer, J.Phys.Chem. 93,3949,(1989)
20. B.Wegewijs, R.M.Hermant, J.W.Verhoeven, M.P. de Haas and J.M.Warman, Chem.Phys.Lett. 168,185,(1990) and references therein
21. H.Beens and A.Weller in: Organic Molecular Photophysics vol.2, p.159 (J.B.Birks ed.), Wiley-Interscience, New York, 1975
22. J.B.Birks, Photophysics of Aromatic Molecules, Wiley-interscience, New York,1971
23. R.S.Mulliken and W.B.Person, Molecular Complexes, Wiley-Interscience, New York,1969
24. A.Weller in : Progress in Reaction Kinetics, vol.1, p.187, (G.Porter ed.), Pergamon, London, 1961
25. see e.g. J.A.Syage and J.Steadman, J.Chem.Phys. 93,2497,(1991)

THEORY OF ELECTRON TRANSFER REACTIONS AND COMPARISON WITH EXPERIMENTS*

R. A. MARCUS and PRABHA SIDDARTH
Noyes Laboratory of Chemical Physics 127-72
California Institute of Technology
Pasadena CA 91125
USA

ABSTRACT. In these lectures, reaction rate theory and the theory of electron transfer (ET) reactions are outlined. The topics discussed include the relation of the potential energy surfaces to free energy curves for ET reactions, classical ET theory for reactions in solution and at interfaces, quantum corrections, the inverted effect, relation to charge transfer spectra, the cross-relation, adiabatic and nonadiabatic ET reactions, electronic matrix elements, relation to other transfer reactions, predictions of and comparison with experiments, and miscellaneous questions.

OUTLINE

* From lectures given by RAM at the NATO Summer School at Aussois, 1991.

E. Kochanski (ed.), Photoprocesses in Transition Metal Complexes, Biosystems and Other Molecules. Experiment and Theory, 49–88.

1. Introduction

Electron transfer (ET) reactions play an important role in many chemical and biological reactions.[1-9] The simplest of all ET reactions is the "self-exchange" reaction, as in the isotopic exchange reactions,

$$Fe(CN)_6^{4-} + Fe^*(CN)_6^{3-} \xrightarrow{k_{11}} Fe(CN)_6^{3-} + Fe^*(CN)_6^{4-} \tag{1}$$

$$Ce(IV) + Ce^*(III) \xrightarrow{k_{22}} Ce(III) + Ce^*(IV) \tag{2}$$

where the asterisk denotes a radioactive isotope. Since the products are chemically equivalent to the reactants, these reactions have a zero value for the standard free energy of reaction ΔG^0. The study of their rates has provided, thereby, direct information on the "intrinsic" factors which control the rate of ET. When two different redox systems are involved, as in

$$Fe(CN)_6^{4-} + Ce(IV) \rightarrow Fe(CN)_6^{3-} + Ce(III) \tag{3}$$

the ET reaction is commonly termed a "cross-reaction." For the latter, the standard free energy of the reaction, ΔG^0, which measures the stability of the products relative to those of the reactants in a precise thermodynamical way, is an additional factor which influences the rate of the ET reaction.

In addition to inorganic ET reactions in solution, studies in the ET field include a wide range of topics, including organic and biological ETs, thermal and photoinduced ETs, charge transfer spectra, and ET reactions at metal electrodes and at other types of interfaces, such as semiconductor-liquid, modified electrode-liquid, polymer-liquid and liquid-liquid interfaces, colloids, and micelles. These areas and their connections are among those depicted in Figure 1. We review some essential features of the theory underlying ET reactions and then compare some of the theoretical predictions with the experimental results.

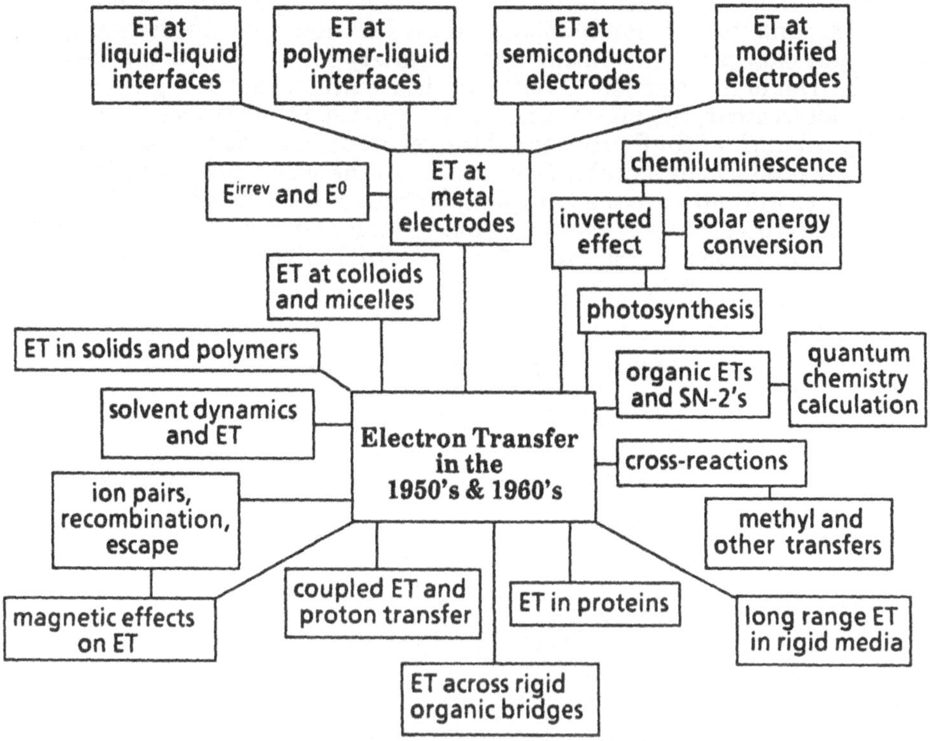

Figure 1. Examples of topics in the electron transfer field.

In general, we consider the ET reaction

$$(Ox)_1 + (Red)_2 \rightarrow (Red)_1 + (Ox)_2 \qquad (4)$$

where the reactants may be moving freely in solution or may be attached to each other so that eq 4 then becomes an intramolecular ET.

2. Theory

2.1 POTENTIAL ENERGY SURFACES AND FREE ENERGY SURFACES

In reactions 1 to 3, the ET reaction is accompanied by changes in the nuclear configurations of the reactants and in the surrounding environment, such as changes in metal-ligand bond lengths in the reactants and changes in the typical orientations of the nearby dipolar solvent molecules. In the case of reaction 1, for example, each Fe^{2+}–OH_2 bond in aqueous solution is about 0.14 Å longer than the Fe^{3+}–OH_2 bond.[10] This latter change imposes a barrier of about 8 kcal mol^{-1} to the reaction.[10] Further, solvent molecules are more oriented around the Fe^{3+} ion than around the Fe^{2+} ion, since the former is more highly charged, contributing approximately 8 kcal mol^{-1} to the free energy barrier of the reaction[10]. Both effects are illustrated in Figure 2. In order for ET to occur, the reactants must approach each other and, at the same time, typical nuclear configurations of the reactants have to evolve toward those of the products by suitable fluctuations in the coordinates. In the present case, these coordinates are the vibrational modes of the reactants, the orientational coordinates of the surrounding solvent molecules, and, to some extent, the vibrational coordinates of the polarized solvent molecules.

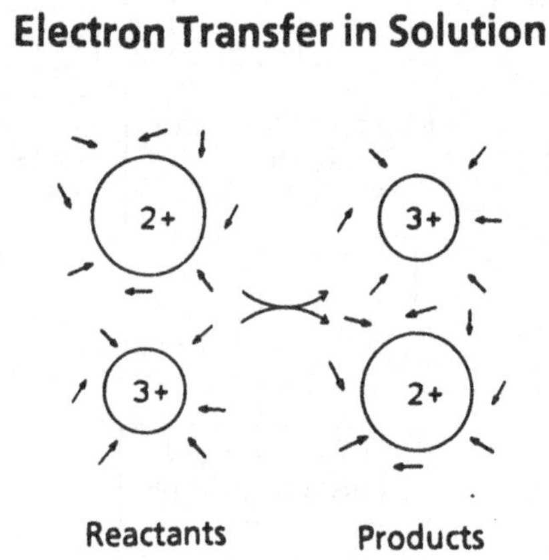

Figure 2. Typical nuclear configurations for reactants, products, and surrounding solvent molecules (cf. ref 11). The longer M-OH_2 bond length in the +2 state is indicated schematically by the larger ionic radius.

To treat the mechanism of ET, it is useful to consider the many-dimensional potential energy surface (PES) for the reactants and the surrounding medium, U_r, and that for the products and surrounding medium, U_p. These PES are a function of many (~thousands) of the solvent's and reactants' coordinates, say N. The intersection of these two PES forms an $(N-1)$-dimensional surface, which, because of the Franck-Condon principle discussed later, serves as the transition state for the reaction. A schematic one-dimensional profile of the two PES and their intersection is given by the solid lines in Figure 3. A somewhat more realistic but still schematic profile is shown by the dashed lines in Figure 3. The latter profile is more complicated, since the dependence of U_r and U_p on the solvational coordinates leads to many local potential energy minima. The dependence of U_r and U_p on the vibrational coordinates, on the other hand, is quadratic to a reasonable approximation for the energies of interest.

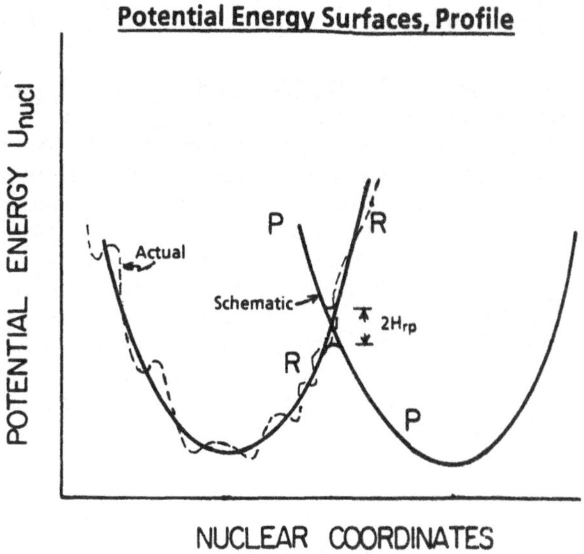

Figure 3. Profile of potential energy surfaces for reactants plus environment, R, and for products plus environment, P. Solid curves: highly schematic. Dashed curves: schematic but more realistic. The typical splitting, $2H_{rp}$, at the intersection of U_r and U_p is exaggerated in the Figure.

It is useful to consider first an intramolecular ET reaction, and the fluctuations in coordinates, at a fixed separation distance R between the electron donor and the acceptor. We subsequently include, in the case of bimolecular reactions, or in the case of an ET reaction between a freely moving reactant and an electrode, the role of the translational coordinates, which permit R to vary. Thus, in the following, the N coordinates first refer to those of a system in which the positions of the reactants, and hence R, are held fixed.

For ET reactions in which there is a weak electronic interaction between the reactants, a reaction coordinate q can be defined globally in a rather precise manner. The method of doing so would take us a little far afield, but is equivalent to using $U_p - U_r$ as a coordinate (cf. ref 12 or an equivalent definition in ref 13). For any given value of q, the system resides on an $(N-1)$-dimensional surface in coordinate space, and the free energy, $G_r(q)$, for the reactants and their surrounding environment can be calculated for any value of q. The expression for $G_r(q)$ contains the value of $\exp[-U_r(q)/k_BT]$, integrated over this $(N-1)$-coordinate surface, and also contains a momentum contribution and a factor h^{N-1}, h being Planck's constant, as in eq B2 of Appendix B. The last two factors, which are omitted here in the interests of notational brevity, cancel when a free energy difference ΔG_r^* is later considered. We thus have

$$\exp\left[-G_r(\mathbf{q})/k_BT\right] = \int \cdots \int \exp\left(-U_r(\mathbf{q})/k_BT\right) dS \tag{5}$$

where dS is a volume element in this $(N-1)$-dimensional space for fixed positions of the translational coordinates of the reactants.

In a linear response approximation, assumed in the theory,[13-15] the contribution of the orientations of the solvent dipoles to changes in the solvent dielectric polarization is proportional to changes in the electric fields, and the approximation is sometimes referred to as the "dielectric unsaturation approximation." In this linear response regime, $G_r(q)$ is then a quadratic function of q. Recent computer simulations have provided further support for the approximation.[16] $G_r(q)$ and $G_p(q)$ can thus be drawn as parabolas, as in Figure 4, even when $U_r(\mathbf{q})$ and $U_p(\mathbf{q})$ are highly anharmonic functions of many of the coordinates. The approximation is much less drastic than assuming that $U_r(\mathbf{q})$ itself is a quadratic function of all the coordinates, collectively denoted here by $\mathbf{q}$.

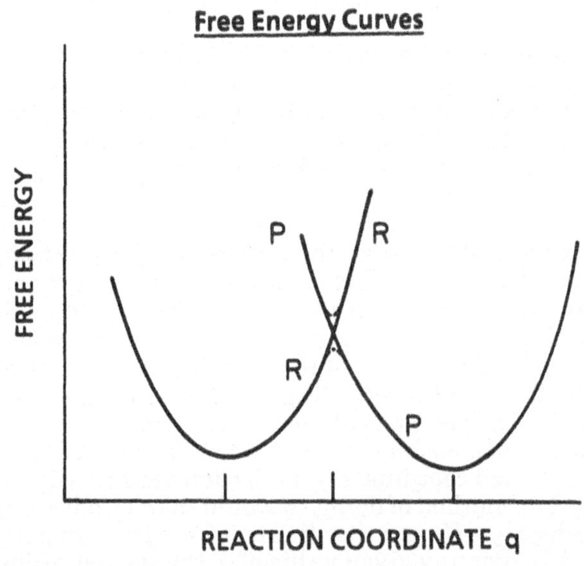

Figure 4. Free energy of reactants plus environment *vs* the reaction coordinate q (R curve), and free energy of products plus environment *vs* q (P curve). The three vertical lines on the abscissa denote, from left to right, q^r, $q^\ddagger$ and q^p.

For electron transfer to occur, one of the conditions which must be satisfied is the Franck-Condon principle. The latter may be described as follows, in terms of a classical description of the nuclear motion: Because the motion of the nuclei is typically slow compared with that of the electrons, the nuclei change neither their instantaneous position nor their instantaneous momenta during an electronic transition. In the present case, this transition is one from the electronic configuration of the reactants to that of the products, and is the electron transfer. The only region in the N-dimensional coordinate space where the above conditions on coordinates and momenta can be satisfied in a *thermal* reaction is where $U_r(\mathbf{q}) = U_p(\mathbf{q})$, *i.e.*, on the $(N-1)$-dimensional surface where the two surfaces U_r and U_p in N-dimensional coordinate space $\mathbf{q}$ intersect. (In the profile of the N-dimensional system in Figure 3, this $(N-1)$-dimensional surface appears as a single point.) An electron transfer in any other region is a vertical transition in the Figure and can only occur upon the absorption or emission of light.

This $(N-1)$-dimensional intersection surface, which is reached by a suitable thermal fluctuation of the coordinates, forms the transition state for this ET reaction. At the intersection, the system can go from the reactants' plus environment's potential energy surface $U_r(\mathbf{q})$, with a profile denoted there by R, to the products' surface P, if some electronic coupling exists between the two reactants. The extent of the coupling is indicated by the splitting $2H_{rp}$ in Figure 3. When H_{rp} is large enough, the system will remain on the lower energy surface, *i.e.*, go from the R surface to the P surface in the transition state region, with a probability of about unity each time the system crosses the intersection surface. The ET reaction is then said to be "adiabatic." If, on the other hand, H_{rp} is very small, for example when the two reactants are far apart, the system has only a small probability of reaching the P surface during this crossing of the TS, and the ET, if it occurs, is then said to be "nonadiabatic." A quantum mechanical-based expression for calculating this probability of transition from the U_r surface to the U_p one at the TS is sometimes employed, using the Landau-Zener expression,[17] as described in Appendix A.

The calculation of the rate constant for an ET reaction thus involves calculating the probability of reaching the transition state multiplied by the probability of going from the R to the P surface in the transition state region. The probability of reaching the transition state can be expressed, in transition state theory, in terms of the free energy required to reach the transition state. We consider first the case where the nuclear motion is treated classically. Quantum mechanical corrections to the following are summarized briefly later.

2.2 CLASSICAL THEORY

We consider either a system in which the donor and acceptor are held fixed at a separation distance R by some rigid molecular framework, or in which they are free to move in solution, approach each other and attain some typical separation distance R in the transition state. At this R we consider the free energy of fluctuations of the vibrational and solvational dipolar coordinates along a reaction coordinate q. There is also an additional free energy term in a bimolecular reaction, the "work" term w_r which is the change in free energy when the reactants fixed at some large separation distance are brought to the separation distance R. This w_r can have both electrostatic and nonelectrostatic contributions. In the interests of brevity of notation, we omit w_r and the corresponding term w_p for the products in the following expressions, although they are normally included in the expressions in the literature.[1,15,18] We first give a quick derivation of an equation for the reorganizational free energy change needed to reach the TS, and then give a more detailed approach which expresses the reorganization parameter λ in terms of molecular properties.

An adiabatic reaction is considered first. In this case the rate constant, written in terms of transition state theory (cf. Appendix A), is

$$k_r = \frac{k_B T}{h} \exp\left(-\Delta G^{\ddagger} / k_B T\right) = \frac{k_B T}{h} \frac{Q^{\ddagger}}{Q} \exp\left(-\Delta E_0^{\ddagger} / k_B T\right) \tag{6}$$

where $Q^{\ddagger}$ is the partition function of the TS, Q is that of the reactants, $\Delta G^{\ddagger}$ is the free energy of activation, and $\Delta E_0{}^{\ddagger}$ is the activation energy at 0° K. In a classical treatment, zero-point energies in $\Delta E_0{}^{\ddagger}$ vanish and the latter becomes $\Delta U^{\ddagger}$, the lowest value of the potential energy in the TS minus that for the reactants, as in eq A3 in Appendix A.

Upon integrating an expression for Q in Appendix B over all coordinates but the reaction coordinate q, a term $\exp[-G(q)/k_B T]$ is obtained. In the linear response approximation $G(q)$ is approximately a quadratic function of q. Combined with another term in Q arising from the momentum p conjugate to q, one factor contributing to Q in eq 6 is a classical harmonic oscillator partition function, $k_B T/h\nu$. Here, ν is the typical frequency associated with motion along q (cf. eq 28 given later). A remaining factor in eq 6 is denoted by $\exp(-\Delta G_r{}^*/k_B T)$. For a unimolecular reaction, one then obtains from eq 6, $k_r = \nu \exp(-\Delta G_r{}^*/k_B T)$, as in Appendix A. Here, $\Delta G_r{}^*$ is the free energy associated with the reorganization (fluctuations) of the solvent dipoles and of the vibrational coordinates needed to reach the TS. For bimolecular reactions, there are additional factors in $Q^{\ddagger}$ and Q associated with the translational motion of the reactants and with their more restricted translational motion in the TS. The net result, derived in Appendix A, is included later in eq 14 for k_r.

The reorganizational free energy $\Delta G_r{}^*$ required to reach the transition state (TS) when the positions of the reactants are held fixed is

$$\Delta G_r{}^* = G_r(q^{\ddagger}) - G_r(q^r) \tag{7}$$

where $G_r(q^{\ddagger})$ is the value of $G_r(q)$ on the intersection surface (the TS) and q^r denotes the value of q for which $G_r(q)$ has a minimum, as in Figure 4. An expression for $\Delta G_r{}^*$ can be obtained using quadratic expressions for the free energies. We have

$$G_r(q) = \frac{k}{2}\left(q - q^r\right)^2 \tag{8a}$$

$$G_p(q) = \frac{k}{2}\left(q - q^p\right)^2 + \Delta G^o \tag{8b}$$

where ΔG^0 is the standard free energy of the reaction, which also equals $G_p(q^p) - G_r(q^r)$. $\Delta G_r{}^*$ is obtained as follows:

On the intersection surface the distribution of the coordinates and momenta is the same for the U_r and U_p surfaces.[14,15] Thus, their entropies are the same for the R and P electronic states at the intersection surface. Similarly, the average of U_r over the coordinates defining the intersection surface equals the average of U_p there. Thus, we have[14]

$$G_r(q^\ddagger) = G_p(q^\ddagger) \tag{9}$$

which together with eqs 7 and 8 yields

$$q^\ddagger = \frac{1}{2}(q^r + q^p) + \frac{\Delta G^0}{k(q^p - q^r)} \tag{10}$$

These equations yield

$$\Delta G_r^* = \frac{\lambda}{4}\left(1 + \frac{\Delta G^0}{\lambda}\right)^2 \tag{11}$$

where

$$\lambda = \frac{k}{2}\left(q^p - q^r\right)^2 \tag{12}$$

Using transition state theory, the net result for the rate constant k_r of the electron transfer reaction can be written as in eq 13a for a unimolecular reaction and in 13b for a bimolecular reaction (cf. Appendix A).

$$k_r = \kappa\nu \exp(-\Delta G_r^*/k_B T) \tag{13a}$$

$$k_r = \kappa\nu v_s \exp(-\Delta G_r^*/k_B T) \tag{13b}$$

where ΔG_r^* is given by eq 11, ν is the frequency associated with the motion along the q-coordinate, and v_s is the effective volume[19] which the reactants occupy in the TS, namely $4\pi R^2/\beta$, where β is the exponent in the dependence of the electron transfer rate on separation distance R (rate varies as $\exp(-\beta R)$, β being about 1 Å^{-1}, depending on the system).[1] The R in v_e is the mean separation distance in the TS. Equation 13b is of the same form as that given by Sutin.[4]

For an adiabatic reaction the $\kappa\nu$ in eq 13 becomes ν, while for a nonadiabatic reaction, it can be shown, as in Appendix A, to be given approximately by eq 14b:

$$\kappa\nu = \nu \qquad \text{(adiabatic)} \tag{14a}$$

$$\kappa\nu = \frac{2\pi}{\hbar}\frac{|H_{rp}|^2}{\left(4\pi\lambda k_B T\right)^{1/2}} \qquad \text{(nonadiabatic)} \tag{14b}$$

The λ appearing in eq 12 is usually termed a reorganization energy and is the sum of λ_i, an inner (*i.e.*, vibrational) contribution, and λ_o, an outer (*i.e.*, solvational, outside any first coordination shell) contribution.[15] The λ_i arises from vibrational motion of the nuclei of the reactants themselves, including those in any first coordination shell, and typically arises from any changes in bond lengths of each species when it is a product, as compared with when it is a reactant (eq 26 below). For example, for ET involving transition metal complexes, this λ_i would include the contribution from the metal atom-to-ligand coordinates. The λ_o includes the contributions from all the rest of the coordinates of the nuclei, which for ET reactions in solution means those of the solvent. We consider next a simple model which provides an estimate of λ.

The vibrations of each reactant are treated as normal modes of vibration. We let q_i denote the *i*th normal mode coordinate, q_i^r its equilibrium value for the reactants, q_i^p that for the products, and k_i the associated force constant $4\pi^2\nu_i^2$, ν_i being the normal mode frequency for that mode.[20] The contribution from the reactants' vibrations to the potential energy of a fluctuation $U_j^{vib}(j=r,p)$ is

$$U_j^{vib} = \frac{1}{2}\sum_i k_i\left(q_i - q_i^j\right)^2 , \qquad j=r,p \tag{15}$$

The ΔG_r^* in eq 13 contains a ΔG_{solv}^* and a ΔU_{vib}^*. (Vibrational entropic terms in this vibrational model can be shown to cancel in ΔG_r^*, a more detailed derivation being given in the literature,[15] a result which can also be inferred using expressions for Q and $Q^\ddagger$ given in Appendix A.)

To express the "outer" contribution to G, we shall treat the solvent as a dielectric continuum and neglect dielectric image effects, which tend to be small for the interaction of the two charges in solution.[21] A more general formulation is again available in refs 14 and 15.[97] If P(r) denotes a particular function of the dielectric orientation polarization of the solvent at point r, then the free energy contribution G_j^{solv} due to these charge-charge and charge-solvent polarization interactions is given by[22]

$$G_j^{solv} = -\frac{1}{8\pi}\left(1-\frac{1}{D_{op}}\right)\int \mathbf{D}_j^2(\mathbf{r})\,d\mathbf{r} - \int \mathbf{P}\cdot\mathbf{D}_j(\mathbf{r})\,d\mathbf{r} + 2\pi c\int \mathbf{P}^2\,d\mathbf{r} \tag{16}$$

where $1/c = 1/D_{op} - 1/D_s$ and $\mathbf{D}_j(\mathbf{r})$ is the electric field associated with the charge distribution ρ_j.

$$\mathbf{D}_j(\mathbf{r}) = -\nabla_r \int \rho_j(\mathbf{r}')/|\mathbf{r}-\mathbf{r}'|\,d\mathbf{r}' , \qquad j=r,p \tag{17}$$

We now have

$$G_r(\mathbf{q},\mathbf{P}) = U_r^{vib}(\mathbf{q}) + G_r^{solv}(\mathbf{P}) \tag{18a}$$

and

$$G_p(\mathbf{q},\mathbf{P}) = U_p^{vib}(\mathbf{q}) + G_p^{solv}(\mathbf{P}) + \Delta G^0 \tag{18b}$$

where **q** now denotes only the totality of the vibrational coordinates q_i rather than all the coordinates.

In order to find the values $P^{\ddagger}(r)$ and $q^{\ddagger}$ in the TS, $G_r(\mathbf{q},\mathbf{P})$ is minimized with respect to variations in P(r) and **q**, subject to the condition that $G_r(\mathbf{q},\mathbf{P})$ and $G_p(\mathbf{q},\mathbf{P})$ are equal at the transition state:

$$G_r(\mathbf{q},\mathbf{P}) = G_p(\mathbf{q},\mathbf{P}) \qquad \text{(TS)} \tag{19}$$

Thus,

$$\delta G_r = 0 \tag{20a}$$

subject to

$$\delta G_r - \delta G_p = 0 \tag{20b}$$

Equation 20 and use of a Lagrangian multiplier m yields, upon introduction of eqs 15, 16, 18 and 19,

$$4\pi c\mathbf{P}(\mathbf{r}) = \mathbf{D}_r(\mathbf{r}) + m\left[\mathbf{D}_r(\mathbf{r}) - \mathbf{D}_p(\mathbf{r})\right] \tag{21}$$

and

$$q_i^{\ddagger} = q_i^r + m\left(q_i^r - q_i^p\right) \tag{22}$$

Introducing these results into eqs 15 and 16, it follows that

$$\Delta G^{*} = \frac{\lambda}{4}\left(1 + \frac{\Delta G^0}{\lambda}\right)^2 \tag{23}$$

where

$$\lambda = \lambda_o + \lambda_i \tag{24}$$

$$\lambda_o = \frac{1}{8\pi c}\int\left(\mathbf{D}_r - \mathbf{D}_p\right)^2 d\mathbf{r} \tag{25}$$

$$\lambda_i = \sum \frac{1}{2} k_i\left(q_i^p - q_i^r\right)^2 \tag{26}$$

If the two reactants can be treated as spherical, then λ_0 is found to be given by[14]

$$\lambda_o = \left(\frac{1}{2a_1} + \frac{1}{2a_2} - \frac{1}{R}\right)\left(\frac{1}{D_{op}} - \frac{1}{D_s}\right) \tag{27}$$

where a_1 and a_2 are the radii of the two reactants, and R is their center-to-center separation distance. Models in which the pair of reactants is approximated by an ellipsoid have also been developed in the literature.[23-25]

The ν in eq 14a is approximately given by[26,27]

$$\nu = \left(\sum_k \lambda_k \nu_k^2 / \lambda\right)^{1/2} \qquad (k = i,o) \tag{28}$$

using the notation in eq 24, or if there are several vibrational frequencies contributing to $\lambda_i\nu_i^2$, the sum in eq 28 is over all of them, $\Sigma_j\lambda_j\nu_j^2 + \lambda_o\nu_o^2$. The often neglected ν_o is associated with the "inertial motion" of the solvent.

From eqs 23 to 27, it can be seen that ΔG^* decreases with increasing a_1 and a_2, decreasing R, increasing D_{op} and decreasing D_s. These observations can be physically interpreted as follows: (a) With increasing ionic radii a_1 and a_2, the ion-solvent interactions decrease so that a smaller reorganization is needed for reaction and hence a smaller λ_o. (b) When the distance between the reactants is decreased, again there is less reorganization, since the somewhat distant solvent molecules can discriminate less between the two charge distributions when R is small, *i.e.*, discriminate less between the charge distribution of the reactants and that of the products. Thereby, a smaller reorganization is needed and therefore a smaller λ_o occurs. (c) With a greater optical dielectric constant D_{op}, the electrostatic fields of the charges are more shielded and therefore the energy difference of ion-solvent interactions for the two charge distributions is less, leading to less reorganization. When D_s equals D_{op}, λ_o is zero, the solvent now being nonpolar.

2.3 ELECTROCHEMICAL ET REACTIONS

Another class of ET reactions is that of electrochemical ET reactions. For the case of ET between an ion or a molecule and a metal electrode, a detailed analysis shows that in the usual region, [$|ne(E - E_0')| < \lambda$, in the notation used below], the transferring electron goes mainly from or into the Fermi level of the metal.[15,26,28,29] Also, now the metal electrode has densely spaced electronic states (a few are indicated schematically in Figure 5).

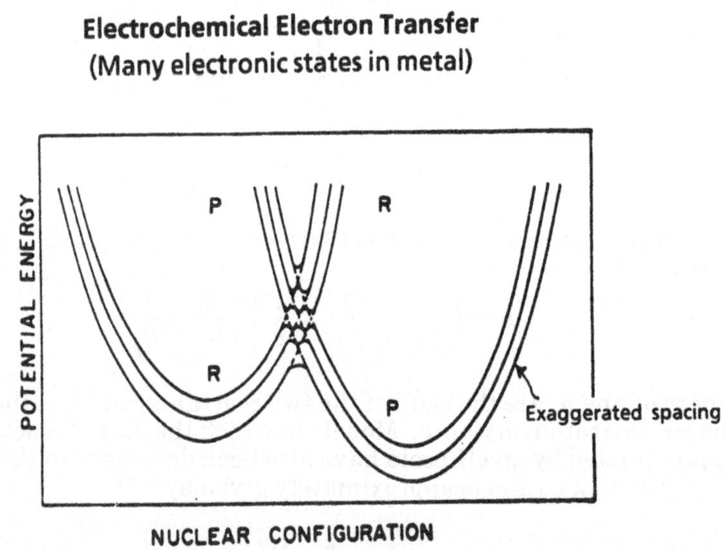

Figure 5. Potential energy profiles for reactants' system R (including electrode and solvent environment) and for products' system P for reaction 32. Several of the continuum of energy levels are depicted, with greatly exaggerated spacing.

Proceeding as before, eq 8 can now be written as

$$G^r(q) = \frac{k}{2}\left(q - q^r\right)^2 \tag{29}$$

$$G^p(q) = \frac{k}{2}\left(q - q^p\right)^2 + ne\left(E - E_0'\right) \tag{30}$$

upon following the arguments given in refs 15 and 28. Here, E is the half-cell potential, E_0' is the "standard" half-cell potential in the prevailing medium, and n is the number of electrons of charge e transferred between reactant and electrode. Thus $n(E - E_0')$, the activation overpotential, plays the same role as ΔG^0 in eq 7.[15,28,30]

In the transition state we have, as before,

$$G^r(q^\ddagger, E_0') = G^p(q^\ddagger, E_0') \tag{31}$$

One ultimately obtains for the rate constant for the electrochemical reaction

$$\text{Ox} + \text{M}(ne) \rightarrow \text{Red} \tag{32}$$

the expression (cf. also Appendix A) given by eq 33a when the reactant moves freely in solution and by eq 33b when it is bound to the electrode.

$$k_{el} = \kappa \nu v_e \, e^{-\Delta G^*_{el}/k_B T} \tag{33a}$$

$$k_{el} = \kappa \nu \, e^{-\Delta G^*_{el}/k_B T} \tag{33b}$$

where k_{el}, the rate constant, is the rate per unit concentration of reactant per unit area of electrode.) In eq 33, $\kappa\nu$ has the same meaning as in eqs 14a and 14b for the adiabatic and nonadiabatic limits, v_e is an effective volume, per unit area of electrode, from which electron transfer can occur ($\simeq \beta^{-1}$, where β was the exponent defined in Section 2.2), and ΔG_{el}^* is given by

$$\Delta G^*_{el} \simeq \frac{\lambda_{el}}{4}\left[1 + \frac{ne\left(E - E_0'\right)}{\lambda_{el}}\right]^2 \tag{34}$$

λ_{el} is given by equations similar to eqs 24 to 26, but now the sum in eq 26 is over the normal modes of the single reactant, and in eq 25 the expressions for $\mathbf{D}_r(\mathbf{r})$ and $\mathbf{D}_p(\mathbf{r})$ now take explicit account of the boundary conditions at the metal-liquid interface, utilizing, for example, the customary image charge description. When the reactant, denoted by 1, is treated as a sphere of radius a_1, λ_o is given for the electrochemical ET by[15,28]

$$\lambda_{o,el} = \left(\frac{1}{2a_1} - \frac{1}{R}\right)(ne)^2\left(\frac{1}{D_{op}} - \frac{1}{D_s}\right) \tag{35}$$

where R is now twice the distance from the center of the reactant to the electrode, in the transition state.

When the two reactants in a bimolecular reaction are in contact in the transition state, the R in eq 27 is approximately equal to $a_1 + a_2$. When the reactant in the electrochemical case is in contact with the metal electrode, in the transition state, the R in eq 35 is approximately $2a_1$. The λ_o for the electrochemical reaction is then seen from eqs 27 and 35 to equal half the λ_o for the self-exchange reaction. Further, the λ_i for the electrochemical reaction is one half the λ_i for the self-exchange reaction, there being only one half the number of vibrational normal modes undergoing changes. Thus the total λ's then satisfy

$$\lambda_{el} \simeq \frac{1}{2}\lambda_{11} \tag{36}$$

However, when there is an adsorbed solvent layer on the electrode which the reactant cannot readily enter, then eq 36 is replaced by the inequality $\lambda_{el} > \lambda_{11}/2$. Consequences of eqs 33 to 36 are given later. If the reactant were bound to the electrode, instead of being free to move in solution, eq 36 would remain applicable only if the dielectric aspects were similar in the two systems and if the R in eq 27 were equal to the R in eq 35.

2.4 PREDICTIONS FROM THEORY

2.4.1 *The inverted effect.* From eq 23, it can be seen that when ΔG^0 is made increasingly negative at constant λ, the free energy barrier ΔG^* initially decreases, and becomes equal to zero when $-\Delta G^0 = \lambda$. Upon making ΔG^0 more negative, $-\Delta G^0$ then exceeds λ, and it is seen from eq 23 that the barrier begins to increase. In summary, the theory predicts that with an increase in $-\Delta G^0$, the rate will increase, reach a maximum, and then decrease, as in Figure 6. The regions where $|\Delta G^0| < \lambda$ and $|\Delta G^0| > \lambda$ are known as the "normal" and "inverted" regions,[13,15] respectively (the regions I and III in Figure 6).

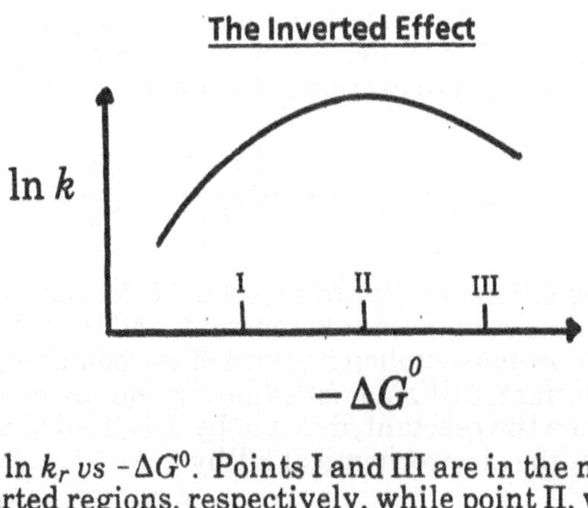

Figure 6. Plot ln k_r *vs* $-\Delta G^0$. Points I and III are in the normal and inverted regions, respectively, while point II, where ln k_r is a maximum, occurs when $-\Delta G^0 = \lambda$.

ΔG^0 can often be made increasingly negative by suitably changing the ligands of a reactant. This decrease in ΔG^0 corresponds to lowering the free energy surface of the products vertically in Figure 4. When this lowering is performed in small decrements, then at some ΔG^0, namely, at $\Delta G^0 \simeq -\lambda$, the intersection of the *R* and *P* free energy curves will occur at the minimum of the *R* curve (curve II in Figure 7). The reaction is then barrierless, *i.e.*, $\Delta G^* = 0$, and the rate constant is a maximum for the given λ (Figure 6). A further lowering of the *P* surface increases the free energy at the intersection and results in the inverted effect on the rate constant (curve III in Figure 7).

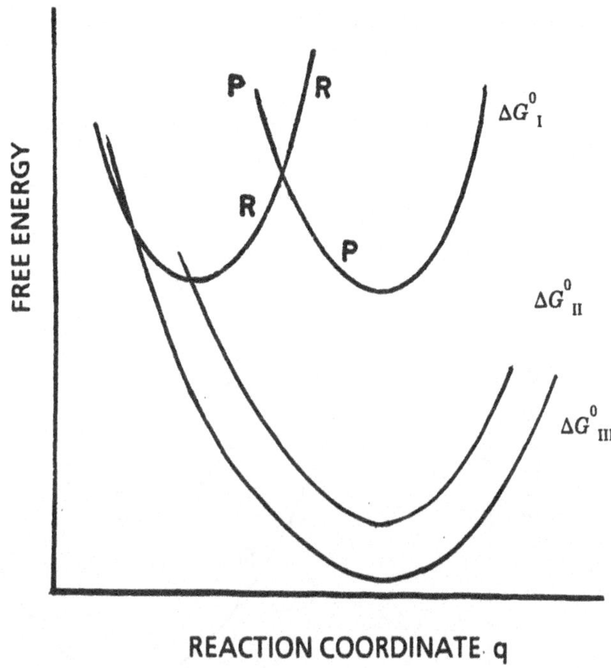

Figure 7. Plot of the free energy *G* versus the reaction coordinate *q*, for reactant (*R*) and product (*P*) systems, for the cases I to III indicated in Figure 6.

Nuclear tunneling yields higher rates than would be predicted by the classical theory alone. The tunneling has a larger effect in the inverted region than in the normal region, because the effective barrier is "narrower" in the latter, as can be seen later in Figure 9. This result has the effect of distorting somewhat the ln k_r *vs* ΔG^0 curve in Figure 7, so that the curve becomes asymmetric when nuclear tunneling occurs. Nevertheless, the maximum in the ln k_r *vs* $-\Delta G^0$ curve still occurs at $-\Delta G^0 \sim \lambda$.[1]

As noted in Section 2.5 and as has been described in ref 92, there is a parallelism between a k_r *vs* $-\Delta G^0$ plot and a charge transfer absorption or emission spectral plot. On one side of the spectral maximum, the correspondence is to the "normal" region, and on the other, to the "inverted" region.

The inverted behavior is sometimes also seen in the preferential formation of an electronically excited product.[31,32] Even though the intersection of the ground electronic state free energy curve *P* of the products and that of the reactants *R* may occur only at high energies in some cases, and so correspond to the reaction's being in the inverted region, an electronically excited-state free energy cuvce P^* of the products may intersect the *R* curve in a more favorable region, as in Figure 8. Then, this electronically excited state may form preferentially. This state can then either emit light, or form other states which emit light, or become deactivated. In this way, the ET reaction has resulted in a chemiluminescence.

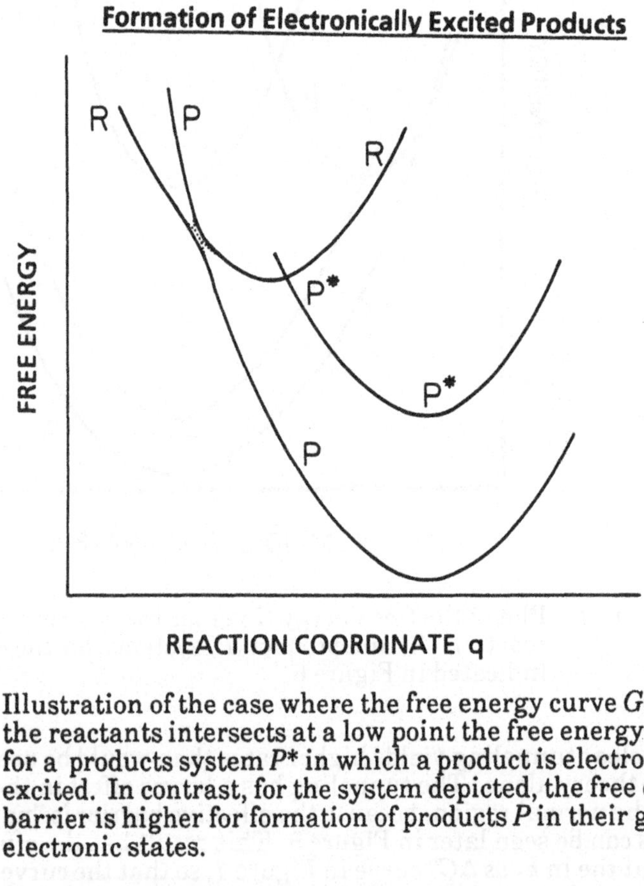

Figure 8. Illustration of the case where the free energy curve $G(q)$ for the reactants intersects at a low point the free energy curve for a products system P^* in which a product is electronically excited. In contrast, for the system depicted, the free energy barrier is higher for formation of products *P* in their ground electronic states.

2.4.2 *The cross-relation.* By noting that λ for a cross-reaction (such as eq 3) is approximately the average of the λ's of the self-exchange reactions (in the present case, eqs 1 and 2),[15]

$$\lambda_{12} = \frac{1}{2}\left(\lambda_{11} + \lambda_{22}\right) \tag{37}$$

the "cross-relation" for rate constants follows from eqs 13 and 23 for the case that the κ in eq 13 is roughly unity for the reactions involved:[15]

$$k_{12} = \left(k_{11}\, k_{22}\, K_{12}\, f_{12}\right)^{1/2} \tag{38a}$$

where $K_{12}=\exp(-\Delta G^0/k_BT)$, the equilibrium constant of the reaction, and f_{12} is a known function of k_{11}, k_{22} and K_{12}:

$$\ln f_{12} = \left(\ln K_{12}\right)^2 / 4 \ln\left(k_{11} k_{22}/\nu^2 \upsilon_s^2\right) \tag{38b}$$

If, instead, the κ in eq 13 were substantially less than unity for the reactions involved, and if it were assumed that $\kappa_{12}\simeq(\kappa_{11}\kappa_{22})^{1/2}$, then eq 38 would again be obtained, with the ν^2 in eq 38b now replaced by $(\kappa_{11}\nu)(\kappa_{22}\nu)$. In practice, f_{12} is usually close to unity.

2.4.3 *Electrochemical relationships.* When the κ's in eqs 13a and 33a are approximately unity, the rate constant k_{ex} for self-exchange in solution, and the electrochemical rate constant $k_{el}{}^{ex}$ corresponding to the so-called exchange current, namely at $E=E_0'$, are related by

$$\left(k_{ex}/A_{ex}\right)^{1/2} \simeq k_{el}^{ex}/A_{el} \tag{39}$$

upon using eq 36. Here, A_{ex} is the pre-exponential factor in k_{ex} (namely $\upsilon_s\nu$, which is roughly equal to 10^{11} to 10^{12} $M^{-1}s^{-1}$) and A_{el} is that in $k_{el}{}^{ex}$ (namely $\upsilon_e\nu$ and which is roughly equal to 10^4 to 10^5 cm s^{-1}).

An important difference, however, between electrochemical ET at metal electrodes and ET in solution is that no inverted effect should occur for the former, since the transferring electron can always go into a high-energy unoccupied level of the metal instead of into a level near the Fermi level, or if the electron transfer is from the metal, it can always come from an occupied level of the metal well below the Fermi level instead of from a level near the Fermi level. In the case of a semiconductor, which has a narrow energy band instead of the very broad one in a metal, the inverted effect again becomes possible.

Another relationship is the dependence of ln k_{el} on the activation overpotential $ne(E-E_0')$, given by eqs 33 and 34. It is analogous to the relation given in eqs 11 and 13 for the dependence of ln k_r on ΔG^0 for a series of homogeneous reactions having a constant λ.

2.5 QUANTUM CORRECTION AND NUCLEAR TUNNELING

In the formulation given above for the rate constant of an ET reaction, the nuclear motion was treated classically. However, there may be some nuclear tunneling through the barrier in Figure 3. The nuclear motion in the ET is described quantum mechanically in such cases . When the reaction is nonadiabatic, the appropriate quantum mechanical expression is obtained using low order time-dependent perturbation theory, namely the Fermi golden rule expression.[29,33] The result is given by

$$k_r = \frac{2\pi}{\hbar} |H_{rp}|^2 (FC) \tag{40}$$

when the reactants are held fixed in position. Here, FC is a "Franck-Condon" factor involving the sum of products of overlap integrals of the vibrational and solvational wavefunctions of the reactants (and solvent) with those of the products, weighted by Boltzmann factors for the given temperature of the system. As before, H_{rp} is the electronic matrix element in Figure 3.

To obtain an expression for FC, one can either attempt to treat all coordinates, vibrational and solvational, quantum mechanically, or treat the higher-frequency vibrations of the inner shell coordinates quantum mechanically and treat the low-frequency modes of the system classically, using for the latter the free energy of solvent reorganization given above.[33] This approach also yields, in the high-temperature limit,[33-35] the expression (cf. Appendix A)

$$k_r = \frac{2\pi}{\hbar} |H_{rp}|^2 \frac{1}{\left(4\pi\lambda k_B T\right)^{1/2}} \exp\left[-\left(\Delta G^0 + \lambda\right)^2 / 4\lambda k_B T\right] \quad \text{(nonadiabatic)} \tag{41}$$

an expression which has also been derived using semiclassical theory.[36] The pre-exponential factor in eq 41 corresponds to the $\kappa\nu$ in eq 13a, for this case that $\kappa << 1$, as already noted in eq 14b.

Several of the quantum expressions which have been introduced for k_r are given below, for the case where the reaction is treated, in addition, nonadiabatically. The simplest of these expressions is the one-frequency model[29]

$$k_r = \frac{2\pi}{\hbar} |H_{rp}|^2 \frac{1}{h\nu} \exp(py - S \coth y)\, I_p(S \operatorname{cosech} y) \tag{42}$$

where $S = \lambda_i / h\nu$, ν is a vibration frequency, regarded as some mean frequency for all vibrations of the reactants and also for the solvent, the solvent being treated as a collection of oscillators, all of this same frequency ν. I_p is a modified Bessel function of the first kind, $p = -\Delta G^0 / h\nu$, and $y = h\nu / 2k_B T$.

A more general quantum expression is one where all high frequency vibrations are treated as having a single frequency ν, while the remaining motion, including that of the solvent, is treated classically. For the case that $h\nu / k_B T$ is so large that the reaction occurs mainly from the lowest vibrational state of the reactants, we have[33]

$$k_r = \frac{2\pi}{\hbar} |H_{rp}|^2 \frac{1}{\left(4\pi\lambda_0 k_B T\right)^{1/2}} \sum_{m=0}^{\infty} \exp\left[-\left(\lambda_0 + \Delta G^0 + mh\nu\right)^2 / 4\lambda_0 k_B T\right] e^{-S} S^m / m! \tag{43a}$$

where m denotes an integer (not to be confused with the m in eq 21). For the more general case where the initial vibrational states may be thermally populated, instead of the system's being only in the lowest vibrational state, eq 43a would be replaced by[33]

$$k_r = \frac{2\pi}{\hbar} |H_{rp}|^2 \frac{1}{\left(4\pi\lambda_0 k_B T\right)^{1/2}} \sum_{m=-\infty}^{\infty} \exp\left[-\left(\lambda_0 + \Delta G^0 + mh\nu\right)^2 / 4\lambda_0 k_B T\right]$$

$$\times \exp(my - S\coth y)\, I_m(S\,\mathrm{cosech}\, y) \qquad (43b)$$

where y has the same significance as in eq 42. At moderate T, eq 43b reduces to eq 43a, while at sufficiently high T, it reduces to eq 41.

We have discussed elsewhere[92] the relationship between the k_r *vs* $-\Delta G^0$ (not ln k_r *vs* $-\Delta G^0$) curve and the plot of charge transfer absorption ($\varepsilon_{\nu_a}/\nu_a$) or emission ($f_{\nu_e}/\nu_e^3$) intensity *vs* frequency, ($\nu_{a,e} - \nu^{max}$) curve. The spectral plot on one side of the maximum ν^{max} corresponds to the "normal" region for k_r, while that on the other side of ν^{max} corresponds to the inverted region.[92] The spectral plot may show evidence of vibrational structure, and it is this structure which corresponds to the different m's in eq 43b. If λ_0 were small enough, eq 43b would display a series of sharp peaks as a function of $-\Delta G^0$, peaks separated by the vibrational energy spacing hν. Instead, there is not merely one mode but others also, treated in eq 43b only by a classical term containing λ_0, and the broadening for each such "line" is given by the Gaussian factor in eqs 43a and 43b. In the regime where most of the broadening is due to fairly high frequencies (but not as high as the principal one, ν in eqs 43), it may be necessary meanwhile to replace the temperature-dependent Gaussian factor in eqs 43 by a more general theoretically-based one containing these vibrations and so showing less temperature dependence.

We comment in Appendix C on the relationship between eqs 42 and 43a when both λ_0 and T both tend to zero, and between eqs 42 and 43b when only λ_0 tends to zero. In eqs 43a and 43b the significance of m is that it is the number of vibrational quanta in the products immediately after the ET minus that in the reactants just before the ET. Equations 43a and 43b reflect the distribution of such m's.

When $h\nu/k_B T$ is large, a linear dependence of ln k_r on ΔG^0 tends to occur in the inverted region, for example as in ref 37, paralleling results on the energy gap law for radiationless transitions.[38] We comment on this linear dependence in Appendix C. There have been a number (*e.g.*, ref 37) of applications of formulae related to or obtained from eq 43, for systems with high frequency vibrations (*e.g.*, the relevant such modes invoked have frequencies around 1000 cm^{-1} to 1500 cm^{-1}). Further, in the inverted region the rate constant k_r tends to become largely temperature-independent when the participating vibrations have high frequencies, as for example in ref 39, because the reaction then largely occurs from the lowest vibrational state of the reactants.

When the quantum effects of the nuclear motion are significant, the k_r is higher than that calculated classically. A useful interpretation can be given in terms of nuclear tunneling. To this end, the Franck-Condon factors are first expressed in terms of semiclassical theory, expressions which in the nuclear tunneling regime are the exponentials given in Figure 9. They may be described as follows: We consider the case of tunneling along a vibrational coordinate x from an initial vibrational state n_r of the reactants to a vibrational state n_p of the products; $n_p - n_r$ is equivalent to the m in eq 43. In Figure 9 the minimum of the vibrational curve U_r^{vib} is higher than that of the U_p^{vib} curve by an amount $mh\nu$. The semiclassical value of the vibrational state-to-state Franck-Condon factor[40] is given in Figure 9, apart from a pre-exponential factor, for two cases, one where the slopes of U_r^{vib} and U_p^{vib} at the intersection have the same signs, and one where they have opposite signs. Each exponential is the semiclassical (also known as WKB) nuclear tunneling probability per encounter of the system with the barrier. In the diagram, the system strikes the barrier once during a period $1/\nu$ of the nuclear motion. In Figure 9, U_r^{vib} and U_p^{vib} are denoted by U_r and U_p.

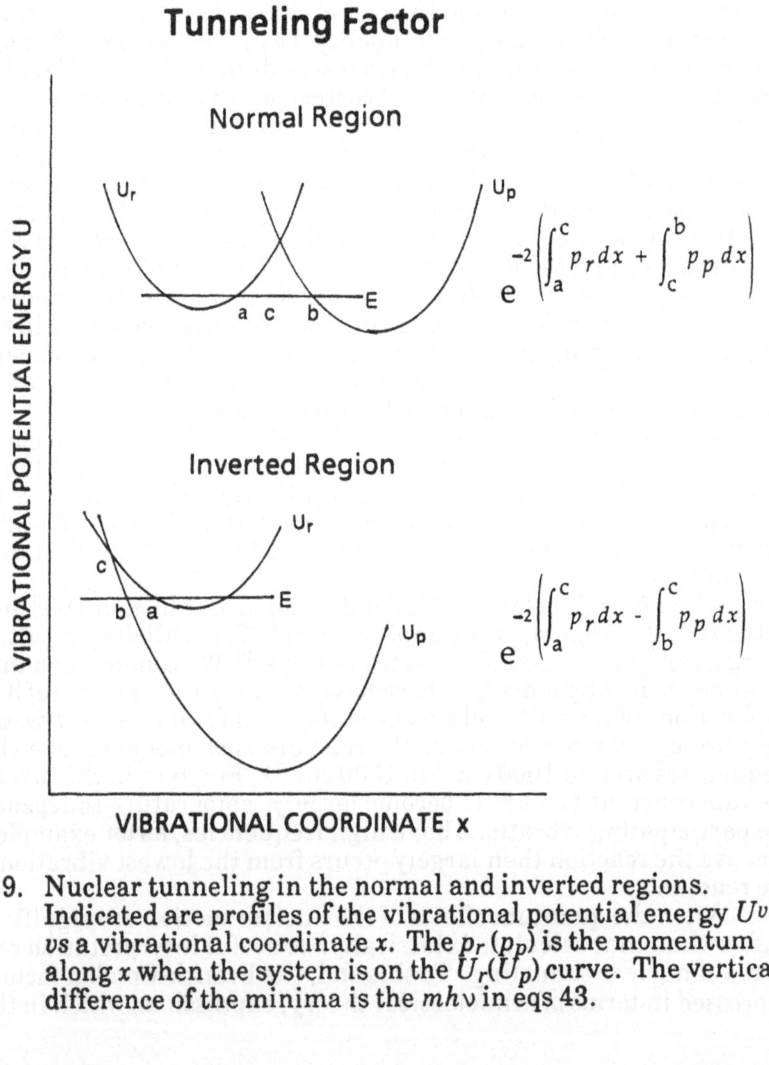

Figure 9. Nuclear tunneling in the normal and inverted regions. Indicated are profiles of the vibrational potential energy U^{vib} *vs* a vibrational coordinate x. The p_r (p_p) is the momentum along x when the system is on the U_r(U_p) curve. The vertical difference of the minima is the $mh\nu$ in eqs 43.

2.6 OTHER

ET theory has also been extended to ET's at various other types of interfaces (*e.g.*, at semiconductor-liquid interfaces at modified electrodes, and at the interface of two immiscible liquids, and at other types of interfaces, involving polymers, for example), charge transfer spectra, and to a variety of other topics, some of which are included in Figure 1. However, in the interests of brevity, a discussion of these topics is omitted here.

3. Comparison of Theory and Experiment

Several applications of the theory outlined in the previous section, are summarized briefly in these notes, though they were described more fully in the lectures themselves.

3.1 RATE CONSTANTS

An example of a comparison of measured self-exchange rate constants with those calculated using eqs 13a and 23 to 27, is given in Table VI of ref 4. The general agreement between theory and experiment is surprisingly good, particularly considering that the rate constants themselves span a wide range of values (from 10^{-9} to 10^{9}).

3.2 SOLVENT EFFECT

The theory predicts a dependence of the ET rates on the solvent dielectric properties via eq 27. Example of a plot of log k versus $(1/D_{op} - 1/D_s)$ for the bis (biphenyl) chromium$^{+/-}$ exchange is given in ref. 41. The plot shows the predicted linear dependence of the logarithm of rate on $(1/D_{op} - 1/D_s)$. However, such a good correlation is not always found, and factors such as ion-pairing, specific solvation effects and strong interaction between reactants can cause unexpected solvent dependences. Additional tests have been made using the solvent dependence of the absorption or emission maximum of charge transfer spectra, e.g., as in the plot[42] for $[(bipy)_2\,Cl\,Ru^{II}\,L\,Ru^{III}\,Cl\,(bipy)_2]^{3+}$, where the ligand L is varied.

3.3 CROSS-RELATION

There have been numerous experimental tests of the cross-relation, eq 38, some of which are shown in Table I of ref 1. The agreement between theory and experiment in most cases is usually good, to within an order of magnitude. Deviations from the cross-relation can result when the driving force is large or when there is a change in mechanism for the self-exchange reactions and the cross-reaction, or when the work terms (not shown in eqs 11 and 13b) do not approximately cancel. In some instances, when a significant discrepancy between the observed and the calculated rate constants exists, this discrepancy has been taken as evidence for the operation of a special mechanism for one of these reactions being compared.

3.4 THE INVERTED EFFECT

The inverted region,[13] depicted in Figure 6, was not observed experimentally for many years.[43] The masking of the inverted effect could have many causes. Experiments that measure the rate of ET by quenching mechanisms may have sometimes given misleading results because quenching by more than one route is possible. (Other alternative routes, besides ET, for quenching are the formation of exciplexes[44] and the competition with other reaction mechanisms.) For ET reactions in solution, the behavior of the rate of ET near the maximum rate can also be masked by diffusion-control.

In recent years, however, the inverted behavior has been experimentally demonstrated in many cases, the first detailed such experiment being that of Miller, Closs and Calcaterra.[45] It involved intramolecular ET for a series of compounds containing two molecular groups with different electron affinities, separated by a rigid bridge B, D-B-A. By varying either of the end groups, a range of ΔG^{0}'s was achieved. Pulse radiolysis was used to generate electrons that were then trapped by the model compounds. The subsequent rate of transfer of electron (corrected for any intermolecular ET), plotted as a function of the standard free energy of the reaction, is shown in Figure 1 of ref 45 and clearly displays the inverted region.

According to the theory, the maximum rate of ET occurs when $-\Delta G^{0} = \lambda$. It would require less energy to reorganize a nonpolar solvent than it would take to reorganize a polar solvent, and so λ would be smaller in the nonpolar case. Therefore, as a further test, the authors repeated their experiments in a nonpolar solvent, iso-octane, and compared their results with the experiments in MTHF. The maximum of the curve tended to be shifted to lower energy in iso-octane as predicted by the theory. Other results which demonstrate the inverted effect include those in refs 46-49.

3.5 RELATION BETWEEN RATE CONSTANTS AND EQUILIBRIUM CONSTANTS

For a series of similar ET reactions (constant λ), a plot of log of the rate constant k_r *vs* the log of the equilibrium constant K is expected to be curved, and the slope α at any particular value of ΔG^{0} is given by (from eqs 11 to 13)

$$\alpha = \frac{1}{2}\left(1 + \frac{\Delta G^{0}}{\lambda}\right) \qquad (44)$$

When $|\Delta G^{0}| << \lambda$, an approximate linear relationship between log k and log K results, with $\alpha \approx 1/2$. The slopes of plots of log k *vs* log K for some reactions are listed in Table III of ref 1, and in many cases a slope close to 0.5 is indeed observed.

3.6 ELECTROCHEMICAL ET REACTIONS

Many comparisons of the self-exchange rates with the rates of the corresponding reactions at an electrode have been made in order to test eq 43. A plot of log k_{ex} *vs* log $k_{el}{}^{ex}$ for some illustrative examples is given in Figure 6.10 of ref 6. The correlation is fairly good, but the electrochemical rates appear to level off at higher rates. This may be a real discrepancy or may be an experimental problem, since at higher electrochemical rates, the reactive contribution to the measured A.C. impedance becomes smaller. There have been some recent attempts to make very small electrochemical devices in order to overcome this difficulty, and work is in progress.[50]

Studies of the effect of activation overpotential on the electrochemical ET rate constant consistent with eq 34 have been described.[51,52]

3.7 ADIABATIC *vs* NONADIABATIC ET

The question of whether or not a particular electron transfer process is adiabatic, *i.e.*, occurs at essentially every crossing of the TS, or nonadiabatic, i.e., rarely occurs during a crossing of the TS, is frequently encountered. It has been possible to answer this question in some specific situations.

This question influences, via the κ in eq 33, the numerical value of A_{el} in eq 39. In a study of an electrochemical ET, the electron transfer "exchange current" and, thereby $k_{el}{}^{ex}$, was measured for the $Ru(NH_3)_6{}^{3+}$ + M(e) → $Ru(NH_3)_6{}^{2+}$ reaction for a variety of metal electrodes M.[53] The nonadiabatic expression for $k_{el}{}^{ex}$ has a term proportional to the density of electronic states in the metal at the Fermi level,[26,29] and that density is, in turn, proportional to a coefficient γ in the temperature dependence of the electronic specific heat of the metal. If the ET reaction at M were adiabatic, however, the $k_{el}{}^{ex}$ would be independent of the density of states and hence of γ. Experimentally, $k_{el}{}^{ex}$ was found to be independent of γ when the latter was varied tenfold by varying the nature of the metal M, for the given redox system.[53]

Reactions which are required to occur over large distances, include those where the donor and acceptor in solution are separated by a long rigid saturated bridge[54-57] or, are far apart in frozen media[58], or when they are far apart in proteins,[59-62] or in an electrochemical system[51] described below. All of these systems have a small matrix element H_{rp} and hence are clearly nonadiabatic. Nonadiabaticity is most clearly tested when, by varying the ΔG^0 in such systems, the value of the maximum rate can be measured (*e.g.*, ref 93), namely at $-\Delta G^0 \sim \lambda$, and eq 13a then used to infer κ. An analogous example, but at an electrode interface *M*, occurred when the activation overpotential $ne(E-E_0')$ was varied for a reaction where the redox agent was bound to a long chain molecule. The latter was part of an ordered monolayer of long chain molecules on the electrode surface.[51] The maximum rate constant (an asymptote in a plot) showed that $\kappa\nu$ was clearly much less than ν and so the ET was clearly nonadiabatic in this case.

For reactions in solution when the reactants are moving freely rather than being held some fixed distance apart the problem of determining adiabaticity *vs* nonadiabaticity has been more difficult. For very fast reactions the high reaction rates which occur when $-\Delta G^0 \sim \lambda$ tend to be masked by the slow and rate controlling diffusion of the reactants toward each other. For slower reactions the estimation of $\kappa\nu$ from the pre-exponential factor of the bimolecular rate constant tends to be masked by other effects, such as the entropy associated with the work terms w_r and w_p (which were omitted for brevity in eq 13b) and any ΔS^* associated with the reorganization in eq 48 given later. This ΔS^* vanishes for isotopic exchange reactions, but the ΔS associated with the w's remains. Some effort has been made to extract the $\kappa\nu$ from such data, *e.g.*, as in the MnO_4^-–MnO_4^{2-} electron exchange reaction where κ appears to be about unity. A value of unity has been used for κ in the comparison noted in Section 3.1 of some absolute reaction rates with those calculated theoretically using eqs 13b and 23-27.[4]

3.8 ELECTRONIC MATRIX ELEMENTS

Recently many experiments have probed the dependence of the electronic matrix element H_{rp}, appearing in eq 40, on the distance and the nature of the medium separating the redox centers.[54-57] A commonly employed technique has been to synthesize compounds with an electron donor D at one end, and an electron acceptor A at the other, with a covalently linked bridge B connecting the two: Here, intramolecular ET can take place from the donor to the acceptor. Usually a series of such compounds are synthesized where the donor and the acceptor groups are kept constant and the bridge length is systematically increased. The rate of ET is measured for the series. In order to infer values of relative electronic matrix elements from these values of rate constants,[1,63] one may proceed as follows:

It was seen in Section 1 that the rate constant k_r of a nonadiabatic ET reaction (small H_{rp}) is given by eq 41 as

$$k_r = \frac{2\pi}{\hbar} |H_{rp}|^2 \frac{1}{\left(4\pi\lambda k_B T\right)^{1/2}} \exp\left(-\Delta G^*/k_B T\right) \tag{45}$$

for an intramolecular process. Equation 45 can be written in the form

$$k_r = A \exp(-E_a/k_B T) \tag{46}$$

where E_a is the activation energy and A the pre-exponential factor.

If the weak dependence of $1/\sqrt{T}$ on T is neglected in eq 45, for purposes of the present lecture, we have

$$E_a \simeq \Delta H^* \tag{47}$$

where ΔH^* is the enthalpy term defined by the standard thermodynamic relation, $\Delta H^* = \partial(\Delta G^*/T) / \partial(1/T)$, since thermodynamic relations apply to these statistical mechanically defined kinetic quantities.

The pre-exponential factor A in eq 46 can then be written as

$$A \simeq \frac{2\pi}{\hbar} |H_{rp}|^2 \frac{\exp(\Delta S^*/k_B)}{\left(4\pi\lambda k_B T\right)^{1/2}} \tag{48}$$

where ΔS^* is equal to $-\partial\Delta G^*/\partial T$.

Using eq 23 for ΔG^*, we have

$$\Delta S^* = \frac{1}{2}\left(1 + \frac{\Delta G^0}{\lambda}\right)\Delta S^0 \tag{49}$$

and

$$\Delta H^* = \frac{\lambda}{4}\left[1 - \left(\Delta G^0/\lambda\right)^2\right] + \frac{\Delta H^0}{2}\left[1 + \left(\Delta G^0/\lambda\right)\right] \tag{50}$$

assuming λ to be independent of temperature.

The ΔS^0 in eq 49 is zero for a self-exchange reaction. We also have $\Delta S^0 \simeq 0$ when the reactants in reaction 4 have charges on $(Ox)_1$ and $(Red)_2$ equal to those on $(Ox)_2$ and $(Red)_1$, respectively, and when, in addition, the radii of the reactants 1 and 2 are about equal. When ΔS^0 is zero, it is seen from eq 49 that $\Delta S^* \simeq 0$. It follows from eq 48 that A then provides a measure of $|H_{rp}|^2$ for the series, there now being no ΔS^* there. Again, if $-\Delta G^0 \approx \lambda$, it is seen from eqs 48 to 50 that $\Delta S^* \simeq 0$ and $\Delta H^* \simeq 0$, $k_r \simeq A$, and that ΔS^* is again absent from eq 48, permitting once more an estimate of $|H_{rp}|^2$. $|H_{rp}|^2$ is also estimated more indirectly from E_a and A using eqs 47 to 50. One alternative route for determining the H_{rp}'s experimentally is to use charge-transfer (intervalence) spectra, upon introducing an approximation for the optical transition dipole matrix element.[64-67] The latter is then proportional both to $|H_{rp}|$ and to the separation distance between the donor and acceptor charges.

One procedure for calculating H_{rp} with theory is to seek the two lowest energy many-electron wavefunctions of the D-B-A system where the electronic charge is equally divided between D and A.[3,5] Then the energy difference of these two delocalized many-electron states equals the $2H_{rp}$ in Figure 3. This procedure can also be implemented more approximately using, instead, a one-electron description in which delocalized orbitals are formed from individual atomic orbitals localized on D and A. The energy difference between two delocalized orbitals equally distributed over D and A again gives H_{rp} but now with the added one-electron approximation.

Alternatively, perturbation theory has been used to estimate H_{rp} by considering properties of the bridge and then treating the interaction of the donor and of the acceptor with the bridge using perturbation theory.[68] In a simple case when the donor and the acceptor are each linked to one atomic orbital of the bridge, H_{rp} is given by[68]

$$H_{rp} = T_D T_A \sum_{\nu} \frac{C_{D\nu} C_{A\nu}}{b_\nu - a} \tag{51}$$

where $T_D(T_A)$ are the matrix elements for interaction between $D(A)$ and the adjacent atomic orbitals of the bridge, $C_{i\nu}$ $(i = D, A)$ is the coefficient of the bridge orbital ν at the point of contact with i, b_ν is the energy of the bridge orbital ν, and a is the energy of the donor orbital. (The donor orbital and acceptor orbital have equal energies in the TS because of the influence of the environment, depicted in Figure 3.) Equation 51 may be extended to more general cases where D and A are both large groups with a number of orbitals connecting $D(A)$ to B.[69] Such a formalism in its one-electron or many-electron form has been termed "superexchange" in the literature, the electron making use of the orbitals of the bridge (both occupied and unoccupied) for electron transfer from D to A (or hole transfer from A to D).

For either of these calculational procedures, many-electron wavefunctions, all-electron SCF models, or semi-empirical methods may be used to calculate the molecular orbitals and the energies required in the expression for H_{rp}. For large systems such as the biological systems to be discussed in the next section, at the present time, semi-empirical methods seem to offer the best alternative. Some examples of calculated values of H_{rp} (using the one-electron extended Hückel theory), along with experimentally determined values of H_{rp}, are given in Tables I-IV of ref 70. Encouraging agreement was found for the relative values of H_{rp} as a function of separation distance in each series.

3.9 EXTENSION TO BIOLOGICAL SYSTEMS

Metalloproteins have been extensively used to examine various aspects of ET reactions.[59-62] One important area of research has been to determine the electronic coupling provided by the protein medium. Both inter- and intramolecular protein electron transfer reactions have been investigated. As an example of the latter, a redox group such as $[Ru(NH_3)_5]^{+3}$ is covalently linked attached to a histidine (His) residue of a protein. Thus, Gray and coworkers have synthesized a series of ruthenium-modified proteins in which the donor and the acceptor are the same, but the ruthenium group is located in different residues of the protein, thus allowing distance and medium effects to be probed. In Figure 2 of ref 71, a myoglobin is illustrated in which there are four surface histidines, all of which can be singly ruthenated and the ET rate between the ruthenium-modified His and the metal porphyrin then measured. From the ratios of the square roots of these rates, ratios of the electronic matrix elements for the various derivatives can be obtained when any distance-dependent effect that may occur in FC, for example in the solvation effect on ΔG^0 and λ, is neglected. These values may be compared with ratios of the H_{rp} values calculated from a superexchange model (as in Table II of ref 72). The order of magnitude agreement between the relative calculated and experimental values is encouraging, and more detailed calculations including more amino acid residues of the proteins are in progress.

The photosynthetic charge transfer system is another area where there has been intense research in the last few years.[73-75] The sequence of reactions is as follows: a) electronic excitation of the special bacteriochlorophyll pair, $BChl_2$; b) transfer of electron from $BChl_2^*$ to a pheophytin, BPh, some 9-10 Å away in 2.8 picoseconds; c) transfer of electron from BPh^- to a quinone Q_A in the next 200 picoseconds.

The protein environment of $BChl_2$, BChl and BPh is largely hydrophobic and therefore gives rise to a small λ_o. The vibrational λ_i is presumably also small since the bond lengths in the reactants (both in the large porphyrin rings of the chlorophyll molecules and the metal-ligand bond lengths) do not undergo much change in bond length upon electron gain or loss. Both aspects lead to an overall small reorganizational term, λ. The driving force of the formation of BPh^- from $BChl_2^*$ also appears to be relatively small, around 0.2 eV, therefore allowing the forward reaction to proceed extremely rapidly, while minimizing loss of energy arising from the absorption of the sunlight. Further, the back ET from BPh^- to form ground state $BChl_2$ appears to be extremely slow, perhaps due in part to an inverted effect, resulting in an efficient charge separation.

Another interesting question regarding these primary steps of photosynthesis that arises is the role of a nearby bacteriochlorophyll molecule

BChl in mediating the fast initial ET (*e.g.*, refs 74-79). Direct ET from $BChl_2^*$ to BPh seems to be somewhat unlikely in view of the large separation between the two pigments. Other possibilities are a mechanism involving $BChl^-$ as an intermediate, or a superexchange mechanism where the BChl serves as a bridge. This topic is under active investigation and discussion.

4. Miscellaneous Questions

In these notes, we have attempted to include remarks on some of the questions raised in or out of class, such as how is the reaction coordinate of an electron transfer reaction is defined (refs 12 and 13), what, in detail, is the Landau-Zener theory for curve crossings (Appendix A), what justification is there for using the same "force constant" for the vibrations of the reactants as for the products (ref 20), how does one proceed from using potential energy surfaces to using free energy surfaces (Appendix A), why does ΔG^0 appear in the nonadiabatic rate expression when the initial Golden Rule formula on which it is based contains instead a ΔE^0 (ref 36), why is there no inverted effect in the electrochemical ET reactions at metal electrodes (Section 2.4), is there evidence for an exciplex formation, in the case of a fluorescence quenching ET reaction, which sometimes masks the inverted effect (ref 44), and what is the relation between long-range electron and triplet energy transfers (ref 94).

In this section, we address several additional questions which were raised.

1. *In the nonadiabatic equation for k_r is the $(4\pi\lambda kT)^{1/2}$ factor part of an electronic factor or a nuclear factor?*

The discussion of the Landau-Zener expression for "curve crossing" in Appendix A shows that this factor appears as a result of the $\kappa(v)$ there, and so arises from the coupling of the electronic and nuclear motions. Thereby, it is to be assigned neither to the one nor to the other, but to both.

2. *When $\lambda \rightarrow 0$, the nonadiabatic expression for k_r becomes infinite. Does this mean that k_r itself becomes infinite?*

A value of $k_r = \infty$ would be a physical impossibility. Rather, when the nonadiabatic ET expression, e.g., eq 41, becomes infinite when $\lambda = 0$, an assumption in which that equation is based has failed. In particular, the transition probability $\kappa(v)$ in Appendix A is required to be small for the pre-exponential factor in eq 41 to be valid. In that Appendix $\kappa(v)$ is shown there to be proportional to $|H_{rp}|^2/\sqrt{\lambda}$. However, $\kappa(v)$ can no longer be small when λ approaches zero, and so a different formalism would be required.

3. *What is the relation, if any, between electron tunneling and nuclear tunneling?*

In Figures 3 and 9 potential energy curves are depicted for the nuclear motion and, as in Figure 9, can be used to describe nuclear tunneling. Not depicted in either of these plots is the many-dimensional surface $V(\mathbf{r}_e)$ for the potential energy for the electrons as a function of the many electronic coordinates $\mathbf{r}_e$. The Schrödinger equation for the electronic wave function, holding the nuclei fixed at positions $\mathbf{q}$ (Born-Oppenheimer approximation), contains this $V(\mathbf{r}_e)$, and, when solved, provides an energy $U(\mathbf{q})$, which serves as the potential energy for the nuclear motion as a function of $\mathbf{q}$ in Figure 3. When the wave function is constrained to reside on the donor, this U becomes the surface U_r, and when it is constrained to reside instead on the acceptor, U becomes the U_p one. In the Schrödinger equation for the electrons, the region between the donor and the acceptor typically has a high potential energy $V(\mathbf{r}_e)$ for the electrons, so high that the electrons usually cannot penetrate that region classically. Nevertheless, ET does occur and, if one wishes to use the term, has occurred by electron tunneling from the donor to the acceptor.

Some measure of this electron "tunneling probability" during a typical period $1/\nu$ of nuclear motion (ν being a typical frequency for that motion) can be obtained as follows: The frequency with which an electron would oscillate between two reactants whose nuclei were held fixed is shown below to be the Bohr frequency, $2H_{rp}/h$, where $2H_{rp}$ is the splitting in Figure 3 due to the electronic interaction of the two reactants. The electron tunneling probability during the time $1/\nu$ is then $2H_{rp}/h\nu$. For example, when $H_{rp} \sim 100\ \mathrm{cm}^{-1}$ and $\nu \sim 400\ \mathrm{cm}^{-1}$, a typical metal-ligand frequency, $2H_{rp}/h\nu \sim 1/2$, and so the electron tunneling probability during a period $1/\nu$ of the nuclear motion is $\sim 1/2$ and the reaction is nearly "adiabatic." In some cases, however, H_{rp} is very small, the electron tunneling probability during the period $1/\nu$ is also therefore very small, and the reaction is said to be "nonadiabatic."

In the case of a donor and acceptor separated by monomeric units in some bridge, H_{rp} decreases exponentially with an increase in the number N of monomeric units. Thereby, it decreases exponentially with increasing donor-acceptor separation distance, when the chain of units is roughly linear. A semi-quantitative explanation is as follows: When there is one molecular orbital (empty in the case of electron transfer and occupied in the case of hole transfer) per unit, quantum mechanical perturbation theory shows that H_{rp} decreases by a factor $V/\Delta E$ per unit, where V is the unit-to-unit electronic matrix element and ΔE is the energy difference between the donor orbital and the unit's orbital. Thus, for N units, H_{rp} decreases by a factor $(V/\Delta E)^N$, *i.e.*, by $\exp(-\beta R/2)$, where $R = Na$, a being the length of a unit, and $\beta = 2a^{-1} \ln(\Delta E/V)$.

For an atom-atom matrix element of ~4 to 5eV, a ΔE of about the same value, and the product of the coefficients of the molecular orbitals on adjacent orbitals being about 1/20 when there are 20 relevant atoms per unit, the V is about (5/20) eV (cf. ref 95). When $a \simeq 5$ Å, β is then calculated to be about 1.3Å^{-1}, which is fairly close to a typical value[70] of $\sim 0.7\ \text{Å}^{-1}$ to $1\ \text{Å}^{-1}$. (When there are 20 atoms contributing to a molecular orbital, each atomic coefficient is about $1/\sqrt{20}$.) The actual situation is more complicated in that a number of orbitals participate rather than merely one per unit, but the same exponential dependence on R prevails.

The connection between the above power rule, $(V/\Delta E)^N$, and eq 51 has been discussed by McConnell[96] for a particular case, in his treatment of triplet energy transfer in chain systems. A similar argument applies to electron transfers.

4. *What is the relation between the splitting* $2H_{rp}$ *and the frequency of the electronic motion?*

This relation can be seen as follows. Suppose that ψ_1 denotes the electronic wave function when the electron is localized on the donor, and ψ_2 when it is localized on the acceptor. The adiabatic wavefunctions, corresponding to the energies of the rounded-off curves at the intersection in Figure 3 would be $(\psi_1+\psi_2)/\sqrt{2}$ and $(\psi_1-\psi_2)/\sqrt{2}$, with energies of U^*-H_{rp} and U^*+H_{rp}. Here U^* is the energy at the intersection $U_r=U_p$. If the electron is originally in the donor at time $t=0$, the electronic wavefunction ψ at any later time is

$$\psi = \frac{1}{2}(\psi_1+\psi_2)\exp\left[-i(U^*-H_{rp})t/\hbar\right]+\frac{1}{2}(\psi_1-\psi_2)\exp\left[-i(U^*+H_{rp})t/\hbar\right] \tag{52}$$

ψ clearly equals ψ_1 at $t=0$. At a time $t=h/4H_{rp}$ it can be seen that $\psi=\psi_2\exp i\theta$, where $\theta=-[(U^*/H_{rp})-1]\,\pi/2$, *i.e.*, that $|\psi|^2=|\psi_2|^2$ and so the electron then resides on the acceptor. After a further time $h/4H_{rp}$, $|\psi|^2$ again equals $|\psi_1|^2$. Thereby, under conditions where the nuclei are fixed in position and the system is at the intersection of U_r and U_p, the electron would oscillate from donor to acceptor and back with a frequency of $2H_{rp}/h$. Since the nuclei move rather than stay fixed, an irreversibility actually occurs, described by the Landau-Zener expression in Appendix A or by more general treatments.

5. *Are the potential energy surfaces parabolic for these and for other reactions?*

We have noted that the PES for ET reactions are only parabolic (harmonic) along the vibrational coordinate axes. Along the solvational coordinates' axes they are highly anharmonic, with various local minima. The free energy plots, on the other hand, including both the solvational and vibrational contributions, are approximately harmonic functions of the reaction coordinate. Other transfer reactions are not represented well by parabolas for reactants and parabolas for products, even for the vibrational motion. In these reactions, a strong interaction of the reactants occurs, one bond is broken and another is formed, and so a quite different model should be used, one which takes account of these effects. One model which was developed leads to an energy barrier, which when phrased in potential energy terms, is given by[80]

$$\Delta U^* = \frac{\lambda}{4} + \frac{\Delta U^0}{2} + \frac{\Delta U^0}{2y}\ln\cosh y \tag{53}$$

where $y=(2\Delta U^0/\lambda)\ln 2$, ΔU^0 being the potential energy of reaction (difference of bond energies of reactants and products), and $\lambda/4$ is the barrier at $\Delta U^0=0$. Further, $\lambda_{12}=\frac{1}{2}(\lambda_{11}+\lambda_{22})$, as in ET reactions.[80] In a region of $|\Delta U^0/\lambda|$ of about 0 to 0.5, this equation is reasonably well approximated by the quadratic equation.[80] However, in marked contrast to the latter, there is no inverted effect: in eq 53, ΔU^* tends to zero monotonically when $\Delta U^0/\lambda$ tends to minus infinity, as expected on physical grounds. It is the difficulty of intersection of the two parabolas in Figure 8 at a conveniently small height which leads to the inverted effect.

6. *Are proton transfers expected to display an inverted effect?*

If the reactants were held so far apart that their interaction would be very weak, a parabolic picture for proton transfers would be adequate. Even in that case, however, the inverted effect would be difficult to observe: the proton levels are closely spaced (~3,000 cm^{-1} apart in OH or CH vibrations), in contrast with the usually wide spacing of electronic energy levels (typically ~30,000 cm^{-1}). The possibility of forming electronically excited states of the reaction products, we have seen in Figure 8, opens new channels for reaction when the formation of the lower electronic state becomes inhibited because of the inverted effect. In the proton transfer case, the new channels would involve, instead, vibrationally excited states of the newly formed O–H, C–H, or N–H proton bond.

Typically, however, proton transfers are not expected to be represented well by the intersection of a pair of parabolas, and a strong interaction behavior, such as that depicted in the response to question 5, would be expected to lead to an equation similar to eq 53 rather than eq 11. On the other hand, in photoexcited proton transfer reactions in isolated systems at sufficiently low temperature (*e.g.*, as in supersonic beams) a resonance might occur between the initial protonic energy level and the final one. Resonances provide the equivalent of a very narrow $\ln k_r$ *vs* $-\Delta G^0$ plot, with a sharp maximum. An equivalent behavior, but for an electronic transition, has been found in electric field controlled resonances in formaldehyde.[81] Perhaps under similar circumstances, an analogous effect can be made to occur for proton transfers.

7. *Is it meaningful to speak of the splitting of the free energy curves, as well as the PES?*

The splitting $2H_{rp}$ of the two PES depends mainly on the separation distance R and so, for a fixed R, is essentially constant along the intersection surface. Because of the relation between the free energy curves in Figure 4 and the PES in Figure 3, the splitting of the $G_r(q)$ and $G_p(q)$ curves is in that case essentially equal to $2H_{rp}$ also. In other cases, if one wished to depict a splitting in Figure 4, that splitting would have to represent the value of H_{rp}, averaged over the intersection surface using a Boltzmann weighting factor.

8. *Does recrossing of the intersection surface occur in the ET reaction?*

Wigner has pointed out that if a "hypersurface" in phase space (*i.e.*, a $2N-1$ dimensional surface in a $2N$-dimensional space which has N coordinates and N momenta as axes) could be found such that there were no recrossing of this hypersurface by classical trajectories of the system during the course of the reaction, classical mechanical transition state theory would be valid.[82] The hypersurface is then the TS. The neglect of such recrossings in TS theory causes the TS calculated rate to be an upper bound to the actual reaction rate.

For an adiabatic reaction in solution, some recrossing probably occurs but typically relatively little, in the calculations of classical trajectories made thus far. The error in TS theory due to neglect of recrossings has been estimated by a large scale numerical calculation of classical trajectories of reactants in a particular S_N2 reaction[83] to be only a factor of about two. This error represents a relatively minor effect in reaction rate theories, which contain other approximations. One main task of such theories is to understand chemical effects which can sometimes vary by many orders of magnitude.

For nonadiabatic reactions, the smallness of $\kappa(v)$ implies that most of the crossings of the $U_r = U_p$ intersection surface will indeed not result in reaction, that is, in a transition from the U_r to the U_p surface. In such a case, after a crossing of the TS, the system will typically remain on the U_r surface, but will ultimately be reflected, and then recross the TS. However, such ineffectiveness to lead to a reaction is taken into account in the calculation of $\kappa(v)$, and no further correction of TS theory due to this source of recrossing is needed in this nonadiabatic case.

9. *How does variational TS theory relate to the TS theory being used?*

Following Wigner's ideas[82] on the implications of recrossings of the hypersurface, the most accurate choice of the hypersurface is the one for which there are the fewest recrossings. It can be shown that such a choice will be the one which gives the lowest calculated reaction rate, and thereby the rate closest to the actual rate. Selecting a TS, and varying it so as to obtain a minimum rate constant, is the essence of variational TS theory. In the case of ET reactions, use of the intersection surface for the TS is expected to provide the most straightforward choice for the TS.

5. Other Topics

There are many topics in the ET field which have been studied experimentally, and, in many cases, theoretically, including some of the topics listed in Figure 1 and not discussed here, such as solvent dynamics, with[84] and without[85] vibrational effects, electron transfer accompanied by rupture of a chemical bond,[86] electron transfer across the interface of two phases, either liquid-liquid,[87] liquid-polymer,[88] or polymer-polymer interfaces,[89] and many others. The field continues to expand in new directions and to offer a challenge to experimentalists and theoreticians alike.

Acknowledgment

It is a pleasure to acknowledge the support of this work by the Office of Naval Research and by the National Science Foundation. A portion of these lectures was written while one of us (R.A.M.) was the Baker Lecturer at Cornell University. He is very pleased to acknowledge the hospitality and discussions of his colleagues there.

Appendix A. Rate Expressions for Adiabatic and Nonadiabatic Reactions

The classical statistical mechanical expression for the rate constant can be written as

$$k_r = \int \cdots \int \dot{q}_1 \, \gamma \exp\left(-H(q_1^{\ddagger})/k_B T\right) dq_2 \ldots dq_N \, dp_1 \ldots dq_N \, / h^N Q \tag{A1}$$

where q_1 denotes the reaction coordinate (denoted by q in the text), $H(q_1^{\ddagger})$ is the sum of the potential energy at the transition state (TS) plus the total kinetic energy (involving all N p_i's), and γ is the probability of a successful electronic transition on reaching the TS (in our case going from the reactants to the products surface there, *i.e.*, going from U_r to U_p). The integral over p_1 is from 0 to $+\infty$ (rate in forward direction), the integrals over the remaining p_i's are from $-\infty$ to $+\infty$, and those over $q_2 \ldots q_N$ are over the full range of these coordinates. Q is the partition function of the reactants

$$Q = \int \cdots \int \exp(-H/k_B T) \, dq_1 \ldots dq_N \, dp_1 \ldots dp_N / h^N \tag{A2}$$

H being the sum of the kinetic and potential energy of the reactants and solvent molecules.

Equation A1 was obtained by using an expression for the probability density of finding the system at $q_1^{\ddagger}$ (per unit length along q_1), multiplying by the velocity $\dot{q}_1$ along the reaction coordinate to obtain the probability flux, then multiplying by γ to obtain the reaction probability flux, and integrating over the coordinates at fixed $q_1 (= q_1^{\ddagger})$ and over the momenta as indicated.

When γ is unity in the range of coordinates and momenta of interest, use of a classical mechanical equation $\dot{q}_i = \partial H / \partial p_i$, and integration over p_1 and over the remaining variables in eq A1 yields the standard TS theory expression for k_r for an adiabatic reaction ($\gamma = 1$), treated classically:

$$k_r = \frac{k_B T}{h} \frac{Q^{\ddagger}}{Q} \exp(-\Delta U^{\ddagger} / k_B T) \tag{A3}$$

where $\Delta U^{\ddagger}$ denotes the lowest potential energy of U_r in the TS minus that of U_r anywhere, and where $Q^{\ddagger}$ is the partition function for the transition state:

$$Q^{\ddagger} = \int \cdots \int \exp\left[-H(q_1 = q_1^{\ddagger}, p_1 = 0)/k_B T\right] dq_2 \ldots dq_N \, dp_2 \ldots dp_N / h^{N-1} \tag{A4}$$

The H in eq A4 is now the sum of the potential energy of the TS, measured relative to the lowest value on the TS, plus the kinetic energy of the TS.

Q is seen from eqs A2 and A4 to contain one coordinate and one momentum, q_1 and p_1, more than $Q^{\ddagger}$. Upon integration in eq A2 over all q_i and p_i but q_1 and p_1, one obtains for Q for a unimolecular reaction a factor given later by eq B2 and denoted by $\exp[-G_r(q_1)/k_B T]$, $G_r(q_1)$ being the "reorganizational" free energy as a function of q_1. It is approximately a quadratic function of q_1, as discussed earlier in the text (linear response approximation). When the integral of this factor over q_1 is combined with the integral over p_1, one obtains a factor which is approximately equal to a classical vibrational partition function, $k_BT/h\nu$, where ν is a typical frequency associated with this motion along the reaction coordinate q_i. If Q' denotes the remaining factor in Q, $Q/(k_BT/h\nu)$, we then have

$$k_r = \nu \frac{Q^\ddagger}{Q'} \exp(-\Delta U^\ddagger / k_B T) \tag{A5}$$

The term $(Q^\ddagger/Q')\exp(-\Delta U^\ddagger/k_BT)$ in eq A5 can be written as the exp $(-\Delta G_r^*/k_BT)$ in the text and is associated, at the separation distance R, with the solvational and vibrational reorganization needed to reach the TS. From eq A5 we then obtain eq 13a of the text for a unimolecular reaction, with $\kappa = 1$.

For bimolecular reactions, three of the coordinates in eq A1 can be chosen to be the center of mass of the two reactants. The contribution of these coordinates and their momenta to the integral in eq A1 just cancels their contribution in eq A2. Three of the remaining coordinates can be chosen to be the relative coordinates of the two reactants. To conform to the definition of a bimolecular reaction rate constant, the integration over these relative coordinates in Q in eq A2 is over a unit volume. If in eq A1 the electron transfer rate has a γ which decreases with separation distance R as exp $[-\beta(R-R_0)]$ and which equals approximately unity at contact ($R = R_0$), integration over the three coordinates in the integral in eq A1 yields a factor $\int 4\pi R^2 \exp[-\beta(R-R_0)]\,dR$, which is well approximated by $4\pi R_0^2/\beta$, since $\beta^{-1} << R$. This term is the v_s in eq 13b in the text. The momentum contribution to Q' and to $Q^\ddagger$ in eq A5 for these translational coordinates cancels. The remaining contributions to $Q^\ddagger$ and Q are the same as those for unimolecular reactions, and yield, as in the unimolecular case, the reorganizational factor exp $(-\Delta G_r^*/k_BT)$. Thereby, one obtains eq 13b of the text with $\kappa = 1$. Under some conditions the R in the TS may not equal the separation distance R_0 at point of contact R_0, and ΔG_r^* and κ are then computed at the R for the TS. (The rate depends on R through κ and through ΔG_r^*, and its maximum could occur at an R somewhat greater than R_0.)

For reactions at an electrode related considerations apply. Now, three of the coordinates in Q and in $Q^\ddagger$ in eq A3 are the translational coordinates of the reactant. In conformity with the definition of k_r for such a reaction, the three coordinates in Q are integrated over a unit volume. Two of them in $Q^\ddagger$ are integrated over a unit area of the electrode, while the third coordinate, the coordinate z, which is perpendicular to the electrode surface, contributes a factor to $Q^\ddagger$ of $\int \exp(-\beta z)dz$, the integral being from $z = 0$, the point of contact with the electrode, to $z = \infty$. Thereby one obtains β^{-1}, which is the v_e in eq 33a in the text. The remaining contributions to eq A3 are the same as above, and so one obtains eq 33a of the text, with $\kappa = 1$.

We turn next to highly nonadiabatic reactions, *i.e.*, when $\gamma << 1$ in eq A1 in the coordinate and momentum region of interest. A one-dimensional approximation to γ is given by the Landau-Zener expression:[17]

$$\gamma = 1 - e^{-\kappa(v)} \tag{A6}$$

where

$$\kappa(v) = \frac{2\pi}{\hbar}\frac{|H_{rp}|^2}{v|s_r - s_p|} \tag{A7}$$

Here, v is the velocity $\dot{q}_1$ along the reaction coordinate at the intersection of the two potential energy curves (at $q_1 = q_1^{\ddagger}$ in our case), and s_r and s_p are the slopes of the potential energy curves at the intersection, $\partial U_r/\partial q_1$ and $\partial U_p/\partial q_1$ in the present case. Strictly speaking, eqs A6 and A7 were derived for a one-coordinate system.[17] If applied, nevertheless, to the present many-coordinate case, these slopes depend on the point $(q_2, ...q_N)$ being crossed at $q_1 = q_1^{\ddagger}$. If some mean value for the slopes is used (since there is an integration in eq A1 over $q_2 ...q_N$), *e.g.*, if we assume U_r to be approximately $\frac{1}{2}k\,(q_1 - q_1^{\,r})^2$ plus terms independent of q_1, and U_p to be $\frac{1}{2}k\,(q_1 - q_1^{\,p})^2$ plus terms independent of q_1, then $|s_r - s_p|$ equals $|k(q_1^{\,r} - q_1^{\,p})|$. If we regard λ as being approximately $\frac{1}{2}k(q_1^{\,r} - q_1^{\,p})^2$, as in the text, then $|s_r - s_p| \simeq \sqrt{(2k\lambda)}$. If we consider the case that $\kappa(v) << 1$ in the region of interest, then in eq A6 $\gamma \simeq \kappa(v)$. The velocity v in the denominator of eq A7 then cancels the $\dot{q}_1$ in eq A1. If the kinetic energy in H in A1 is regarded as being $\frac{1}{2}p_1^{\,2}/\mu_1$ plus terms independent of p_1, integration of eq A1 then yields

$$k_r = \frac{2\pi}{\hbar}\frac{|H_{rp}|^2}{\left(4\pi\lambda k_B T\right)^{1/2}}\frac{Q^{\ddagger}}{Q'}\exp\left(-\Delta U^{\ddagger}/k_B T\right) \tag{A8}$$

upon once again extracting the factor $k_BT/h\nu$ from Q and also using the relation $\sqrt{(k/\mu)} = 2\pi\nu$. Two ν's then cancel.

The term $(Q^{\ddagger}/Q')\exp(-\Delta U^{\ddagger}/k_BT)$ in eq A8 equals $\exp(-\Delta G_r^*/k_BT)$ for unimolecular reactions as before, while for bimolecular reactions and electrode reactions in which the reactant is not bound to the electrode, it equals $v_s \exp(-\Delta G_r^*/k_BT)$ and $v_e \exp(-\Delta G_r^*/k_BT)$, respectively, where v_s and v_e were defined above. The $|H_{rp}|$ in eq A8 now denotes the matrix element when the two reactants (or the reactant and the electrode) are in contact.

In this way, in the case of very small γ, eqs 13 and 33 of the text are obtained, but with $\kappa\nu$ being given in each case by eq 14b in the text. For intermediate values of γ, one could either integrate eq A1 using eq A6, or simply interpolate approximately by using eq 14b when the right-hand side of eq 14b is less than ν and then using eq 14a otherwise.

Physical insight into eq A7 in terms of the "frequency" of the electronic motion in the TS, obtained as $2H_{rp}/h$ in Section 4, namely the separation of the adiabatic energy levels in the TS in Figure 3 divided by h, and of an effective frequency for the nuclear motion in the vicinity of the "crossing point" is given in ref 90.

Appendix B. The Reorganizational Free Energy G(*q*)

The expression for $G(q)$, the free energy of reorganization as a function of q, is considered next. (q is the reaction coordinate denoted by q_1 in Appendix A.) We consider unimolecular reactions first. Using classical statistical mechanics, we have

$$\exp[-G(q)/k_BT] = \int \dots \int \exp[-H(q_1 = q, p_1 = 0)/k_BT]\, dq_2 \dots dq_N dp_2 \dots dp_N / h^{N-1} \tag{B1}$$

The kinetic energy term in H in eq B1 is of the form $\frac{1}{2}\Sigma_{i,j} g^{ij} p_i p_j$. With this quadratic form, a standard expression[91] can be used to evaluate the integral in eq B1 over p_2 to p_N. If the determinant of the g^{ij}'s, for i and j not equal to 1, is denoted by $1/g'$, then we have upon integration of the p_i's

$$\exp[-G(q)/k_BT] = \frac{\left(2\pi k_BT\right)^{(N-1)/2}}{h^{(N-1)}} \int \cdot \int \exp[\ U(q_1 = q)/k_BT]\, dS \tag{B2}$$

where dS denotes the $N-1$-dimensional volume element $\sqrt{g'}dq_2 \dots dq_N$. Apart from a prefactor, eq B2 is the same as eq 5 in the text, upon introducing r subscripts.

Where the reaction being considered is bimolecular or when it involves an ET between an ion in solution and an electrode, there are also translational coordinates to be considered, as noted in Appendix A. The $G(q)$ given by eq B1 or B2 is then intended to be the value when the reactants are held fixed at some separation distance R, and is a function of that R. In this case, some of the $N-1$ coordinates would be the translational coordinates of the reactants and the $G(q)$ would be defined with eq B1, but using a number of coordinates less than the $N-1$ there, less by the number of translational coordinates of the reactants, six in the bimolecular case, and three in the case of the electrode reaction.

Appendix C. Comparison of Quantum Expressions and Remarks on ln k_r *vs* $-\Delta G^0$ Asymmetry

It is instructive to examine the limiting situation for eqs 42 and 43 when $\lambda_o \to 0$ and when both $T \to 0$ and $\lambda_o \to 0$, and to see how the two expressions become identical there. At sufficiently low temperatures, reaction occurs only from the lowest vibrational energy level (no other level is populated), a situation which is obtained in eq 42 by letting $T \to 0$. Equation 42 then becomes

$$k_r = \frac{2\pi}{\hbar} |H_{rp}|^2 e^{-S} S^m / \Gamma(m+1) h\nu \qquad (T=0) \tag{C1}$$

where $\Gamma(m+1)$ is the Gamma function ($= m!$ when m is an integer) and $m = -\Delta G^0/h\nu$, defined here for $m \geq 0$.

Equation 43a can be expected to approach eq C1, and eq 43b to approach eq 42, only when the λ_o in eqs 43a and 43b is made to approach zero, since eqs C1 and 42 contain no λ_o contribution. When λ_o approaches zero, the function $(4\pi\lambda_o k_BT)^{-1/2} \exp[-(\lambda_o + \Delta G^0 + mh\nu)^2/4\lambda_o k_BT]$ becomes extremely small, when regarded as a function of $\Delta G^0 + mh\nu$, except at $\Delta G^0 + mh\nu = 0$. (The effect of the λ_o in the denominator of the exponent dominates that in the pre-exponential factor.) Indeed, in the limit of $\lambda_o \to 0$, this function becomes a Dirac delta function $\delta(\Delta G^0 + mh\nu)$. Thereby, eq 43a in this limit becomes

$$k_r = \frac{2\pi}{\hbar} |H_{rp}|^2 \sum_{m=0}^{\infty} \delta(\Delta G^0 + mh\nu)\, e^{-S}\, S^m / \Gamma(m+1) \tag{C2}$$

where the $m!$ in eq 43a has been written as $\Gamma(m+1)$. The derivation of the nonadiabatic expressions, eqs 42 and 43a, is based on the use of Fermi's Golden Rule, which in turn considers transitions from the initial state to a continuum of final states which are quasi-degenerate with the initial state and so preserve conservation of energy in the reaction step. In eq C2, this degeneracy can only occur if $h\nu$ were regarded as small enough that the sum over m becomes a sum over a continuum, *i.e.*, becomes an integral over dm. If we then use the well-known relation that $\delta(\Delta G^0 + mh\nu)$ equals $(1/h\nu)\,\delta([\Delta G^0/h\nu] + m)$, and the integral over m is evaluated, it follows that eq C2 reduces to eq C1. Thus, at low temperatures and at $\lambda_o = 0$, eqs 42 and 43a become the same expression, as expected. A similar argument shows that, at any temperature, when λ_o tends to zero, eq 43b reduces to eq 42.

In concluding this Appendix, we comment briefly on the dependence of expressions such as eq C1 on the temperature in the inverted region. Introducing Stirling's formula for $\Gamma(m+1)$, $(m/e)^m \sqrt{(2\pi m)}$, it is seen that $\ln k_r$ varies with m as $m \ln(Se/m) - \ln\sqrt{(2\pi m)}$. Since a logarithmic dependence on m is a rather weak one, the principal dependence of $\ln k_r$ on m, and hence on $-\Delta G^0$, is seen to be a linear one, with a slope $\gamma = \ln(\lambda e/|\Delta G^0|)$. Thus, the dependence of $\ln k_r$ on $-\Delta G^0$ in this quantum treatment is essentially linear rather than quadratic.

References

1. Marcus, R. A. and Sutin, N. (1985) Biochim. Biophys. Acta., **811**, 265.
2. J. R. Bolton, N. Mataga and G. McLendon (eds.)(1991) Adv. Chem. Ser., **228**, assorted articles.
3. Newton, M. D. and Sutin, N. (1984) Ann. Rev. Phys. Chem., **35**, 437.
4. Sutin, N. (1983) Prog. Inorg. Chem., **30**, 441.
5. Newton, M. D. (1991) Chem. Rev., **91**, 767.
6. Cannon, R. D. (1980) Electron Transfer Reactions, Butterworths, London.
7. Eberson, L. (1987) Electron Transfer Reactions in Organic Chemistry, Springer, New York.
8. Fox, M. A. and Chanon, M. (eds.) (1988) Photoinduced Electron Transfer, Elsevier, New York. 4 vols.
9. Twigg, M. V. (ed.) (1991) Mechanisms of Inorganic and Organometallic Reactions, vol. 7, Chaps. 1 and 2, and earlier volumes.
10. Brunschwig, B. S., Logan, J., Newton, M. D. and Sutin, N. (1980) J. Am. Chem. Soc., **102**, 5798.
11. Sutin, N. (1988) Pure & Appl. Chem., **60**, 1817.
12. King, G. and Warshel, A. (1990) J. Chem. Phys., **93**, 8682.
13. Marcus, R. A. (1960) Disc. Faraday Soc., **29**, 21.
14. Marcus, R. A. (1956) J. Chem. Phys., **24**, 966.

15. Marcus, R. A. (1965) J. Chem. Phys., **43**, 679.
16. Kuharski, R. A., Bader, J. S., Chandler, D., Sprik, M., Klein, M. S. and Impey, R. W. (1988) J. Chem. Phys., **89**, 3248; Hwang, J.-K. and Warshel, A. (1987) J. Am. Chem. Soc., **109**, 715.
17. Landau, L. (1932) Phys. Z. Sowjetunion, **1**, 88; **2**, 46; Zener, C. (1932) Proc. Roy. Soc., **137A**, 696; **140A**, 660. See also ref 90 below.
18. Marcus, R. A. (1964) Ann. Rev. Phys. Chem., **16**, 155.
19. Marcus, R. A. (1981) Int. J. Chem. Kinetics, **13**, 865.
20. Since the value of k_i will differ before and after the ET, it is often approximated as a particular average of the value for the reactants' *i*th mode, k_i^r and the value for the products' *i*th mode, k_i^p: $k_i = 2k_i^r k_i^p/(k_i^r + k_i^p)$. This step has been called "symmetrizing" the PES. Calculations have been made showing that this approximation. (Marcus, R. A. (1987) in A. E. Hansen, J. Avery and J. P. Dahl (eds), Understanding Molecular Properties, Reidel, Boston, p. 229.)
21. Marcus, R. A. (1965) J. Chem. Phys., **43**, 58.
22. Marcus, R. A. (1965) J. Chem. Phys., **43**, 3477, Appendix I.
23. Marcus, R. A. (1965) J. Chem. Phys., **43**, 1261.
24. Cannon, R. D. (1977) Chem. Phys. Lett., **49**, 299.
25. Brunschwig, B. S., Ehrenson, S. and Sutin, N. (1986) J. Phys. Chem., **90**, 3657.
26. Dogonadze, R. R. (1971) in Hush, N. S. (ed), Reactions of Molecules at Electrodes, Wiley, New York, Chap. 3, Eq. 201.
27. Sumi, H. (1980) J. Phys. Soc. Jpn., **49**, 1701; Sumi, H. and Marcus, R. A. (1986) J. Chem. Phys., **84**, 4894.
28. Marcus, R. A. (1977) in P. A. Rock (ed.), Special Topics in Electrochemistry, Elsevier, New York, p. 161; Marcus, R. A. (1959) Can. J. Chem. **37**, 155.
29. Levich, V. G. and Dogonadze, R. R. (1959) Dokl. Acad. Nauk SSSR, **124**, 123; Dogonadze, R. R., Kuznetsov, A. M. and Vorotyntsev, M. A. (1972) Phys. Stat. Sol. B, **54**, 125, 425; Levich, V. G. (1970) in Eyring, H. (ed.), Physical Chemistry, An Advanced Treatise, Academic, New York, Vol. 9, Chap. 12.
30. Marcus, R. A. (1963) J. Phys. Chem., **67**, 853, 2889.
31. Marcus, R. A. (1965) J. Chem. Phys., **43**, 2654; (1970) **52**, 2803.
32. Wallace, W. L. and Bard, A. J. (1979) J. Phys. Chem., **83**, 1350.
33. Kestner, N. R., Logan, J. and Jortner, J. (1974) J. Phys. Chem., **78**, 2148; Ulstrup, J. and Jortner, J. (1975) J. Phys. Chem., **63**, 4358; Efrima, S. and Bixon, M. (1976) Chem. Phys., **13**, 447; Siders, P. and Marcus, R. A. (1981) J. Am. Chem. Soc. **103**, 741, 748.
34. Marcus, R. A. and Siders, P. (1982) J. Phys. Chem., **86**, 622.
35. Warshel, A. (1982) J. Phys. Chem., **86**, 2218.
36. Marcus, R. A. (1984) J. Chem. Phys., **81**, 4494.
37. Kober, E. M., Casper, J. V., Lumpkin, R. S. and Meyer, T. J. (1986) J. Phys. Chem. **90**, 3722; Chen, P., Duesing, R., Graff, D. K. and Meyer, T. J. (1991) *ibid.* **95**, 5850.
38. Engleman, R. and Jortner, J. (1970) Mol. Phys. **18**, 145.
39. Liang, N., Miller, J. R. and Closs, G. L. (1990) J. Am. Chem. Soc. **112**, 5353.

40. *E.g.*, Landau, L. D. and Lifshitz (1958) Quantum Mechanics, Pergamon Press, New York, eq 51.6, and analogous arguments for the case where the slopes are of opposite sign in the tunneling region; Child, M. S. (1991) Semiclassical Mechanics With Molecular Application, Clarendon Press, New York, eq (5.120) with $\alpha=0$ there.
41. Li, T. T.-T. and Brubaker Jr., C. H. (1981) J. Organomet. Chem., **216**, 223.
42. Powers, J. M. and Meyer, T. J. (1980) J. Am. Chem. Soc., **102**, 1289; Callahan, R. W. and Meyer, T. J. (1976) Chem. Phys. Lett., **39**, 82.
43. Rehm, D. and Weller, A. (1970) Isr. J. Chem., **8**, 259; Nagle, J. K., Dresick, W. J. and Meyer, T. J. (1975) J. Am. Chem. Soc., **97**, 2909; Scandola, F. and Balzani, V. (1979) J. Am. Chem. Soc., **101**, 6140.
44. Kikuchi, K., Takahashi, Y., Katagiri, T., Niwa, T. and Hoshi, M. (1991) Chem. Phys. Lett., **180**, 403; Kikuchi, K., Takahashi, Y., Hoshi, M., Niwa, T., Katagiri, T. and Miyashi, T. (1991) J. Phys. Chem. **95**, 2378.
45. Miller, J. R., Closs, G. L. and Calcaterra, L. T. (1984), J. Am. Chem. Soc., **106**, 3047.
46. Gould, I. R., Mueller, L. J. and Farid, S. (1991) Z. Phys. Chem., **170**, 143; Gould, I. R., Ege, D., Moser, J. E. and Farid, S. (1990) J. Am. Chem. Soc., **112**, 4290, and references cited therein.
47. Asahi, T., Mataga, N., Takahashi, Y., Miyashi, T. (1990) Chem. Phys. Lett., **171**, 309, and references cited therein.
48. Gaines, G. L., O'Neil, N. P., Svec, W. A., Niemczyk, M. P. and Wasielewski, M. R. (1990) J. Am. Chem. Soc., **13**, 719, and references cited therein.
49. Fox, L. S., Kozik, M., Winkler, J. R. and Gray, H. B. (1990) Science **247**, 1069.
50. Penner, R. M., Heben, M. J., Longin, T. L. and Lewis, N. (1990) Science, **250**, 1118; private communication.
51. Chidsey, G. E. (1991) Science, **251**, 919.
52. Savéant, J. M. and Tessier, D. (1982) Faraday Disc. Chem. Soc., **74**, 57;Weaver, M. J. and Hupp, J. T. (1982) Am. Chem. Soc. Symp. Series, **198**, 181.
53. Iwasita, T., Schmickler, W., and Schultze, J. W. (1985) Ber. Bunsenges. Phys. Chem., **89**, 138.
54. Isied, S. S., Vassilian, A., Magnuson, R. H. and Schwarz, H. A. (1985) J. Am. Chem. Soc., **107**, 7432; Vassilian, A., Wishart, J. F., van Hemelyrck, B, Schwarz, H. and Isied, S. S. (1990) J. Am. Chem. Soc., **112**, 7278.
55. Closs, G. L., Calcaterra, L. T., Green, N. J., Penfield, K. W. and Miller, J. R. (1986) J. Phys. Chem., **90**, 3673.
56. Penfield, K. W., Miller, J. R., Paddon-Row, M. N., Cotsaris, E., Oliver, A. M. and Hush, N. S. (1987) J. Am. Chem. Soc., **109**, 5061.
57. Wasielewski, M. R., Niemczyk, M. P., Johnson, D. G., Svec, W. A. and Minsek, D. W. (1989) Tetrahedron, **45**, 4785.
58. Beitz, J. V. and Miller, J. R. (1979) J. Chem. Phys., **71**, 4579.
59. Isied, S. S. (1984) Prog. Inorg. Chem., **32**, 443.
60. Bowler, B. E., Raphael A. L. and Gray, H. B. (1990) Prog. Inorg. Chem., **38**, 259.
61. Sykes, A. G. (1991) Adv. Inorg. Chem., **36**, 377.

62. Hoffmann, B. M., Natan, M. J., Nocek, J. M. and Wallin, S. A. (1991) Struct. Bond., **75**, 85.
63. Isied, S. S., Vassilian, A., Wishart, J. F., Creutz, C., Schwarz, H. A. and Sutin, N. (1988) J. Am. Chem. Soc., **110**, 635.
64. Hush, N. S. (1967) Prog. Inorg. Chem., **8**, 391.
65. Hopfield, J. J. (1977) Biophys. J., **18**, 311.
66. Creutz, C. (1983) Prog. Inorg. Chem., **30**, 1.
67. Stein, C. A., Lewis, N. A. and Seitz, G. J. (1982) J. Am. Chem. Soc., **104**, 2596.
68. McConnell, H. M. (1961) J. Chem. Phys., **35**, 508.
69. Larsson, S. (1981) J. Am. Chem. Soc., **103**, 4034.
70. Siddarth, P. and Marcus, R. A. (1990) J. Phys. Chem., **94**, 2985.
71. Axup, A. W., Albin, M., Mayo, S. L., Crutchley, R. J. and Gray, H. B . (1988) J. Am. Chem. Soc., **110**, 435.
72. Siddarth, P. and Marcus, R. A. (1990) J. Phys. Chem., **94**, 8430.
73. For a recent review, see Kartha, S. Das, R. and Norris, J. R. (1991), in H. Sigel (ed.), Metal Ions in Biological Systems, **27**, 323.
74. Breton, J. and Vermeglio, A. (eds.) (1988) The Photosynthetic Bacterial Reaction Center--Structure and Dynamics, NATO ASI Series A: Life Sciences 149, Plenum, New York.
75. Michel-Beyerle, M. E. (ed.) (1985) Antennas and Rection Centers of Photosynthetic Bacteria, Springer, Berlin.
76. Holzapfel, W., Finkele, U., Kaiser, W., Oesterhelt, D., Scheer, H., Stilz, H. U. and Zinth, W. (1990) Proc. Natl. Acad. Sci. U.S.A., **87**, 5168.
77. Kirmaier, C. and Holten, D. (1990) Proc. Natl. Acad. Sci. U.S.A., **87**, 3552.
78. Plato, M., Mobius, K., Michel-Beyerle, M. E., Bixon, M. and Jortner, J. (1988) J. Am. Chem. Soc., **110**, 7279.
79. Marcus, R. A. (1988) Isr. J. Chem. **28**, 205.
80. Marcus, R. A. (1968) J. Phys. Chem. **72**, 891.
81. Polik, W. F., Guyer, D. R. and Moore, C. B. (1990) J. Chem. Phys. **92**, 3453.
82. Wigner, E. (1938) Trans. Faraday Soc. **34**, 29.
83. Bergsma, J. P., Gertner, B. J., Wilson, K. R. and Hynes, J. T. (1987) J. Chem. Phys. **86**, 1356.
84. Sumi, H. and Marcus, R. A. (1986) J. Chem. Phys. **84**, 4894; Nadler, W. and Marcus, R. A. (1987) *ibid.* **86**, 3906.
85. Susman, L. D. (1980) Chem. Phys. **49**, 295; Helman, A. B. (1982) *ibid.* **65**, 271; Calef, D. F. and Wolynes, P. G. (1983) J. Phys. Chem. **87**, 3387; van der Zwan, G. and Hynes, J. J. (1985) *ibid.* **89**, 4181.
86. Savéant, J.-M. (1987) J. Am. Chem. Soc. **109**, 6788.
87. Marcus, R. A. (1990) J. Phys. Chem. **94**, 4152, 7742; Marcus, R. A. (1991) J. Phys. Chem. **95**, 2010; Geblewicz, G. and Schiffrin, D. J. (1988) J. Electroanal. Chem. **244**, 27.
88. Leidner, C. R. and Murray, R. W. (1984) J. Am. Chem. Soc. **106** 1606.
89. Leidner, C. R. and Murray, R. W. (1985) J. Am. Chem. Soc. **107**, 551; Jernigan, J. C. and Murray, R. W. (1990) *ibid.*, **112**, 1034.
90. Kauzmann, W. (1957) Quantum Chemistry, Academic Press, New York, p. 539.

91. Bellman, R. (1960) Introduction to Matrix Analysis, McGraw Hill, New York, p. 96.
92. Marcus, R. A. (1989) J. Phys. Chem. **93**, 3078.
93. Moser, C. C., Keske, J. M., Warncke, K., Farid, R. S. and Dutton, P. L. (1992) Nature **355**, 796.
94. Johnson, M. D., Miller, J. R., Green, N. S. and Closs, G. L. (1989) J. Am. Chem. Soc. **111**, 3751.
95. Siddarth, P. and Marcus, R. A. (1992) J. Phys. Chem. **96**, 000.
96. McConnell, H. M. (1960) J. Chem. Phys. **33**, 115.
97. An alternative discussion of the solvent reorganization in terms of charges and potentials instead of in terms of electric fields and polarization vectors is given in the first citation of ref. 28.

CALCULATION OF THE LIFETIME OF MOLECULES IN ELECTRONICALLY EXCITED STATES

S. D. PEYERIMHOFF
Institut für Physikalische und Theoretische Chemie
Universität Bonn
Wegelerstraße 12
D-5300 Bonn 1

ABSTRACT. It is shown that quantum chemical calculations are able to calculate lifetimes of small molecules in electronically excited states with an accuracy which is comparable to that of experimental determination. The advantage of the calculations is, that they can treat processes from the nanosec scale up to lifetimes of minutes in principle by the same procedure, and that this is possible for radiative and radiationless transitions. Calculations are furthermore able to give insight into the mechanism of the various processes.

1. Introduction

Quantum chemical ab initio calculations have become an important tool for studying small molecules in their ground and electronically excited states [1]. Their main advantage is that they are able to compute the entire potential hypersurface, from the potential minimum of the given state up to the various dissociation channels. In particular interactions between various states, which are important for photochemical processes, are directly obtainable in quantum chemical calculations.

In addition to the energy of an excited species its lifetime is of great importance. Since quantum chemical calculations generally deal with isolated molecules, collisional desactivation is not directly treated in such an approach. The main desactivation processes determining the lifetime are then: (1) radiative transitions, which must be divided into two categories from a theoretical point of view: the spin- and dipole-allowed processes belong to the first category and are quite fast (lifetime in the nanosec range) while the spin- or dipole-forbidden transitions belong to the second category and allow lifetimes which can range from microseconds to minutes or even hours. (2) non-radiative transitions due to intersystem-crossing or internal conversion (or generally interaction of electronic states of different or the same spin and spatial symmetry), whereby various mechanism for the interaction are possible; frequent mechanisms are spin-orbit coupling, non-adiabatic effects (i.e. coupling by the kinetic energy operator) or tunneling, for example.

The present lecture will discuss the basic procedures for the quantum mechanical computation of the various desactivation processes and will give a number of simple examples. More details of the computational approaches, and ample illustrations for the theoretical investigation of more complex systems can be found in the literature.

E. Kochanski (ed.), Photoprocesses in Transition Metal Complexes, Biosystems and Other Molecules. Experiment and Theory, 89–101.

2. Calculation of excited states of molecular systems

There are basically two ingredients in modern ab initio calculations, which determine the accuracy of the results: (1) AO basis set and (2) the theoretical approach. The most important requirements will be summarized in what follows.

2.1. AO BASIS SET REQUIREMENTS

For atoms in their ground state standard AO basis sets of different flexibility are listed in books [2] or are even directly included in the library of modern program packages [3]. Depending on the property to be computed, a simple double-zeta type (6-31G or similar) basis may suffice (for geometries), or a more extensive one, including polarization and correlation orbitals, must be employed (energy differences, dipole moments).
For the treatment of excited states additional functions must often be added, in particular if dissociation occurs into excited atomic states. The AO basis set tables in the literature unfortunately do not include such atomic functions for excited atoms (higher principal quantum numbers or higher angular momentum states). If Rydberg states are involved, as is quite often the case in photochemistry, the corresponding diffuse functions representing such states must also be included in the AO basis. Tables for the exponents of such functions are available.
In summary, the AO basis set has to be defined carefully, with an eye on the electronic states one wants to consider. For realistic results the basis has to be at least of double-zeta plus polarization quality and has to include some function to account for electron correlation in addition to those, which are required to describe the actually excited orbitals (spectroscopic orbitals).

2.2. COMPUTATIONAL PROCEDURE

The well-known SCF procedure describes the system basically as a product of independent particles (obeying the Pauli Principle, however). Since the movement of the particles is not entirely independent of one another, i.e. there is a correlation in their movement, which is not accounted for in the independent particle approach, an error results in the energy of even the best SCF (Hartree-Fock) function, and this is generally referred to as correlation energy. This error can be considered to result in the main from a pairwise interaction of electrons, and it clearly changes if electron pairs are broken as in the dissociation or electronic excitation. As a result the correlation energy error is generally different for the various electronic states (the difference can be larger than 1 eV), and in order to obtain reliable excitation energies, excited state treatments have to be able to account for the electron correlation, at least for the difference in electron correlation in the various states. The SCF approach is not sufficient.

The most general wavefunction which takes correlation into account is the configuration interaction wavefunction

$$\Psi_{CI}=c_0\Phi_0+c_1\Phi_1+c_2\Phi_2+\ldots \tag{1}$$

in which the ϕ_i are excited configurations as allowed by the AO basis. Generally, the most important ϕ_i terms are single and double excitations with respect to the dominant (physically interpretable) configuration (because the Hamiltonian possesses no higher than two-particle interactions).

If potential energy surfaces are described in an area further away from their minimum, there is often more than one configuration which is dominant. A typical example is the O_2 molecules. At its ground state minimum the dominant configuration (over 90 %) is $3\sigma_g^2 1\pi_u^4 1\pi_g^2$. Since $3\sigma_g$ is essentially made up from pz+pz on the two oxygens, $1\pi_u$ from px+px or py+py, and the antibonding $3\sigma_u$ and $1\pi_g$ MO's from the corresponding negative linear combinations, all orbitals tend to become isoenergetic with increasing distance between the two oxygens. As a result, the CI wavefunction at large R has equal contributions (always 12.5 %) from the configurations $3\sigma_g^2 1\pi_u^4 1\pi_g^2$, $3\sigma_u^2 1\pi_u^4 1\pi_g^2$, $3\sigma_g^2 1\pi_u^2 1\pi_g^4$, $3\sigma_u^2 1\pi_u^2 1\pi_g^4$ and 50 % from $3\sigma_g 3\sigma_u 1\pi_u^3 1\pi_g^3$.

This example makes clear that there can be several "important" terms in the expansion of eq. (1). The general type of CI approach to date is therefore one, in which single and double excitation configurations not only with respect to one, but relative to the most important, the so-called "reference configurations", are considered. The short hand notation is "multi-reference-CI", MR-CI or MRD-CI (to indicate that only up to double-excitations are included). For the proper description of excited states such multi-reference CI wavefunctions are generally required. All examples to be discussed are based on a special such MRD-CI variant [4]: all single and double excitations with respect to the reference configurations are generated (this can be 10^5-10^8, referred to as MRD-CI space), but only a selection therefrom (based on an energy-lowering criterium) is actually used to make up the Hamiltonian matrix (between 5000 and 30000) and the wavefunction, while the contribution of the remaining configurations to the total energy is considered in a perturbation-like treatment. Hence the wavefunction corresponds to a truncated MRD-CI expansion, the energy corresponds very nearly to the entire MRD-CI space (and often times the full CI, i.e. the expansion considering all and not only double excitations, is also estimated). Table 1 shows a typical example of such a convergence pattern.

TABLE 1. Convergence pattern for a CI calculation on the BH ground state (at R_e).

Size of sec. equation	relative error in the total energy (a.u.)
1	0.102260
586	0.005208
7308	0.000874
10389	0.000285
13880	0.000151
17049	0.000053
132686	0.000000 $\hat{=}$ -25.2276274 hartree

The errors in such calculations, employing reasonably-sized AO basis sets, are in the order of 0.1 - 0.2 eV for transition energies, a few degrees and a few hundreths of an Å for the geometry of the minima on the potential hypersurface. Various examples can be found in Ref. [1].

3. Calculation of Radiative Lifetimes

The probability for spontaneous emission from the initial state Ψ_n to the final state Ψ_m is given by the Einstein transition probability A_{nm} [5]. This quantity can be obtained in quantum mechanical calculations from

$$A_{nm}=\frac{64\pi^4}{3h}\nu_{nm}^3|R_{nm}|^2 \tag{2}$$

in which

$$R_{nm}=<\Psi_n|\hat{O}|\Psi_m> \tag{3}$$

Since there is an exponential decrease in the number of systems in state n, i.e.

$$N_n=N_n^0\exp[-A_{nm}]t, \tag{4}$$

after the time

$$\tau=1/A_{nm} \tag{5}$$

the number of systems left in state n is reduced to 1/e of the initial number. This time is considered the mean life (simply "lifetime") of the state n. If transitions are possible to many states m, the lifetime then corresponds to $\tau = 1/\Sigma_m A_{nm}$.

The Einstein transition probability for emission A_{nm} is directly related [5] to the corresponding quantity B_{mn} which determines absorption:

$$B_{mn}=(8\pi hc\nu_{nm}^3)^{-1}A_{nm} \tag{6}$$

Frequently the strength of an absorption is expressed in terms of the oscillator strength f_{mn}.

$$f_{mn}=(m_e hc^2\nu_{nm}/\pi e^2)B_{mn}, \tag{7a}$$

or

$$f_{mn}=\frac{2}{3}\Delta E(m,n)|R_{nm}|^2 \tag{7b}$$

if atomic units are used throughout.

Hence the lifetime is proportional to the inverse of the oscillator strength. The quantitative determination of τ or f requires the knowledge of the energy difference $\Delta E(m,n)$ or ν_{mn} be tween the states, the computation of which has been discussed in Sect. 2, and the evaluation of the quantum mechanical expression according to eq. (3).

3.1. SPIN- AND DIPOLE-ALLOWED TRANSITIONS

For a spin and dipole-allowed transition the operator $\hat{O}$ in eq. (3) is simply the dipole operator; Ψ_n and Ψ_m have the same spin multiplicity. In the Born-Oppenheimer approximation the total wavefunction is written as a product of the electronic part (evaluated at a fixed position Q of the nuclei) and a nuclear (vibrational) part:

$$\Psi_i = \Psi_e(Q,\vec{r})\chi_v(\vec{Q}) \tag{8}$$

Insertion into eq. (3) gives

$$R_{e'v'e''v''} \sim |\int_{\vec{Q}} \chi_{v'}^* \left\{ \int_{\vec{r}_i} \Psi_{e'}^* \, |\Sigma e\vec{r}_i| \Psi_{e''} d\vec{r}_i \right\} \chi_{v''} d\vec{Q}|^2 \tag{9a}$$

if the standard notation, i.e. double prime for the lower and a single prime for the upper state, is chosen.

The term in braces is generally referred to as the electronic transition moment $R_{e'e''}$.

$$R_{e'e''} = \langle \Psi_{e'} | \hat{O} | \Psi_{e''} \rangle \tag{10}$$

It can easily be evaluated by employing MRD-CI wavefunctions for $\Psi_{e'}$ and $\Psi_{e''}$ respectively. It will generally depend on the nuclear coordinates, i.e. the point of the potential energy surface for which the MRD-CI wavefunction has been computed. If $R_{e'e''}$ has been calculated for a number of nuclear geometries, an analytical fit for $R_{e'e''}(\vec{Q})$ can be made, and the expression of eq. (9a) can be computed as

$$R_{e'v'e''v''} \sim |\langle \chi_{v'} | R_{e'e''}(\vec{Q}) \rangle \chi_{v''} \rangle_Q|^2 \tag{9b}$$

This procedure is the correct way of computing the transition matrix element. The well-known Franck-Condon approximation assumes a constant value for $R_{e'e''}$ rather than its dependence on Q and leads to the simplified form

$$R_{e'v'e''v''} \sim |\ \bar{R}_{e'e''}|^2 \ |\langle \chi_{v'} | \chi_{v''} \rangle|^2 \tag{11}$$

with the square of the vibrational overlap named Franck-Condon factors.

In many instances $R_{e'e''}$ shows only a weak dependence on the nuclear coordinate so that the Franck-Condon approximation is adequate. This is not the case in areas of the avoided crossing of states, in which the electronic wavefunctions change very much with nuclear positions. The approximation is also questionable in the area of higher symmetry points. An obvious example is the $X^1\Sigma^+(1^1A') \leftarrow {}^1\Delta(2^1A')$ transition in HCN [6]. At linear nuclear arrangement, which has the best vibrational overlap, $R_{e'e''}$ is zero by symmetry. It becomes sizeable only with increasing bending angle HCN, so that a realistic transition probability can only be obtained from eq. (9).

A large number of calculated transition probabilities, oscillator strengths, radiative lifetimes obtained from such theoretical MRD-CI wavefunctions or variants thereof are contained in the literature [7]. Generally, lifetimes are obtained within at least a factor of two compared to the experimentally derived data, which oftentimes also show considerable error limits. Typical results are contained in Table 2, for Rydberg-like (NH_3) and Valence-shell transitions.

TABLE 2. Calulated data for transition probabilities compared with the corresponding measured value [7,8].

Molecule	Transition	Calc.	Exptl.
NH_3	X - A	f_e=0.089	0.13, 0.079, 0.088, 0.0696
N_2	$C^3\Pi_u$-$B^3\Pi_g$	τ_e=32 ns	36.6±0.5, 40.5±1.3, 34.8±12
SiH	$A^2\Delta$-$X^2\Pi$	τ_0=662 ns	534±23
C_2	$A^1\Pi_u$-$X^1\Sigma_g^+$	τ_0=10.7 μs	11.4±2.2
cyclohexadiene	S_0-S_1	f_e=0.105	0.11

Fig. 1 shows a comparison between the measured inner-shell excitation spectrum of SiH_3F and the calculated location of the peaks and their relative intensities [9]. I feel the agreement is convincing, and the definitive assignment of peaks is possible on the basis of the quantum chemical calculations for energies and transition probabilities.

Fig. 1. Experimental inner-shell spectrum (Δ) of SiH_3F and calculated spectrum. The calculated intensity is indicated by vertical lines, the peak positions are relative to the calculated silicon 1s ionization limit indicated by the dashed vertical line. The K edge is not known from experiments.

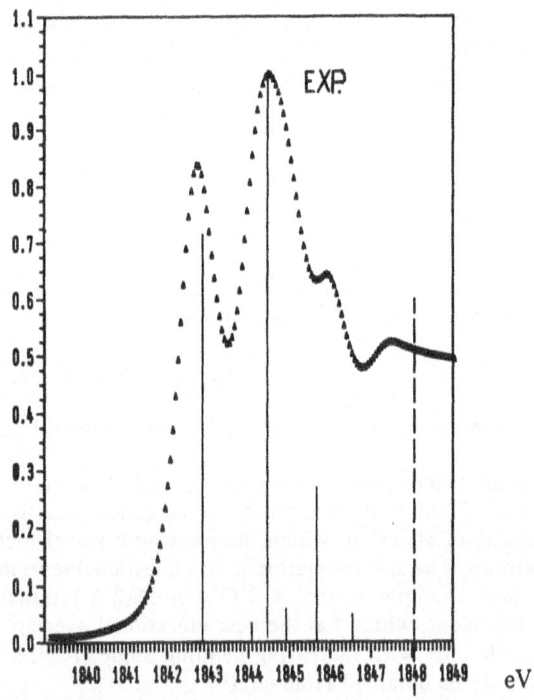

Somewhat more complicated is the calculation of the lifetime of inner-shell excited molecules, since their lifetime generally is determined by the combination of an L to K shell transition and the ejection of an Auger electron, whereby this process can be in competition with dissociation. Methods for describing such a process in molecules have also been developed [10], but detailed comparative experimental investigations are still lacking.

3.2. SPIN- AND DIPOLE-FORBIDDEN TRANSITIONS

In principle, the procedures for treating spin and dipole forbidden transitions are the same as those described in Sect. 3.1 - they are only more complicated.
The simple electric dipole operator has to be replaced by further terms representing the magnetic dipole and quadrupole operators

$$A_{nm}=\frac{4}{3}\alpha^3(\Delta E)^3\left\{ |<n|\vec{R}|m>|^2+\frac{1}{4}\alpha^2|<n|\vec{L}+2\vec{S}|m>|^2+\frac{3}{40}\alpha^2(\Delta E)^2|<n|\vec{R}x\vec{R}|m>|^2 \right\} \tag{12}$$

(all values in atomic units, α being the fine structure constants), and the MRD-CI wavefunctions are not eigenfunctions of S^2 and S_z any more. Formally the singlet and triplet functions (or those of other multiplicities) can be written according to perturbation theory as

$$\begin{aligned} "^1\Psi_e" &= {}^1\Phi+\Sigma a_k{}^3\Phi_k+\Sigma b_k{}^1\Phi_k \\ "^3\Psi_e" &= {}^3\Phi+\Sigma a_k'{}^1\Phi_k+\Sigma b_k'{}^3\Phi_k \end{aligned} \tag{13}$$

whereby the magnitude of the "contamination" of the pure spin function by other multiplicities (and spatial symmetries) is given by the spin-orbit interaction

$$a_k=<^1\Phi|\hat{H}_{SO}|^3\Phi_k>/\Delta E(k) \tag{14}$$

with similar formulae for the other coefficients.

Instead of employing perturbation theory, the functions of eq. (13) can also be obtained in a variational-type treatment [11].

The radiative lifetime of a state decaying by a quadrupole transition, as for example by the $b^1\Sigma_g^+$ - $a^1\Delta_g$ transition in the O_2 molecule, is obtained in the same manner as has been discussed in the previous section with the exception that the electric dipole operator in eq. (9) is replaced by the electric quadrupole term. The computation is also more sensitive to a correct description of the energy separation between the two states involved, because ΔE enters into the formula in the 5th power rather than in the third as before (eq. 12). In this connection it should also be mentioned that the rotatory strength, involving not only the electric dipole moment operator in eq. (3), but also the magnetic dipole $\Sigma(\vec{r}_k\times\nabla_k)$, can also be computed for the circular dichroism spectra, as has been shown quite recently, for example, for R-methyloxirane [12].

For transitions between states of different spin multiplicity the expansions of eq. (13) have to be inserted into eq. (3) (or the corresponding electronic part eq. (10)). In this case the transition moment

$$R_{nm} = \Sigma_k \frac{<n|H_{so}|k>}{\Delta E} <m|\hat{O}|k> + \Sigma_k \Sigma_l \frac{<m|H_{so}|k>}{\Delta E} <k|\hat{O}|l> \frac{<n|H_{so}|l>}{\Delta E} \tag{15}$$

consists of a sum of contributions between states Φ_m and Φ_K (in the short hand notation m and k) or Φ_k and Φ_ℓ of the same spin multiplicty, whereby the weighting factor is determined by the spin orbit coupling and the energy difference between states. There is certainly the question about convergence of the series, i.e. how many states k and ℓ have to be considered. In computations one often takes the approach of "chemical convergence", i.e. one assumes convergence if the addition of a few more states does not change the result.

Examples are contained in Table 3.

TABLE 3. Computed radiative lifetimes (in sec unless specified otherwise) for various molecules in excited states which cannot combine with lower states by dipole- or spin-allowed processes [13,14].

Molecule	$b^1\Sigma^+$	$a^1\Delta$	$b^1\Sigma^+(\rightarrow a^1\Delta)$
O_2	11	4330	720
exptl.	12	3980	400 within a factor of two
S_2	3.4	350	780
exptl.	?	?	?
SO	0.013	0.45	450
exptl.	0.007	?	?
SeO	4.7 ms	1.1	325
exptl.	1.4;>0.8;0.55	?	?
NF	0.015	2.0	
exptl.	0.015;0.022	1;5.6	
NCl	0.0025	1.1	428
exptl.	0.0006	0.002	?
AsH	0.35 ms	22 ms	130
exptl.	?	?	?

The advantage of the calculations is that they produce not only numbers, but give more detailed insight into the processes. In O_2, for example, the main perturber states are as follows [14]:

$X^3\Sigma_g^-$ ($m_s = 0$)	perturber:	$^1\Sigma_g^+$, $^3\Pi_g$
$X^3\Sigma_g^-$ ($m_s = \pm 1$)	perturber:	$^3\Pi_g$ and $^1\Pi_g$
$a^1\Delta_g$ ($m_s = \pm 2$)	perturber:	$^3\Pi_g$ ($m_s = \pm 1$)
$b^1\Sigma_g^+$ ($m_s = 0$)	perturber:	$^3\Sigma_g^-$ ($m_s = 0$), $^3\Pi_g$

The desactivation of the $b^1\Sigma_g^+$ state can occur via three possibilities:

- $b^1\Sigma_g^+ \rightarrow a^1\Delta_g$, in an electric quadrupole transition corresponding to a lifetime of 720 sec as given in Table 3.

- $b^1\Sigma_g^+ \rightarrow X^3\Sigma_g^-$ ($m_s = 0$), the difference in quadrupole moments of the $b^1\Sigma_g^+$ and $X^3\Sigma_g^-$ states determines the transition probability this process would correspond to a lifetime of 6 x 10^6 sec.

- $b^1\Sigma_g^+$ - $X^3\Sigma_g^-$ ($m_s = \pm 1$) is the dominant desactivation process corresponding to the lifetime of 11 sec. The process corresponds to the magnetic dipole transition (the spin part of the L + 2S operator in eq. 12) between the perturber $^3\Sigma_g^-$ ($m_s = 0$) of $b^1\Sigma_g^+$ and the $X^3\Sigma_g^-$ ($m_s = \pm 1$) state.

  The desactivation of the $a^1\Delta_g$ occurs as follows:

- $a^1\Delta_g \rightarrow X^3\Sigma_g^-$ ($m_s = 0$) is an electric quadrupole transition between $^1\Delta_g$ and the $^1\Sigma_g^+$ perturber of the $X^3\Sigma_g^-$ state and corresponds to a lifetime of 5 x 10^7 sec.

- $a^1\Delta_g \rightarrow X^3\Sigma_g^-$ ($m_s = \pm 1$) results from a magnetic dipole operator (orbital part L in eq. 12) between $^1\Delta_g$ and the $^1\Pi_g$ perturber of $X^3\Sigma_g^+$ as well as between the perturber $^3\Pi_g$ of $^1\Delta_g$ and $X^3\Sigma_g^-$. This is the dominant process resulting in the lifetime given in Table 3.

The analysis for the other molecules can be made along the same lines. The shortening of the $b^1\Sigma^+$ lifetime from O_2 to S_2, for example, is a consequence of the larger spin-orbit coupling in the heavier system which produces larger contributions of the perturber states. The same argument holds for the trend from SO to SeO or NF to NCl and to some extent also AsH. The massive decrease of the $b^1\Sigma^+$ lifetime from O_2 to SO is due to the loss of gerade and ungerade symmetry so that the perturber states allow electric dipole transitions rather than magnetic dipole processes [14]. The lifetime of the $a^1\Delta$ state follows the same lines as discussed for $b^1\Sigma^+$. The lifetime corresponding to a desactivation of the $b^1\Sigma^+$ state to the $a^1\Delta$ is always of the same magnitude in the different molecules since it is caused by the same physical process (quadrupole transition).

In summary it can be seen from these examples that the transition probability for spin-forbidden processes can also be reliably obtained from such quantum mechanical calculations. The accuracy does not seem to be lower than for spin-allowed processes (τ within a factor of two). This in not surprising since the total transition probability is obtained as a sum of terms involving spin-allowed processes. The two other ingredients: spin-orbit coupling to determine the weight of the perturber, can be obtained within an error smaller than 10 %; the convergence behavior of the perturbation approach is more difficult to estimate and can be avoided by variational procedures which are starting to become in use for heavier systems.

4. Radiationless Transitions

Non-radiative transitions can be considered to occur as a consequence of potential surface interactions. The actual location of the crossing is quite important. The proper calculation of the relative energies of various states is therefore even more critical in order to obtain reliable lifetimes than it is in the evaluation of the intensities for radiative transitions.

The most simple approach to obtain the probability for a radiationless transition is to employ the Fermi-Wentzel golden rule. The line width is obtained as

$$\Gamma_{v'} = 2\pi \left| \int_{Q} \chi^{*}_{v'}(Q) \left\{ \int \Phi^{*}_{e} | \hat{H}_{int} | \Phi_{e'} d\vec{r}_{i} \right\} \chi_{k}(E) dQ \right|^{2} \tag{16}$$

whereby $\chi_{v'}$ is the bound vibrational level and χ_k the corresponding level lying in the continuum of the second electronic, generally repulsive, state. The predissociation rate is

$$k^{pre} = const\ \Gamma \tag{17}$$

whereby const = $2\pi\gamma c$ if γ is the atomic unit of energy expressed in cm^{-1} and c the velocity of light in cm/s. Furthermore

$$\tau^{pre} = 1/k^{pre} \tag{18}$$

It is obvious that the technical ingredients for calculating radiationless transitions are the same as already discussed in the previous sections, provided the operator for the interaction H_{int} is known. The most frequently occurring interaction mechanisms are coupling by spin-orbit interaction (intersystem crossing) and coupling by the nuclear kinetic energy operator (internal conversion), generally referred to as non-adiabatic coupling (rotational coupling, radial coupling).

An example for the first type is the predisscociation of the $C^2\Pi$ state of the NO molecule [15] by the $^4\Pi$ state as indicated in Fig. 2. Table 4 presents the numerical values. Experiments

TABLE 4. Predissociation data for the $C^2\Pi$ state of NO for certain vibrational levels.

v'	$k^{pre}(10^8 s^{-1})$	τ^{pre}(ns)	$\Gamma^{pre}(cm^{-1})$
0	0.8	12.5	0.0004
1	42.0	0.27	0.023
2	200	0.05	0.11

deduce a value of k^{pre} = 1.06 x 10^8 s^{-1} for the $C^2\Pi_{1/2}$ state (J = 3.5), which is very much in line with our theoretical numbers. The calculations also find the predissociation process distinctly faster than the radiative transition for which they yield 70 ns for v' = 0 (compared to an experimentally derived value of 90 ns, again for $C^2\Pi_{1/2}$, v' = 0, J = 3.5).

Fig. 2. Calculated potential curves important for the predissociation of the $2^2\Pi$ ($C^2\Pi$) state of NO.

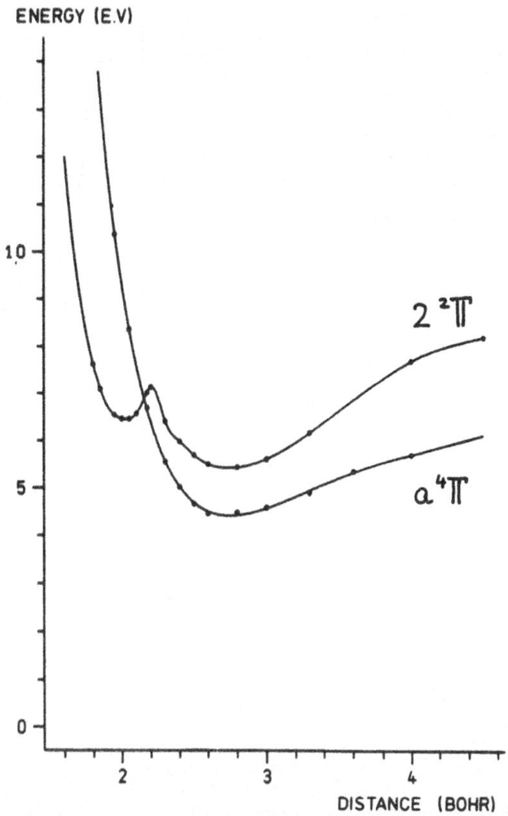

An example for the second type, in which the kinetic energy operator is responsible for the coupling, is seen in Fig. 3 for ArH [16]. At small internuclear separation there are interactions between the various states of $^2\Sigma^+$ symmetry, in particular between the almost repulsive $X^2\Sigma^+$ ground state and the first two bound states $A^2\Sigma^+$ and $3^2\Sigma^+$. The coupling elements $\langle X^2\Sigma^+|\partial/\partial R|A^2\Sigma^+\rangle$ and those between the other states can be computed as a function of the internuclear separation R. Table 5 shows some of the results.

TABLE 5. Line widths (in cm^{-1}) and corresponding lifetimes (in ns) for the $A^2\Sigma^+$ state of ArD.

		v' = 0	v' = 1	v' = 2
Γ	calc.	0.037	0.118	0.215
Γ	exp.	<0.05	$\simeq$0.1	
τ	calc.	0.14	0.045	0.025

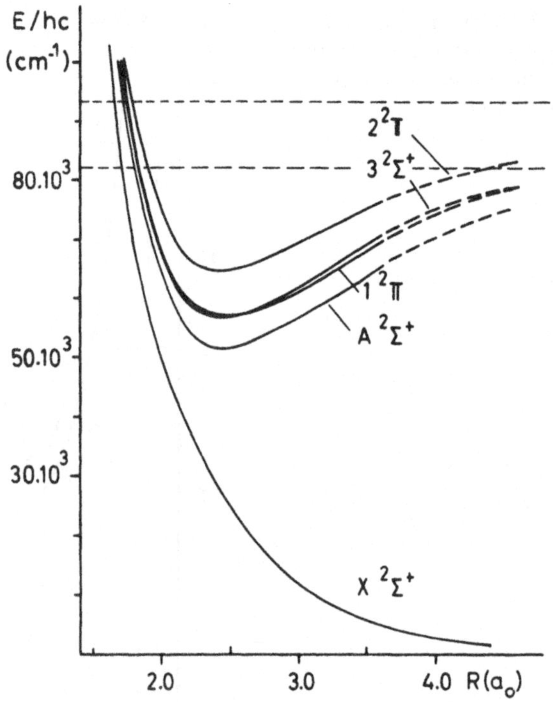

Fig. 3. Calculated potential curves for the ground state and the four lowest excited states of ArH. Full curves: spline fit through ab initio points [16]. Dashed curve: analytical continuation to asymptotic limit (as help to the eye). The dashed horizontal lines indicate the dissociation limits, i.e. the average of H(2s,2p)+Ar correlating with $A^2\Sigma^+$, $1^2\Pi$ and $3^2\Sigma^+$ and the limit Ar(^{3}P)+H correlating with $2^2\Pi$.

The calculations seem to predict similar values as experiments. The computed value for the radiative lifetime for the $A^2\Sigma^+ \rightarrow X^2\Sigma^+$ transition is 130 ns. The calculated predissociation lifetime is always shorter than 0.1 ns and as such much shorter than the radiative lifetime. This result supports the experimental assumption [17] that the lowest $^2\Sigma^+$ state predissociates before it can radiate.

5. Summary

The quantum chemical calculations for the lifetime of a molecule in its excited state decaying via radiative or radiationless processes seem to reach the same accuracy as experimental determinations, provided AO basis sets which include at least some polarization functions and multi-reference type treatments (accounting for at least part of electron correlation) are employed. The majority of theoretical studies have so far been undertaken for small molecules, primarily because for these the most accurate experimental data are available for comparison. So far the best values for transition probabilities in spin-forbidden processes seem to stem from calculations.

Extension of the treatments to larger systems are certainly possible, as long as the wavefunctions of the various states are available. The higher density of electronic states, the greater possibilities for various interactions and the requirement to have reliable values for the relative energetic location of states in larger molecules will add to the complexity of the theoretical treatment of larger molecules.

References

[1] See for example: Bruna, P. J. and Peyerimhoff, S. D. (1987), "Excited State Potentials", in "Ab initio Methods in Quantum Chemistry", Vol. 1, ed. Lawley, K. P., John Wiley & Sons, p. 1

[2] Poirier, R., Kari, R. and Csizmadia, I. G. (1985), "Handbook of Gaussian Basis Sets", Elsevie, for example

[3] for example in the Program System GAUSSIAN 90 or in TURBOMOLE

[4] Buenker, R. J. and Peyerimhoff, S. D. (1983), in "New Horizons of Quantum Chemistry", eds. Löwdin, P. O. and Pullmann, B., Reidel, Dordrecht, p. 183

[5] for example: Herzberg, G. (1950), "Molecular Spectra and Molecular Structure", Vol. 1, van Nostrand, p. 20, 382

[6] Perić, M., Peyerimhoff, S. D. and Buenker, R. J. (1977) Can. J. Chem. 55, 3664

[7] Peyerimhoff, S. D. (1984) Faraday Symp. Chem. Soc. 19, 63, and references therein

[8] a. Runau, R., Peyerimhoff, S. D. and Buenker, R. J. (1977) J. Mol. Spectry 68, 253

b. Shih, S. K., Butscher, W., Buenker, R. J. and Peyerimhoff, S. D. (1978) Chem. Phys. 29, 241

c. Lewerenz, M., Bruna, P. J., Peyerimhoff, S. D. and Buenker, R. J. (1983) Mol. Phys. 49, 1

d. Chabalowski, C. F., Peyerimhoff, S. D. and Buenker, R. J. (1983) 81, 57

e. Share, P. E., Kompa, K. L., Peyerimhoff, S. D. and van Hemert, M. C. (1988) Chem. Phys. 120, 411

[9] Koch, A., this laboratory

[10] Schimmelpfennig, B., Nestmann, B. and Peyerimhoff, S. D., submitted to J. Phys. B

[11] Yarkony, D. R. (1988) J. Chem. Phys. 89, 7324, and references therein

[12] Carnell, M., Peyerimhoff, S. D., Breest, A., Gödderz, K. H., Ochmann, P. and Hormes, J. (1991) Chem. Phys. Lett 180, 477

[13] a. Matsushita, T., Klotz, R., Marian, C. M. and Peyerimhoff, S. D. (1987), Mol. Phys. 62, 1385

b. Bettendorff, M., Klotz, R. and Peyerimhoff, S. D. (1986) Chem. Phys. 110, 315

c. Matsushita, T., Marian, C. M., Klotz, R. and Peyerimhoff, S. D. (1987) Can. J. Phys. 65, 155

[14] Klotz, R., Marian, C. M., Peyerimhoff, S. D., Heß, B. A. and Buenker, R. J. (1984) Chem. Phys. 89, 223

[15] de Vivie-Riedle, R., van Hemert, M. C. and Peyerimhoff, S. D. (1990) J. Chem. Phys. 92, 3613

[16] van Hemert, M. C., Dohmann, H. and Peyerimhoff, S. D. (1986) Chem. Phys. 111, 55

[17] Johns, J. W. C. (1970) J. Mol. Spectry 36, 488

RELATIVISTIC EFFECTS IN MOLECULAR CALCULATIONS

S. D. Peyerimhoff
Institut für Physikalische und Theoretische Chemie
Universität Bonn
Wegelerstraße 12
D-5300 Bonn 1

ABSTRACT. Two types of relativistic effects, which are important in molecules involving heavy atoms, are discussed: the kinematic effect resulting in a contraction of s and expansion of d shells, and the zero-field splitting due to spin-orbit interaction, which can be larger than the difference between electronic levels. The kinematic effects can be taken into account by employing the spin-free no-pair operator, the spin-orbit splitting can be evaluated by the operator in the Breit-Pauli form. Various examples for molecules such as AuH, NiH and CuO demonstrate the effect of the relativistic contributions on the properties of ground and excited molecular states.

1. Introduction

Relativistic effects become important for molecules involving heavy atoms, such as transition metal compounds. There are two major areas, in which the relativistic effects become particularly effective:

(1) In the classical picture the electron mass changes if the electron in the Bohr orbit approaches velocities comparable to the speed of light, and thus the radii of the various shells shrink for the innermost electrons. As a consequence of this contraction and hence stronger nuclear screening, the outer shells with high ℓ quantum numbers, in particular the d and f shells, become higher in energy and more expanded in space. This leads to a different charge distributions in the various atomic and molecular shells and will be referred to as the kinematic relativistic effect, which can be understood in a qualitative manner.

(2) Spin-orbit interaction is considerable in systems with large nuclear charges, since it is essentially proportional to Z and r^{-3} (r denoting the electronic coordinate), and thus this effect plays a major role in the zero-field splitting of electronic levels of heavier compounds. Since electronic states corresponding to different d and s occupation of the transition metal atoms are relatively close in energy, their energetic separation may be comparable with spin-orbit splittings, and as a result mixing of the spin-orbit components originating from different electronic states might become important.

It is clear, that correlation effects also play a major role in the treatment of states of different s and d occupation, so that multi-reference type CI treatments as discussed in the previous chapter on the lifetime of electronically excited states, are essential for a proper

E. Kochanski (ed.), Photoprocesses in Transition Metal Complexes, Biosystems and Other Molecules. Experiment and Theory, 103–111.

description of transition metal compounds.
In this lecture the two different relativistic effects outlined above will be discussed and exemplified for various systems.

2. Kinematic Relativistic Effects

The most common way to take the relativistically modified charge density into account is the use of properly adjusted pseudopotentials [1]. Another, more basic approach are Dirac-Hartree-Fock calculations [2,3]. Just as in non-relativistic molecular calculations it is important to go beyond a Hartree-Fock approach in order to include correlation effects. This is possible today by employing standard MRD-CI type calculations with a relativistically modified Hamiltonian, as derived by B. Heß [4]. This spin-free no-pair operator has the standard form of one- and two-particle terms (in natural units):

$$H=\Sigma_i H(i)+\sum_{i<j} V(i,j) \tag{1}$$

The one-particle operators are given by

$$H(i)=E_i+V_{eff}(i), \tag{2}$$

in which

$$E_i=\sqrt{p_i^2+m^2} \tag{3}$$

and

$$V_{eff}(i)=A_i(V_{ext}(i)+R_iV_{ext}(i)R_i)A_i-W_1(i)E_iW_1(i)-\frac{1}{2}\left\{W_1^2(i),E_i\right\} \tag{4}$$

The abbreviations in the effective one-electron potential are

$$A_i=\sqrt{(E_i+m)/2E_i} \tag{5}$$

$$R_i=(\sigma_i\cdot p_i)/E_i+m \tag{6}$$

and the kernel of the integral operator W_1 is

$$W_1(p_i,p_i')=A_i(R_i-R_i')A_i'\frac{V_{ext}(p_i,p_i')}{E_i+E_i'} \tag{7}$$

The unmodified two-electron term in eq. (1) is generally employed.

This operator has the following advantages:

- it can be derived from quantum electrodynamics
- the binding energies for one-electron atoms and (multiple) ions agree to order $(Z\alpha)^3$ with those of the Dirac equation
- it approaches the correct non-relativistic limit, may be treated variationally and avoids the shortcomings of molecular four-component methods in basis set expansion techniques
- it can be used in standard electronic structure calculations, whereby the computational effort for implementing the relativistic corrections (in essence modifying the one-electron integrals) is negligible.

In the framework of our investigations this operator has the great advantage, that relativistic and non-relativistic calculations can be carried out by the same method, so that a comparison of results will unambiguously show the influence of relativity. The results to be discussed are from the group of B. Heß.

A typical system in which the influence of the kinematic relativistic effects becomes apparent is the gold atom [5]. The ground and the first two excited states, possessing the following electron configurations

^{2}S: [Xe] $4f^{14}\,5d^{10}\,6s^1$, ^{2}D: [Xe] $4f^{14}\,5d^9\,6s^2$ and ^{2}P: [Xe] $4f^{14}\,5d^{10}\,6p^1$

are considered. The results of Table 1 show very drastically the large difference between the relativistic and non-relativistic approach.

TABLE 1. Comparison of calculated transition energies, ionization potentials and electron affinities (all in eV) for the gold atom as obtained from relativistic and non-relativistic calculations.

	Non-relativistic		Relativistic		Exptl.
	SCF	CI	DHF	CI	
^{2}S-^{2}D	5.13	5.03	1.86	1.70	1.74
				1.59[a)]	
^{2}S-^{2}P	2.71	3.46	4.24	5.17	4.95
				4.99[a)]	
IP	5.92	6.91	7.67	9.05	9.22
EA	0.10	1.02	0.66	1.97	2.31
				2.28[b)]	

[a)]ACPF calculation
[b)]CASSCF MO for Au^-

In the non-relativistic calculation the ^2S-^2D transition shows the larger energy difference than the ^2S-^2P, whereas this situation is already reversed at the SCF type (Dirac Hartree-Fock DHF) level in the relativistic treatment. The reason for this can easily be understood on a qualitative basis considering Fig. 1. In the relativistic treatment the s-type shells are contracted, the d shells expanded and therefore higher in energy. The 5d→6s gap in the non-relativistic treatment (^2S-^2D) is larger than 6s→6p (^2S-^2P), and therefore such calculations give even the wrong ordering of states. The agreement between relativistic CI type values and the experimental transition energies is within standard error limits. The influence of the relativistic effects on electron affinities and ionization potentials is also quite apparent.

Fig. 1. Energy of important orbitals in the Au atom in the relativistic and non-relativistic treatment. The occupation in the ground state is indicated.

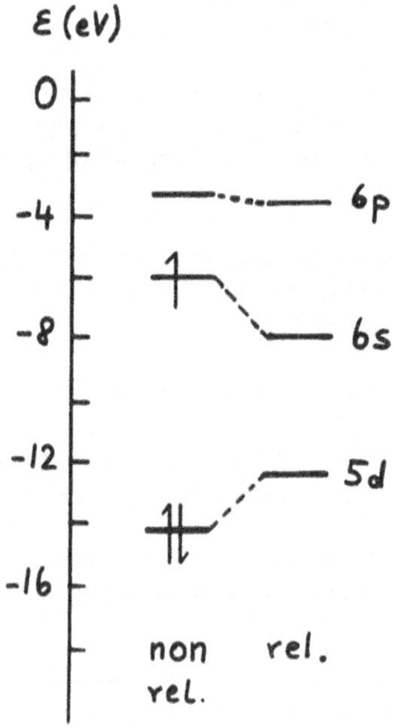

The relativistic effects do not only influence the electronic spectrum, but also binding properties as seen from the results of Table 2, which shows a comparison of potential curve data for AuH obtained in the relativistic and non-relativistic approach [6]. The bond length

TABLE 2. Calculated parameters for AuH.

	non-relativistic	relativistic	exptl.
R_e(pm)	172	152	152.38
ω_e(cm^{-1})	1693	2381	2305
B_e(cm^{-1})	5.69	7.36	7.24
D_e(eV)	2.19	3.33	3.36

is shortened by 20 pm by the relativistic effects, the potential well becomes much deeper as seen by the dissociation energy (which is increased by 50 %) and the much higher zero-point frequency. Again, the effects can qualitatively be interpreted by the different participation of d orbitals (enhanced hybridization) in the bonding.

In summary the examples show quite convincingly that kinematic relativistic effects are of considerable importance and cannot be overlooked if compounds involving heavy atoms are studied.

3. Spin-Orbit Interaction

In the framework of ab initio calculations the spin-orbit operator employed consists of the one- and two-electron Breit-Pauli operators [7] given as

$$H_{SO}=\frac{1}{4m^2c^2}\Sigma_\alpha\Sigma_i\left(\frac{Z_\alpha}{r_{i\alpha}^3}\sigma_i\,(r_{i\alpha}\times p_i)\right)-\frac{1}{4m^2c^2}\sum_i\sum_{j\neq i}\left(\frac{1}{r_{ij}^3}(r_{ij}\times p_j)\,(\sigma_i+2\sigma_j)\right) \tag{8}$$

Our experience indicates that spin-orbit splittings, at least up to molecules containing 3d electrons, can be obtained quite reliably by employing this operator in a perturbation treatment, whereby the zero-order functions are CI expansions and result from calculations employing the relativistic no-pair operator or the standard electronic Hamiltonian, if light elements are to be considered. Typical calculated spin-orbit splittings given in Table 3 show that the computed values represent the spectroscopically measured data quite well.

TABLE 3. Calculated spin-orbit splittings (in cm^{-1}).

System	calc.	exptl.
$HF^+(^2\Pi)$	283	292.8
$HCl^+(^2\Pi)$	640	648.17
$HBr^+(^2\Pi)$	2609	2652.8
Cu (^2D_g)	-2006	-2042
Ni (^3D_g)	-1506	-1508
Pd (^3D_g)	-3420	-3530
NiH ($X^2\Delta$)	-1014	-1012
first order only	-1202	

The importance of considering the spin-orbit splitting is exemplified for NiH. All data are taken from the work of C. Marian [8]. Fig. 2 shows the calculated potential energy curves for the lowest-lying states $^2\Delta$, $^2\Sigma^+$ and $^2\Pi$ of NiH obtained from a standard calculation, employing relativistic corrections via the no-pair operator, but not including any spin-orbit effects. The $^2\Delta$ state is seen to be the ground state, but the two excited states are very close in energy.

Fig. 2. Calculated potential energy curves for the three lowest-lying electronic states of NiH without considering spin-orbit effects.

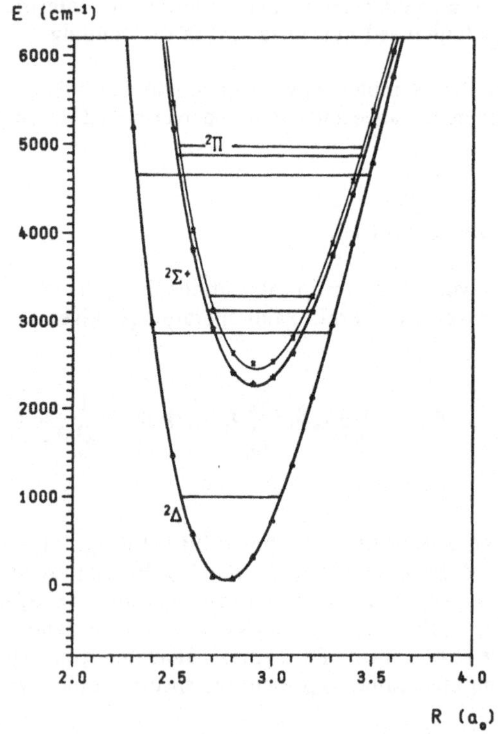

If the spin-orbit contribution is evaluated, the splitting of the $^2\Delta_{5/2}$ and $^2\Delta_{3/2}$ components is larger than the original $^2\Sigma^+$ - $^2\Pi$ energy difference in Fig. 2. The similar statement holds for the two $^2\Pi_{3/2}$ and $^2\Pi_{1/2}$ components, one of which lies now higher and the other lower than the $^2\Sigma^+$ state. At the same time it is seen from Fig. 3 that there are now various vibrational levels with the same Ω quantum number, which are able to interact or to perturb each other if their energetic location is approximately equal. The Ω=5/2 vibronic levels are essentially unperturbed and stem from the $^2\Delta_{5/2}$ state. Among the Ω=3/2 levels the $^2\Pi_{3/2}$ (v=0) and the $^2\Delta_{3/2}$ (v=1) are strongly mixed due to their proximity in energy so that the vibronic function v=4 is written in terms of the unperturbed terms as $^2\Pi$ (0.73 v=0, -0.20 v=1) + $^2\Delta$ (0.25 v=0, 0.60 v=1), for example. Generally the mixing for higher levels becomes more complex so that an assignment to the unperturbed vibrational states becomes difficult. In this connection it is noteworthy, that the vibronic mixing can of course be different for the deuterated species ND, because the difference in reduced mass causes a different vibrational level structure. Details can be found in the original reference [8].

Fig. 3. Calculated potential energy curves of the lowest NiH states by taking into account spin-orbit effects.

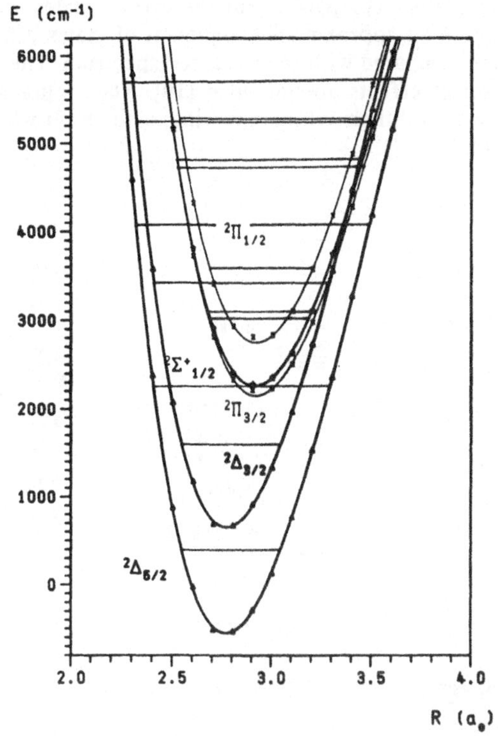

The comparison with experiment is shown in Fig. 4. The interpretation of the experimental lines is greatly facilitated by the theoretical study [8] and further experiments are under way [9] to find some of the levels predicted in Fig. 4 by the quantum chemical calculations.

Fig. 4. Calculated and measured vibronic spectrum for NiH (J = 5/2). The levels on the right-hand side of each of the Ω columns (dashed lines) are calculated, those on the left-hand side are the measured levels. For Ω = 0.5 the experimental levels are split into e and f components; Λ doubling has not been considered in the theoretical study and hence only single levels appear.

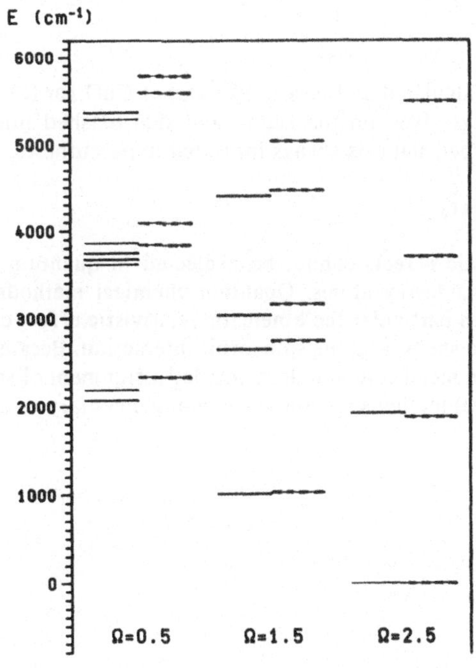

The last example is a comparison for the various states of CuO (Fig. 5). Without the spin-orbit interaction an approximate assignment of states can be undertaken if the calculated intensities are compared with the measured spectrum. The more realistic picture is obtained if the spin-orbit effects are included [10]. The figure shows also the large number of electronic states and their vibronic mixing, a situation which makes definitive assignments quite difficult.

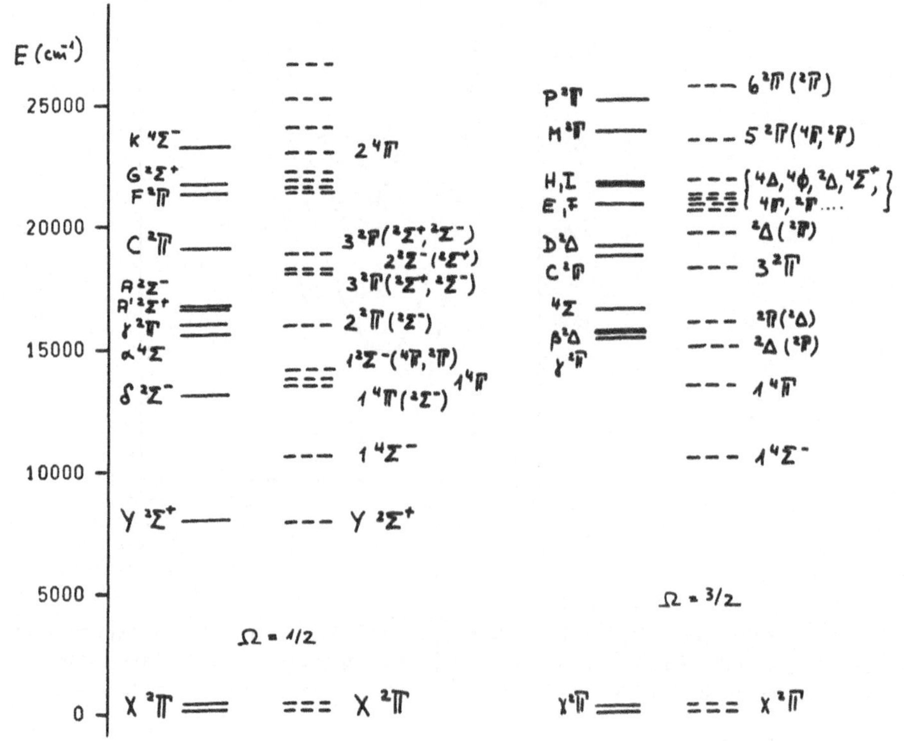

Fig. 5. Calculated and measured states of CuO for Ω = 1/2 and Ω = 3/2. The calculated data are always given on the right hand side (dashed lines), the contributions of the various unperturbed states is always indicated in parentheses. Heavy mixing between states is seen.

4. Summary

Relativistic effects cannot be neglected in quantum mechanical calculations for systems containing heavy atoms. Quantum chemical methods are available to date to treat such effects, in particular the kinematic relativistic effect causing the contraction and expansion of atomic shells, and the spin-orbit interaction. Because of the complexity of the spectra of transition metal compounds, a detailed assignment of states can probably only be made if the theoretical methods are employed complementary to experimental studies.

References

[1] see for example Hay, P. J., Wadt, W. R., Bobrowicz, F. W. (1978) J. Chem. Phys. 69, 984;
Krauss, M., Stevens, W. J., Basch, H. J. (1985) J. Comp. Chem. 6, 287;
Roos, R. B., Ermler, W. C. (1985) J. Phys. Chem. 89, 5202
Igel, G., Wedig, U., Dolg, M., Fuentealba, P., Preuss, H., Stoll, H. and Frey, R. (1984) J. Chem. Phys. 81, 2737

[2] Desclaux, J. P. and Pyykkö, P. (1976) Chem. Phys. Lett. 39, 300

[3] McLean, A. D., Lee, Y. S. (1982) in "Current Aspects of Quantum Chemistry", ed. Carbo, R., Elsevier, Amsterdam

[4] Heß, B. A. (1986) Phys. Rev. A 33, 3742; (1985) Phys. Rev. A 32

[5] Jansen, G., and Heß, B. A. (1989) Chem. Phys. Lett. 160, 507

[6] Jansen, G., and Heß, B. A. (1989) Z. Phys. D 13, 363

[7] Moss, R. E. (1973), "Advanced Molecular Quantum Mechanics", Chapman and Hall, London

[8] Marian, C. M. (1990) J. Chem. Phys. 93, 1176

[9] Urban, W., Bonn University

[10] Hippe, D., Peyerimhoff, S. D., J. Chem. Phys., submitted for publication

SPECTROSCOPIC TECHNIQUES FOR ORGANOMETALLIC INTERMEDIATES

J.J. TURNER
Department of Chemistry
University of Nottingham
Nottingham UK
NG7 2RD

ABSTRACT. An understanding of organometallic photochemistry often requires a knowledge of intermediates and excited states. This article outlines how various spectroscopic techniques might be applied to obtain this information.

1. Introduction

A typical, simple organometallic photoreaction [1] is:

$$Mo(CO)_6 + PPh_3 \xrightarrow[\text{heptane}]{h\nu} Mo(CO)_5(PPh_3) + CO$$

It is known that this reaction proceeds via dissociation of the hexacarbonyl to form the square pyramidal $Mo(CO)_5$ species, which then adds the phosphine. Such a simple scheme may however disguise some extremely complex photochemisty. For instance:

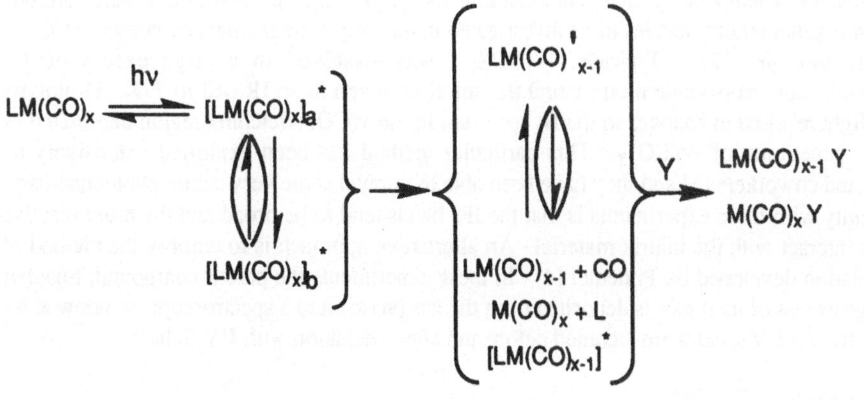

E. Kochanski (ed.), Photoprocesses in Transition Metal Complexes, Biosystems and Other Molecules. Experiment and Theory, 113–123.

In this scheme the ground state molecule $LM(CO)_x$ (where L is a ligand such as a phosphine or pyridine) absorbs light and is promoted to a particular excited state $[LM(CO)_x]_a^*$, which may well decay to - or be in equilibrium with - one or more excited states such as $[LM(CO)_x]_b^*$. From one of the excited states, photochemistry may proceed, typically by loss of one of the ligands, producing intermediates such as $LM(CO)_{x-1}$ or $M(CO)_x$ which in the simplest case will react with another potential ligand Y to give $LM(CO)_{x-1}Y$ or $M(CO)_xY$. There may however be intermediate stages involving reaction to, or equilibrium with, other intermediates or structures, eg $LM(CO)'_{x-1}$. Finally the initial intermediate may not be produced in its ground state, eg $[LM(CO)_{x-1}]^*$.

For a photocatalytic reaction such as,

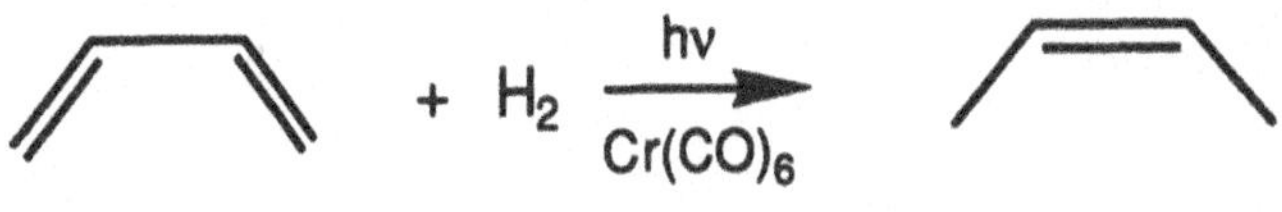

it is clear that the intermediate stages will be very complicated involving species such as $Cr(CO)_5$, $Cr(CO)_3(diene)(H_2)$ (??) etc.

Such intermediates are likely to be highly reactive and hence, like most excited states, will be short-lived under conventional conditions. Special spectroscopic techniques are necessary. These can range from one extreme where the intermediate is "frozen" at very low temperature to another where the reaction is probed at room temperature on timescales down to a few femtoseconds. This chapter will describe, with selected examples, some of these experimental methods, and consider ways of investigating excited states.

2. Matrix Isolation

2.1. METHODS

The object of the matrix isolation technique is to trap a reactive intermediate in a solid inert environment at low temperature, and hence examine its spectroscopic properties at leisure. Among the earliest organometallic species to be investigated in this way were the hexacarbonyls, $M(CO)_6$ (M = Cr, Mo, W) [2]. Typically, $W(CO)_6$ was dissolved in a large excess of 1:4 methylcyclohexane-isopentane mixture and the solution frozen in an IR cell to 77K. Photolysis with UV light resulted in changes in the IR spectrum in the ν(CO) stretching region interpreted as due to the generation of $W(CO)_5$. This particular method has been exploited extensively by Wrighton and coworkers [3] and they have been able to unravel some very subtle photochemistry. The difficulty with these experiments is that the IR bands tend to be broad and the more reactive fragments interact with the matrix material. An alternative approach is to employ the method of matrix isolation developed by Pimentel [4]. In these experiments the parent compound, together with a large excess of inert gas, is deposited from the gas phase on to a spectroscopic window at 4 - 10K. The IR and UV spectra are recorded before and after irradiation with UV light.

2.2. AN EXAMPLE - $Fe(CO)_4$

The photochemistry of $Fe(CO)_5$ provides a good early example of the application of this method to organometallic chemistry [5]. Figure 1 shows the IR spectrum in the ν(CO) region before and after photolysis of $Fe(CO)_5$ in a Ne matrix at 4K [5a].

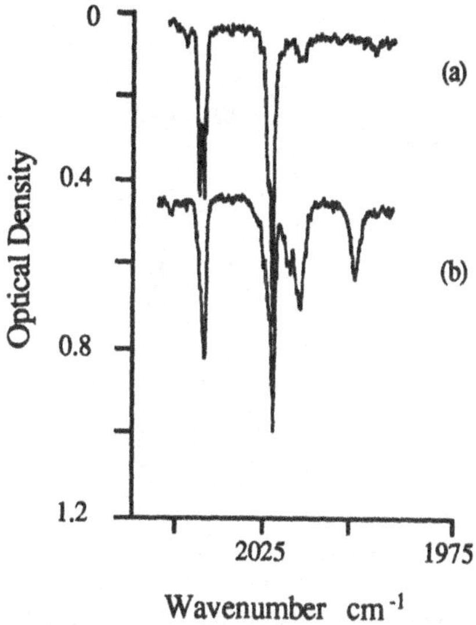

Figure 1. IR spectra of $Fe(CO)_5$ in a Ne matrix at 4K after (a) spray on, (b) 1 min UV photolysis.

Simple group theory predicts that in the ν(CO) region, the trigonal bipyramidal $Fe(CO)_5$ should show two IR active bands of symmetry a_2'' and e'. The weak bands in the spectrum are due to the small percentage of molecules containing ^{13}CO; the high frequency band is split by 'matrix effects' [4]. If photolysis leads to formation of $Fe(CO)_4$ one can envisage this fragment adopting several possible structures, with appropriate IR activity as shown:

Structure	T_d	D_{4h}	C_{3v}	D_{2d}	C_{2v}
ν(CO) IR activity	t_2	e_u	$2a_1+e$	b_2+e	$2a_1+b_1+b_2$

The photoproduct appears to have three IR bands but this may be misleading since band overlap may occur; alternatively one - or more - bands may be too weak to observe. Further information is required and this is provided by the use of isotopes combined with the Energy Factored Force Field [6]. This demonstrates conclusively [5b] that in an Ar (or Ne) matrix, $Fe(CO)_4$ adopts a C_{2v} structure with bond angles determined from the intensities of the IR bands, as shown in figure 2(a).

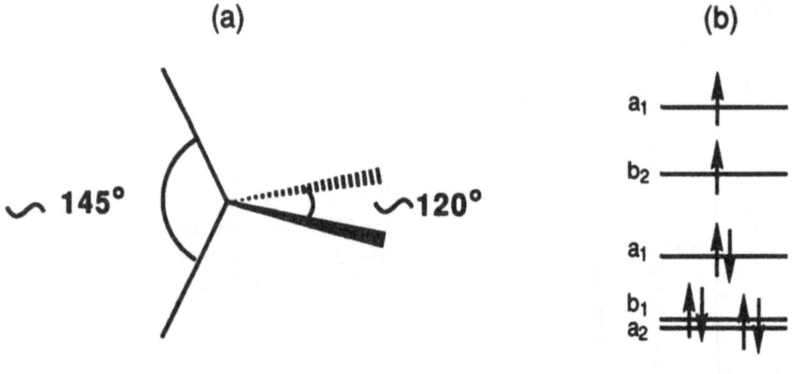

Figure 2.

(a) Shape of $Fe(CO)_4$ determined from matrix IR studies [5b]

(b) Order of levels in C_{2v} structure of $Fe(CO)_4$ [7,8,9]

This structure, surprising at first, is precisely what is expected on the basis of angular overlap [7], extended Hückel [8] and ab initio calculations [9]. Moreover this structure requires the fragment to have two unpaired electrons (see figure 2b). Poliakoff and colleagues [10], employing magnetic circular dichroism combined with matrix isolation, confirmed this by demonstrating $Fe(CO)_4$'s paramagnetism. In matrix experiments [5b] it was shown that the $Fe(CO)_4$ fragment is capable of interacting sufficiently strongly with CH_4 (or Xe) to distort the structure leading, according to theory, to a singlet structure:

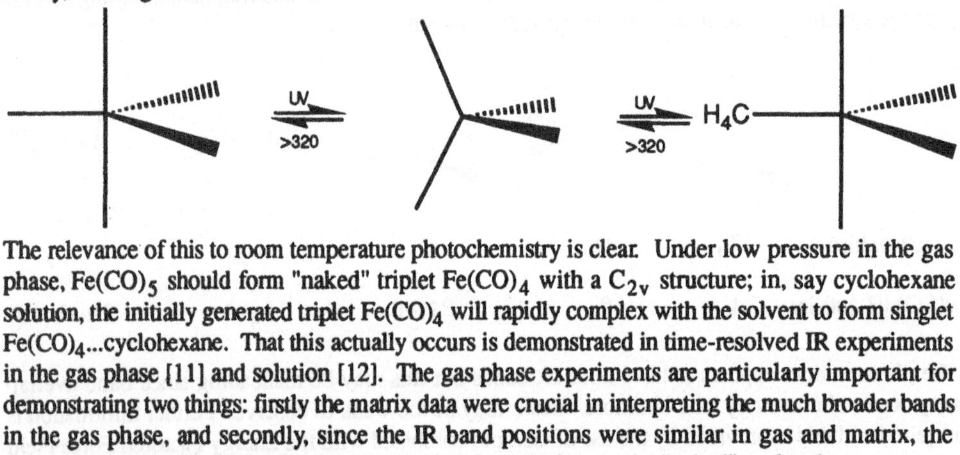

The relevance of this to room temperature photochemistry is clear. Under low pressure in the gas phase, $Fe(CO)_5$ should form "naked" triplet $Fe(CO)_4$ with a C_{2v} structure; in, say cyclohexane solution, the initially generated triplet $Fe(CO)_4$ will rapidly complex with the solvent to form singlet $Fe(CO)_4$...cyclohexane. That this actually occurs is demonstrated in time-resolved IR experiments in the gas phase [11] and solution [12]. The gas phase experiments are particularly important for demonstrating two things: firstly the matrix data were crucial in interpreting the much broader bands in the gas phase, and secondly, since the IR band positions were similar in gas and matrix, the unreactive matrix (Ne or Ar) imposes no structural constraints on the "naked" molecule.

Matrix isolation can thus provide detailed information about the structure and subtle photochemistry of highly reactive, even coordinatively unsaturated, or "naked", organometallic intermediates. The difficulty however is that no information is obtained about kinetics. Next therefore we shall consider a novel way of studying species which are somewhat more stable than those requiring the matrix method.

3. Low temperature solvents

In a low-temperature solvent it is likely that moderately unstable species will be stabilised sufficiently to permit study with conventional techniques. Perhaps the most revealing spectroscopic technique for organometallic species has been infrared [13]. The most appropriate solvents will therefore have to be inert and transparent to IR radiation, ie liquid noble gases [LNG]. Such solvents have long been used for spectroscopic studies of stable molecules [14]; it is relatively recently that they have found application for organometallic photochemistry [15]. A straightforward example involves the photochemistry of $Fe(CO)_2(NO)_2$ in liquid krypton doped with N_2 [16]. As photolysis proceeds, the three IR chromophores ν(NN), ν(CO) and ν(NO) provide a very direct way of monitoring the reaction from $Fe(CO)_2(NO)_2$ to $Fe(N_2)(CO)(NO)_2$ to $Fe(N_2)_2(NO)_2$.

Since the species are being studied in a liquid environment it is possible to monitor kinetics and, by measuring kinetic change with temperature, to extract thermodynamic information eg $Ni(CO)_3(N_2)$ [17]. A novel compound examined by the LNG technique is $Cr(CO)_5(H_2)$, produced by the photolysis of $Cr(CO)_6$ in liquid xenon doped with H_2; this species has a lifetime at -50C of several hours and it proved possible to detect the extremely weak ν(H-H) IR band of the "sideways" bonded (η^2-H_2) group [18]. This type of interaction, first demonstrated by Kubas [19], is of importance in catalytic cycles involving hydrogen [20].

Finally, at its critical point, xenon has a pressure and temperature of 58 atmos. and 16.9°C respectively. It is possible therefore to construct apparatus for studying organometallic photochemistry in a supercritical solvent; this has some interesting applications [21].

The low-temperature solvent method thus permits the study of novel photogenerated species; a limiting factor is that the species have to be relatively stable which therefore precludes the examination of highly reactive intermediates, including coordinatively unsaturated species.

The two methods oulined so far refer to species under rather unusual circumstances; of course we need to know about their behaviour under more conventional conditions. If the species are unstable this means very rapid spectroscopy and the first question is what techniques are available for this and what relevance are the low temperature experiments.

4. Fast Time-resolved Spectroscopy

4.1. METHODS

The most popular experiments have involved flash generation of an intermediate and examination by UV/visible spectroscopy. The diffculty about these experiments is that the UV/visible spectra of organometallics in solution are generally broad and featureless. This means that assignment has had to rely on kinetic techniques rather than spectroscopic [22]. One approach is to use the UV/vis data from matrix experiments in which the characterisation has relied on IR spectra. A different, and perhaps more satisfactory approach, is to monitor the behaviour of the intermediates by fast IR spectroscopy. Unfortunately, largely because of detector problems, fast time-resolved infrared (TRIR) is in practice much more difficult than the corresponding UV/vis experiments.
Fortunately the ν(CO) bands of metal carbonyls are extremely intense; this enabled the Mülheim

group to obtain TRIR, at room temperature, of the $M(CO)_5$ species in cyclohexane [23]. They employed a conventional flashlamp as the photolyis source and for detection a globar, monochromator and fast IR detector.

Since then a number of groups have developed TRIR apparatus, mostly employing laser photolysis (eg excimer) and IR laser detection (CO or diode lasers) [24]. In all these experiments detection is based on a "point-by-point" approach. That is, following the laser flash, the IR signal is monitored at a fixed frequency and the change in intensity with time recorded; the experiment is repeated at a different IR frequency, and in this way it is possible to build up a complete spectrum in the ν(CO) region as a function of time.

4.2. AN EXAMPLE - $[CpFe(CO)_2]_2$

The photochemistry of $[CpFe(CO)_2]_2$ ($Cp = \eta^5\text{-}C_5H_5$) provides an illustration of the use of TRIR techniques. Flash photolysis with UV/Visible detection [25] established that the compound dissociates via two paths, suggested to be:

$$[CpFe(CO)_2]_2 \longrightarrow Cp_2Fe_2(CO)_3 \text{ (long-lived)} + CO$$

$$[CpFe(CO)_2]_2 \longrightarrow 2\ CpFe(CO)_2 \text{ (short-lived)}$$

Matrix isolation [26,27], particularly using ^{13}CO isotopes [26], showed that photolysis leads to $CpFe(\mu\text{-}CO)_3FeCp$ with the three bridging CO groups arranged symmetrically about the Fe-Fe bond. These experiments also illustrate one particular limitation of the matrix technique. Undoubtedly UV photolysis also generates the radicals, but in the enclosed matrix "cage", these readily recombine so that they do not appear in the IR spectrum. Room temperature flash photolysis combined with IR detection [28] does however show clearly that both $CpFe(\mu\text{-}CO)_3FeCp$ and $CpFe(CO)_2$ are produced.

4.3. ULTRAFAST STUDIES

The time resolution of the TRIR technique described above is limited by the detection system to about 100 ns. This timescale is perfectly adequate for most "chemical" processes but may well be inadequate for the primary steps in the photochemistry. To study such processes, complex laser mixing techniques are needed, where the IR signal is "upconverted" to the UV/vis region where detection can be much faster. There have been several reports of such equipment but only two applications to organometallic chemistry [29,30] . The difficulty is that the IR bands tend to be rather broad because of complications due to vibrational relaxation, and there are no matrix data to confirm the assignment. In a recent ps/fs experiment [29] on $[CpFe(CO)_2]_2$, it is clear that the spectrum changes over the timescale of the experiment until it is recognisably that of $CpFe(CO)_2$ [28]. However it is difficult to differentiate the genuine early stages of the photodissociation from vibrational relaxation effects.

So far we have not mentioned Time-resolved Resonance Raman spectroscopy (TR^3), which has played such a fundamental role in monitoring fast processes in biological systems. There are very few reports of such experiments with organometallics, partly because the molecules are so photosensitive.

5. Excited states

5.1. INTRODUCTION

To understand the whole photochemical pathway it is necessary to probe the nature of the excited states. Figure 3 schematically illustrates the ground state, one excited state, and some spectroscopic connections between them.

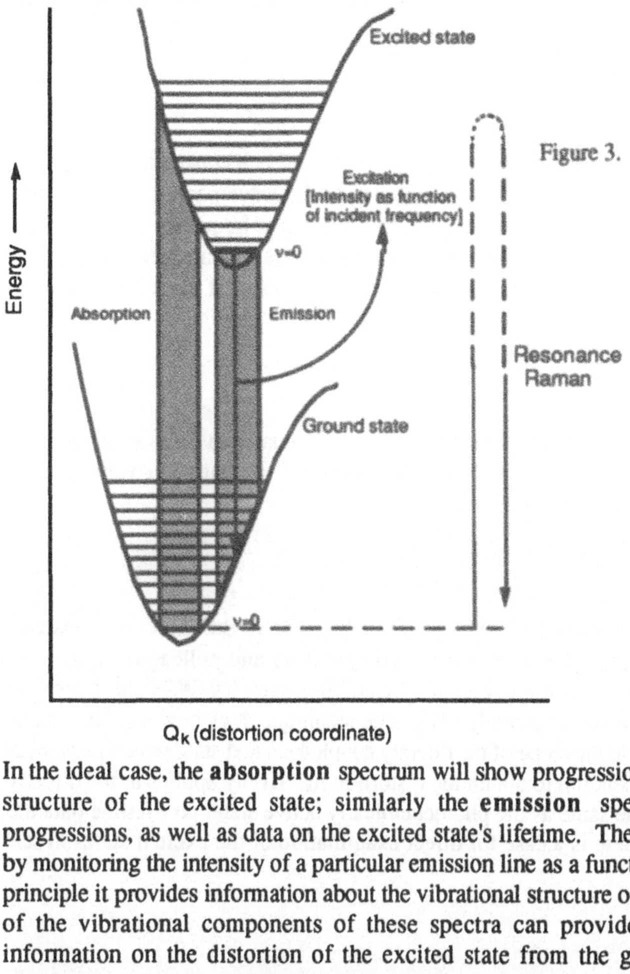

Figure 3. Schematic representation of Potential Energy Curves

In the ideal case, the **absorption** spectrum will show progressions associated with the vibrational structure of the excited state; similarly the **emission** spectrum will show ground state progressions, as well as data on the excited state's lifetime. The **excitation** spectrum is obtained by monitoring the intensity of a particular emission line as a function of the incident wavelength; in principle it provides information about the vibrational structure of the excited state. The intensities of the vibrational components of these spectra can provide, via Franck-Condon analysis, information on the distortion of the excited state from the ground state ie Δ_k along the Q_k coordinate. However it is only in a few cases, and at very low temperatures, that sufficient fine structure is observed to provide informaton on the structure of the excited state [31]. The low-temperature matrix technique can demonstrate highly resolved spectra. Thus photolysis of Cp_2ReH_2 in a N_2 matrix generates Cp_2Re for which structured absorption, emission and excitation spectra have been recorded [32]. Low temperatures can also be obtained in free-jet expansion experiments and very recently Reber and Zink [33] have observed a well resolved excitation spectrum involving the first charge transfer excited state of bis(hexafluoroacetylacetonate)Pt(II) in a molecular beam. Making some assumptions about the potential function, it proved possible to estimate the changes in length of several bonds on going from ground to excited state [33].

5.2. RESONANCE RAMAN - TIME DEPENDENT APPROACH

Resonance Raman (RR) spectra require coupling to the excited state involved in the transition. Hence in principle RR spectra should provide similar information to electronic spectra, and Franck-Condon analysis of Raman intensities should allow determination of the structure of excited states [34]. However for large molecules such analysis can be extremely complicated and a rather different way of looking at such spectra has been developed by Heller [35]. This is based on the relationship between spectroscopy in the frequency and time domains. Thus for absorption spectroscopy:

$$I(\omega) = C\omega \int_{-\infty}^{\infty} e^{-i\omega t} \langle\phi|\phi(t)\rangle dt$$

where I is the absorption intensity as a function of frequency (ω), C is a constant and $\langle\phi|\phi(t)\rangle$ measures the overlap between the wave packet at time t = 0 and the wave packet at time t. The connection with the excited state distortion arises because, for a particular frequency ω_k, the overlap depends on Δ_k and ω_k. A similar formula can be derived for Resonance Raman spectroscopy, and with certain simplifying assumptions, the relationship between the intensity of the different normal modes can be expressed:

$$I_1/I_2 = (\Delta_1^2\,\omega_1^2)/(\Delta_2^2\,\omega_2^2)$$

In other words, the ratio of the intensities of two RR bands is directly related to the distortions along the two normal coordinates Q_1 and Q_2. To obtain actual, as opposed to relative Δ_k values, some calibration is necessary and this is provided by the equation:

$$2\sigma^2 = \Sigma_k \Delta_k^2 \omega_k^2$$

where $8\sigma^2$ is the square of the **absorption** bandwidth at 1/e of the peak height. This development has been exploited for organometallic systems, particularly by Zink and colleagues [36]. For instance it has been shown that the photoactive state of $W(CO)_5$(pyridine) has W-N and trans C-O bonds lengthened by 0.18 and 0.12 Å respectively [37]. The technique does however have some limitations: the theory really refers to the slope of the directly coupled excited state which means that there may be a considerable extrapolation to obtain the distortion Δ_k; it only applies to the directly coupled state which may not be the same as the photochemically active state; no lifetime data are obtained. All this suggests that there is a case for direct examination of the excited states in real time.

5.3. DIRECT OBSERVATION

The obvious way to obtain structural information is by direct observation of the vibrational spectrum of the excited state. There have been many TR^3 experiments on a wide range of inorganic systems. In the long lived ^{3}MLCT excited state of $[Ru(bpy)_3]^{2+}$, the RR spectrum indicates that bpy$^-$ is formed, ie that the formula of the excited state can be written $[Ru^{3+}(bpy)_2\,bpy^-)]^{2+}$ [38]. Estimation of the individual bond length changes requires a sophisticated normal coordinate analysis [39]. For "$Pt_2(pop)_4^{4-}$", the Pt-Pt stretching vibration is higher in frequency in the excited state than in the ground state [40], which matches molecular orbital models of the bonding. However there have been very few TR^3 experiments on organometallics. One example is $ClRe(CO)_3$(bpy) where there where there is clear evidence from emission [41] and TR^3 experiments [42] that the long-lived excited state involves - as with $[Ru(bpy)_3]^{2+}$ - charge transfer from metal to bpy ligand. However the ν(CO) bands of the excited state were not resonance enhanced and hence could provide no information on the excited state. Moreover because of the strong emission, it has proved impossible to obtain RR spectra of the molecule in the ground state [43]. On the other hand for $W(CO)_4$(diimine) complexes Zink has obtained TR^3 spectra which demonstrate resonance enhancement for one of the ν(CO) modes which shows an upward shift of some 50 cm^{-1} [44].

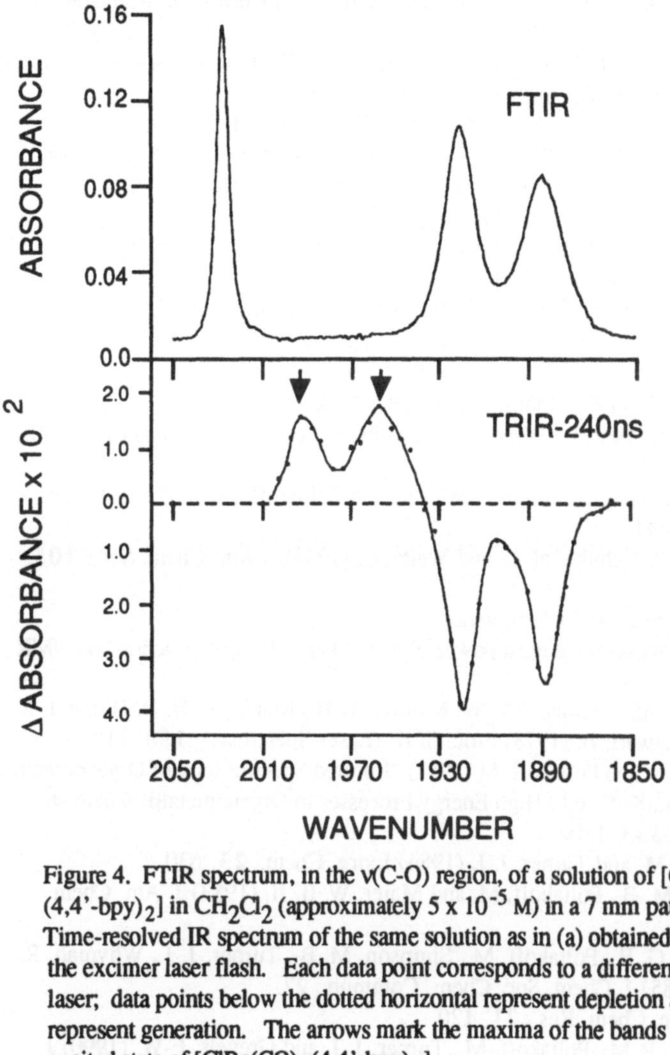

Figure 4. FTIR spectrum, in the ν(C-O) region, of a solution of $[ClRe(CO)_3$-$(4,4'\text{-bpy})_2]$ in CH_2Cl_2 (approximately 5 x 10^{-5} M) in a 7 mm path length cell. Time-resolved IR spectrum of the same solution as in (a) obtained 240 ns after the excimer laser flash. Each data point corresponds to a different line of the CO laser; data points below the dotted horizontal represent depletion and those above represent generation. The arrows mark the maxima of the bands assigned to the excited state of $[ClRe(CO)_3(4,4'\text{-bpy})_2]$.

Since the TR^3 spectra show some limitations it would be valuable to obtain IR spectra of the excited states. The first report of such experiments [45], on $[ClRe(CO)_3(4,4'\text{-bpy})_2]$, showed that in the excited state there is an increase in frequency of two of the ν(CO) bands of about 60 cm^{-1} (figure 4) [46] and that the decay of the excited state IR signal matched previous emission studies [48].

The upward shift in ν(CO) frequencies in this compound and in $W(CO)_4$(diimine) [44] is due to charge transfer to bpy ligand which leaves the metal oxidised; this diminishes back bonding to the CO groups and hence raises the frequency. More recently it has been possible, using TRIR, to follow the whole photochemistry of $W(CO)_5$(4-cyanopyridine), and monitor the conversion of the excited state to dissociation products [49].

6. References

[1] Geoffroy, G. L. and Wrighton, M. S. (1979) Organometallic Photochemistry, Academic Press, New York.

[2] Stolz, I. W., Dobson, G. R. and Sheline, R. K. (1962) J. Am. Chem. Soc., **84**, 3589; (1963) J. Am. Chem. Soc., **85**, 1013. For further details see next chapter.

[3] e.g. Young, K. M. and Wrighton, M. S. (1990) J. Am. Chem. Soc., **112**, 157.

[4] Hitam, R. B., Mahmoud, K. A. and Rest, A. J. (1984) Coord. Chem. Rev. **55**, 1; Almond, M. J. and Downs, A. J. (1989) "Spectroscopy of Matrix Isolated Species", in Clark, R. J. H. and Hester, R. E. (eds.), Advances in Spectroscopy, **17**, 1.

[5] (a) Poliakoff, M. and Turner, J. J. (1973) J. Chem. Soc. Dalton, 1351; (b) (1974) J. Chem. Soc. Dalton, 2276; (c) Poliakoff, M. (1978) Chem. Soc. Rev., 527.

[6] for a description of the application of the Energy Factored Force Field to transition metal carbonyls see Braterman, P. S. (1975) Metal Carbonyl Spectra, Academic Press.

[7] Burdett, J. K. (1974) J. Chem. Soc. Faraday II, **70**, 1599.

[8] Elian, M. and Hoffmann, R. (1975) Inorg. Chem., **14**, 1058.

[9] Daniel, C., Bernard, M., Dedieu, A., Wiest, R. and Veillard, A. (1984) J. Phys. Chem., **88**, 4805.

[10] Barton, T. J., Grinter, R., Thomson, A. J., Davis, B. and Poliakoff, M. (1977) J. Chem. Soc. Chem. Commun., 841.

[11] Ouderkirk, A., Wermer, P., Schultz, N. L. and Weitz, E. (1983) J. Am. Chem. Soc., **105**, 3354.

[12] Grevels, F-W. (1991) Lecture at this Conference.

[13] Of course for **stable** compounds the most powerful methods are X-ray diffraction and NMR spectroscopy.

[14] Bulanin, M. O. (1973) J. Mol. Struct., **19**, 59; Beattie, W. H., Maier, W. B. II, Holland, R. F., Freund, S. M. and Stewart, B. (1978) Proc. SPIE (Laser Spectrosc.), **158**, 113.

[15] Turner, J. J., Healy, M. A. and Poliakoff, M (1987) "Infrared Spectroscopy of Organometallic Intermediates" in Suslick, K. S. (ed.) High Energy Processes in Organometallic Chemistry, ACS Symposium Series **333**, 110.

[16] Gadd, G. E., Poliakoff, M. and Turner, J. J. (1984) Inorg. Chem., **23**, 630.

[17] Turner, J. J., Simpson, M. B., Poliakoff, M. and Maier, W. B. II, (1983) J. Am. Chem. Soc., **105**, 3898.

[18] Upmacis, R. K., Gadd, G. E., Poliakoff, M., Simpson, M. B., Turner, J. J., Whyman, R. and Simpson, A. F. (1985) J. Chem. Soc. Chem. Commun., 27.

[19] Kubas, G. J. (1988) Acc. Chem. Res., **21**, 120.

[20] Jackson, S. A., Hodges, P. M., Poliakoff, M., Turner, J. J. and Grevels, F-W. (1990) J. Am. Chem. Soc., **112**, 1221; Hodges, P. M., Jackson, S. A., Jacke, J., Poliakoff, M., Turner, J. J. and Grevels, F-W. (1990) J. Am. Chem. Soc., **112**, 1234.

[21] Howdle, S. M., Healy, M. A. and Poliakoff, M. (1990) J. Am. Chem. Soc. **112**, 4804.

[22] e.g. Meyer, T. J. and Caspar, J. V. (1985) Chem. Rev., **85**, 187.

[23] Hermann, H., Grevels, F-W., Henne, A. and Schaffner, K.(1982) J.Phys. Chem., **86**, 5151.

[24] For a review of experiments up to about the end of 1985, including those in the gas phase where the photochemistry is different from solution, see: Poliakoff, M. and Weitz, E. (1986) Adv. Organomet. Chem., **25**, 277.
[25] Caspar, J. V. and Meyer, T. J. (1980) J. Am. Chem. Soc., **102**, 7794.
[26] Hooker, R. H., Mahmoud, K. A. and Rest, A. J. (1983) J. Chem. Soc. Chem. Commun., 1022.
[27] Hepp, A. F., Blaha, J. P., Lewis, C. and Wrighton, M. S. (1984) Organometallics, **3**, 174.
[28] (a) Moore, B. D., Simpson, M. B., Poliakoff, M. and Turner, J. J. (1984) J. Chem. Soc. Chem. Commun., 972; (b) for a recent paper with up-to-date data see: Dixon, A. J., George, M. W., Hughes, C., Poliakoff, M. and Turner, J. J. (1991), J. Am. Chem. Soc., in press.
[29] Moore, J. N., Hansen, P. A. and Hochstrasser, R. M. (1987) Chem. Phys. Letters **138**, 110; (1989) J. Am. Chem. Soc., **111**, 4563; Anfinrud, P. A., Han, C-H., Lian, T. and Hochstrasser, R. M. (1991) J. Phys. Chem., **95**, 574.
[30] Spears, K. G., Zhu, X., Yang, X. and Wang, L. (1988) Optics Commun., **66**, 167; Wang, L., Zhu, X. and Spears, K. G. (1988) J. Am. Chem. Soc., **110**, 8695; (1989) J. Phys. Chem., **93**, 2.
[31] For example, Yertsin, H., Otto, H., Zink, J. I. and Gliemann, G. (1980) J. Am. Chem. Soc., **102**, 951.
[32] Bell, S. E. J., Hill, J. N., McCamley, A. and Perutz, R. N. (1990) J. Phys. Chem., **94**, 3877.
[33] Reber, C. and Zink, J. I. (1991) 9th Int. Symp. Photochem. and Photophys. Coord. Compounds, O-8.
[34] Clark, R. J. H. and Dines, T. J. (1986) Angew. Chem. Int. Ed. Engl., **25**, 131.
[35] Heller, E. J., Sundberg, R. L. and Tannor, D. (1982) J. Phys. Chem., **86**, 1822.
[36] Zink, J. I. and Shin, K-S. K. (1991) Adv. Photochem., **16**, 119.
[37] Tutt, L. and Zink, J. I. (1986) J. Am. Chem. Soc., **108**, 5830.
[38] Bradley, P. G., Kress, N., Hornberger, B. A., Dallinger, R. F. and Woodruff, W. H. (1981) J. Am. Chem. Soc., **103**, 7441.
[39] Strommen, D. P., Mallick, P. K., Danzer, G. D., Lumpkin, R. S. and Kincaid, J. R. (1990) J. Phys. Chem., **94**, 1357; Mallick, P. K., Strommen, D. P. and Kincaid, J. R. (1990) J. Am. Chem. Soc., **112**, 1686; Danzer, G. D. and Kincaid, J. R. (1990) J. Phys. Chem., **94**, 3976.
[40] Che, C-M., Butler, L. G., Gray, H. B., Crooks, R. M. and Woodruff, W. M. (1983) J. Am. Chem. Soc., **105**, 5492.
[41] Wrighton, M. S. and Morse, D. L. (1974) J. Am. Chem. Soc., **96**, 998.
[42] Smothers, W. K. and Wrighton, M. S. (1983) J. Am. Chem. Soc., **105**, 1067.
[43] Stufkens, D. J. (1991) personal communication.
[44] Perng, J-H. and Zink, J. I. (1990) Inorg. Chem., **29**, 1158.
[45] Glyn, P., George, M. W., Hodges, P. M. and Turner, J. J. (1989) J. Chem. Soc. Chem. Commun., 1655.
[46] Recent experiments [47] with an IR diode laser show that the high frequency band occurs at about 2055 cm^{-1}. The shifts raise interesting questions about the electron distribution in the excited state.
[47] Johnson, F. P. A., George, M. W. and Turner, J. J. (1991) unpublished observations.
[48] Giordano, P. J. and Wrighton, M. S. (1979) J. Am. Chem. Soc., **101**, 2888.
[49] Glyn, P., Johnson, F. P. A., George, M. W., Lees, A. J. and Turner, J. J. (1991) Inorg. Chem., **30**, 3543.

PHOTOCHEMISTRY INVOLVING $Cr(CO)_5$ - STILL SOME PUZZLES

J. J. TURNER
Department of Chemistry
University of Nottingham
Nottingham UK
NG7 2RD

ABSTRACT. Although the photochemistry of $Cr(CO)_6$, including the generation of $Cr(CO)_5$ and examination of its properties, has been the subject of many investigations, the final word has by no means been said and this is an appropriate moment and context to review the position. Where appropriate we shall mention the photochemistry of $Mo(CO)_6$ and $W(CO)_6$.

1. Introduction

In the early 1960s several workers [1,2,3] postulated the existence of metal pentacarbonyls ($M(CO)_5$: M = Cr,Mo,W) in the solution photochemistry of $M(CO)_6$:

$$M(CO)_6 \overset{h\nu}{\Rightarrow} M(CO)_5 + CO$$
$$M(CO)_5 + L \Rightarrow M(CO)_5L$$

The first question was whether $M(CO)_5$ could be identified and its structure determined; two obvious possible structures are square pyramid (C_{4v}) or trigonal bipyramid (D_{3h}). The first indication came from experiments in low-temperature glasses. Following experiments with $W(CO)_6$ [4], Stolz et al [5] dissolved $Cr(CO)_6$ in 1:4 isopentane-methylcyclohexane, cooled the solution in an IR cell to -180C and photolysed with UV light. IR bands appeared at 2088(w), 1955(s) and 1928(m) cm^{-1} and were assigned to the $\nu(CO)$ stretching bands of $Cr(CO)_5$ with C_{4v} symmetry (a_1, e and a_1 respectively); the D_{3h} structure should have **two** IR active $\nu(CO)$ bands. Shortly afterwards El-Sayed [6] suggested that the yellow colour obtained on photolysis of $Mo(CO)_6$ in 1:1 ether-isopentane was due $Mo(CO)_5$, but Dobson [7] presented convincing arguments that photolysis of $M(CO)_6$ species in such a media produced weak complexes of $M(CO)_5$ with ether. Dobson also made the important observation that in the pure hydrocarbon solvent, photolysis yielded a yellow colour which disappeared when the irradiation light was turned off; the colour was much longer-lived at low temperatures. Dobson also speculated that the IR spectrum of the $Mo(CO)_5$ fragment [5] might be due, not to "naked" $Mo(CO)_5$, but to "a weak complex between $Mo(CO)_5$ and an extremely poor donor, pehaps the nitrogen under which the experiment was carried out" [8]. Although, as we shall see, this perceptive explanation is wrong in detail, it is quite correct in principle.

E. Kochanski (ed.), Photoprocesses in Transition Metal Complexes, Biosystems and Other Molecules. Experiment and Theory, 125–140.

The low temperature work was extended to frozen gas matrices by Turner and colleagues [9] who photolysed $W(CO)_6$ in solid argon at 20K. From the pattern of the IR bands of the photoproduct they confirmed the probable C_{4v} structure of $W(CO)_5$; they also noted that $W(CO)_5$ had absorption bands at 239 and 436 nm and commented that "it was hard to imagine Ar acting as a ligand", hence in principle removing the problem of any weak interactions with $W(CO)_5$. More detailed matrix studies [10] provided IR and UV/vis data for all the $M(CO)_5$ fragments; in particular it was noted that the visible absorption band of $Cr(CO)_5$ in **Ar** was assigned to an absorption at 542 nm, compared with a corresponding absorption at 485 nm in a **pentane-methylcyclohexane glass** [11]. This difference is striking; it led immediately to an investigation [12] of the behaviour of $Cr(CO)_5$ in a **methane** matrix at 20K on the argument that the two hydrocarbon environments might be similar. Sure enough, the visible absorption band of $Cr(CO)_5$ in CH_4 is at 492 nm, ie quite close to the glass value. On the assumption that Ar is a non-interacting species (but see below), it was proposed that the shift from Ar to CH_4 is a measure of the interaction between CH_4 and the empty coordination site of $Cr(CO)_5$. It was therefore suggested that a similar interaction might occur with alkane solvents at room temperature. Hence (part of) the room temperature photochemistry could be:

$$Cr(CO)_6 \overset{h\nu}{\Rightarrow} \underset{(542\ nm)}{Cr(CO)_5} + solvent(S) \Rightarrow \underset{(483\ nm)}{Cr(CO)_5 ... S}$$

Thus on photolysis of parent hexacarbonyl the initial product is a "naked" $M(CO)_5$ fragment which, in the case of Cr, absorbs at 542 nm; however in the solvent environment this is practically instantaneously converted to a weak $M(CO)_5$...solvent complex which, in the case of Cr, absorbs at ca. 480-490 nm. Hence Dobson's yellow colour was probably due to interaction of $M(CO)_5$ with the "innocuous" hydrocarbon solvent. This was therefore one of the earliest pieces of evidence for arrested C-H activation. Since the interaction is weak, there is an equilibrium set up with the naked fragment and in the presence of a "strong" ligand (L), the naked fragment reacts to give $Cr(CO)_5L$.

At about the same there were attempts to observe $Cr(CO)_5$ in solution by flash photolysis techniques [13,14]. It was shown subsequently however [15,16] that the earlier experiments were misinterpreted because of the presence of solvent impurities which react with $Cr(CO)_5$. In pure cyclohexane, $Cr(CO)_5$ has a visible absorption band with a maximum at 503 nm, ie at about the wavelength expected for $Cr(CO)_5$...cyclohexane, **not** "naked" $Cr(CO)_5$ [15]. With a cyclohexane solution saturated under 1 atmosphere of CO, the half-life of the solvated species was 25 us., before reaction with CO and re-formation of $Cr(CO)_6$.

This left several problems:

(i) Proof of the structure of $Cr(CO)_5$; clearly, relying on simple group theory - three IR active ν(CO) bands for C_{4v}, two for D_{3h} - is unsatisfactory.

(ii) The origin and significance of the shift in the visible absorption band.

(iii)The magnitude of the $Cr(CO)_5$...Q interactions, and its influence on the chemistry.

(iv) The primary photochemical steps.

These problems will be addressed in the following sections.

2. The structure and spectroscopy of $Cr(CO)_5$

2.1. MATRIX DATA

The Energy Factored (or Cotton-Kraihanzel) Force Field (EFFF) has been shown to be a powerful method of understanding the ν(CO) vibrations of metal carbonyls [17]. Unfortunately there are frequently more force constants than frequencies; for instance a C_{4v} $M(CO)_5$ fragment has five k(CO) force constants but, as we saw above, only three IR active frequencies. The dilemma is solved by the use of isotopes. In the EFFF approximation, the G matrix is diagonal in the reduced masses of the CO groups; it is thus a straighforward matter to calculate the spectrum of isotopically (^{13}C or ^{18}O) substituted molecules, provided the C-O force constants are known. Conversely if a complex ν(CO) isotopic pattern is observed, in principle a suitable set of force constants can be found to generate a theoretical spectrum which matches the experimental. This is important because now there are in general more frequencies than force constants, and hence the confidence in the resulting force field is measured by the degree of fit between theoretical and experimental spectrum. In addition the intensities of ν(CO) bands usually follow closely the local dipole moment approach so that information on geometrical structure is also obtained.

In fact by matrix photolysis at 20K of ^{13}CO-enriched $M(CO)_6$ parent molecules, the structure of each of the $M(CO)_5$ species was proved to be C_{4v}, with axial/equatorial ("droop") bond angles near to 90^0 [18]. Equally importantly it was demonstrated [19] that, although the IR spectra hardly changed in frequency, the position of the maximum of the visible absorption band of $Cr(CO)_5$ was much more sensitive to the matrix material than previously observed:

Matrix	Ne	SF_6	CF_4	Ar	Kr	Xe	CH_4
Band max (nm)	624	560	547	533	518	492	489
(kcm^{-1})	16.0	17.9	18.3	18.8	19.3	20.3	20.4

This effect was so striking that it was necessary to demonstrate that it was not a "generalised matrix effect". Photolysis was performed in **mixed** matrices, eg Ne/2%Xe at 4K. In this matrix $Cr(CO)_5$ showed **two** visible bands at approximately 628 and 487 nm, and displayed substitutional photochemistry:

$$\underset{(628\text{ nm})}{Cr(CO)_5...Ne} \underset{h\nu/432\text{ nm}}{\overset{h\nu/618\text{ nm}}{\rightleftharpoons}} \underset{(487\text{ nm})}{Cr(CO)_5...Xe}$$

Since the overwhelming environment is always Ne, this experiment demonstrated conclusively that the shift in visible band is caused by interaction with the empty site on the metal fragment [19]. The substitution photochemistry observed even at 4K requires some mobility on the part of the $Cr(CO)_5$ fragment - as we shall see, in section 5.3, this provides an important clue to the detailed photobehaviour of $Cr(CO)_6$.

Some authors have said recently that the frequency of the visible absorption band is a direct measure of the interaction energy between the fragment and matrix material. It is important to point out that this is probably not so. Careful examination of the IR data show that, although the force fields of, say $Cr(CO)_5$...Ne and $Cr(CO)_5$...Xe, are very similar, the IR intensities suggest different "droop" angles. This leads to the interpretation of the shift shown in Figure 1; although there is **some** change in ground state energy with matrix the major part of the shift comes from the sensitivity of the excited state to bond angle [20].

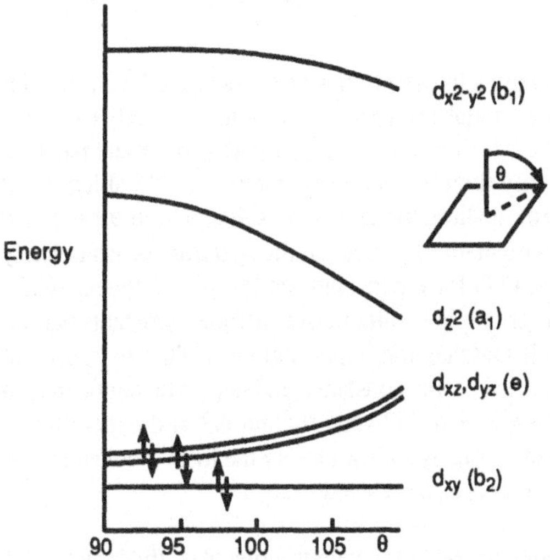

Figure1.
Orbital energies of d^6 $M(CO)_5$ C_{4v} fragment as a function of droop angle θ.
The energy of the transition ($b_2^2 e^4 \rightarrow b_2^2 e^3 a_1^1$) is very sensitive to θ.

However it is still arguable that the matrix environment is sufficiently perturbing, even in Ne, to cause distortion of the fragment away from its "true" ground state structure. This is an important point since on generation of $Cr(CO)_5$ by condensation of Cr atoms and CO/Ar the ν(CO) IR spectrum [21] indicates a D_{3h} structure which changes to C_{4v} on warming the matrix [22]. Conclusive structural evidence comes from gas phase work.

2.2. GAS PHASE DATA

Breckenridge and Sinai probed the visible absorption spectra of transients following the laser flash photolysis (355 nm) of $Cr(CO)_6$ in the gas phase [23]. With Ar or CH_4 as buffer gases, short-lived transients were observed with broad absorptions at 500-530 nm; by analogy with the matrix data [19], these were assigned to $Cr(CO)_5$...Ar and $Cr(CO)_5$...CH_4. In these experiments [23], when $Cr(CO)_6$ was flashed alone there was evidence only for $Cr(CO)_5Cr(CO)_6$. However in later experiments [24], with either $Cr(CO)_6$ alone or with He as buffer gas, it was possible to assign an absorption at 620 nm to "naked" $Cr(CO)_5$; this is very close to $Cr(CO)_5$ in a Ne matrix (623 nm) and suggests that in Ne the interaction must be negligible and hence that the ground state structure is indeed C_{4v}. However with Ar, methane or propane as buffer gases there was no evidence for $Cr(CO)_5$...Q (Q = Ar, CH_4 or C_3H_8) which seems to contradict the earlier work [23]. There is little doubt that effectively "naked" $Cr(CO)_5$ is observed in the gas but the position on the weak complexes is obscure. Further informatiom is provided by Time-resolved IR (TRIR) experiments.

Weitz et al [25] reported that XeF laser (351 nm) photolysis of $Cr(CO)_6$ in an Ar buffer gas generates a species whose kinetic behaviour is consistent with $Cr(CO)_5$, and which has IR bands at 1980 and 1948 cm^{-1}. The t_{1u} band of $Cr(CO)_6$ is observed at 2000.4 cm^{-1} in the gas phase [26] and at 2003 cm^{-1} in a Ne matrix [19,27]; the gas to matrix shift is very small and is much smaller than for other matrices. Thus the best comparison of gas and matrix data for $Cr(CO)_5$ will involve a Ne matrix:

Molecule	Environment	Frequencies(cm^{-1})	Assignment
$Cr(CO)_6$	gas	2000.4	t_{1u}
	Ne matrix	2003	t_{1u}
$Cr(CO)_5$	Ne matrix	1978	e
		1945	a_1
	Ne(2%Ar) matrix	1972	e ($Cr(CO)_5$...Ar)
		1941	a_1(„)
	Ar gas	1980	e (?)
		1948	a_1(?)

There is no doubt that gas phase TRIR detects square pyramidal $Cr(CO)_5$ but the data cannot really distinguish between naked $Cr(CO)_5$ and $Cr(CO)_5$...Ar. On the other hand, by monitoring TRIR spectra following the flash photolysis of $W(CO)_6$ in a mixture of ethane and Ar, Rayner et al [28] detected the conversion of the initially generated $W(CO)_5$ (ca. 1980 and 1942 cm^{-1}) to $W(CO)_5...C_2H_6$ (1982 and 1948 cm^{-1}). Interestingly they were unable to detect any transient bands due to $W(CO)_5...CH_4$. There do not seem to be equivalent $Cr(CO)_6$ experiments, but it seems reasonable to suppose that $Cr(CO)_5$...Q (Q = CH_4 or Ar) would be undetectable, as suggested by the later UV/Visible experiments [24]. These TRIR experiments provide valuable information on the weak interactions of $Cr(CO)_5$ (see section 4.2)

2.3. THEORY OF $Cr(CO)_5$ STRUCTURE

The unusual geometry of some of the metal carbonyl fragments in matrices prompted early semi-empirical extended Huckel molecular orbital (EHMO) calculations [29]. For low-spin d^6 $M(CO)_5$ (M = Cr, Mo, W), the favoured structure is indeed square pyramidal with a "droop" angle of about 90°; however the lowest energy structure of the triplet configuration is predicted to be trigonal bipyramidal. Later calculations confirmed these conclusions [30,31]. Unfortunately EHMO calculations cannot accurately estimate the energy difference between states of different multiplicity. However *ab initio* calculations [32,33] show that for $Cr(CO)_5$ the $^1A_1$ state of the C_{4v} structure lies some 9-10 kcal mol^{-1} below the $^3A_2'$ state of the D_{3h} structure, ie the lowest energy form of $Cr(CO)_5$ is the singlet C_{4v}.

The simplest explanation of the variation of the absorption band with matrix in terms of bond angle change has been mentioned above. Another approach [33] invokes the spatial extent of the excited state a_1 orbital, whose lobe extends into the vacant sixth coordination site. The **destabilization** resulting from the open-shell orbital interaction with the closed-shell solvent will increase in the order Ne < Ar < Kr < Xe in agreement with the absorption data. Extensive *ab initio* calculations [34,35] suggest that dispersion energy makes a very significant contribution to the interaction between $M(CO)_5$ species and noble gases but it has proved extremely difficult to calculate reliable numbers for the bond energy.

In view of the great interest in C-H activation [36] and agostic interactions [37], more intriguing is the interaction between $Cr(CO)_5$ and saturated hydrocarbons. Employing a variety of matrix techniques, including ^1H isotopic enrichment, several attempts have been made to determine the specific interaction geometry between $Cr(CO)_5$ and CH_4 [38]. Unfortunately none have provided clear evidence but it is reasonable to assume that one C-H bond is being stretched as in Crabtree's model [36], although there are steric effects which are very important [39]. We shall reum to the energies of such interactions shortly.

3. $Cr(CO)_5$ in solution

3.1. UV/VIS DETECTION

The matrix data suggest that the interaction between $Cr(CO)_5$ and CF_4 is less than that between $Cr(CO)_5$ and CH_4. This in turn suggests that perfluoroalkane solvents will interact less than alkane solvents and this should have a dramatic effect on the kinetics. This was elegantly demonstrated by Kelly and colleagues [40,41], who compared the flash photolysis behaviour of $Cr(CO)_6$ in perfluoromethylcyclohexane (Fcyc). with the previous results in cyclohexane [16]. In each case a transient is produced within the time resolution of the apparatus (ca. 5 ns), in cyclohexane at 503 nm [16] but in Fcyc. at 620 nm; moreover on flashing CO-saturated solutions, the transients re-formed $Cr(CO)_6$, but at very different rates:

$$Cr(CO)_6 \overset{h\nu}{\Rightarrow} \text{"}Cr(CO)_5\text{"} \overset{CO}{\Rightarrow} Cr(CO)_6$$

cyclohexane	503 nm	$k_2 = 3 \times 10^6\ dm^3\ mol^{-1}\ s^{-1}$	[16]
perfluoromethylcyclohexane	620 nm	$k_2 = 3 \times 10^9$,,	[40]

The reason for this 1000-fold difference in rates in clearly that the interaction between $Cr(CO)_5$ and cyclohexane is much stronger, so that displacement by CO is much slower. This is also strikingly demonstrated in an experiment with cyclohexane-doped Fcyc (10^{-2}M). The initially generated signal at 620 nm rapidly decays as a band at ca. 510 nm grows in, due to formation of $Cr(CO)_5$...cyclohexane. From the rate of decay as a function of cyclohexane concentration the rate constant can be evaluated [40]:

$$Cr(CO)_6 \Rightarrow \text{"}Cr(CO)_5\text{"} + cyc \Rightarrow Cr(CO)_5...cyc$$

perfluoromethylcyclohexane $k_2 = 2 \times 10^9\ dm^3\ mol^{-1}\ s^{-1}$

The coincidence of the visible bands of "$Cr(CO)_5$" in gas [24], Ne matrix [19] and Fcyc [40,41] might suggest that the interactions are the same and hence essentially zero. It is also important to note that in Fcyc, without CO, there is rapid formation of dimer $Cr(CO)_5Cr(CO)_6$ by reaction of the fragment with parent. The dimer is not seen in cyclohexane presumably because of the formation of $Cr(CO)_5$...cyc.

These experiments are certainly consistent with the solution generation of either naked or very weakly complexed $Cr(CO)_5$. However this interpretation relies on broad visible bands and clearly IR would be more structurally informative.

3.2. IR DETECTION

The simplest experiment is to attempt to generate $Cr(CO)_5$ in a fluid solution at low temperature where the $Cr(CO)_5$ will complex with some substrate sufficiently strongly to permit conventional IR spectroscopy. As mentioned above, when discussing early flash photolysis experiments, impurities in hydrocarbon solvents can determine the chemistry. Thus in "fairly" pure methylcyclohexane at -78C, photolysis of $Cr(CO)_6$ yields ν(CO) IR bands at 1953 and 1927 cm^{-1} which in a few minutes change to 1940 and 1908 cm^{-1} (plus 2073 cm^{-1}, the precursor of which was too weak to see) [42]. The 1953 and 1927 cm^{-1} bands are close to the bands observed in early ([2088], 1955, 1928 cm^{-1} [4]) and later ([2087], 1953, 1925 cm^{-1} [43]) low-temperature glass experiments. These bands are assigned to $Cr(CO)_5$...MCH. In the fluid solution this species converts into $Cr(CO)_5$...impurity with bands at 2073, 1940 and 1908 cm^{-1}, the impurity probably being aromatic or an olefin or even H_2O. Employing a special purification technique [44] for the methylcyclohexane, it was possible to generate $Cr(CO)_5$...MCH which was stable for an hour at -78C before slowly converting back to $Cr(CO)_6$.

Perhaps even more striking is that photolysis of $Cr(CO)_6$ in either liquid Xe or Xe-doped liquid Kr at -98C produces a species with a lifetime of about 2 s and with IR bands at 2086, 1960 and 1934 cm^{-1} [45]. This species is clearly $Cr(CO)_5$...Xe since $Cr(CO)_5$ in a Xe matrix has bands at 2086, 1956 and 1929 cm^{-1}. This experiment illustrates the value of the matrix visible absorption data; the bands of $Cr(CO)_5$ in Xe and CH_4 are at 482 and 489 nm respectively and the implication is that the weak complexes $Cr(CO)_5...CH_4$ and $Cr(CO)_5$...Xe should not be very different in stability.

However by the same argument as presented earlier, definite proof of the species present in solution requires isotopic IR data. In elegant experiments employing fast time-resolved IR spectroscopy, Schaffner and colleagues [46] were able to observe the ν(CO) bands of $Cr(CO)_5$...cyc, but the picture was obscured by the presence of impurities. In later experiments [47], $Cr(CO)_5$...cyc was shown to have bands at 1960 and 1937 cm^{-1} (the expected high frequency band at ca. 2090 cm^{-1} was too weak to observe). Moreover starting with ^{13}CO-enriched $Cr(CO)_6$, and employing EFFF arguments assisted by Timney's method [48], the structure of $Cr(CO)_5$...cyc was conclusively shown to be C_{4v} as expected. The rate constant for the back reaction with CO was 3.6 x 10^6 dm^3 mol^{-1} s^{-1}, compared with Kelly's value [16] from visible flash photolysis of 3 x 10^6 dm^3 mol^{-1} s^{-1}. An interesting feature of these experiments was that unless the cyclohexane was highly pure the transient reacted to give $Cr(CO)_5...H_2O$. This type of experiment has also been important in observing unstable species such as $Cr(CO)_5(N_2)$ [49] and $Cr(CO)_5(H_2)$ [50], based on IR spectra obtained in either matrices [8] or liquid noble gases [51,52].

4. The $Cr(CO)_5$...Q Bond Energies

Much has been said so far about the spectroscopic and kinetic evidence for the interaction between $Cr(CO)_5$ and, particularly, saturated hydrocarbons. We now consider important measurements of this interaction in both solution and gas phase.

4.1. SOLUTION PHOTOACOUSTIC MEASUREMENTS

In photoacoustic calorimetry applied to $Cr(CO)_6$, a laser pulse dissociates $Cr(CO)_6$ and a microphone attached to the cell monitors thermal changes generated as acoustic waves. Thus, for example, ΔH can be obtained for the reaction:

$$Cr(CO)_6 \xrightarrow[\text{alkane}]{h\nu} Cr(CO)_5...alkane + CO$$

Assuming that $Cr(CO)_6$ and $Cr(CO)_5$...alkane have similar solvation energies then the difference between ΔH for the above reaction and ΔH for the gas phase reaction,

$$Cr(CO)_6 \Rightarrow Cr(CO)_5 + CO$$

will be a measure of the $Cr(CO)_5$...alkane bond energy.

Employing a laser pyrolysis technique, Smith and colleagues [53] obtained the gas phase dissociation energy values for $M(CO)_6$ (M = Cr, Mo and W; 36.8, 40.5 and 46.0 kcal mol^{-1} respectively). Using the photoacoustic technique, first Peters [54] and then Burkey [55] have measured the solution ΔH values in several hydrocarbons:

Solvent	heptane	pentane	isooctane	cyclohexane
ΔH(kcal mol^{-1})	27.0[54]	27.9[55]	25.8[55]	24.2[55]
	27.2[55]			

Thus for $Cr(CO)_5$...alkane, interaction is of the order of 10 kcal mol^{-1} (ie, ca. 37 - 27), which is a very substantial number. If this is really a measure of agostic interaction, then, given that the visible absorption bands of $Cr(CO)_5$...Xe and $Cr(CO)_5...CH_4$ are so close, it will be fascinating to measure the dissociation energy of $Cr(CO)_5$...Xe which can have no agostic interaction.

4.2. GAS PHASE ENERGETICS BY TRIR

As mentioned above (section 2.2) gas phase TRIR experiments can provide energetic data on $M(CO)_5$/alkane interactions. Unfortunately there are no data for Cr, but those for W are highly relevant.

From the kinetic behaviour of the decay of "naked" $W(CO)_5$ and the growth of $W(CO)_5$...alkane it was possible to measure both the equilibrium constants, and their variation in temperature for the reactions [28]:

$$W(CO)_5 + alkane \Leftrightarrow W(CO)_5...alkane$$

Hence values of the dissociation energy were obtained:

Alkane	ethane	pentane	cyclohexane	(CH_3F)	
$W(CO)_5$/alkane energy (kcal mol^{-1})	7.4	10.6	11.6	(11.2)	(± ca.2-3)

The value for $W(CO)_5$...pentane (10.6 kcal mol^{-1}) is close to the value for $Cr(CO)_5$...pentane (8.9 kcal mol^{-1}) obtained by completely different photoacoustic calorimetry [55]. What was very striking about the gas phase TRIR experiments was that the interaction energies of $W(CO)_5$ with CH_4 and CF_4 appeared to be too low to see any signal due to $W(CO)_5...CH_4$ or $W(CO)_5...CF_4$; hence it was only possible to put an upper limit on the bond energies, <5 kcal mol^{-1}. It is not clear why $W(CO)_5...CH_4$ and $W(CO)_5...C_2H_6$ (and hence presumably the corresponding Cr species) should be so different. Further experiments in solution are needed.

5. $Cr(CO)_5$ - picosecond and femtosecond studies

The solution flash photolysis experiments described so far have been limited to a time resolution of, at best, ca. 5 ns (UV/visible detection) and 10 µs (IR detection). These show that on these timescales the first detectable species is $Cr(CO)_5$...solvent. There is strong evidence from matrix experiments [56] that the initial stages of the photochemistry are rather complicated: we shall return to this shortly, but there have been several attempts to observe the early stages with better time resolution. We now consider some of these experiments.

5.1. UV/VISIBLE EXPERIMENTS

Peters and colleagues [57] concluded that in both cyclohexane and THF, the solvated species ($Cr(CO)_5$...S) is produced in less than 25 psec. Simon and Xie [58] showed that the $Cr(CO)_5$...cyclohexane complex formed in a time less than the resolution of their equipment - 0.8 ps; a longer time (2.8 ps) was observed for $Cr(CO)_5$...MeOH, reflecting a longer solvent reorganisation time. In very revealing experiments, Joly and Nelson [59] irradiated $Cr(CO)_6$ in methanol with 100 femtosec pulses (at 308 nm) and probed subsequent absorption at several wavelengths, 410, 440, 480 and 500 nm. The interpretation of the data is difficult because the absorption of the excited state of $Cr(CO)_6$ overlaps that of the initial product but the conclusion is that the initially excited $Cr(CO)_6$ ($^1T_{1g}$) is purely dissociative and forms "naked" $Cr(CO)_5$ in about 300 fs; this then reacts with solvent to form $Cr(CO)_5$(MeOH) with a risetime of about 1.6 ps, similar to the observations of Simon and Xie [58]. The data "show absorption which is strongest to the blue of 480 nm after the initial dissociation is complete and before the solvent coordinates". This might at first sight be surprising since the band of naked $Cr(CO)_5$ is expected at around 620 nm; we return to this very important point shortly. Another significant observation in this work is that the visible band of $Cr(CO)_5$(MeOH) shifts to the blue over a period of some 70 ps and it is suggested that this may be because $Cr(CO)_5$(MeOH) is initially formed vibrationally hot and cools over this period. This effect has also been observed by Lee and Harris [60]. The transient absorption of $Cr(CO)_5$...cyclohexane was probed at 622 nm which is at the red end of the band maximum of ca.500 nm; the signal at 622 decays with a time constant of ca. 21 ps as the absorption maximum shifts to the blue. The obvious explanation of this is that $Cr(CO)_5$...cyclohexane is formed vibrationally hot and hence the electronic transition to the excited state is at lower energy. As the molecule cools the band maximum shifts to the blue; probing in the blue at 500 nm shows the corresponding grow-in, with a time-constant of ca. 17 ps which is considerably longer than the <0.8 ps of [58]. Similar results were obtained in THF [60], except that at 622 nm the decay is faster in THF than in cyclohexane (9 v. 21 ps), presumably because the maximum absorption of $Cr(CO)_5$...THF is to the blue of $Cr(CO)_5$...cyclohexane (450 v. 500 nm) and hence at 622 nm the THF complex is vibrationally hotter by 2220 cm^{-1} than the cyclohexane complex. Confirmation of this comes from probing the $Cr(CO)_5$...THF at longer wavelength - ie at a wavelength associated with an even hotter molecule - when the decay rate dropped to 4 ps.

These results can be summarised:

	hν			
$Cr(CO)_6$	=>	naked $Cr(CO)_5$	=>	$Cr(CO)_5$...solvent
	<0.5 ps	(480 nm)	<2ps	(λ_{max} shifts to blue in 4 to 70 ps depending on degree of vibrational heating)

There are two immediate questions:
(i) Is the vibrational cooling expected?
(ii)Why does naked $Cr(CO)_5$ absorb at ca.480 nm and not ca.620 nm?

5.2. VIBRATIONAL RELAXATION

Direct measurement of vibrational relaxation (ie cooling) of metal carbonyls has been done in elegant experiments by Heilwell and colleagues [61]. A metal carbonyl in solution is irradiated in one of its ν(CO) IR bands with an intense very short pulse of IR: this excites the molecule from v = 0 to v = 1. The rate of return to v = 0 is monitored by a second IR source. In this way it has been shown that the lifetimes of "hot" $Cr(CO)_6$ in CCl_4 and n-hexane are 440 and 145 ps respectively. Hence it is not surprising that "hot" $Cr(CO)_5$...Q takes many ps to cool. Other evidence comes from gas phase experiments.

As described above, pulsed laser photolysis of metal carbonyls in the gas phase produces carbonyl fragments. Yardley [62] was the first to note, for $Fe(CO)_5$, that a single photon can generate lower fragments than $Fe(CO)_4$. This observation has been confirmed using TRIR on this [63] and other carbonyls, particularly those of Cr, Mo and W [64]. The degree of fragmentation depends on the laser energy; the results for $Cr(CO)_6$ can be summarised [65]:

Laser (wavelength nm)	XeF(351)	XeCl(308)	KrF(248)	KrCl(222)	ArF(193)
Fragments	5 (4)	5 4	5 4 (3)	4 3	4 3 (2)

Here 5, 4, 3 and 2 represent $Cr(CO)_5$, $Cr(CO)_4$, $Cr(CO)_3$ and $Cr(CO)_2$ respectively, **bold** and () represent most important and least important species. This variation follows from the excess energy in one photon at each laser frequency. (By contrast in solution only $Cr(CO)_5$ is produced because of the very rapid dissipation of the excess energy.) Thus a great deal of energy is dumped into the fragments and, assuming this is manifest in population of high vibrational levels of the ν(CO) stretching modes, anharmonicity will ensure that immediately after the flash the ν(CO) bands will be broad, and that as the fragments cool the bands will narrow and shift in frequency to the blue. This is exactly what is observed, depending on the conditions, over a period of the order of a microsecond.

Further evidence of the importance of vibrational excitation is provided by Hopkins's [68] beautiful picosecond time-resolved resonance Raman (TR^3) experiments. Two 5 ps pulses at 266 nm are employed: the first dissociates $Cr(CO)_6$ in cyclohexane; the second probes the RR spectrum of $Cr(CO)_5$ (naked or solvated). **Stokes** signals (ie v = 0 to v = 1 via a virtual excited state) at 381 cm^{-1} (M-C stretch) and 1935 cm^{-1} (C-O stretch) grow in with a time scale of approximately 100 ps. More importantly an **anti-Stokes** signal (v = 1 to v = 0) appears at 381 cm^{-1}, within the time-resolution of the instrument, and decays away, again over 100 ps. This compares closely with the direct IR measurement of the v = 1 to v = 0 decay for $Cr(CO)_6$ in n-hexane [61]. The longer decay time observed in TR^3 compared with visible detection (ie 100 v. 4-22 ps) is presumably due to the fact that TR^3 monitors only v = 1 to v = 0, whereas the visible experiments monitor decay from much higher vibrational levels.

TR^3 at 266 nm should be equally sensitive for detection of both "naked" $Cr(CO)_5$ and $Cr(CO)_5$...cyclohexane. Moreover the matrix experiments suggest that, although the visible band is sensitive to coordination in the empty site, the IR spectrum - at least of the v(CO) bands - is little affected. Thus TR^3 will not distiguish between them, but in the light of the UV/vis experiments it is likely that the experiment is probing the vibrational relaxation of the **solvated** species.

The experiments have concentrated on $Cr(CO)_6$ as a source of $Cr(CO)_5$; presumably relaxation problems would be different if the fragment were produced from a different source. There are some very relevant comparative experiments with $W(CO)_5L$ species (L = CO, pyridine, piperidine) [69,70,71]. With $W(CO)_6$ in cyclohexane the $W(CO)_5$...S species is produced in its vibrationally relaxed ground state in a time similar to $Cr(CO)_5$...S from $Cr(CO)_6$. However the relaxed $W(CO)_5$...S is produced from the pyridine and piperidine complexes in a shorter time, probably less than 5 ps. The situation is complex, but one possible explanation [72] is that the difference in vibrational relaxation time is related to the amount of energy deposited in the dissociating ligand, CO, pyridine or piperidine. With the organic ligands, more energy may be removed and hence less left in the $W(CO)_5$ fragment. Clearly the relaxation problem merits further close study if the role of the solvent is to be fully understood.

5.3. VISIBLE BAND OF 'NAKED' $Cr(CO)_5$

Before turning to fast IR experiments in solution we need to consider why the naked $Cr(CO)_5$ absorbs at ca. 480 nm. Interestly the most direct answer comes from matrix experiments. Experiments on $Cr(CO)_6$ and $Cr(CO)_5$ in matrices with polarised photolysis light and polarised spectroscopy, in both the UV/visible and IR, were best interpreted on the (simplified) scheme [20,56]:

$$\underset{^1A_{1g}}{Cr(CO)_6} \overset{h\nu}{\Rightarrow} \underset{^1T_{2g}/^1T_{1g}}{[Cr(CO)_6]^*} \Rightarrow \underset{^1E\,(C_{4v})}{[Cr(CO)_5]^*} + CO \Rightarrow \underset{^1E'(D_{3h})}{Cr(CO)_5} \Rightarrow \underset{^1A_1\,(C_{4v})}{Cr(CO)_5}$$

That is, following excitation of $Cr(CO)_6$, CO is ejected and $Cr(CO)_5$ is produced with a C_{4v} structure but in an **excited** electronic state [73]. This species achieves the ground state by travelling downhill in energy through the D_{3h} structure, ie via a Berry pseudorotation as shown in figure 2 [56]. Since the D_{3h} structure can rearrange to C_{4v} via three paths the empty coordination site of the $Cr(CO)_5$ fragment will either finish facing the way it starts or in two other directions. In a mixed Ne/Xe matrix this would generate $Cr(CO)_6$, $Cr(CO)_5$...Ne and $Cr(CO)_5$...Xe. Figure 2 also explains the mixed matrix photochemistry in Ne/2%Xe (section 2.1; reference [19]); the ground state C_{4v} fragment (e.g. $Cr(CO)_5$...Ne) is excited via the ^1E state to the D_{3h} structure and from there to one of three orientations, hence possibly coming to rest opposite either a Xe or a Ne atom. Since $Cr(CO)_5$...Ne and $Cr(CO)_5$...Xe have different visible absorption bands the exchange can be driven one way or the other by light of appropriate wavelength. This behaviour is a possible explanation for the quantum yield for $M(CO)_6$ solution photochemistry being close to 2/3 [74]; this proposal has been supported by recent experiments [75,76].

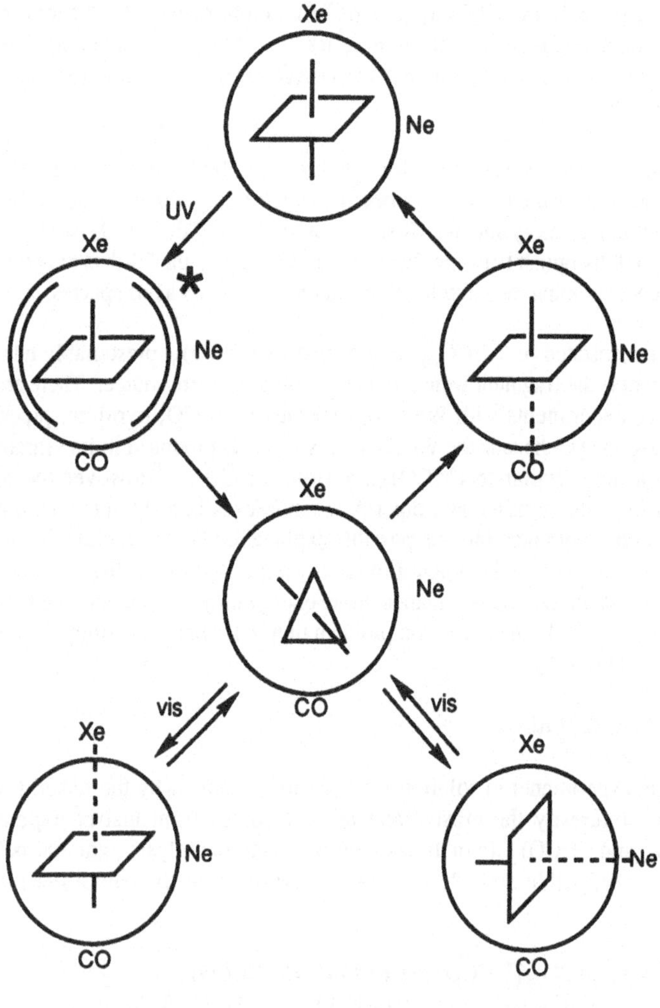

Figure 2. Schematic representation of photochemical behaviour of $Cr(CO)_5$ fragment in a mixed (Ne/Xe) low temperature matrix.
* represents the C_{4v} fragment in the excited ^1E state.

The fact that the $Cr(CO)_5$ is claimed to be produced in an electronic excited state has been the subject of some controversy [77], but the *ab initio* calculations support this behaviour [33]. The calculations also indicate that the visible absorption band of the C_{4v} ground state (ie $^1A_1 \rightarrow {}^1E$) is at longer wavelength than that of the excited C_{4v} state ($^1E \rightarrow {}^1B_2$). Immediately this suggests that the band observed at 480 nm [59] is due to C_{4v} $Cr(CO)_5$ in a ^1E excited state. It could also in principle be an absorption of the D_{3h} structure but these absorptions are calculated to be at longer wavelength.

The whole of the above discussion has assumed a photochemical process entirely through singlet states. If the photochemistry involves triplet states then in the matrix interpretation it makes almost no difference except that the final step from D_{3h} to C_{4v} ground state will not be a smooth progression but a triplet to singlet relaxation. However it has been claimed [78] that the solution photochemistry of $M(CO)_6$ molecules is only consistent with a triplet mechanism. Since the absorption in $Cr(CO)_6$ at 308 nm is undoubedly to a ^1T state, there would have to a very rapid intersystem crossing to the triplet manifold to match the observations in the fs experiments [59]. The absorption at 480 nm would then be either from the ^3E state of the C_{4v} structure or the $^3A_2'$ state of the D_{3h} geometry.

5.4. PICOSECOND IR STUDIES

Ideally fast time-resolved IR should be able to probe the very early stages of the photochemistry and distinguish between C_{4v} avd D_{3h} structures. The problem is that the fragment is likely to be vibrationally very hot and hence the IR bands will be broad, making distinction between the options difficult. Flash photolysis (266 nm) of $Cr(CO)_6$ in cyclohexane was monitored on timescales down to about 20 ps at different ν(CO) frequencies. A signal at 1980 cm^{-1} grew in over 100 ps and this was originally assigned to naked $Cr(CO)_5$ [79]. This would have contradicted the visible detection experiments. However more recent work [80] suggests that this band is dominated by vibrationally hot $Cr(CO)_6$, presumably formed from geminate recombination of $Cr(CO)_5$ with CO. A signal at 1961 cm^{-1}, assigned to $Cr(CO)_5$...cyclohexane, grew in over about 150 ps, reasonably consistent with the other spectroscopic observations. To reduce the problem of high vibrational heating, Spears and colleagues have performed an ingenious experiment [81], bearing some relation to those of Langford [70, 71, 72]. $Cr(CO)_6$ was dissociated in the usual way with a 20 ps 300 nm pulse to form $Cr(CO)_5$...cyclohexane, which on the timescales of these experiments is very long-lived. Since the bond dissociation energy of this species is only ca. 10 kcal mol^{-1} and it has a a broad absorption band centred at ca. 500 nm, a second pulse at 532 nm readily dissociates this complex. A signal at 1970 cm^{-1} grows in at least as fast as the response time and then decays over 15±5 ps. Since the shift from gas to cyclohexane for the t_{1u} mode of $Cr(CO)_6$ is ca. 14 cm^{-1} it is assumed that the same shift would apply to naked $Cr(CO)_5$ and hence its e mode should be at 1986 - 14 = 1972 cm^{-1}. Hence the 1970 cm^{-1} was assigned to naked $Cr(CO)_5$ in its ground vibrational state. In view of the relaxation measurements, it is rather surprising if naked, **cold** $Cr(CO)_5$ is produced in <5 ps. However if Langford's suggestion [72] is correct, this might occur because the hydrocarbon molecule which comes off $Cr(CO)_5$...S carries with it most of the excess energy. Other transient signals were observed in these experiments and were interpreted as vibrationally excited $Cr(CO)_5$...cyclohexane formed via a different path than from naked cold $Cr(CO)_5$. Clearly the exact details of the stages from excited $Cr(CO)_6$ to vibrationally cold $Cr(CO)_5$...cyclohexane is complex and is not yet clear. Picosecond IR experiments with $Cr(CO)_6$ in perfluoroheptane [82] suggest that there is considerable vibrational overlap between species that might exist on this reaction coordinate. Experiments that might shed some light on this would be those combining the low-temperature matrix technique with time-resolved IR and we have such experiments planned.

6. Conclusion

The developing story of the $Cr(CO)_5$ fragment is a fascinating one. A host of experimental techniques have been applied to its behaviour. Although the chemistry of $Cr(CO)_5$ at timescales of a nanosecond and longer are well understood, there are still unanswered questions about the very fast processes. No doubt the $Cr(CO)_5$ fragment will continue to be a source of novel science.

7. References

[1] Strohmeier, W. (1961) Chem. Ber., **94**, 3337.
[2] Massey, A. G. and Orgel. L. E. (1961) Nature **191**, 1387.
[3] Dobson, G. R., El-Sayed, M. A., Stolz, I. W. and Sheline, R. K. (1962) Inorg. Chem., **1**, 526.
[4] Stolz, I. W., Dobson, G. R. and Sheline, R. K. (1962) J. Am. Chem. Soc., **84**, 3589.
[5] Stolz, I. W., Dobson, G. R. and Sheline, R. K. (1963) J. Am. Chem. Soc., **85**, 1013.
[6] El-Sayed, M. A. (1964) J. Phys. Chem., **68**, 433.
[7] Dobson, G. R. (1965) J. Phys. Chem., **69**, 677.
[8] In fact photolysis of $Cr(CO)_6$ species in an Ar matrix doped with N_2 gives a quite specific species $Cr(CO)_5(N_2)$ which has different spectral properties from $Cr(CO)_5$/hydrocarbon: Burdett, J. K., Downs, A. J., Gaskill, G. P., Graham, M. A., Turner, J. J. and Turner, R. F. (1978) Inorg. Chem., **17**, 523.
[9] Graham, M. A., Rest, A. J. and Turner, J. J. (1970) J. Organomet. Chem., **24**, C54.
[10] Graham, M. A., Poliakoff, M. and Turner, J. J. (1971) J. Chem. Soc. (A), 2939.
[11] Boylan, M. J., Braterman, P. S. and Fullerton, A. (1971) J. Organomet. Chem., **31**, C29. This paper also demonstrated that the claim to detect D_{3h} $Mo(CO)_5$ on melting a glass containing C_{4v} $Mo(CO)_5$ was "spurious".
[12] Graham, M. A., Perutz, R. N., Poliakoff, M. and Turner, J. J. (1972) J. Organomet. Chem., **34**, C34.
[13] McIntyre, J. A. (1970) J. Phys. Chem., **74**, 2403.
[14] Nasielski, J., Kirsch, P. and Wilputte-Steinert, L.(1971) J. Organomet. Chem., **29**, 269.
[15] Kelly, J. M., Hermann, H. and Koerner von Gustorf, E. (1973) J. Chem. Soc. Chem. Commun., 105.
[16] Kelly, J, M., Bent, D. V., Hermann, H., Schulte-Frohlinde, D. and Koerner von Gustorf, E. (1974) J. Organomet. Chem., **69**, 259.
[17] Braterman, P. S. (1975) Metal Carbonyl Spectra, Academic Press.
[18] Perutz, R. N. and Turner, J. J. (1975) Inorg. Chem., **14**, 262.
[19] Perutz, R. N. and Turner, J. J. (1975) J. Am. Chem. Soc., **97**, 4791.
[20] Turner, J. J., Burdett, J. K., Perutz, R. N. and Poliakoff, M. (1977) Pure Appl. Chem., **49**, 271.
[21] Kundig, E. P. and Ozin, G. A. (1974) J. Am. Chem. Soc., **96**, 3820.
[22] Huber, H., Kundig, E. P., Ozin, G. A. and Poe, A. J. (1975) J. Am. Chem. Soc., **97**, 308.
[23] Breckenridge, W. H. and Sinai, N. (1981) J. Phys. Chem., **85**, 3557.
[24] Breckenridge, W. H. and Stewart, G. M. (1986) J. Am. Chem. Soc., **108**, 364.
[25] Seder, T. A., Church, S. P., Ouderkirk, A. J. and Weitz, E. (1985) J. Am. Chem. Soc., **107**, 1432.
[26] Jones, L. H., McDowell, R. S. and Goldblatt, M. (1969) Inorg. Chem., **8**, 2349.
[27] Perutz, R. N. (1974) Ph.D. Thesis, University of Cambridge.
[28] Ishikawa, Y-o., Brown, C. E., Hackett, P. A. and Rayner, D. M. (1988) Chem. Phys. Letters, **150**, 506; Brown, C. E., Ishikawa, Y-o., Hackett, P. A. and Rayner, D. M. (1990) J. Am. Chem. Soc., **112**, 2530.
[29] Burdett, J. K. (1974) J. Chem. Soc. Faraday Trans. 2, **70**, 1599.
[30] Elian, M. and Hoffmann, R. (1975) Inorg. Chem., **14**, 1058.

[31] Pensak, D. A. and McKinney, R. J. (1979) Inorg. Chem., **18**, 3407.
[32] Demuynck, J., Strich, A. and Veillard, A. (1977) Nouveau J. Chem., **1**, 217.
[33] Hay, P. J. (1978) J. Am. Chem. Soc., **100**, 2411.
[34] Demuynck, J., Kochanski, E. and Veillard, A. (1979) J. Am. Chem. Soc., **101**, 3467.
[35] Rossi, A., Kochanski, E. and Veillard, A. (1979) Chem. Phys. Letters, **66**, 13.
[36] Crabtree, R. H. (1985) Chem. Rev., **85**, 245.
[37] Green, M. L. H. and O'Hare, D. (1985) Pure Appl.Chem., **57**, 1897.
[38] Various unpublished experiments in Nottingham.
[39] Saillard, J-Y. and Hoffmann, R. (1984) J. Am. Chem. Soc., **106**, 2006.
[40] Bonneau, R. and Kelly, J. M. (1980) J. Am. Chem. Soc., **102**, 1220.
[41] Kelly, J. M., Long, C. and Bonneau, R. (1983) J. Phys. Chem., **87**, 3344.
[42] Tyler, D. R. and Petrylak, D. P. (1981) J. Organomet. Chem., **212**, 389.
[43] Boylan, M. J., Black, J. D. and Braterman, P. S. (1980) J. Chem. Soc. Dalton, 1646.
[44] Murray, E. C. and Keller, R. N. (1969) J. Org. Chem., **34**, 2234.
[45] Simpson, M. B., Poliakoff, M., Turner, J. J., Maier, W. B. II and McLaughlin, J. G. (1983) J. Chem. Soc. Chem. Commun., 1355.
[46] Hermann, H., Grevels, F-W., Henne, A. and Schaffner, K. (1982) J. Phys. Chem., **86**, 5151.
[47] Church, S. P., Grevels, F-W., Hermann, H. and Schaffner, K. (1985) Inorg. Chem., **24**, 418.
[48] Timney, J. A. (1979) Inorg. Chem., **18**, 2502.
[49] Church, S. P., Grevels, F-W., Hermann, H. and Schaffner, K. (1984) Inorg. Chem., **23**, 3830.
[50] Church, S. P., Grevels, F-W., Hermann, H. and Schaffner, K. (1985) J. Chem. Soc. Chem. Commun., 30.
[51] Maier, W. B. II, Poliakoff, M., Simpson, M. B. and Turner, J. J. (1980) J. Chem. Soc. Chem. Commun., 587; Turner, J. J., Simpson, M. B., Poliakoff, M., Maier, W. B. II, and Graham. M. A. (1983) Inorg. Chem., **22**, 911.
[52] Upmacis, R. K., Gadd, G. E., Poliakoff, M., Simpson, M. B., Turner, J. J., Whyman, R. and Simpson, A. F. (1985) J. Chem. Soc. Chem. Commun., 27.
[53] Lewis, K. E., Golden, D. M. and Smith, G. P. (1984) J. Am. Chem. Soc., **106**, 3905.
[54] Yang, G. K., Peters, K. S. and Vaida, V. (1986) Chem. Phys. Letters, **125**, 566. (Note that an earlier paper - Bernstein, M., Simon, J. D. and Peters, K. S. (1983) Chem. Phys. Letters. **100**, 241 - had incorrect data.)
[55] Morse, J. M., Parker, G. H. and Burkey, T. J. (1989) Organometallics, **8**, 2471.
[56] Burdett, J. K., Grzybowski, J. M., Perutz, R. N., Poliakoff, M., Turner, J. J. and Turner, R. F. (1978) Inorg. Chem., **17**, 147.
[57] Simon, J. D. and Peters, K. S. (1983) Chem. Phys Letters, **98**, 53; Welch, J. A., Peters, K. S. and Vaida, V. (1982) J. Phys. Chem., **86**, 1941.
[58] Simon, J. D. and Xie, X. (1986) J. Phys. Chem., **90**, 6751. See also later papers on solvents where there is competition between the alkane and non-alkane ends of the molecules - Xie, X. and Simon, J. D. (1990) J. Am. Chem. Soc., **112**, 1130, and references therein.
[59] Joly, A. G. and Nelson, K. A. (1989) J. Phys. Chem., **93**, 2876.
[60] Lee, M. and Harris, C. B. (1989) J. Am. Chem. Soc., **111**, 8963.
[61] Heilwell, E. J., Cavanagh, R. R. and Stephenson, J. C. (1987) Chem. Phys. Letters, **134**, 181.

[62] Nathanson, G., Gitlin, B., Rosan, A. M. and Yardley, J. T. (1981) J. Chem. Phys., **74**, 361, 370.
[63] Seder, T. A., Ouderkirk, A. J. and Weitz, E. (1986) J. Chem. Phys., **85**, 1977.
[64] Reviewed in Weitz, E. (1987) J. Phys. Chem., **91**, 3945.
[65] Ishikawa, Y-o., Brown, C. E., Hackett, P. A. and Rayner, D. M. (1990) J. Phys. Chem., **94**, 2404, based on data in this reference, and references [25], [66] and [67].
[66] Fletcher, T. R. and Rosenfeld, R. N. (1985) J. Am. Chem. Soc., **107**, 2203.
[67] Seder, T. A., Church, S. P. and Weitz, E. (1986) J. Am. Chem. Soc., **108**, 4721.
[68] Yu, S-C., Xu, X., Lingle, R. and Hopkins, J. B. (1990) J. Am. Chem. Soc., **112**, 3668.
[69] Langford, C. H., Moralejo, C. and Sharma, D. K. (1987) Inorg. Chim. Acta, **126**, L11.
[70] Moralejo, C., Langford, C. H. and Sharma, D. K. (1989) Inorg. Chem., **28**, 2205.
[71] Moralejo, C. and Langford, C. H. (1991) Inorg. Chem., **30**, 567.
[72] Langford, C. H. private communication.
[73] $^1T_{2g}$ and $^1T_{1g}$ states correlate with C_{4v} states of $^1E + {}^1B_2$ and $^1E + {}^1A_2$ symmetry respectively; in either case the ^1E state is of lower energy. The ^1E state correlates with the $^1A_1'$ and $^1E'$ states of D_{3h} structure with $^1E'$ having the lower energy; the $^1E'$ state correlates with both ^1E and $^1A_1$ (ie excited and ground states) with C_{4v} structure so the overall effect is energy downhill. See Daniel, C., Benard, M., Dedieu, A. and Veillard, A. (1984) J. Phys. Chem., **88**, 4805.
[74] Nasielski, J. and Colas, A. (1975) J. Organomet. Chem., **101**, 215.
[75] Wieland, S. and van Eldick, R. (1990) J. Phys. Chem., **94**, 5865.
[76] Nayak, S. K. and Burkey, T. J. (1991) Organometallics, **10**, 3745.
[77] See the discussion at the end of Turner, J. J. and Poliakoff, M. (1983) ACS Symp. Ser. No. **200**, 35.
[78] Nasielski, J. and Colas, A. (1978) Inorg. Chem. **17**, 237.
[79] Wang, L., Zhu, X. and Spears, K. G. (1988) J. Am. Chem. Soc., **110**, 8695.
[80] Wang, L., Zhu, X. and Spears, K. G. (1989) J. Phys. Chem., **93**, 2.
[81] Sprague, J. R., Arrivo, S. M. and Spears, K. G. (1991) J. Phys. Chem., in press.
[82] Spears, K.G., Wang, L., Zhu, X. and Arrivo, S.M. (1990) SPIE, **1209**, 32.

Photochemistry of Organo–metal Carbonyls: Stereochemical and Catalytic Aspects

F.–W. GREVELS
Max–Planck–Institut für Strahlenchemie
Stiftstraße 34–36
D–4330 Mülheim an der Ruhr
Germany

ABSTRACT. Multiple photosubstitution of CO ligands in Group 8 and Group 6 transition metal carbonyls is investigated in the context of photocatalytic processes, such as olefin isomerization and diene hydrogenation or hydrosilylation. Emphasis is placed on four topics. (i) Quantum yields and photokinetics: the photochemical conversion of $Fe(CO)_5$, $Ru(CO)_5$, and $W(CO)_6$ into multi–substituted derivatives (using trimethyl phosphite and/or *E*–cyclooctene as the incoming ligand) is monitored by means of quantitative IR–spectroscopy; quantum yields for the individual steps are evaluated on the basis of the appropriate photokinetic formalism, which accounts for mutual internal light filtering and includes both consecutive and parallel reactions, if necessary. (ii) $M_3(CO)_{12}$ (Fe, Ru, Os) cluster photochemistry. (iii) Catalytic aspects: as specific examples, the $Fe(CO)_3$ and the $Cr(CO)_3$ groups are recognized as the repeating units in the catalytic cycles of alkene isomerization and diene hydrogenation or hydrosilylation, respectively; photogenerated labile complexes containing the respective $M(CO)_3$ units are employed as reservoir complexes for investigating the catalytic processes separated from the photo–induction period. (iv) Structure and bonding: the single–faced π–acceptor character of olefin ligands is the key factor, which governs the structures and stabilities of olefin–substituted metal carbonyls in the trigonal–bipyramidal (d^8) and octahedral (d^6) geometries and, thus, largely determines the catalytic activities of such compounds; for mechanistic studies on labile type olefin complexes advantage is taken of the exceptional coordination properties of *E*–cyclooctene, which is quite a unique olefin in this respect.

1. Introduction

Carbon monoxide is among the most versatile auxiliary ligands in organometallic chemistry. Although the CO ligand is a strong π–acceptor, it shows remarkable flexibility in its demand for π–back donation from low–valent metals and, thus, tolerates a wide range of other ligands with varying σ–donor/π–acceptor properties. Transition metal carbonyl fragments not only accomodate almost every kind of organic moieties (with even and odd numbers of electrons, such as mono–, di–, and trienes, arenes, and alkyl, allyl, dienyl and trienyl groups), but also act as templates in many organometallic reactions, both stoichiometric and catalytic in nature.

Photochemical techniques have their regular place in the repertoire of methods in organo–metal carbonyl chemistry [1]. It is commonly observed that, compared with their ground state properties, the reactivity of such compounds toward loss of CO (and/or other ligands) is greatly enhanced upon electronic excitation to a ligand field excited state. Thus,

E. Kochanski (ed.), Photoprocesses in Transition Metal Complexes, Biosystems and Other Molecules. Experiment and Theory, 141–171.

UV–vis irradiation of metal carbonyls in the presence of potential ligands and substrate molecules frequently is the method of choice for the synthesis of substituted derivatives or for the generation of catalytically active species [2]. As an important point to note, the photochemical events themselves in most cases show little, if any, temperature dependence, whereas subsequently occurring thermal steps may be slowed down considerably upon lowering the temperature. In consequence of this, photochemical reactions are particularly well suited for the preparation of thermally labile complexes, which would not survive the more drastic conditions of processes initiated by heating in the dark. Highly reactive intermediates, too unstable for isolation, may be accumulated in low–temperature solutions and matrices, such that spectroscopic characterization becomes feasible [3].

Multiple CO photosubstitution in Group 6 and Group 8 metal carbonyls by C=C units or other ligands is among the topics dealt with in detail in this article, whereby emphasis is placed on the photokinetics and evaluation of quantum yields, the identification of reactive intermediates, the structural features of the products, and on their reactivity and catalytic activity.

2. Photoreactions of Group 8 Metal Carbonyls

2.1. PHOTOSUBSTITUTION OF CO GROUPS IN PENTACARBONYLIRON

The photochemistry of pentacarbonyliron is fascinating with regard to the multifariousness of synthetic and catalytic applications and from the mechanistic point of view as well. A wide range of substituted $Fe(CO)_{5-n}L_n$ derivatives, with structures varying between trigonal–bipyramidal and square–pyramidal, is readily generated upon extended photolysis of $Fe(CO)_5$ in the presence of the respective ligand(s). The initial step involves loss of CO from the $^3E'$ excited state of $Fe(CO)_5$ with formation of the $Fe(CO)_4$ fragment [4,5], the C_{2v} structure of which has been elucidated on the basis of its CO stretching vibrational pattern in the gas phase [6], in low–temperature matrices [7], and in solution [8,9]. In the absence of suitable ligands L this fragment reacts with excess $Fe(CO)_5$ to form $Fe_2(CO)_9$ with a quantum yield near unity [10].

2.1.1. *Multiple Substitution of CO by a Phosphorus Ligand.* In the trigonal–bipyramidal derivatives $Fe(CO)_{5-n}L_n$ *n*–donor ligands such as phosphines or phosphites preferentially occupy an axial position. Complexes of this type are readily formed upon irradiation of $Fe(CO)_5$ in the presence of the respective ligand (Scheme 1).

There is some evidence [11] that the photoreaction of $Fe(CO)_5$ with phosphorus ligands surprisingly yields both $Fe(CO)_4L$ and $Fe(CO)_3L_2$ as primary products. Indeed, this early report is confirmed by the observation that single–flash excitation (309 nm) of $Fe(CO)_5$ in the presence of excess trimethyl phosphite yields a mixture of $Fe(CO)_4L$ (35%), $Fe(CO)_3L_2$ (57 % !), and $Fe(CO)_2L_3$ (8 %). This result deserves particular attention, as it suggests that the di–substituted and tri–substituted derivatives are formed following the absorption of a single photon by the parent $Fe(CO)_5$!

This led us to investigate the system in more detail [12]. Figure 1 shows the formation of the substituted $Fe(CO)_{5-n}[P(OCH_3)_3]_n$ derivatives (n = 1, 2, 3) and the disappearance of the parent complex $Fe(CO)_5$ upon irradiation (λ = 302 nm) in the presence of a 20–fold

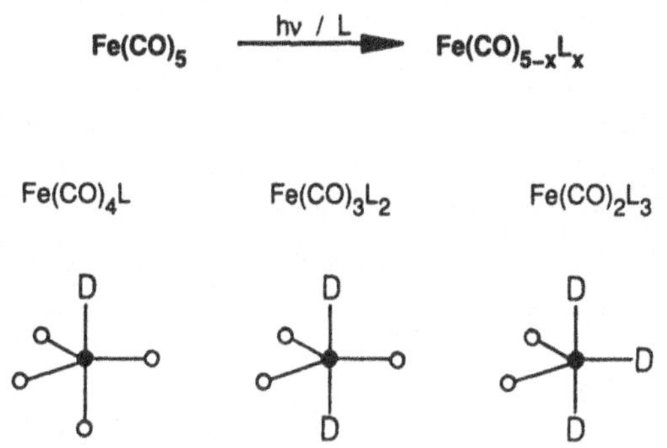

Scheme 1

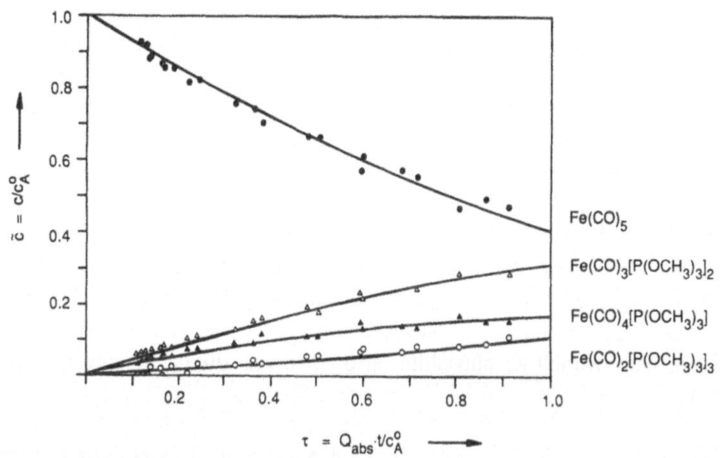

Figure 1: Multiple substitution of CO groups in $Fe(CO)_5$ by trimethyl phosphite (20–fold excess) upon irradiation at λ = 302 nm (the curves are computed on the basis of the six quantum yields given in Scheme 2 and the ε–values of the four complexes).

excess of trimethyl phosphite, monitored by means of quantitative IR spectroscopy. In a series of complementary experiments, starting with $Fe(CO)_4[P(OCH_3)_3]$ and $Fe(CO)_3[P(OCH_3)_3]_2$ under analogous conditions of irradiation, the secondary photosubstitution reactions were separately studied. In summary, the results clearly demonstrate that both consecutive and parallel formation of the multi–substituted complexes takes place (Scheme 2). The set of six quantum yields is evaluated on the basis of the following kinetics, which account for mutual internal light filtering ($x_i = \varepsilon_i \cdot c_i / \Sigma(\varepsilon_j \cdot c_j)$, "molar light fraction") by all four complexes involved in the system:

$$-dc_A/dt = (\Phi_{AB} + \Phi_{AC} + \Phi_{AD}) \cdot x_A \cdot Q_{abs}$$
$$dc_B/dt = [\Phi_{AB} \cdot x_A - (\Phi_{BC} + \Phi_{BD}) \cdot x_B] \cdot Q_{abs}$$
$$dc_C/dt = [\Phi_{AC} \cdot x_A + \Phi_{BC} \cdot x_B - \Phi_{CD} \cdot x_C] \cdot Q_{abs}$$
$$dc_D/dt = [\Phi_{AD} \cdot x_A + \Phi_{BD} \cdot x_B + \Phi_{CD} \cdot x_C] \cdot Q_{abs}$$

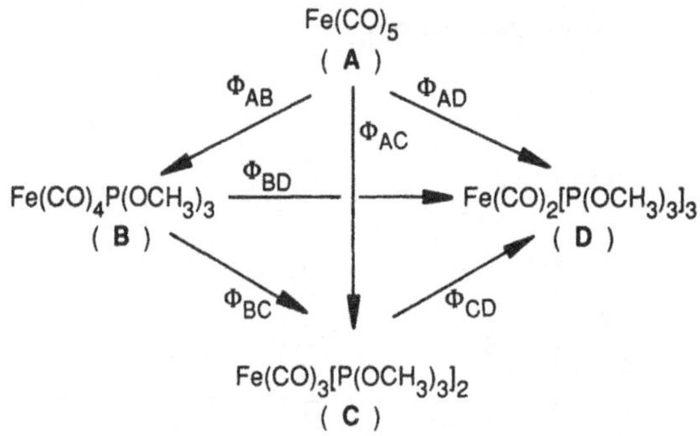

$\Phi_{AB} = 0.29$
$\Phi_{AC} = 0.43$
$\Phi_{AD} = 0.06$
$\Phi_{BC} = 0.54$
$\Phi_{BD} = 0.13$
$\Phi_{CD} = 0.43$

$\Phi_{-A} = 0.78$

Scheme 2

Integration of these equations yields the concentrations of the starting material and the three products [13] as implicit functions of the integrated light absorption [13], the molar absorbance data of the four compounds, and the six quantum yields. By using an iterative computer program the quantum yields are evaluated from a sufficiently large number of experimental data sets, whereupon a plot of the concentrations *vs* the integrated light absorption can be obtained (solid curves in Figure 1, which demonstrate an excellent fit). The ratio $\Phi_{AB} : \Phi_{AC} : \Phi_{AD}$ is in accord with the product ratio observed in the single flash experiment, thus indicating that the di-substituted complex, $Fe(CO)_3[P(OCH_3)_3]_2$, is indeed the predominant product resulting from the absorption of a single photon by $Fe(CO)_5$. Moreover, the secondary photoreaction of the tetracarbonyl complex $Fe(CO)_4[P(OCH_3)_3]$ with trimethyl phosphite also involves direct di-substitution to an albeit lesser extent.

2.1.2. *Multiple Substitution of CO by an Olefin.* Olefin-substituted carbonyliron complexes, like their counterparts with *n*-donor ligands, are conveniently prepared photochemically from $Fe(CO)_5$ (Scheme 3). As a common feature of the two types of compounds the coordination geometries are essentially trigonal-bipyramidal. However, in contrast to the

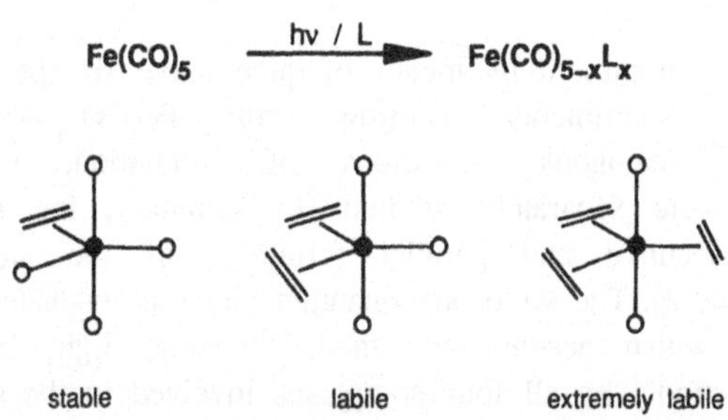

Scheme 3

preference of *n*–donor ligands for the axial positions, mono–olefins occupy the coordination sites in the equatorial plane [14], which is dictated by their single–faced π–acceptor character [15]. Moreover, the iron–olefin bonds in the series of $Fe(CO)_{5-n}(\eta^2\text{-olefin})_n$ complexes show a dramatic decrease in stability for $n \geq 2$ [16,17], apparently because the metal fails to meet the demand of the olefins for π–back donation. The lability of $Fe(CO)_3(\eta^2\text{-alkene})_2$ complexes is essential to the alkene isomerization with carbonyliron catalysts *(vide infra)* [11,16–19], which involves the $Fe(CO)_3$ group as the repeating unit in the catalytic cycle. However, quantitative investigations into the photochemical generation of such compounds are severely hampered by partial decomposition due to the low stability.

Fortunately, one particular olefin (*E*–cyclooctene = ECO) is available, which forms exceptionally stable complexes [20,21] with a wide variety of transition metals and, thus, allows the photochemical conversion of $Fe(CO)_5$ into the mono– and di–substituted $Fe(CO)_3(\eta^2\text{-olefin})_2$ complexes to be monitored by means of IR–spectroscopy on a quantitative level (Figure 2) [22]. Further photosubstitution of CO by a third olefin is possible, but even with *E*–cyclooctene the $Fe(CO)_2(\eta^2\text{-olefin})_3$ complex is too unstable for detection under the conditions employed in this study. The liberated carbon monoxide, if not rigorously removed, reacts with $Fe(CO)_2(\eta^2\text{-ECO})_3$ at ambient temperature to replace one of the three olefin ligands. Thus, $Fe(CO)_3(\eta^2\text{-ECO})_2$ appears to be the ultimate product resulting from the photoreaction of $Fe(CO)_5$ with excess *E*–cyclooctene (Scheme 4). The quantum yields are evaluated from the experimental data by a procedure [12] analogous to that applied to the reaction with trimethyl phosphite (*vide supra*), but appropriately simplified.

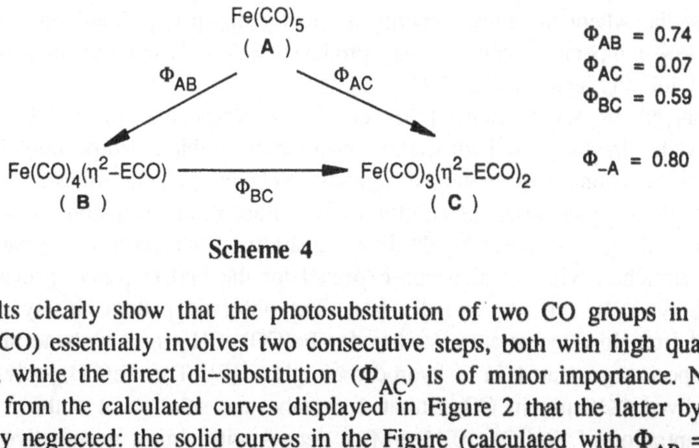

Scheme 4

The results clearly show that the photosubstitution of two CO groups in $Fe(CO)_5$ by the olefin (ECO) essentially involves two consecutive steps, both with high quantum yields (Φ_{AB}, Φ_{CB}), while the direct di–substitution (Φ_{AC}) is of minor importance. Nevertheless, it is evident from the calculated curves displayed in Figure 2 that the latter bypath cannot be completely neglected: the solid curves in the Figure (calculated with $\Phi_{AB} = 0.74$, $\Phi_{AC} = 0.07$, and $\Phi_{BC} = 0.59$) fit the experimental data points distinctly better than the dotted curves (calculated with $\Phi_{AB} = 0.81$, $\Phi_{AC} = 0$, and $\Phi_{BC} = 0.59$ [22]). Accordant with this result, single–flash excitation (309 nm) of $Fe(CO)_5$ in the presence of *E*–cyclooctene yields $Fe(CO)_4(\eta^2\text{-ECO})$ as the far predominant product, but in addition ca. 10 % $Fe(CO)_3$–$(\eta^2\text{-ECO})_2$ is also formed.

2.1.3. *Mechanistic Aspects.* The large proportion of initial di–substitution in the photoreaction of $Fe(CO)_5$ with a phosphorus ligand [11,12] is still somewhat difficult to

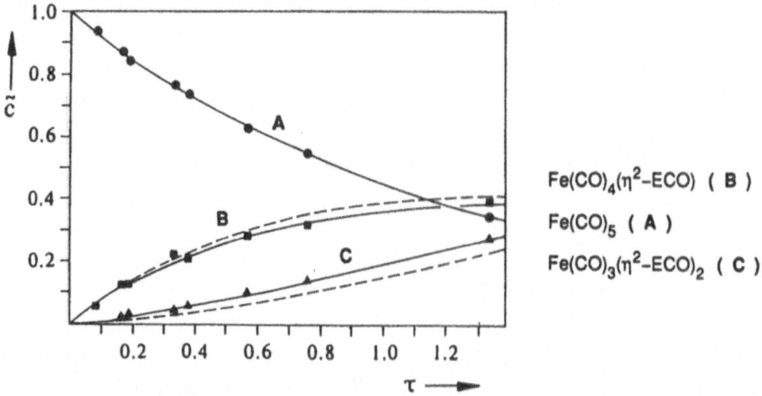

Figure 2: Multiple substitution of CO groups in $Fe(CO)_5$ by *E*–cyclooctene (ECO; 20 to 40–fold excess) upon irradiation at λ = 302 nm (the solid curves are computed on the basis of the three quantum yields given in Scheme 4 and the ε–values of the three complexes).

rationalize. It might be argued that irradiation at wavelengths around 300 nm provides sufficient energy (ca. 95 kcal/mol) for breaking more than one metal–CO bond, i.e., to generate a mixture of $Fe(CO)_3$ and $Fe(CO)_4$ fragments, which then could be trapped by the ligand. However, this would not explain the entirely different results obtained with the olefin as the trapping agent. Moreover, multiple loss of CO from $Fe(CO)_5$ following the absorption of a single photon has only been observed in the gas phase [6], but not in condensed media, where the excess energy is rapidly dissipated. Continuous irradiation of $Fe(CO)_5$ in low–temperature matrices has produced $Fe(CO)_3$, but this involves secondary photolysis of initially generated $Fe(CO)_4$ [7].

Time–resolved IR spectroscopy [23] (cf. [3]) is employed for the detection of the solvated $Fe(CO)_4$ fragment in hydrocarbon solution at ambient temperature [8,9] and for monitoring its reactions with potential ligands (Scheme 5). The transient IR spectrum displayed in Figure 3 is observed shortly (3 μs) after flash excitation (λ = 309 nm) of $Fe(CO)_5$ in cyclohexane solution [9]. It shows, apart from depletion of the parent complex, the four CO stretching vibrational bands expected for the $Fe(CO)_4$·solv species with a C_{2v} structure [A_1 (vw), B_1 (st), A_1 (m), B_2 (st)]. The observed pattern closely resembles the data reported from low–temperature studies for $Fe(CO)_4 \cdot CH_4$ in a methane matrix and for the related species generated in a hydrocarbon glass [7]. The $Fe(CO)_4$·solv fragment is rather short–lived (≤10 μs) in CO–saturated cyclohexane solution at ambient temperature. At low concentration of $Fe(CO)_5$ (0.15 mM) it decays with almost quantitative regeneration of the parent complex, but the formation of $Fe_2(CO)_9$ as a permanent product competes effectively as the concentration of $Fe(CO)_5$ increases (≥ 1 mM).

The latter process is completely suppressed in the presence of *E*–cyclooctene or trimethyl phosphite, both of which react even more rapidly than CO with the photogenerated $Fe(CO)_4$·solv fragment, as indicated by the almost instantaneous appearance of the respective $Fe(CO)_4L$ derivatives. However, in the case of trimethyl phosphite

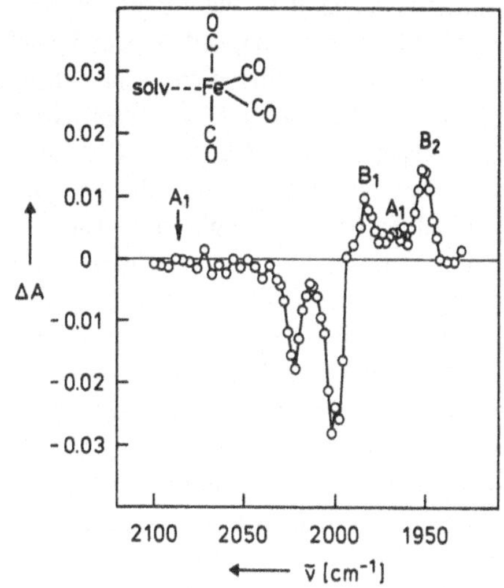

Figure 3: Transient IR spectrum observed shortly (ca. 3 μs) after flash photolysis (λ = 309 nm) of $Fe(CO)_5$ (0.15 mM) in CO–saturated cyclohexane solution at 20 °C.

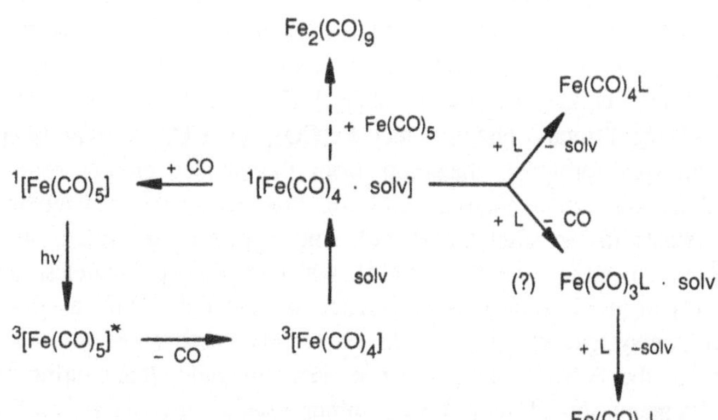

Scheme 5

further spectral changes, slower by 4 to 5 orders of magnitude, are much in evidence, which indicate the ultimate formation of the di–substituted product, $Fe(CO)_3$(trimethyl phosphite)$_2$. Unfortunately, the spectral information is not sufficient to identify the intermediate(s) involved in this latter reaction, but it is clear that the initially formed $Fe(CO)_4$(trimethyl phosphite) remains unaffected. One might speculate about dimerization of the tetracarbonyliron fragment to form $Fe_2(CO)_8$ as an intermediate, which then could undergo unsymmetrical cleavage while taking up two ligands yielding $Fe(CO)_5$ and $Fe(CO)_3L_2$. However, if so, the production of $Fe(CO)_4L$ and $Fe(CO)_3L_2$ in ca. 1:1.5 ratio would place un upper limit of ca. 0.6 on the quantum yield for the disappearance of $Fe(CO)_5$, which is not in accord with the experimental data (Φ_{-A} = 0.78, Scheme 2).

Moreover, it seems unlikely from the kinetic point of view that a major proportion of the $Fe(CO)_4$·solv species should dimerize before being trapped by the ligand L, the more so since L is present in large excess. Hence one is forced to assume that the missing link to the di-substituted product is a mononuclear species. An attractive proposal involves competitive displacement of CO or solvent from $Fe(CO)_4$·solv (Scheme 5), depending on the nature of the incoming ligand. Solvent displacement by the phosphorus donor ligand with formation of the axially substituted $Fe(CO)_4L$ complex implicates a rearrangement of the $Fe(CO)_4$ skeleton from C_{2v} to C_{3v} structure. It may well be that at some stage along the reaction coordinate CO detachment becomes equally favourable, particularly since the M–CO bond dissociation energy quoted for $Fe(CO)_4$ is not very large [7]. Subsequently occurring displacement of the solvent from the resulting $Fe(CO)_3L$·solv by a second ligand L would be straightforward. By contrast, displacement of the solvent from $Fe(CO)_4$·solv by the olefin is possible without any significant distortion of the $Fe(CO)_4$ geometry, such that the olefin approaching the metal centre finds optimum conditions for metal(d_π) $\rightarrow$ olefin(π^*) back donation. Hence it seems plausible that in this case CO substitution plays only a minor role.

Worthwhile to note in this context, the nature of the incoming ligand does not affect the quantum yield for the disappearance of $Fe(CO)_5$, which amounts to ca. 0.8 in the reactions with both trimethyl phosphite (Scheme 2: Φ_{-A} = 0.78) and *E*–cyclooctene (Scheme 4: Φ_{-A} = 0.80). This finding is in line with the above proposal. Moreover, the data confirm that the primary photolytic dissociation of CO from $Fe(CO)_5$ is indeed a very efficient process, in accord with the early reports on the photochemical conversion of the parent pentacarbonyl into $Fe_2(CO)_9$ [10]. The lowest ligand field excited triplet state ($^3E'$) of $Fe(CO)_5$ is considered as the photoactive state [4,5], which evolves directly to the ground state ($^3B_2$) of the primary photoproduct $Fe(CO)_4$ (+ CO). It may be populated either directly by the spin-forbidden transition from the singlet ground state ($^1A_1'$) of $Fe(CO)_5$ or, more likely, through intersystem crossing from its singlet counterpart ($^1E'$) or from higher excited states in the singlet manifold. The respective transitions are difficult to locate in the UV–vis spectrum of $Fe(CO)_5$ [24], which is almost featureless around the calculated positions [4] in the 300 nm region. Relaxation from the $^1E'$ to the ground state, in competition with intersystem crossing, could be largely responsible for the observed deviation from unity of the $Fe(CO)_5$ disappearance quantum yield. Recapturing of CO by the photogenerated fragment should play a very minor role, if any, in view of the large excess of the incoming ligands.

Gas phase kinetic studies [6] revealed that the reaction of CO with $Fe(CO)_4$ is slower by nearly three orders of magnitude, compared with other metal fragments such as, e.g., $Cr(CO)_5$. This previous finding contrasts with the very fast reaction of $Fe(CO)_4$·solv with CO (or other ligands) in cyclohexane solution, which compares well with kinetic data reported for the decay of $Cr(CO)_5$·solv in CO–saturated cyclohexane. Hence it seems that the reaction of the solvated species, unlike that of 'naked' $Fe(CO)_4$ in its triplet ground state ($^3B_2$), is a spin–allowed process. This implies, in accordance with a previous suggestion made for $Fe(CO)_4{\cdot}CH_4$ [7a], that the solvation of the photogenerated fragment involves triplet to singlet conversion (Scheme 5) with retention of the C_{2v} structure. This is somewhat surprising since an alkane can hardly be looked upon as a good π–acceptor suited for stabilizing the b_2 HOMO of the d^8 $Fe(CO)_4$ fragment.

The quantum yield Φ_{BC} for the conversion of $Fe(CO)_4(\eta^2$–ECO) into

$Fe(CO)_3(\eta^2\text{-}ECO)_2$ (Scheme 4) shows a marked wavelength dependence [22], as it increases from 0.59 at 302 nm to 0.82 at 254 nm. This points toward the involvement of two different ligand field excited states, related to the doubly degenerate $^3E'$ state of $Fe(CO)_5$, with differing probabilities for CO *vs* olefin detachment from $Fe(CO)_4(\eta^2\text{-}ECO)$. It is evident from the remarkable high quantum yields at either wavelength that CO dissociation distinctly predominates, since loss of the olefin would bring about no net reaction.

2.1.4. *Photoinduced Catalysis of Alkene Isomerization.* The photocatalytic isomerization of alkenes with carbonyliron complexes [11,16–19] (Scheme 6) is known to be a very efficient process. Quantum yields exceeding unity by 2 or 3 orders of magnitude are observed when $Fe(CO)_5$ is used as the nominal catalyst. This is consistent with the notion

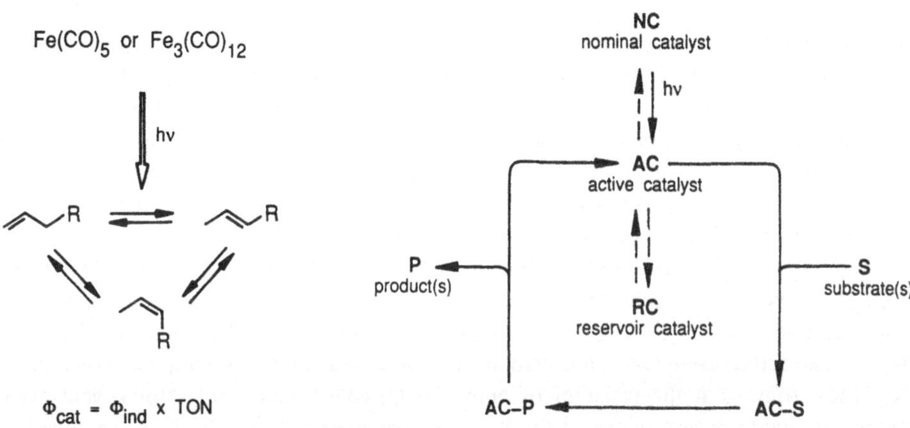

Scheme 6

of a photo-induced catalytic process, which involves the photochemical generation of the active species from an inactive precursor, while the catalytic cycle as such does not require the action of light. In such a case the quantum yield for the overall process (Φ_{cat}) depends on both the photoinduction quantum yield and the turnover number, the product of which (Φ_{ind}×TON) may or may not be larger than unity. The identity of the actual catalyst involved in the alkene isomerization and its properties (lifetime, turnover number, turnover rate) have been controversely discussed over the years. Meanwhile it has been established [16,17], in agreement with a previous proposal [11], that the $Fe(CO)_3$ unit plays a key role as the repeating unit in the catalytic cycle. In detail, photosubstitution of two CO groups in $Fe(CO)_5$ by the alkene generates an $Fe(CO)_3(\eta^2\text{-alkene})_2$ complex acting as a reservoir catalyst, the lability of which is a prerequisite for the catalytic activity (the stable *E*–cyclooctene complex, *vide supra*, is a unique exception in this class of compounds). Metal–alkene dissociation provides the vacant coordination site required for the double bond migration via an $Fe(CO)_3(\eta^3\text{-alkenyl})H$ intermediate, followed by alkene'/alkene exchange to complete the catalytic cycle (Scheme 7).

Quantitative studies suffer from the lability of the $Fe(CO)_3(\eta^2\text{-alkene})_2$ type complex, which gives rise to side reactions leading to inactive products such as $Fe(CO)_3(\eta^4\text{-diene})$ or $Fe(CO)_4(\eta^2\text{-alkene})$. In consequence of this it proves rather difficult to quantify the

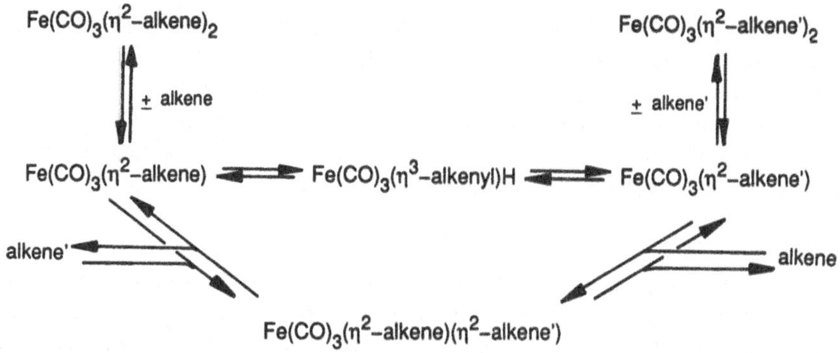

Scheme 7

generation and the decay of the active species, which under continuous irradiation at ambient temperature are interwoven in time with the catalytic process. The obvious solution to this problem is to perform the photochemical step(s) at reduced temperature, low enough to completely suppress both the catalytic process and the decay of the active catalyst. This way the labile $Fe(CO)_3(\eta^2\text{-alkene})_2$ complex accumulates to sufficiently high concentrations, such that, after switching off the light source and warming the solution to an appropriate temperature, the catalysis can be monitored at leisure.

In neat 1-pentene at -60 °C, for example, $Fe(CO)_5$ is largely converted (ca. 70 %) into $Fe(CO)_3(\eta^2\text{-1-pentene})_2$ upon extended irradiation. No 1-pentene isomerization occurs at this temperature, but ca. 500 turnovers (with a turnover rate in the order of 10 min^{-1}, based on the active complex concentration) are observed after warming the solution to 0 °C [16b]. Aiming at the isolation of pure $Fe(CO)_3(\eta^2\text{-1-pentene})_2$, efforts were made to remove the large excess of the olefin at low temperature and to crystallize the complex, but did not meet with success. Apparently some isomerization occured during the protracted work-up procedure and no crystalline material could be obtained from the $Fe(CO)_3(\eta^2\text{-1/2-pentene})_2$ mixture of complexes.

Such difficulties are avoided by using a cycloalkene as the olefinic ligand. Thus, extended irradiation of a solution of $Fe(CO)_5$ and *Z*-cyclooctene (ZCO) in hexane at -40 °C results in the formation of $Fe(CO)_3(\eta^2\text{-ZCO})_2$, which is labile in solution above -35 °C, but nevertheless can be isolated in high yield (80-90 %) as a crystalline material [16]. The X-ray structure analysis (Figure 4) [25] shows the trigonal-bipyramidal geometry expected for this type of complex. It closely resembles that of $Fe(CO)_3(\eta^2\text{-ECO})_2$ [21c], although the distances between the metal and the equatorial olefin ligands are slightly larger. Owing to its lability in solution the complex is destined for using it as a versatile source of the $Fe(CO)_3$ unit, not only for synthetic purposes [16,26], but also for investigations into the kinetics of the tricarbonyliron-catalyzed alkene isomerization [16].

Terminal alkenes, such as 1-pentene, readily replace the two *Z*-cyclooctene ligands at temperatures above -30 °C. Extensive isomerization of excess 1-pentene, with turnover numbers approaching 2000, occurs upon warming to -20 °C or above. Clearly, $Fe(CO)_3(\eta^2\text{-ZCO})_2$ serves as a pre-formed reservoir catalyst in the tricarbonyliron-catalyzed alkene isomerization, thus providing an alternative access to the catalytic cycle [16],

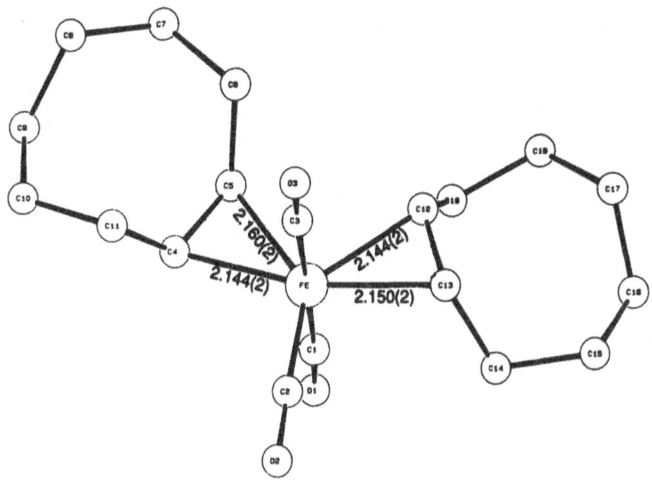

Figure 4: Structure of $Fe(CO)_3(\eta^2$–Z–cyclooctene$)_2$ in the crystal [25].

well suited for mimicking the photoinduced catalysis. The obvious advantage is to start with a well–defined concentration of the activated species, such that reliable kinetic data can be obtained. The catalytic activity lasts for days at –20 °C, hours at 0 °C, and ca. 15 min at +20 °C.

The isomerization of neat 1–pentene, displayed in the diagram on the right in Figure 5, ultimately approaches the thermodynamic equilibrium mixture of the three pentene isomers. The catalysis can be halted by cooling to dry–ice temperature and continues to proceed with the same velocity upon warming up again. The turnover rate is constant (ca. 12 min^{-1} at 0 °C) up to ca. 80 % conversion, thus indicating that the reverse processes are negligible up to this stage. The repeating unit (Scheme 7) evidently discriminates between the different pentene isomers. It picks out the terminal alkene from the reaction mixture and prefentially releases the more weakly bound isomers with an internal C=C bond. Similar results are obtained with 1–hexene, as shown in Figure 5 in the diagram on the left. Again the turnover rate (ca. 18 min^{-1} at 0 °C) is constant up to high conversion and, noteworthy, the products are formed in constant ratio from the very beginning. This finding leads to the conclusion that the double bond migration via the various $Fe(CO)_3(\eta^3$–alkenyl)H intermediates (see Scheme 7, which, for the sake of clarity, shows only one such species) is fast compared with the subsequent exchange of the isomerized alkene' for another 1–alkene substrate molecule.

More detailed investigations into the kinetics of the 1–pentene isomerization reveal that the reaction rates are proportional to the $Fe(CO)_3(\eta^2$–ZCO$)_2$ catalyst concentration (i.e., the turnover rate is constant), but independent of the substrate concentration (as long as the conversion is below 70–80 %). The turnover rate shows a marked temperature dependence (0.5 min^{-1} at –20 °C, 12 min^{-1} at 0 °C, 175 min^{-1} at +20 °C), from which the activation energy (22 kcal/mol) is evaluated. This data, together with the other findings, is consistent with the notion that a (nearly) constant fraction of the tricarbonyliron $Fe(CO)_3$ complex is actively employed in the catalytic cycle, which, moreover, involves a metal–alkene dissocation (release of the isomerized alkene' in Scheme 7) as the rate determining step.

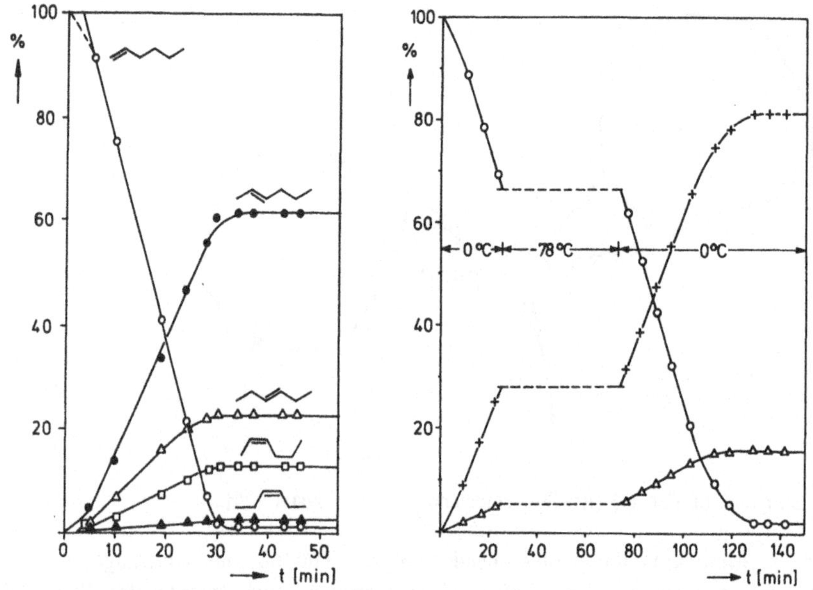

Figure 5: Catalytic 1–alkene isomerization with $Fe(CO)_3(\eta^2\text{–}ZCO)_2$. – The diagram on the right shows the conversion of neat 1–pentene (9.1 M) at 0 °C into *cis*– and *trans*–2–pentene (12.1 mM complex concentration). The catalysis is interrupted by cooling to dry–ice temperature [16]. – The diagram on the left shows the conversion of neat 1–hexene (8.0 M) at 0 °C into the 2– and 3–hexene isomers (16 mM complex concentration) [16b].

Concerning the photocatalytic quantum yield, $\Phi_{cat} = \Phi_{ind} \times TON$, it seems justified to take the conversion of $Fe(CO)_5$ into the stable complex $Fe(CO)_3(\eta^2\text{–}ECO)_2$ as a model for the generation of the labile $Fe(CO)_3(\eta^2\text{–alkene})_2$ reservoir catalyst. This process essentially involves two photochemical steps. In consequence of this, the overall quantum yield increases gradually with increasing conversion of $Fe(CO)_5$, but remains well below 0.5 due to both internal light filtering and the deviation from unity of the particular steps. At ca. 60 % conversion of $Fe(CO)_5$, for example, the overall quantum yield for the formation of $Fe(CO)_3(\eta^2\text{–}ECO)_2$ amounts to ca. 0.2 (Figure 2). Taking this data as Φ_{ind} together with a turnover number of 2000 one obtains $\Phi_{cat} = 0.2 \times 2000 = 400$, which is a good agreement with experimental values [11,19] determined for the photoinduced catalysis with continuous irradiation of $Fe(CO)_5$ in 1–pentene.

2.2. MULTIPLE PHOTOSUBSTITUTION OF CO GROUPS IN PENTACARBONYL-RUTHENIUM.

Quantitative investigations into the photochemistry of $Ru(CO)_5$ are fraught with some difficulties arising from the moderate thermal stability of this complex [27]. Thus, following its photochemical generation from $Ru_3(CO)_{12}$ and carbon monoxide in solution at ambient temperature, $Ru(CO)_5$ slowly undergoes thermal substitution of one CO group by a phosphorus donor ligand. This has to be taken into account as a correction term in the photokinetics.

Upon extended irradiation in the presence of excess trimethyl phosphite $Ru(CO)_5$ is converted into a mixture of substituted $Ru(CO)_{5-n}L_n$ complexes (n = 1, 2, 3), as monitored by means of quantitative IR spectroscopy [28]. Up to high conversion (60–80 %) the material balance shows no significant deficit. The evaluation of quantum yields for the individual steps is based on the photokinetic formalism previously applied to the related reactions of $Fe(CO)_5$ (*vide supra*), appropriately modified to account for the above-mentioned thermal bypath. For the sake of accuracy, the data for the secondary photosubstitution of $Ru(CO)_4L$ and $Ru(CO)_3L_2$ are taken from complementary experiments using these complexes as the starting materials.

In striking contrast to the behaviour of $Fe(CO)_5$, the multiple photosubstitution of CO groups in $Ru(CO)_5$ essentially proceeds in a step-by-step fashion (Scheme 8). Primary disubstitution, if occurring at all, is within the experimental margin of error. This observation demonstrates that caution has to be exercised in drawing conclusions by

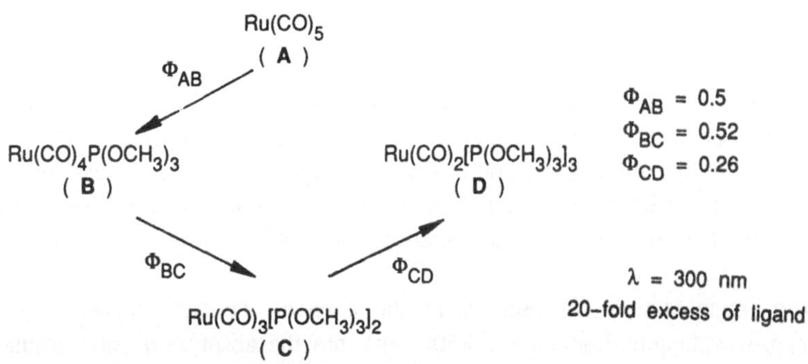

Scheme 8

analogy. Mechanistic studies, including wavelength dependent measurements and flash photolysis experiments with time-resolved IR detection, are currently underway, but the reason for the different behaviour of the carbonyliron and -ruthenium complexes is not yet clear. As a common feature, the quantum yields in both systems are decreasing in going to the higher substituted derivatives, which apparently reflects competitive CO and donor ligand photodetachment.

Multiple photosubstitution of CO by olefin ligands has been observed with $Ru(CO)_5$ [20b] and $Os(CO)_5$ [29]. These reactions apparently proceed via a sequence of consecutive steps, whereby the formation of higher substituted products is favoured by shorter wavelengths of irradiation. However, more detailed mechanistic information is lacking and quantum yield data are not yet available.

2.3. PHOTOFRAGMENTATION OF THE $M_3(CO)_{12}$ CLUSTERS IN THE PRESENCE OF OLEFINS.

The wavelength-dependent photochemistry of the group 8 metal $M_3(CO)_{12}$ cluster complexes has received considerable attention. A vast amount of information on the nature of the electronic transitions, primary photoproducts and other reactive intermediates, quantum yields and photokinetics has been accumulated over the years [30]. Of the two types of photoreactions, CO substitution and fragmentation of the trinuclear framework, the latter predominates upon irradiation into the characteristic long-wavelength absorptions of the clusters.

Mononuclear olefin-substituted carbonylruthenium complexes are most conveniently prepared from $Ru_3(CO)_{12}$ [30–32] (Scheme 9). Irradiation through a cut-off filter ($\lambda \geq 370$ nm) into the long-wavelength absorption around 395 nm in the presence of, e.g., methyl acrylate results in the exclusive formation of $Ru(CO)_4(\eta^2$–methyl acrylate), which can be

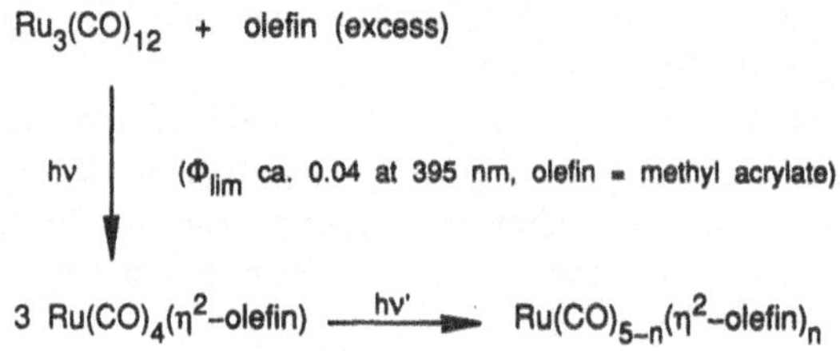

Scheme 9

isolated in high chemical yield [31]. Under these conditions the products are screened from further irradiation. Secondary photosubstitution with formation of $Ru(CO)_{5-n}(\eta^2\text{–olefin})_n$ complexes ($n \geq 2$) [17,33][20b] occurs upon irradiation through solidex glass or quartz. Worthwhile to note, the $Ru(CO)_4(\eta^2$–olefin) complexes are generally less stable than the iron analogues and, thus, may serve as a source of the $Ru(CO)_4$ unit in ligand exchange reactions [31].

Unsymmetric fragmentation is observed in the case of the iron cluster $Fe_3(CO)_{12}$, which upon long-wavelength irradiation ($\lambda \geq 500$ nm) into the absorption band around 603 nm in the presence of ethene [30d] or *E*–cyclooctene [12][22b] yields $Fe(CO)_5$, $Fe(CO)_4$–(η^2–olefin), and $Fe(CO)_3(\eta^2\text{–olefin})_2$ in nearly 1:1:1 ratio (Scheme 10). The observation

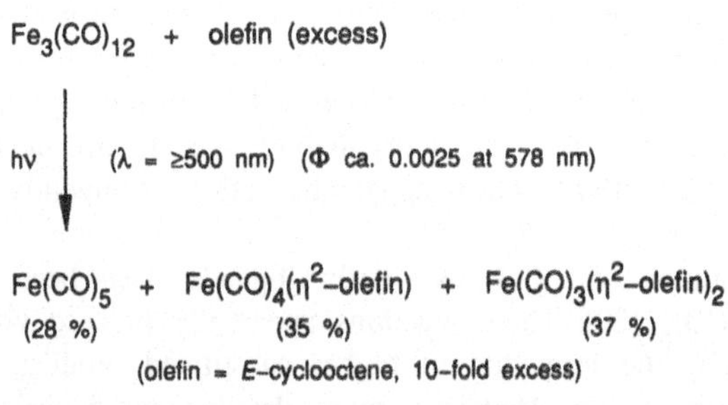

Scheme 10

of the latter type of complex as a primary product is an important finding with respect to the photocatalytic alkene isomerization with $Fe_3(CO)_{12}$ as the nominal catalyst, as it provides clear evidence [30d] that the catalytic cycle involves the same active species otherwise generated from $Fe(CO)_5$ upon short-wavelength irradiation (*vide supra*).

Quantitative studies [12][22b] again are facilitated by using *E*–cyclooctene (ECO) as the olefin ligand. It turns out that the product ratio approaches 1:1:1 at low concentration of ECO, whereas a large excess favours the formation of $Fe(CO)_3(\eta^2\text{-ECO})_2$ at the expense of $Fe(CO)_5$. It seems that this is due to the competitive generation of the substituted cluster $Fe_3(CO)_{11}(\eta^2\text{-ECO})$, which temporarily appears in the reaction mixture with albeit marginal stationary concentration, but nevertheless has been isolated as a pure substance [12]. The electronic absorption spectrum of $Fe_3(CO)_{11}(\eta^2\text{-ECO})$ shows a long-wavelength maximum in close neighbourhood to that of the parent $Fe_3(CO)_{12}$ (Figure 6). Irradiation into this region yields $Fe(CO)_3(\eta^2\text{-ECO})_2$ (42 %) and $Fe(CO)_4(\eta^2\text{-ECO})$

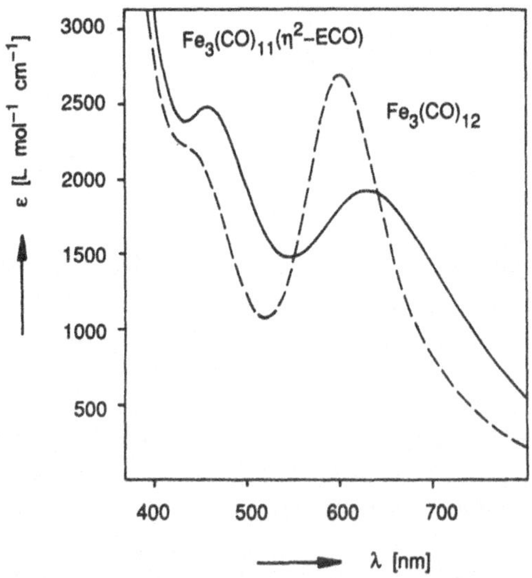

Figure 6: Electronic absorption spectra of $Fe_3(CO)_{12}$ and $Fe_3(CO)_{11}(\eta^2\text{-}E\text{-cyclooctene})$ [12] in hexane.

(50 %), but little $Fe(CO)_5$ (8 %). This photofragmentation pattern closely parallels that of the related phosphine-substituted cluster $Fe_3(CO)_{11}[P(C_6H_5)_3]$, which upon similar long-wavelength photolysis in the presence of excess ECO is converted into a nearly 1:1:1 mixture of $Fe(CO)_4[P(C_6H_5)_3]$, $Fe(CO)_4(\eta^2\text{-ECO})$, and and $Fe(CO)_3(\eta^2\text{-ECO})_2$ (Φ = 0.0035 at 578 nm), but yields no $Fe(CO)_5$.

Photofragmentation of the osmium cluster $Os_3(CO)_{12}$ in the presence of olefins, again with filtered radiation ($\lambda \geq 370$ nm), leads to mono- and dinuclear products, $Os(CO)_4(\eta^2\text{-olefin})$ and $Os_2(CO)_8(\mu\text{-}\eta^1,\eta^1\text{-CHR-CHR})$ (Scheme 11) [34]. The diosmacyclobutane complexes undergo reversible ring cleavage, both thermally and photochemically, thus providing access to a variety of exchange products including diosmacyclobutene systems. The transient existence of $Os_2(CO)_8$ in hydrocarbon solution at ambient temperature has been demonstrated by means of flash photolysis in combination with time-resolved IR spectroscopy [35].

A comprehensive discussion of the mechanistic aspects of the trinuclear cluster photofragmentation would be beyond the scope of this short paragraph. Briefly, there is accumulating evidence [30] that a non-radical reactive isomer of the parent $M_3(CO)_{12}$ with a vacant coordination site (resulting from terminal → bridging CO rearrangement) is responsible for the long-wavelength photochemistry. This species should coordinate the

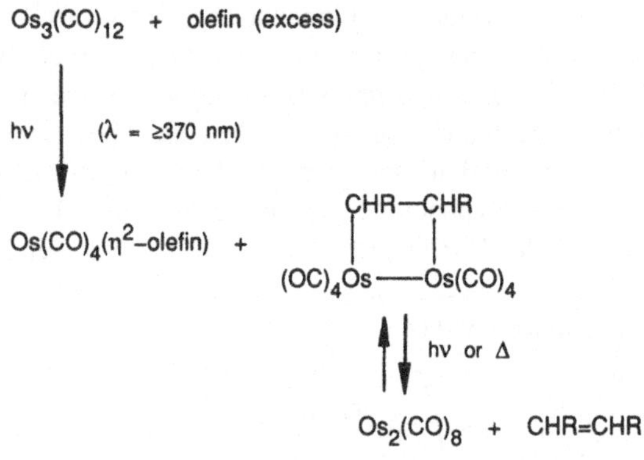

Scheme 11

olefin and subsequently undergo either loss of CO (associative substitution being the overall result) or metal–metal bond cleavage to form a mono– and a dinuclear fragment. Competetive reversion to the ground state parent $M_3(CO)_{12}$ would account for the relatively low quantum yields. Further breakdown of the dinuclear fragment with concomitant olefin coordination would complete the reaction sequence in the cases of iron and ruthenium, while the $Os_2(CO)_8(\mu\text{-}\eta^1,\eta^1$–CHR–CHR) complex is a stable product. An isomer of the latter compound, probably with a CO–bridge and a terminal olefin, has been observed as a relatively long–lived transient species in the above mentioned flash photolysis experiments [35].

3. Olefin–substituted Group 6 Metal Carbonyls

Sequential substitution of two CO groups upon extended irradiation of Group 6 metal carbonyls in the presence of alkenes has been reported as early as three decades ago [36]. However, for a long time it seemed commonly accepted that the resulting $M(CO)_{6-n}(\eta^2\text{-olefin})_n$ complexes (n = 1, 2) at best are moderately stable, such that their characterization almost exclusively relied on IR spectroscopic data. Thus, it is not surprising that much more attention was paid to the photochemistry of Group 6 metal carbonyls with *n*–donor ligands, which are generally more stable and well suited for detailed investigations into their spectroscopic and photochemical properties [1]. Nonetheless, the photochemistry of the complexes with η^2–olefin ligands [37], like that of η^4–diene and η^6–triene compounds [38], has some synthetic potential, which is worth being exploited. Moreover, the stereochemical and catalytic aspects are of great interest.

3.1. STRUCTURE AND BONDING

Low thermal stability is in fact a characteristic property of most of the monosubstituted $M(CO)_5(\eta^2$–olefin) derivatives (M = Cr, Mo, W) [39–42], the complexes of *E*–cyclooctene being again the remarkable exception for all three metals [40,41]. Some of the labile $M(CO)_5(\eta^2$–olefin) complexes can nevertheless be prepared as pure, crystalline materials.

$$M(CO)_6 + \text{olefin} \xrightarrow{h\nu} M(CO)_5(\eta^2\text{-olefin}) + CO$$

$$M(CO)_5(\eta^2\text{-olefin}) + L \longrightarrow M(CO)_5L + \text{olefin}$$

Scheme 12

They serve as versatile sources of the respective $M(CO)_5$ moities (Scheme 12) in ligand exchange reactions [40a][41–44]. The compounds $Cr(CO)_5(\eta^2$–*Z*–cyclooctene) [40a][41], $Mo(CO)_5(\eta^2$–norbornene) [43], and $W(CO)_5(\eta^2$–cyclopentene) [42] are particularly useful in this respect. Specific applications include the preparation of selectively labelled $M(CO)_5(^{13}CO)$ and, e.g., the synthesis of diazadiene complexes in which the L–L ligand is coordinated to the $M(CO)_5$ unit in a monodentate fashion [43,45].

By contrast, and quite surprising at first, high thermal stability is a recurrent feature in the many di–substituted complexes of type *trans*–$M(CO)_4(\eta^2$–olefin$)_2$ (M = Cr, Mo, W) [39,41,46,47], which have been prepared with simple alkenes, cycloalkenes, and α,β–unsaturated esters. Thus, for example, *trans*–$Cr(CO)_4(\eta^2$–ethene$)_2$ [46], previously considered to be unstable even in low–temperature xenon solution [48], proved to be stable at room temperature and above not only in crystalline form, but also in solution and in the gas phase, if rigorously purified and freed from labile by–products.

More recently, the first representatives of the isomeric *cis*–$M(CO)_4(\eta^2$–olefin$)_2$ type of complex were discovered, which by far are the least stable ones in the whole series and could be isolated only with *E*–cyclooctene [47,49].

The above trends in stability can be rationalized on a qualitative level in terms of competitive demand of the CO and olefin ligands for π–back donation from the metal, which seems to be the predominant factor for the M–olefin bond strength [50]. The X–ray diffraction structure analysis of *trans*–$W(CO)_4(\eta^2$–methyl acrylate$)_2$ [39], one of the first examples of this type of compounds, reveals a perpendicular orientation of the *trans*–positioned olefins (Figure 7). Five other *trans*–$M(CO)_4(\eta^2$–olefin$)_2$ complexes, which have been examined by means of X–ray diffraction, invariable show the same structural

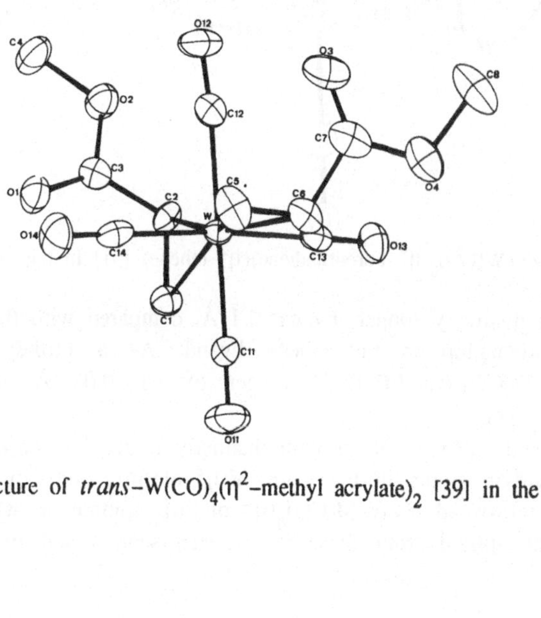

Figure 7: Structure of *trans*–$W(CO)_4(\eta^2$–methyl acrylate$)_2$ [39] in the crystal.

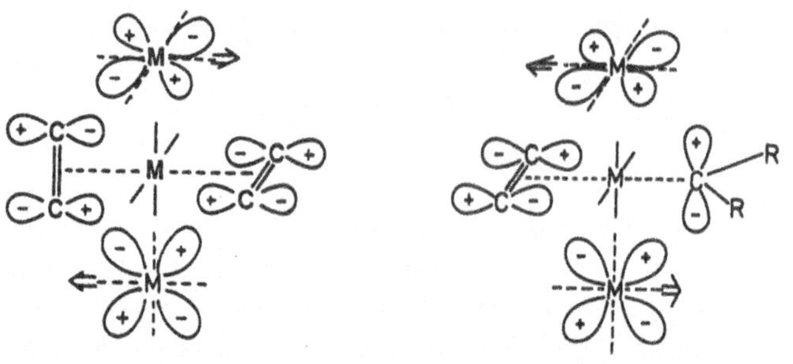

Figure 8: Trans–orthogonal orientation of the olefin and carbene acceptor orbitals in *trans*–$M(CO)_4(\eta^2$–olefin$)_2$ and *trans*–$M(CO)_4(\eta^2$–olefin)(carbene) complexes.

feature [41,42,47]. Because of their single–faced π–acceptor character the two olefins in this orientation avoid competition for π–back donation from the metal (Figure 8). Moreover, such competition with the CO ligands is minimized. By contrast, a CO group (with its *two* orthogonal π–acceptor orbitals) situated in *trans*–position to an olefin will strongly rival for π–back donation and, thus, weaken the metal–olefin bond. This effect is clearly demonstrated by the structure of *mer*–$W(CO)_3(\eta^4$–norbornadiene$)(\eta^2$–ethene) [51], which has a *trans*–$M(\eta^2$–C=C$)_2$ unit and a *trans*–$M(\eta^2$–C=C)(CO) unit in the same molecule (Figure 9). The metal–carbon distance of the norbornadiene double bond in

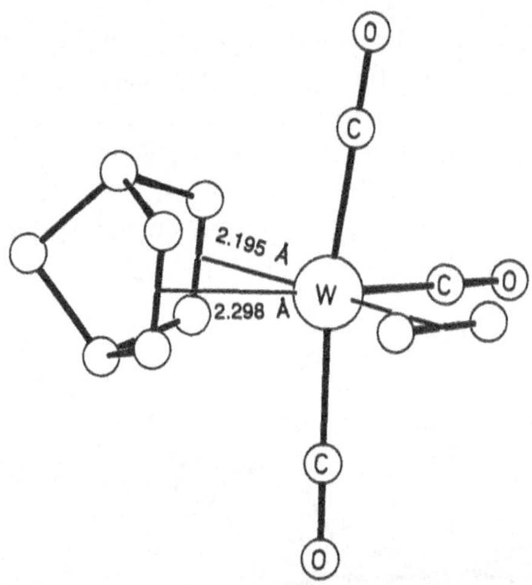

Figure 9: Structure of *mer*–$W(CO)_3(\eta^4$–norbornadiene$)(\eta^2$–ethene) [51] in the crystal.

trans–position to CO is distinctly longer, by ca. 0.1 Å, compared with the other one in the *trans*–orthogonal orientation to the ethene ligand. As a further example, the metal–olefin bond in $W(CO)_5(\eta^2$–ECO) is longer by ca. 0.08 Å, compared with *trans*–$W(CO)_4(\eta^2$–ECO$)_2$ [41].

Noteworthy, these trends also go along with distinctly larger ^{13}C–NMR coordination shifts of the olefinic carbon atoms in the *trans*–$M(\eta^2$–olefin$)_2$ units in *mer*–$M(CO)_3$–$(\eta^4$–norbornadiene$)(\eta^2$–olefin) and *trans*–$M(CO)_4(\eta^2$–olefin$)_2$ complexes, which invariably resonate ca. 25–35 ppm upfield from those in the *cis*–isomers and in the $M(CO)_5$–

(η^2–olefin) derivatives [39,41,46,47,51].

Furthermore, the electronic absorption spectra of the latter two classes exhibit a marked bathochromic shift of the features in the ligand field region, compared with *trans*–$M(CO)_4(\eta^2\text{–olefin})_2$ and $M(CO)_6$ as well (Figure 10). Clearly, this is also attributable to the single–faced π–acceptor character of the olefin ligands, since one of the

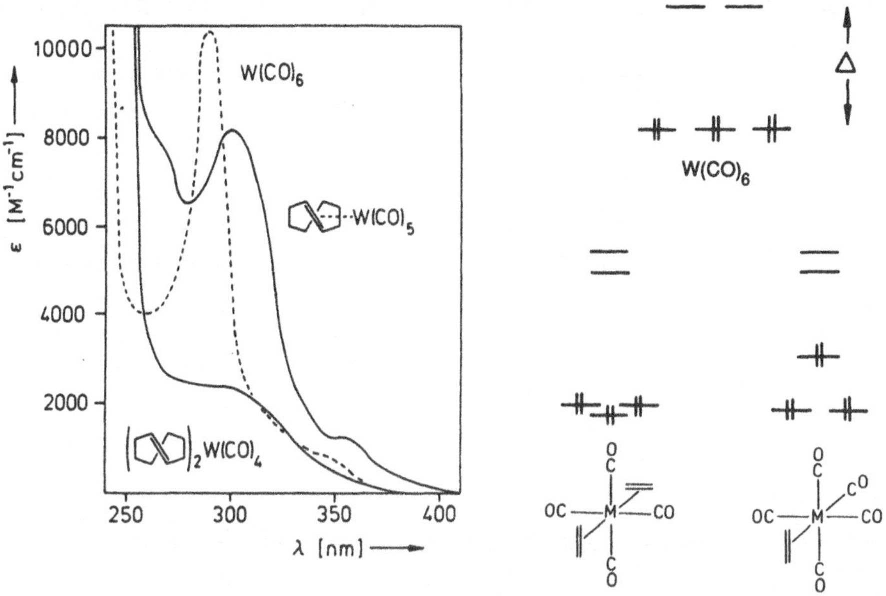

Figure 10: Electronic absorption spectra in hexane and qualitative d–orbital schemes of $W(CO)_6$, $W(CO)_5(\eta^2\text{–ECO})$, and $W(CO)_4(\eta^2\text{–ECO})_2$ (ECO = *E*–cyclo–octene).

occupied metal d_π orbitals (not interacting with the olefin) should be distinctly less stabilized. It is interesting to note in this context that photoelectron spectroscopic data [52] reveal accidental degeneracy of the two low–lying d_π orbitals in an $M(CO)_5(\eta^2\text{–olefin})$ complex and of all three d_π orbitals in a *trans*–$M(CO)_4(\eta^2\text{–olefin})_2$ complex, which strongly suggests that the π–acceptor ability of the olefin π^* orbital compares well with the CO π^* orbitals.

Scheme 13 summarizes the different types of structures with one, two or three C=C units coordinated to the various Group 6 carbonylmetal fragments, which have been observed and characterized so far. In all cases, *E*–cyclooctene forms the most stable monoolefin–metal bonds. With respect to the particular metals, the following trend in the stabilities of the compounds is observed:

$$Cr < Mo < W$$

As a concluding remark in this section it should emphasized that the above bonding considerations for the $M(CO)_4(\eta^2\text{–olefin})_2$ systems are equally valid for $M(CO)_4(\eta^2\text{–olefin})(\text{carbene})$ complexes [53]. The ultimately stable products formed upon irradiation of $M(CO)_5(\text{carbene})$/olefin mixtures show a *trans*–coplanar arrangement of the olefin and carbene units, i.e., a *trans*–orthogonal orientation of the two single π–acceptor orbitals, unless chelate restraints command the less favourable *cis*–structure.

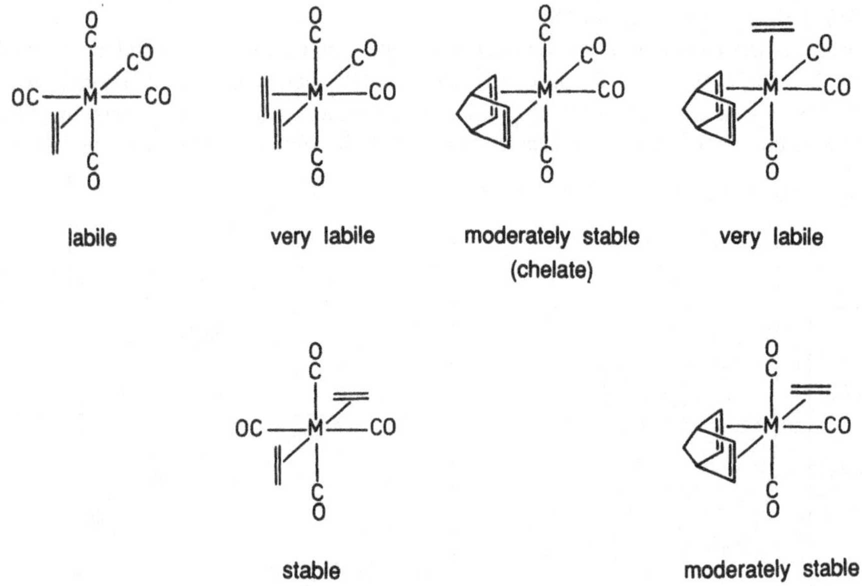

Scheme 13

3.2. SEQUENTIAL PHOTOSUBSTITUTION OF TWO CO GROUPS IN $M(CO)_6$ BY AN OLEFIN

Because of its superior bonding properties, *E*–cyclooctene is selected for monitoring the conversion of $W(CO)_6$ into the substituted $W(CO)_5(\eta^2$–olefin) and $W(CO)_4(\eta^2$–olefin$)_2$ complexes on a quantitative level.

Concerning the photolytic CO dissociation from a monosubstituted Group 6 $M(CO)_5L$ complex it has been calculated that on both the excited and ground state surfaces a *cis*–vacant $M(CO)_4L$ fragment should be more stable than the isomeric species with a vacant coordination site in *trans*–position to L [5,54]. In accord with this theoretical prediction the *cis*–vacant $W(CO)_4(\eta^2$–ECO) fragment is indeed generated, together with a substantial amount of $W(CO)_5$, as a primary photoproduct of $W(CO)_5(\eta^2$–ECO). This has been observed in solution at ambient temperature by means of flash photolysis in combination with time–resolved IR spectroscopy and in low–temperature matrices with conventional IR spectroscopy as well. Hence, the formation of the ultimately stable *trans*–$M(CO)_4(\eta^2$–olefin$)_2$ complex is expected to involve the *cis*–isomer as a key intermediate. This is in fact observed when the photoreaction of $W(CO)_5(\eta^2$–ECO) with excess *E*–cyclooctene is followed by means of quantitative IR spectroscopy (Figure 11) [49]. It is difficult to isolate *cis*–$W(CO)_4(\eta^2$–ECO$)_2$ from this solution, but the complex is readily accessible via independent routes involving ligand exchange of a labile $W(CO)_4$ complex. Again, the quantum yields are evaluated on the basis of the photokinetic formalism and the iterative procedure previously applied to the photosubstitution reactions of $Fe(CO)_5$ (*vide supra*). The excellent fit of the computed curves with the experimental data illustrates the reliability of the results.

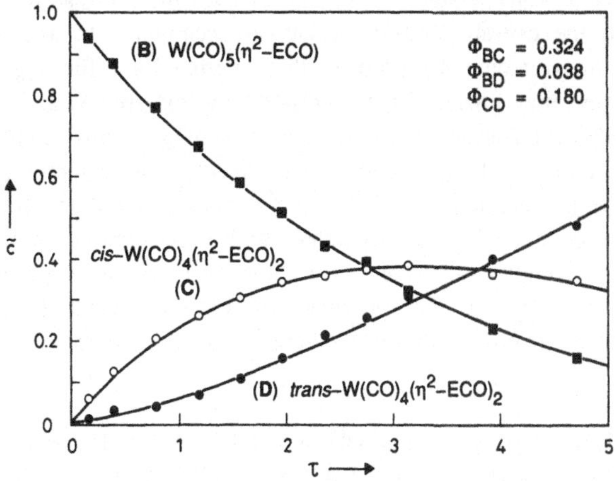

Figure 11: Photochemical conversion of $W(CO)_5(\eta^2\text{-ECO})$ (B) into *cis*- and *trans*-$W(CO)_4(\eta^2\text{-ECO})_2$ (C, D) upon irradiation at λ = 365 nm in the presence of excess *E*-cyclooctene (ECO) [49] (the curves are computed on the basis of the quantum yields Φ_{BC}, Φ_{BD}, and Φ_{CD} and the ε-values of the three complexes given in Scheme 14).

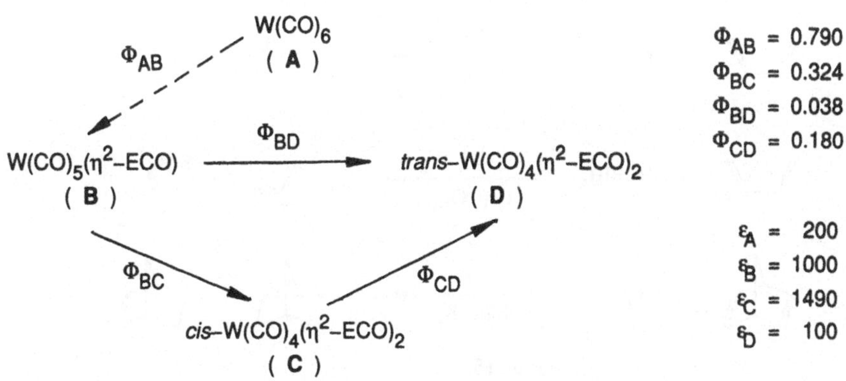

Scheme 14

$W(CO)_5(\eta^2\text{-ECO})$ and *cis*-$W(CO)_4(\eta^2\text{-ECO})_2$ appear as intermediate products, when $W(CO)_6$ is irradiated under analogous conditions in the presence of excess *E*-cyclooctene (Scheme 14). Because of their relatively high absorbances, compared with $W(CO)_6$ and the ultimately formed *trans*-$W(CO)_4(\eta^2\text{-ECO})_2$, the two intermediate products do not accumulate to a high level of concentration in the overall process. The initial photosubstitution of a CO group in $W(CO)_6$ occurs with a quantum yield (Φ_{AB} = 0.79), which compares well with published data [55] for the analogous reaction with a donor ligand. There is no indication that primary di-substitution takes place. Moreover, the direct conversion of $W(CO)_5(\eta^2\text{-ECO})$ into the *trans*-disubstituted product (Φ_{2D}) plays, at best, a minor role. Furthermore, it is important to note that the *cis* → *trans* isomerization of $W(CO)_4(\eta^2\text{-ECO})_2$ is exclusively a photochemical process. This finding, in accordance with previous considerations [46], characterizes the *cis*–disubstituted complex as the 'bottle neck' in the overall reaction: it has to accumulate to a certain level of concentration in order to overcome the problem that internal light filtering by the other components in the mixture may prevent light absorption by this species and, consequently, hinder the *cis* → *trans* isomerization. Based on this reasoning it now readily understood why, for example, the synthesis of *trans*–$Cr(CO)_4(\eta^2\text{-ethene})_2$ [46] requires cooling of the solution to $-50\ ^0C$. The intermediate *cis*–$Cr(CO)_4(\eta^2\text{-ethene})_2$ is too unstable at ambient temperature for being accumulated to a sufficiently high concentration. Moreover, it not surprising that this labile complex shows up as a short–lived transient species in the gas phase photolysis of $Cr(CO)_6$ in the presence of excess ethene [56], while the thermodynamically stable *trans*–$Cr(CO)_4(\eta^2\text{-ethene})_2$ is not observed under those conditions.

3.3. PHOTOCATALYTIC DIENE HYDROGENATION AND HYDROSILYLATION WITH GROUP 6 METAL CARBONYLS: TRACKING THE ACTIVE SPECIES

Conjugated dienes are selectively *cis*–1,4–hydrogenated or –hydrosilylated upon irradiation of a carbonylchromium catalyst in the presence of the substrates [57–61]. Norbornadiene (NBD) yields two hydrogenated products: norbornene (1,2–addition of H_2) and nortricyclene (*homo*–1,4–addition of H_2) in ca. 1:3 ratio [62–65] (Scheme 15). The

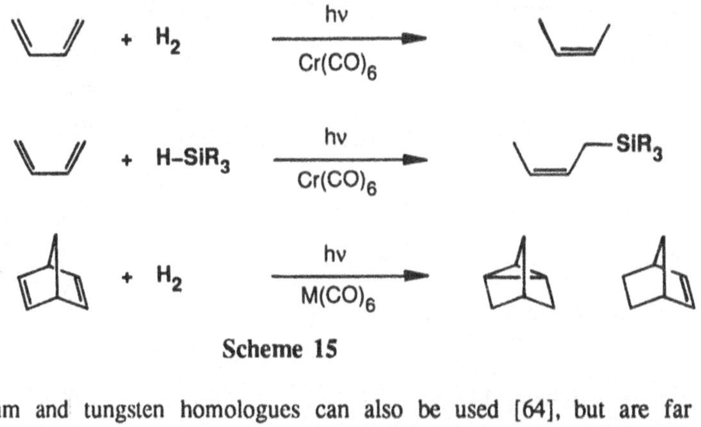

Scheme 15

molybdenum and tungsten homologues can also be used [64], but are far less reactive compared with the carbonylchromium catalyst. Both types of addition are also observed in the hydrosilylation of NBD [42,47], but here the 1,2–addition is far predominant.

When starting with a metal hexacarbonyl, the photo–induction period of these processes involves the formation of the respective $M(CO)_4(\eta^4\text{-diene})$ complexes, but further irradiation is required to generate a vacant coordination site for taking up H_2 or H–SiR_3, which then can be transferred onto the organic substrate. Over the past two decades much effort was made in order to elucidate the nature of the actual catalyst. Most recent results [66,67] indicate that the $M(CO)_3$ moiety acts as the repeating unit, thus disfavouring the previous proposal of an $M(CO)_4(\eta^2\text{-diene})$ species with a vacant coordination site *cis* to the η^2–coordinated diene, suited for mediating the 1,2–addition.

3.3.1. *Spectroscopic Studies.* Low–temperature matrix experiments show that photolytic CO dissociation from $Cr(CO)_4(\eta^4$–1,3–diene) complexes results in the formation of the respective *fac*–$Cr(CO)_3(\eta^4$–1,3–diene) fragments [66a]. Similar experiments with $Cr(CO)_4$–(η^4–NBD) reveal that in this case two isomeric fragments are formed, *mer*– and *fac*–$Cr(CO)_3(\eta^4$–NBD) [66b] which are identified on the basis of their characteristic CO stretching vibrational patterns (Figure 12). Complementary studies [67], including flash

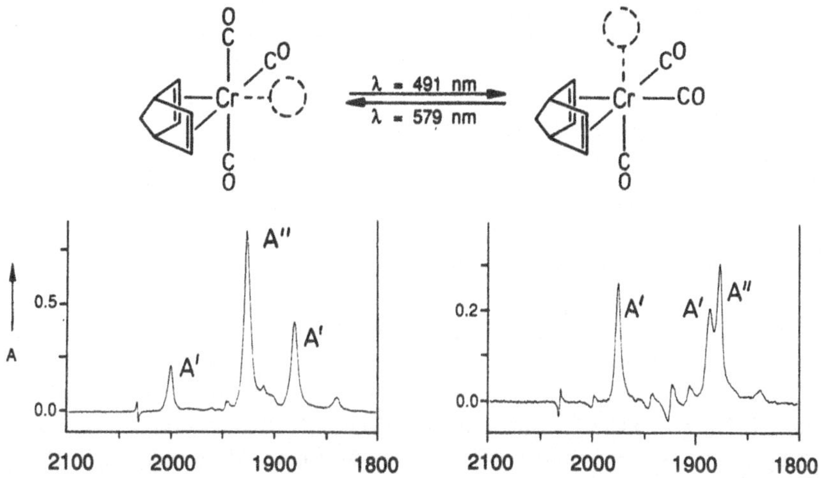

Figure 12: Infrared CO stretching vibrational patterns of *mer*–$Cr(CO)_3(\eta^4$–NBD) (left) and *fac*–$Cr(CO)_3(\eta^4$–NBD) (right), generated from $Cr(CO)_4(\eta^4$–NBD) in a methane matrix upon photolysis with λ = 313 nm, followed by secondary irradiation with λ = 579 nm and 491 nm, respectively. Note that residual absorptions of the starting material and the respective other fragment isomer are removed by means of computer–assisted subtraction.

photolysis experiments in combination with time–resolved IR spectroscopy, provide further evidence for the photolytic generation of the *mer*– and *fac*–$M(CO)_3(\eta^4$–NBD) fragments from $M(CO)_4(\eta^4$–NBD). Subsequent addition of H_2 to the vacant coordination site, followed by transfer to the coordinated diene, should lead to the observed hydrogenation products (Scheme 16).

One important factor has been ignored until recently, *viz.*, that in competition with H_2 or H–SiR_3 the diene, present in large excess, could occupy the vacant coordination site and thus might inhibit the catalytic process [47][67b]. In fact, this has been observed in the cases of the molybdenum and tungsten catalysts, where complexes of type *mer*–$M(CO)_3(\eta^4$–NBD)(η^2–NBD) were detected and isolated [47][67b]. This finding provides a satisfactory rationale for the relatively low photocatalytic activity of these two metals. It seems likely that the isomers with a *fac*–$M(CO)_3$ skeleton are also formed in these systems, but are too unstable for detection (due to the CO group in *trans*–position to the η^2–coordinated NBD, cf. Scheme 13). However, the *E*–cyclooctene complex *fac*–$W(CO)_3(\eta^4$–NBD)(η^2–ECO) is stable enough for isolation [47]. It is generated from $W(CO)_4(\eta^4$–NBD) upon irradiation at –30 °C in the presence of ECO and isomerizes, both thermally and photochemically, with formation of the more stable *mer*–isomer.

The chromium complex *mer*–$Cr(CO)_3(\eta^4$–NBD)(η^2–NBD), expected to be present in

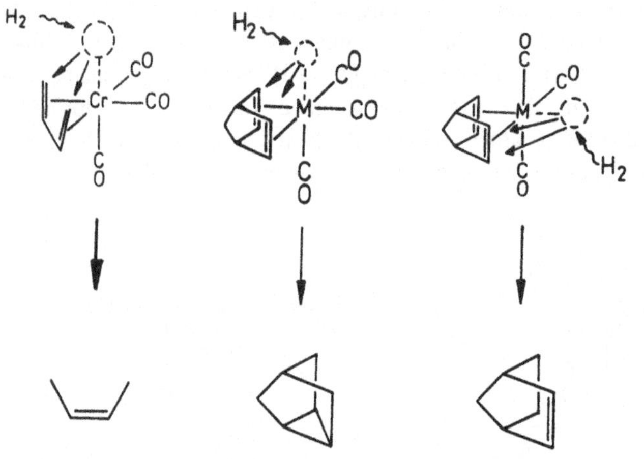

Scheme 16

the photocatalytic solutions, eluded detection by means of IR spectroscopy, even in low-temperature solution. However, two other *mer*-$Cr(CO)_3(\eta^4\text{-NBD})(\eta^2\text{-olefin})$ compounds, with ECO and ethene as the olefin ligands, are accessible on preparative scale. Not unexpectedly, the ECO complex *mer*-$Cr(CO)_3(\eta^4\text{-NBD})(\eta^2\text{-ECO})$ is stable at ambient temperature and thus renders it feasible to determine the quantum yield ($\Phi = 0.18$ at 300 nm and 0.11 at 365 nm) for the formation of this type of complex [47].

3.3.2. *The Use of mer-*$Cr(CO)_3(\eta^4\text{-NBD})(\eta^2\text{-ethene})$ *as a Reservoir Catalyst.* The ethene complex *mer*-$Cr(CO)_3(\eta^4\text{-NBD})(\eta^2\text{-ethene})$ is prepared from $Cr(CO)_4(\eta^4\text{-NBD})$ by irradiation in ethene-saturated pentane solution at −50 °C and is isolated as a pure, crystalline substance [68]. In solution at −10 °C or above the ethene ligand is gradually displaced by other ligands, such that this complex can be used as a source of the $Cr(CO)_3(\eta^4\text{-NBD})$ fragment.

Most interestingly, it acts as a catalyst for the hydrogenation of excess NBD without irradiation, whereby it exactly reproduces the results of the photoinduced catalysis with respect to the product distribution under varying conditions. This includes the stereospecific *endo*-deuteration and a significant H/D isotope effect when D_2 is used instead of H_2 under atmospheric pressure (Scheme 17). With a 200-fold excess of NBD substrate, ca. 100 turnovers take place, until catalysis ceases after about 2 h, due to gradual decomposition to $Cr(CO)_4(\eta^4\text{-NBD})$ and traces of $Cr(CO)_6$. These findings confirm the proposed role of the $Cr(CO)_3$ group as the repeating unit in the catalytic cycle. Combining this turnover number with an estimated quantum yield in the order of 0.1 for the photogeneration of the active species (taken from the reaction with ECO, see above) one could make an assesssment of the quantum yield for the photoinduced catalytic hydrogenation: $\Phi_{cat} = \Phi_{ind} \times TON = 10$. This value is somewhat higher than previously published experimental data (1.9 [63]), probably because the *in situ* generated catalyst reacts with liberated CO to re-form the inactive $Cr(CO)_4(\eta^4\text{-NBD})$.

One might have expected that the *mer*-$Cr(CO)_3$ structure is retained in the active catalyst, such that 1,2-hydrogenation would be favoured (cf. Scheme 16), at least

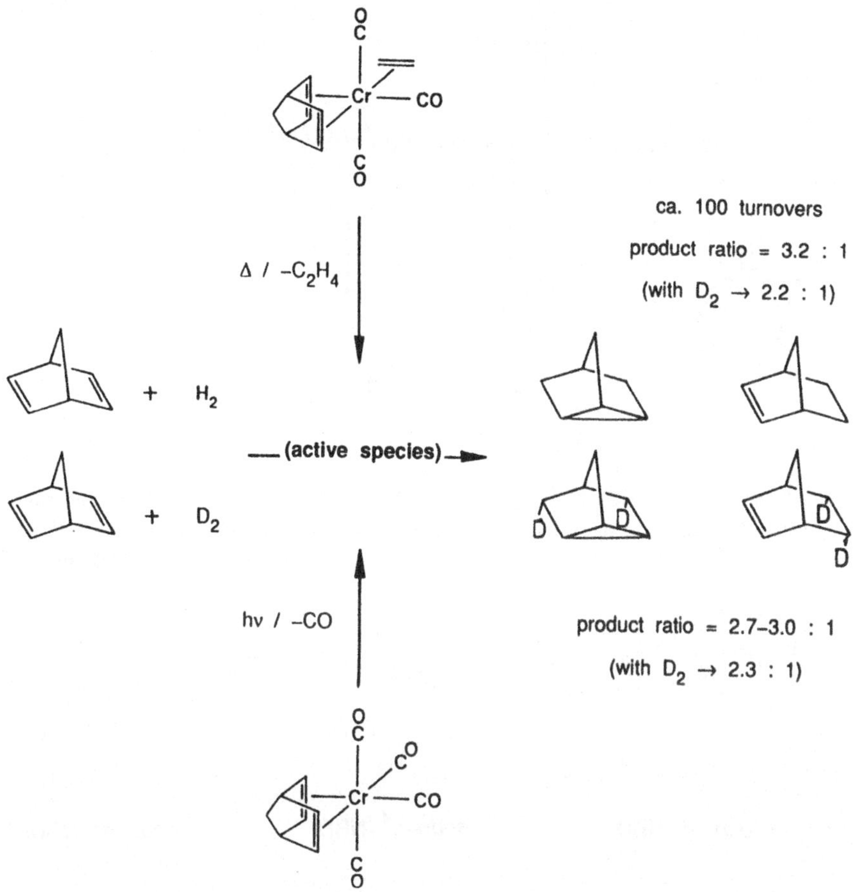

Scheme 17

temporarily. However, the two hydrogenation products, NTC and NBD, are formed in constant ratio from the very beginning. Hence it seems likely that the catalysis involves a rapid interconversion of the *mer–* and *fac–*$Cr(CO)_3$ structures. Accordingly, when excess isoprene is used as the substrate instead of NBD, this conjugated diene is *cis*–1,4–hydrogenated. This implicates that the stereochemistry of CO loss from $Cr(CO)_4$–(η^4–1,3–diene) or $Cr(CO)_4(\eta^4$–NBD) in the photo–induction period has no influence on the overall outcome of the photocatalytic process. Certainly, this is quite a disappointing result for the ambitious photochemist in search of stereoselective photochemistry as a synthetic tool.

As a surprising result, *exo*–d_2–NBN is formed in addition to *endo*–d_2–NBN and d_2–NTC, when the deuteration of norbornadiene with the *mer*–$Cr(CO)_3(\eta^4$–NBD)–(η^2–ethene) catalyst is performed under 60 bar pressure of D_2 (Scheme 18). Logically, the *exo*–deuteration must involve η^2–coordination of NBD from the *exo*–side. It seems possible that D_2, present in high concentration, blocks two coordination sites and thus hinders *endo*–η^4–coordination of NBD. If so, the supposed binding of NBD from the *exo*–side would become feasible. In accordance with this consideration, norbornene undergoes further hydrogenation with formation of norbornane at the end of the catalytic

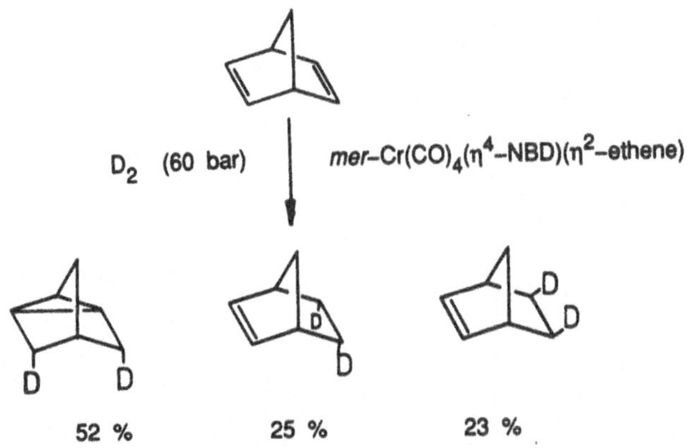

Scheme 18

hydrogenation of NBD under high pressure. This is not observed under atmospheric pressure. In conclusion, a third catalyst species, formulated as $Cr(CO)_3$(*exo*–η^2–NBD)$(D_2)_2$ (Scheme 19), is proposed to be present in the catalytic system under high pressure of D_2 (or H_2).

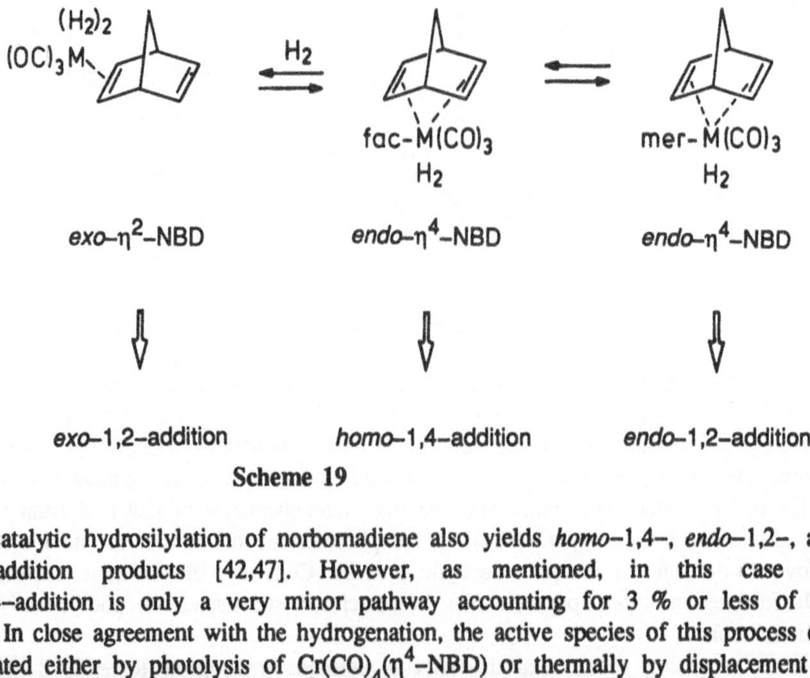

Scheme 19

The catalytic hydrosilylation of norbornadiene also yields *homo*–1,4–, *endo*–1,2–, and *exo*–1,2–addition products [42,47]. However, as mentioned, in this case the *homo*–1,4–addition is only a very minor pathway accounting for 3 % or less of the products. In close agreement with the hydrogenation, the active species of this process can be generated either by photolysis of $Cr(CO)_4$(η^4–NBD) or thermally by displacement of ethene from *mer*–$Cr(CO)_3$(η^4–NBD)(η^2–ethene), Scheme 20, thus again providing clear evidence of the role of the $Cr(CO)_3$ group as the repeating unit in the catalytic cycle. As a further indication of the close mechanistic relationship, the ratio of *exo*– and *endo*–1,2–addition increases dramatically with higher concentration of the silane substrate. Moreover, secondary hydrosilylation with ultimate production of bis(silyl)norbornane derivatives takes place.

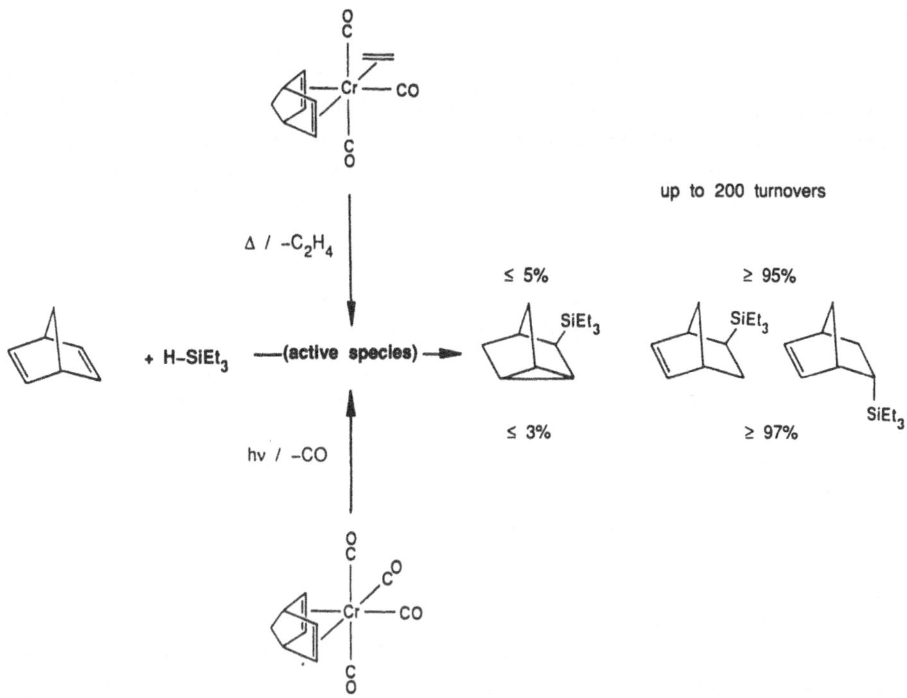

Scheme 20

Concluding remarks

As a main concern the two lectures presented in this article are intended for demonstrating how to gain an insight into the concatenation of events in organometallic photoreactions and photocatalytic processes such as olefin isomerization, hydrogenation, hydrosilylation, etc.. It is certainly appealing (and necessary) to employ sophisticated techniques, inevitably photophysical in nature, in order to go back on the time scale of a photoprocess as far as possible, even to the first femto- and picoseconds following the absorption of the photon. However, for achieving a comprehensive understanding of an organometallic photoreaction in all its complexity it is also necessary (and not less appealing) to employ more conventional techniques aiming, for example, at the isolation of reactive compounds or the analysis of ultimate products. The latter aspects, although possibly less attractive for the excited state enthusiast, are nevertheless essential for a successful application of photochemical methods in organometallic chemistry.

Acknowledgements

I wish to thank my colleagues and students for fruitful collaboration. Their names are cited in the list of references, which also includes a substantial amount of unpublished results. Part of this work was supported by the Alexander von Humboldt-Stiftung (through stipends and other support to S. Özkar and J. Takats) and the EEC (contract SC1-0007-C), which is gratefully acknowledged.

References

[1] Background information on the entire field, as it developed over the decades, is given by (a) W. Strohmeier, *Angew. Chem.* **1964**, *76*, 873–881; *Angew. Chem. Int. Ed. Engl.* **1964**, **3**, 730; (b) E. Koerner von Gustorf and F.–W. Grevels, *Fortschr. Chem. Forsch. (Top. Curr. Chem.)* **1969**, *13*, 366–450; (c) M. Wrighton, *Chem. Rev.* **1974**, *74*, 401; (d) G.L. Geoffroy and M.S. Wrighton, *Organometallic Photochemistry*, Academic Press, New York 1979.

[2] There are several excellent reviews, e.g., (a) M.S. Wrighton, D.S. Ginley, M.A. Schroeder, and D.L. Morse, *Pure Appl. Chem.* **1975**, *41*, 671; (b) L. Moggi, A. Juris, D. Sandrini, and M.F. Manfrin, *Rev. Chem. Intermed.* **1981**, *4*, 171; (c) R.G. Salomon, *TetraheJron* **1983**, *39*, 485.

[3] Turner, J. J., *Spectroscopic Techniques for Organometallic Intermediates*, this volume, pp. 113–123.

[4] C. Daniel, M. Bénard, A. Dedieu, R. Wiest, and A. Veillard, *J. Phys. Chem.* **1984**, *88*, 4805.

[5] Veillard, A., *Photochemistry of Organometallics: Quantum Chemical Approach*, this volume, pp. 173–216.

[6] E. Weitz, *J. Phys. Chem.* **1987**, *91*, 3945.

[7] (a) M. Poliakoff and J. J. Turner, *J. Chem. Soc., Dalton Trans.* **1974**, 2276–2285; (b) M. Poliakoff, *Chem. Soc. Rev.* **7** (1978) 527; (c) M. Poliakoff and E. Weitz, *Acc. Chem. Res.* **20** (1987) 408; (d) M. Poliakoff, *J. Chem. Soc., Dalton Trans.* **1974**, 210–212.

[8] S.P. Church, F.–W. Grevels, H. Herrmann, J.M. Kelly, W.E. Klotzbücher, and K. Schaffner, *J. Chem. Soc. Chem. Commun.* **1985**, 594.

[9] F.–W. Grevels, J. M. Kelly, W. E. Klotzbücher, and K. Schaffner, unpublished results.

[10] (a) J. Dewar and H. O. Jones, *Proc. Roy. Soc. (A)* **1905**, *76*, 558–577; (b) O. Warburg and E. Negelein, *Biochem. Z.* **1929**, *204*, 495–499.

[11] M.A. Schroeder and M.S. Wrighton, *J. Am. Chem. Soc.* **1976**. *98*, 551.

[12] H. Angermund, A.K. Bandyopadhyay, F.–W. Grevels, F. Mark, and K. Schappert, *8th International Symposium on the Photochemistry and Photophysics of Coordination Compounds, Santa Barbara / California* (August 1989), Abstract C–20; full paper in preparation.

[13] The concentrations of the complexes are displayed in Figures 1, 2, and 11 as reduced quantities $\tilde{c}_i = c_i/c^o_A$. Light absorption is measured by means of a modified version of an electronically integrating actinometer (W. Amrein, J. Gloor, and K. Schaffner, *Chimia* **1974**, *28*, 185–188), which compensates for incomplete absorption of light in the sample cell; the integrated light absorption is given in the Figures as the reduced quantity $\tau = Q_{abs} \cdot t/c^o_A$.

[14] C. Krüger, B. L. Barnett, and D. Brauer in E. A. Koerner von Gustorf, F.–W. Grevels, and I. Fischler (Eds.), *The Organic Chemistry of Iron*, Vol. 1, Academic Press, New York, **1978**, pp. 1–112.

[15] (a) J. Demuynck, A. Strich, and A. Veillard, *Nouv. J. Chim.* **1977**, *1*, 217; (b) T.A. Albright, J.K. Burdett, and M.–H. Whangbo, *Orbital Interactions in Chemistry*, Wiley–Interscience, New York 1985, Chapt. 19.

[16] (a) H. Fleckner, F.–W. Grevels, and D. Hess, *J. Am. Chem. Soc.* **1984**, *106*, 2027–2032; (b) H. Fleckner, Doctoral Dissertation, Universität Duisburg, 1987.

[17] Y.–M. Wuu, J.G. Bentsen, C.G. Brinkley, and M.S. Wrighton, *Inorg. Chem.* **1987**, *26*, 530–540.

[18] (a) F. Asinger, B. Fell, and G. Collin, *Chem. Ber.* **1983**, *96*, 716–735; (b) F. Asinger, B. Fell, and K. Schrage, *Chem. Ber.* **1965**, *98*, 372–380; ibid. 381–386.

[19] (a) J.C. Mitchener and M.S. Wrighton, *J. Am. Chem. Soc.* **1981**, *103*, 975–977; (b) R.L. Whetten, K.–J. Fu, and E.R. Grant, *J. Am. Chem. Soc.* **1982**, *104*, 4270–4272.

[20] (a) A.C. Cope, C. R. Ganellin, H.W. Johnson (jr.), T.V. van Auken, and H.J.S. Winkler, *J. Am. Chem. Soc.* **1963**, *85*, 3276–3279; (b) R.G. Ball, G.–Y. Kiel, J. Takats, C. Krüger, E. Raabe, F.–W. Grevels, and R. Moser, *Organometallics* **1987**, *6*, 2260–2261.

[21] (a) M. von Büren and H.–J. Hansen, *Helv. Chim. Acta* **1977**, *60*, 2717–2722; (b) M. von Büren, M. Cosandey, and H.–J. Hansen, *Helv. Chim. Acta* **1980**, *63*, 892–898; (c) H. Angermund, F.–W. Grevels, R. Moser, R. Benn, C. Krüger, and M.J. Romaõ, *Organometallics* **1988**, *7*, 1994–2004.

[22] (a) H. Angermund, A.K. Bandyopadhyay, F.–W. Grevels, and F. Mark, *J. Am. Chem. Soc.* **1989**, **111**, 4656–4661; (b) H. Angermund, Doctoral Dissertation, Universität Duisburg, 1986.

[23] For a description of the time–resolved IR instrumentation used in the Max–Planck–Institut für Strahlenchemie (Mülheim an der Ruhr) see: K. Schaffner and F.–W. Grevels, *J. Mol. Struct.* **1988**, *173*, 51–55.

[24] M. Dartiguenave, Y. Dartiguenave, and H. B. Gray, *Bull. Soc. Chim. France* **1969**, 4223–4225.

[25] C. Krüger, F.–W. Grevels, et al., unpublished results.

[26] (a) E. Delgado, J. Hein, J. C. Jeffery, A. L. Ratermann, F. G. A. Stone, and L. J. Farrugia, *J. Chem. Soc., Dalton Trans.* **1987**, 1191–1199; (b) W.–Y. Zhang, D. J. Jakiela, A. Maul, C. Knors, J. W. Lauher, P. Helquist, and W. Enders, *J. Am. Chem. Soc.* **1988**, *110*, 4652–4660; (c) R. H. Fong and W. H. Hersh, *Organometallics* **1988**, 7, 794–796; (d) P. Binger, B. Biedenbach, R. Schneider, and M. Regitz, *Synthesis*, **1989**, 960–961; (e) P.–J. Colson, M. Franck–Neumann, and M. Sedrati, *Tetrahedron Letters* **1989**, *30*, 2393–2396; (f) P. Eilbracht, C. Hittinger, K. Kufferath, and G. Henkel, *Chem. Ber.* **1990**, *123*, 1079–1087; (g) P. Eilbracht, C. Hittinger, K. Kufferath, A. Schmitz, and H.–D. Gilsing, *Chem. Ber.* **1990**, *123*, 1089–1095.

[27] (a) R. Huq, A. J. Poë, and S. Chawla, *Inorg. Chim. Acta* **1980**, *38*, 121–125; (b) W. R. Hastings, M. R. Roussel, and M. C. Baird, *J. Chem. Soc., Dalton Trans.* **1990**, 203–205.

[28] (a) F.–W. Grevels, F. Mark, and J. Schrickel, unpublished results; (b) J. Schrickel, Doctoral Dissertation in preparation, Universität Duisburg.

[29] G.–Y. Kiel, J. Takats, and F.–W. Grevels, *J. Am. Chem. Soc.* **1987**, *109*, 2227–2229.

[30] (a) R. C. Austin, R. S. Paonessa, P. J. Giordano, and M. S. Wrighton, *Adv. Chem. Ser.* **1978**, *No. 168*, 189–214; (b) D. R. Tyler, R. A. Levenson, and H. B. Gray, *J. Am. Chem. Soc.* **1978**, *100*, 7888–7893; (c) A. J. Poë and C. V. Sekhar, *J. Am. Chem. Soc.* **1986**, *108*, 3673–3679; (d) J. G. Bentsen and M. S. Wrighton, *J. Am. Chem. Soc.* **1987**, *109*, 4518–4530; (e) J. G. Bentsen and M. S. Wrighton, *J. Am. Chem. Soc.* **1987**, *109*, 4530–4544; (f) N. M. J. Brodie, R. Huq, J. Malito, S. Markiewicz, A. J.

Poë, and V. C. Sekhar, *J. Chem. Soc., Dalton Trans.* **1989**, 1933–1939; (g) J. A. DiBenedetto, D. W. Ryba, and P. C. Ford, *Inorg. Chem.* **1989**, *28*, 3503–3507. These papers provide profound information on the state of the art concerning the mechanistic aspects of $M_3(CO)_{12}$ photochemistry. The references cited therein may also serve as a comprehensive bibliography concerning previous relevant papers published by the authors and other researchers in the field.

[31] (a) F.-W. Grevels, J. G. A. Reuvers, and J. Takats, *J. Am. Chem. Soc.* **1981**, *103*, 4069–4073; (b) F.-W. Grevels, J. G. A. Reuvers, and J. Takats, *Inorg. Synth.* **1986**, *24*, 176–180.

[32] B. F. G. Johnson, J. Lewis, and M. V. Twigg, *J. Organomet. Chem.* **1974**, *67*, C75–C76.

[33] F.-W. Grevels, J. G. A. Reuvers, and J. Takats, *Angew. Chem.* **1981**, *93*, 475–477; *Angew. Chem. Int. Ed. Engl.* **1981**, *20*, 452–460.

[34] (a) M. R. Burke, J. Takats, F.-W. Grevels, and J. Takats, *J. Am. Chem. Soc.* **1983**, *105*, 4092–4093; (b) J. Takats, *Polyhedron* **1988**, *7*, 931–941.

[35] F.-W. Grevels, W. E. Klotzbücher, F. Seils, K. Schaffner, and J. Takats, *J. Am. Chem. Soc.* **1990**, *112*, 1995–1996.

[36] I.W. Stolz, G.R. Dobson, and R.K. Sheline, *Inorg.Chem.* **1963**, *2*, 1264–1267.

[37] (b) F.-W. Grevels, J. Jacke, W.E. Klotzbücher, S. Özkar, and V. Skibbe, *Pure Appl. Chem.* **1988**, *60*, 1017–1024.

[38] C. G. Kreiter, *Adv. Organomet. Chem.* **1986**, *26*, 297–375.

[39] F.-W. Grevels, M. Lindemann, R. Benn, R. Goddard, and C. Krüger, *Z. Naturforsch. B* **1980**, *35*, 1298–1309.

[40] (a) F.-W. Grevels and V. Skibbe, *J. Chem. Soc. Chem. Commun.* **1984**, 681–683; (b) J.K. Klassen and G.K. Yang, *Organometallics* **1990**, *9*, 874–876.

[41] V. Skibbe, Doctoral Dissertation, Universität Duisburg, 1985.

[42] D. Chmielewski, Doctoral Dissertation in preparation, Universität Duisburg.

[43] (a) C. Kayran, Doctoral Dissertation, Middle East Technical University, Ankara / Turkey; (b) F.-W. Grevels, C. Kayran, and S. Özkar, manuscript in preparation.

[44] (a) L. Weber and D. Wewers, *Chem. Ber.* **1985**, *118*, 541–550; (b) U. Feldhoff, F.-W. Grevels, R. P. Kreher, K. Angermund, and C. Krüger, *Chem. Ber.* **1986**, *119*, 1919–1930; L. Weber, D. Bungardt, A. Müller, and H. Bögge, *Organometallics* **1989**, *8*, 2800–2804.

[45] (a) B. S. Creaven, F.-W. Grevels, and C. Long, *Inorg. Chem.* **1989**, *28*, 2231–2234; (b) B. S. Creaven, C. Long, R. A. Howie, G. P. McQuillan, and J. Low, *Inorg. Chim. Acta* **1989**, *157*, 151–152.

[46] (a) F.-W. Grevels, J. Jacke, and S. Özkar, *J. Am. Chem. Soc.* **1987**, *109*, 7536–7537.

[47] J. Jacke, Doctoral Dissertation, Universität Duisburg, 1989.

[48] M.F. Gregory, S.A. Jackson, M. Poliakoff, and J.J. Turner, *J. Chem. Soc. Chem. Commun.* **1986**, 1175–1177.

[49] F.-W. Grevels, J. Jacke, F. Mark, S. Özkar, and V. Skibbe, full paper in preparation.

[50] For a quantitative treatment involving ab initio SCF and CAS SCF calculations see: C. Daniel and A. Veillard, *Inorg. Chem.* **1989** *28*, 1170–1173.

[51] F.-W. Grevels, J. Jacke, P. Betz, C. Krüger, and Y.-H. Tsay, *Organometallics* **1989**, **8**, 293–298.

[52] D. Stufkens, private communication.

[53] (a) K. Angermund, F.–W. Grevels, C. Krüger, and V. Skibbe, *Angew. Chem.* **1984**, *96*, 911–913; *Angew. Chem. Int. Ed. Engl.* **1984**, *23*, 904–905; (b) C. Alvarez, A. Pacreau, A. Parlier, H. Rudler and J.–C. Daran, *Organometallics* **1987**, *6*, 1057–1064.
[54] C. Daniel and A. Veillard, *Nouv. J. Chim.* **1986**, *10*, 83–90.
[55] J. Nasielski and A. Colas, *Inorg. Chem.* **1978**, *17*, 237–240.
[56] B.H. Weiller and E.R. Grant, *J. Am. Chem. Soc.* **1987**, *109*, 1252–1253.
[57] (a) J. Nasielski, P. Kirsch and L. Wilputte–Steinert, *J. Organomet. Chem.* **1971**, *27*, C13–C14; (b) G. Platbrood and L. Wilputte–Steinert, *Tetrahedron Lett.* **1974**, 2507–2508.
[58] M. Wrighton and M.A. Schroeder, *J. Am. Chem. Soc.* **1973**, *95*, 5764–5765.
[59] (a) G. Platbrood and L. Wilputte–Steinert, *J. Organomet. Chem.* **1974**, *70*, 407–412; (b) G. Platbrood and L. Wilputte–Steinert, *J. Mol. Catal.* *1*, 265–273.
[60] I. Fischler, M. Budzwait and E.A. Koerner von Gustorf, *J. Organomet. Chem.* **1976**, *105*, 325–330.
[61] M.S. Wrighton and M.A. Schroeder, *J. Am. Chem. Soc.* **1974**, *96*, 6235–6237.
[62] G. Platbrood and L. Wilputte–Steinert, *Bull. Soc. Chim. Belg.* **1973**, *82*, 733–735.
[63] G. Platbrood and L. Wilputte–Steinert, *J. Organomet. Chem.* **1974**, *70*, 393–405.
[64] D.J. Darensbourg, H.H. Nelson (III), and M.A. Murphy, *J. Am. Chem. Soc.* **1977**, *99*, 896–903.
[65] (a) M.J. Mirbach, D.Steinmetz, and A. Saus, *J. Organomet. Chem.* **1979**, *168*, C13–C15; (b) M.J. Mirbach, T.N. Phu, and A. Saus, *J. Organomet. Chem.* **1982**, *236*, 309–320.
[66] (a) W. Gerhartz, F.–W. Grevels, W.E. Klotzbücher, E.A. Koerner von Gustorf, and R.N. Perutz, *Z. Naturforsch. B* **1985**, *40*, 518–523; (b) F.–W. Grevels, J. Jacke, W.E. Klotzbücher, K. Schaffner, R.H. Hooker, and A.J. Rest, *J. Organomet. Chem.* **1990**, *382*, 201–224.
[67] (a) S.A. Jackson, P.M. Hodges, M. Poliakoff, J.J. Turner, and F.–W. Grevels, *J. Am. Chem. Soc.* **1990**, *112*, 1221–1233; (b) P.M. Hodges, S.A. Jackson, J. Jacke, M. Poliakoff, J.J. Turner, and F.–W. Grevels, *J. Am. Chem. Soc.* **1990**, *112*, 1234–1244.
[68] D. Chmielewski, F.–W. Grevels, J. Jacke, and K. Schaffner, *Angew. Chem.* **1991**, *103*, 1361–1363; *Angew. Chem. Int. Ed. Engl.* **1991**, *30*, 1343–1345.

PHOTOCHEMISTRY OF ORGANOMETALLICS : QUANTUM CHEMICAL APPROACH

Alain Veillard
UPR 139 du CNRS
Institut Le Bel, 4, Rue Blaise Pascal
67000 STRASBOURG, France

ABSTRACT. A number of photochemical reactions of organometallics have been studied theoretically during the last years. Potential energy curves and potential energy surfaces have been obtained for some typical reactions from ab initio CI calculations using CASSCF wavefunctions as reference wavefunctions. State correlation diagrams have also been derived for other photochemical reactions. In a number of cases the photoactive excited state has been identified and the reaction path has been elucidated. Only a limited number of mechanisms seems to be operative, two for the reactions leading to the elimination of a closed-shell fragment (carbon monoxide or molecular hydrogen for instance) and one for the homolytic process leading to radical fragments. In most cases the reaction proceeds on more that one potential energy curve or potential energy surface, the system passing from one curve (or surface) to the next one through intersystem crossing or internal conversion.

1. INTRODUCTION

Organometallics have a rich and interesting photochemistry which has been extensively used in the last ten years to generate unsaturated and very reactive species[1]. As pointed out by Meyer, "this extensive photochemistry has generally been based only on product and quantum yield studies. There is little insight in this area into excited-state dynamics, detailed photochemical mechanisms, or the nature of the excited state or states responsible for the photochemistry"[2].

Until recently, current understanding of the photochemical reactions of organometallics has been based on molecular orbital diagrams coupled with an

E. Kochanski (ed.), Photoprocesses in Transition Metal Complexes, Biosystems and Other Molecules. Experiment and Theory, 173–216.

analysis in terms of the bonding and antibonding character of the orbitals involved[1]. It is usually assumed that photodissociation results from exciting an electron from a bonding to the corresponding antibonding orbital. A typical example is $Mn_2(CO)_{10}$, with the homolysis of the metal-metal bond resulting from the excitation to the $^1B_2$ state which corresponds to the promotion of an electron from the σ to the σ^* orbital (σ and σ^* denote the molecular orbitals which are respectively bonding and antibonding with respect to the metal-metal bond)[1]. This type of analysis is conceptually appealing since very simple and certainly useful, but it suffers from some drawbacks :

i) it does not explain easily the existence of concurrent photochemical reactions at a given wavelength. This difficulty may be circumvented by pointing out[1,3] that, for instance in $Re(CO)_5CH_3$ and related systems, the $d_{z^2}(\sigma^*)$ orbital is not only antibonding with respect to the $Re\text{-}CH_3$ bond but also with respect to the Re-CO bond, so that carbonyl dissociation might compete with the homolysis of the metal-methyl bond. The same explanation would hold for the concurrent reactions of $HCo(CO)_4$ [4] and of $Mn_2(CO)_{10}$ [5], although the coefficients of the carbonyl ligands turn out to be rather small in the σ^* orbital.

ii) a related difficulty concerns for instance the photosubstitution reaction of the complexes $Fe(CO)_3$(R-DAB) (R-DAB = 1, 4diaza - 1,3-butadiene) with the loss of a carbonyl ligand observed upon an internal transition of the Fe-(R-DAB) metallacycle[6].

iii) there are a number of reports in the literature of a photochemical reaction being observed upon excitation at different wavelengths[3,7,8]. This seems difficult to interpret on the basis that the photocleavage of a bond results from exciting an electron from the bonding to the antibonding orbital, i.e. from excitation into a single electronic state.

iv) photolysis of the metal-metal bond in $Mn_2(CO)_{10}$ is observed upon excitation into the $^1B_2$ state (corresponding to the $\sigma \rightarrow \sigma^*$ excitation). The potential energy curve (with respect to the metal-metal distance) of this state $^1B_2$ is not dissociative. It should be similar to the potential energy curve for the state $^1\Sigma_u^+$ of H_2, which is bonding and dissociates to ionic products, not to radicals[9].

v) photolysis of the metal-hydrogen bond in $HMn(CO)_5$ occurs upon irradiation at 51800 cm^{-1} [10]. Calculations indicate that the singlet state corresponding to the $\sigma \rightarrow \sigma^*$ excitation lie at much higher energy, above 60000 cm^{-1} [11].

The idea that the photochemical cleavage of a bond is linked to the excitation of an electron from a bonding to the corresponding antibonding orbital is in general too simple to account for the mechanism of the photochemical reactions (cf. below for a further discussion of the validity of this rationale). Another theoretical approach to the mechanism of photochemical reactions is to compute the potential energy surfaces (PES) or potential energy curves (PEC) which connect the reactants to the primary products. Then it becomes possible to get some idea of the path followed by the system from the reactants to the products (usually with some assumptions, concerning for instance the occurence of internal conversion or intersystem crossing). Potential energy surfaces (or potential energy curves) have been reported recently for a few photochemical reactions of organometallics and have allowed a better understanding of the mechanism of these reactions[12-15]. When the calculation of the potential energy surfaces (or curves) exceeds the computational facilities, one can restrict the calculations to a number of remarkable points on the surfaces and then generate the corresponding state correlation diagrams[16-20].

The first part of this chapter is devoted to a survey of the theoretical methods used. The next section deals with the calculation of excited states and excitation energies. The following section summarizes the theoretical results obtained for different classes of photochemical reactions corresponding to :

a) the loss of a carbonyl ligand ;

b) the photosubstitution of metal carbonyls ;

c) the elimination of molecular hydrogen from a dihydride ;

d) the elimination of molecular hydrogen from dihydrogen complexes ;

e) the photolysis of a metal-hydrogen bond ;

f) the photolysis of a metal-metal bond.

A particular interest has been paid to systems like $HCo(CO)_4$ and $HMn(CO)_5$ which undergo concurrent photoreactions either at a unique wavelength or at different wavelengths[4,10,21].

2. METHODOLOGY

A progress in the understanding of the mechanism of the photochemical reactions of organometallics was the use of state correlation diagrams[17-20]. In this approach, one identifies first the ground state and low-lying excited states of the reactants and of the primary products, on the basis of either a qualitative energy level scheme or ab initio calculations. One assumes that some symmetry element is retained during the course of the reaction. Then one sets up a state correlation diagram, based on the spin and symmetry properties, between the reactants and the primary products (rules for the construction of a state correlation diagram may be found in Ref. 16). A photochemical reaction is expected to occur either if the excited molecule goes directly on a single surface (and without any appreciable barrier) to the products or if internal conversion may be achieved easily between the excited state and the ground state surfaces, usually as a consequence of an avoided crossing. State correlation diagrams have helped to identify the symmetry of the excited states involved in a number of photochemical reactions (for instance the $^3E'$ state in the photodissociation of $Fe(CO)_5$[17], the $^3A_1$ state in the photolysis of the Co-H bond in $HCo(CO)_4$ [17] and the

$^3B_2$ state in the photodissociation of $Mn_2(CO)_{10}$ [19]. However they suffer from the following limitations :

i) they are not very useful to identify the spin multiplicity of the excited states involved, since efficient intersystem crossing is relatively common for organometallics[22-24] ;

ii) they do not say anything about the relative stability of the reactant and products and ignore the energy barriers which do not result from avoided crossings ;

iii) in the absence of a wavefunction, one has to make some assumption on the formal oxidation states of the atoms and this is sometimes risky. For instance we assumed in our earlier work[17] that the Co atom in $HCo(CO)_4$ is d^8, in accordance with the common belief[25,26] that this is an hydride compound. This resulted in an avoided crossing between the two states $^1A_1$. Subsequent work[12] has shown that the Co-H bond in this compound should be described as covalent (in agreement with more recent descriptions of the bonding[26,27]), and that the potential energy curve of the $^1A_1$ ground state does not show any avoided crossing.

State correlation diagrams are just a mere approximation to the potential energy surfaces (PES) which connect the reactants to the primary photochemical products. The calculation of these PES represents a more realistic approach, and in principe this could be achieved through a variety of quantum chemical approaches. In reality the situation is less favorable and for instance none of the semi-empirical approaches seems very useful. The Extended Hückel theory, which has been extremely powerful for mapping PES of organometallics in their ground state[28], is not very appropriate for excited states since it ignores the spin. The INDO/S CI formalism yields excitation energies which are not always accurate[29]. Other semi-empirical methods like MNDO and AM1 appear to be rather oriented towards ground-state properties[30] and have not been parameterized for transition metals so far. Density functional methods encounter some problems in the description of bond making and bond breaking and in the treatment of excited states[31]. In the hierarchy of *ab initio* methods, SCF theory does not describe properly the dissociation of covalent bonds and yields only qualitative results for the excited states of organometallics[32] while configuration interaction (CI) based on SCF reference wavefunctions shows a slow convergence of the results as a function of the length of the CI expansion, resulting sometimes in relatively inaccurate excitation energies[33]. So far the best approach, which has been extensively tested[12-14,29,34,35] seems to perform contracted CI calculations[36] based on CASSCF (complete active space SCF) reference wavefunctions[37]. It has been found that this type of calculations, which is referred as MR-CISD (multireference CI with all single and double excitations within a given space) with a CASSCF reference wavefunction, represents a good approximation to a full CI calculation[38].

The right way to calculate potential energy surfaces for a number of electronic states consists of carrying out for each electronic state a CASSCF calculation followed by contracted CI (CCI) calculations[39]. However this method becomes too expensive for most organometallics. The approach which has been used in a number of cases[12,13,14,34] consists of performing a CASSCF calculation for a particular state and using this CASSCF reference wavefunction for the CCI calculations of all the states of interest. The choice of this particular state was dictated by the following considerations. The ground state is not the best candidate, since the CCI calculations would be biased in favour of this state (leading to overestimated excitation energies[34]). Whenever possible, a better choice corresponds to a state with an occupation number of one for the orbitals which are empty in the ground state but populated in the low-lying excited states. Some details regarding these CASSCF calculations are given in Table 1. For $Fe(CO)_5$, the occurence of symmetry breaking solutions in the early CASSCF calculations imposed the use of a CASSCF wavefunction optimized for the ground state[34].

Table 1. The CASSCF calculations

	State	Main configuration[a]	Active space[a]	Nb. of electrons in the active space	Ref.
$Cr(CO)_6$	$^7A_{1g}$	$d_{xy}^1\, d_{xz}^1\, d_{yz}^1\, d_{x^2-y^2}^1\, d_{z^2}^1\, 4s^1$	3d, 4s, 4dπ	6	52
$HMn(CO)_5$	$^5A_2$	$d_{xy}^2\, d_{\pi}^2\, \sigma^2\, \sigma^{*1}\, d_{x^2-y^2}^1$	$3d, 4d_{xy}, 4d\pi, \sigma, \sigma^*$	8	13,34
$H_2Fe(CO)_4$	$^5A_1$	$\sigma_g^2\, d_{y^2-z^2}^1\, \sigma_g^{*1}\, \sigma_u^1\, \sigma_u^{*1}$	$3dy^2\text{-}z^2, \sigma_g, \sigma_g^*, \sigma_u, \sigma_u^*, 2\sigma_g$	6	14
$HCo(CO)_4$	$^3A_1$	$d_{\pi}^4\, d_{\delta}^4\, \sigma 1\, \sigma^* 1$	$3d_\delta, 4d_\delta, \sigma, \sigma^*$	6	12

[a]σ and σ* denote the m.o.'s which are bonding and antibonding with respect to the metal-hydrogen bond.

The *ab initio* potential energy curves are obtained from the CCI calculations. For each electronic state, two CCI calculations are performed at a given point of the surface : the first one with one reference configuration corresponding to the calculated state, the second one being a multireference calculation including all the configurations that appear with a coefficient larger than a given threshold (usually equal to 0.08) in the first monoreference CI wavefunction (the number of reference states being usually less than ten). The electrons correlated are usually the 3d electrons and those of the metal-hydrogen bonds. The CI calculations are of the SD type, including single and double excitations to all virtual orbitals except the counterparts of the carbonyl 1s and of the metal 1s, 2s and 2p orbitals. The number of configurations ranges from a few tens of a thousand up to one million but this number is reduced to at most a few thousands by the contraction. In a number of cases one had to rely on SCF wavefunctions as reference wavefunctions for the CI calculations[15,40]. This was the case for early calculations on $Fe(CO)_5$ [40] and for large systems like the metalloporphyrins[15] (CASSCF calculations being precluded by the size of the system).

The potential energy curves calculated are cross-sections of the many-dimensional potential energy surfaces, obtained with a number of restrictive assumptions :

- the bond lengths are kept fixed (except for the bond which dissociates) and are usually fixed to the experimental values[12,14,40] ;

- the highest possible symmetry is maintained along the reaction path. For instance, for the photodissociation of $HCo(CO)_4$, it has been assumed that C_{3v} symmetry is retained along the reaction paths corresponding to the dissociation of the Co-H and Co-$(CO)_{axial}$ bonds[12]. C_{2v} symetry was maintained along the reaction path corresponding to the dissociation of an equatorial ligand from $Fe(CO)_5$ since the product $Fe(CO)_4$ is of C_{2v} symetry (it was possible to show through state correlation diagrams that qualitatively similar conclusions would be reached by considering either the dissociation of an equatorial ligand under the lower symetry C_s or the dissociation of an axial ligand)[40].

- the geometry may be slightly idealized : for instance the same metal-carbon bond length was used for the axial and equatorial carbonyl ligands in $Fe(CO)_5$ [40].

- additional assumptions are sometimes needed to keep the computations within reasonable limits. For instance, it was assumed that, during the dissociation of a carbonyl ligand from $Fe(CO)_5$, the rearrangement of the fragment $Fe(CO)_4$ will be similar for all the electronic states. This rearrangement was calculated at the SCF level along the potential energy surface $^1A_1$ corresponding to the ground state $^1A'_1$ of $Fe(CO)_5$, _i.e._ for the reaction of thermal elimination[40]. For the photodissociation of $HCo(CO)_4$, the rearrangement is defined by the angle τ between the axial carbonyl ligand and one equatorial ligand. Since the value of τ is close to 100° for both the reactant $HCo(CO)_4$ and the product $Co(CO)_4$, this value has been retained for all the points along the reaction path[12].

- relativistic effects are not included (for this reason, PES have been computed so far only for systems with a metal of the first transition series where these effects are comparatively less important[41,42]).

Technical details regarding the choice of the basis set, the reference wavefunction and the extent of configuration interaction may be found in the original papers[12-15,40].

The potential energy surfaces obtained from the calculations should satisfy the following conditions :

i) they should reproduce correctly the sequence and energetics of the excited states of the reactant. This point will be discussed in detail in the next paragraph.

ii) the structure and the electronic ground state (and possibly the lowest excited states) of the products should be correctly described. Very often, the products of a _primary_ photoreaction are too unstable for conventional studies and their spectrum and structure has not always been elucidated with certainty. Nevertheless the comparison between the theoretical predictions and the experimental data has been possible for a limited number of systems. Theoretical calculations yield for $Fe(CO)_4$ a $^3B_2$ ground state with C_{2v} symetry[40]. Experimentally, $Fe(CO)_4$ has been shown to be paramagnetic[43] and the C_{2v} structure deduced from the infrared spectrum[44] is very close to the theoretical structure. The ground state of $HCo(CO)_3$ has been theoretically assigned to the closed-shell state $^1A_1$ of the C_{3v} structure[35], with experimental evidence for the existence of $HCo(CO)_3$ in two different configurations including the one of C_{3v} symetry[45].

iii) the relative stability of the reactant and products should be correctly described by the calculations (this will insure that the potential energy curves have the right slope). The dissociation of a carbonyl ligand from $Fe(CO)_5$ was calculated to be endothermic by 43 kcal/mole[40], a value which is somewhat too high in the light of more recent theoretical[46] and experimental[47] determinations. The theoretical value for bond dissociation energy of the Co-H bond in $HCo(CO)_4$ is 44 kcal/mole[12] *vs.* an experimental value of 57 kcal/mole[48]. The bond dissociation energy for the Co-C_{axial} bond was calculated as 22 kcal/mole but this value is probably underestimated[12,49] (an experimental value is lacking).

3. THE CALCULATION OF EXCITED STATES.

3.1. General remarks.

One major difficulty in the calculations of excited states is due to the fact that many organometallics are characterized by the occurence of a large number of excited states within a relatively narrow energy region. For instance 18 excited states were located between 33000 and 51000 cm^{-1} for $Fe(CO)_5$ [40]. Further difficulty arises when one wants to compare the calculated excitation energies to experimental values, since most often the electronic spectra of organometallics are poorly resolved (certainly as a consequence of this high density of states). These electronic spectra show usually a broad band (corresponding generally to a charge transfer transition) with one or several shoulders (corresponding to the ligand field LF or d → σ* excitations). Since the excitations which are responsible for the photochemical reactions studied here are the LF, d → σ* and σ → σ* excitations, most of the theoretical studies did not include the metal-to-ligand charge-transfer (MLCT) excited states corresponding to the d → π^{*}_{ligand} excitations or the Rydberg type excited states (corresponding to the 3d → 4s, 4p, 4d excitations). In what follows, we restrict ourselves to *ab initio* calculations of excited states in relation with the study of photochemical reactions.

None of the theoretical studies of the excited states presented here has included a calculation of the oscillator strengthes. This is certainly unfortunate, since in some cases their estimate could help in assigning the bands observed in the experimental absorption spectrum. A typical example corresponds to $HCo(CO)_4$, with the experimental absorption spectrum beginning at about 36000 cm^{-1} and only one resolved, intense band at 44000 cm^{-1} [4]. The lowest singlet state is calculated to be a ^1E state corresponding to a d_δ → σ* excitation[12], which is neither a pure d → d band nor a pure (metal to hydrogen) charge transfer band. It is presently unclear whether the band at 44000 cm^{-1} corresponds to this ^1E state or to a (metal to carbonyl) charge transfer state (in this case, the band corresponding to the $^1A_1 \rightarrow {}^1E$ excitation must be relatively weak and hidden below the intense band at 44000 cm^{-1})[12,35].

3.2. Excited states of $Cr(CO)_6$.

$Cr(CO)_6$ is one organometallic molecule with a relatively well resolved electronic spectrum[50]. The theoretical excitation energies[51,52] are compared in Table 2 to the corresponding experimental values. The sequence of excited states is repro-

Table 2 - Experimental and theoretical excitation energies (in cm^{-1}) for $Cr(CO)_6$

Electronic Excitation	Theoretical[a]	Theoretical[b]	Experimental
$^1A_{1g} \rightarrow {}^1T_{1g}$ LF	37900	34490	29500-31500
$^1A_{1g} \rightarrow {}^1T_{1u}$ MLCT	41650	-	35700
$^1A_{1g} \rightarrow {}^1T_{2g}$ LF	44930	42410	38850
$^1A_{1g} \rightarrow {}^1T_{1u}$ MLCT	47000	-	43600

[a]CI calculations with a SCF reference wavefunction[52].

[b]CI calculations with a CASSCF reference wavefunction[51].

duced correctly by the calculations, but the theoretical excitation energies based on a SCF reference wavefunction are too high by up to one eV. There is a sizeable improvement in the LF excitation energies when the reference wavefunction is a CASSCF wavefunction rather than a SCF wavefunction.

3.3. Excited states of $Fe(CO)_5$.

Using a CASSCF reference wavefunction optimized for the ground state, the LF excited states $^1E'$ and $^1E''$ were calculated respectively at 33400 and 34100 cm^{-1} [34]. As judged from the results obtained for the LF states of $HMn(CO)_5$ with different reference wavefunctions[34], these excitation energies are probably overestimated by about 5000 cm^{-1}. This would put the LF state ^1E at about 28000 - 29000 cm^{-1} [34], in good agreement with the experimental spectrum which shows a very weak absorption in the region of 28600 cm^{-1} [53] and with the fact that the photochemistry starts in this region[53,54]. More recently the $^1E'$ LF state has been calculated at 29100 cm^{-1} from CI calculations based on the use of CASSCF reference wavefunctions optimized for each electronic state[55].

3.4. Excited states of $HMn(CO)_5$

The excitation energies calculated by Veillard et al.[34] and by Daniel[13,56] are shown in Table 3. The experimental spectrum of $HMn(CO)_5$ in the gas phase shows

Table 3 - Theoretical excitation energies (in cm^{-1}) for $HMn(CO)_5$

Electronic excitation	Ref.[34]	Ref.[13,56]
$^1A_1 \rightarrow {}^3E$ LF	25200	23900
$^1A_1 \rightarrow {}^3E\ d_\pi \rightarrow \sigma^*$	-	32900
$^1A_1 \rightarrow {}^3A_2$ LF	27600	33020
$^1A_1 \rightarrow {}^1A_2$ LF	33300 (36700^a)	38500
$^1A_1 \rightarrow {}^1E$ LF	33700^b	33600^b
$^1A_1 \rightarrow {}^1E\ d_\pi \rightarrow \sigma^*$	42300^b	43100^b
$^1A_1 \rightarrow {}^3A_1\ \sigma \rightarrow \sigma^*$	46900	61700
$^1A_1 \rightarrow {}^1B_2\ d_\delta \rightarrow \sigma^*$	53000	-
$^1A_1 \rightarrow {}^1A_2$ MLCT	53600	-
$^1A_1 \rightarrow {}^1E$ MLCT	-	53900^b
$^1A_1 \rightarrow {}^1A_1\ \sigma \rightarrow \sigma^*$	61800^b	-

[a] with a CASSCF reference wavefunction optimized for each electronic state

[b] Symmetry allowed

one broad band around 46700 cm^{-1} with two shoulders at 34500 and 51300 cm^{-1}, these three features having been assigned as MLCT transitions[57]. According to the results of Table 3, the transition at 34500 cm^{-1} should rather correspond to the LF excitations. This assignment is supported by the low value of the intensity in the experimental spectrum, a general feature of the LF transitions. The intense band at 46700 cm^{-1} corresponds probably to several transitions such as the $d_\pi \rightarrow \sigma^*$ and MLCT transitions (the allowed transition $^1A_1 \rightarrow {}^1E$ corresponding to the $d_\pi \rightarrow \sigma^*$ excitation is computed at 42300 cm^{-1}).

3.5. Excited states of $HCo(CO)_4$.

The theoretical excitation energies are given in Table 4. In the experimental

Table 4 - Theoretical excitation energies (in cm-1) for $HCo(CO)_4$

Electronic excitation	Ref.[12]	Ref.[35]
$^1A_1 \rightarrow {}^3E$ $d_\delta \rightarrow \sigma^*$	25900	24250[b]
$^1A_1 \rightarrow {}^3A_1$ $\sigma \rightarrow \sigma^*$	34600	-
$^1A_1 \rightarrow {}^1E$ $d_\delta \rightarrow \sigma^*$	36000[a]	34000-36000[a,c]

[a]Symmetry allowed.

[b]With a CASSCF reference wavefunction optimized for each electronic state.

[c]Estimated

spectrum absorption begins around 36000 cm^{-1}, the only resolved feature being a band at 44000 cm^{-1} [4]. There are two possible ways to fit the theoretical results with the experimental data[35]. The first one considers that the band at 44000 cm^{-1} corresponds to a MLCT transition and that the $d_\delta \rightarrow \sigma^*$ absorption is hidden below,somewhere between 36000 and 44000 cm^{-1}. However Sweany argues[58] that weaker singlet-singlet absorption in this region should have been observable. The second one considers that the band at 44000 cm^{-1} corresponds to the $d_\delta \rightarrow \sigma^*$ excitation. It seems presently difficult to choose between these two possibilities.

3.6. Excited states of $H_2Fe(CO)_4$.

The theoretical excitation energies for some of the symmetry allowed transitions have been calculated by Daniel[14] and are reported in Table 5. The electronic spectrum of this dihydride is characterized by two sets of low-lying states denoted $b^{1,3}A_1$ and $a^{1,3}B_2$ between 26000 cm^{-1} and 40000 cm^{-1}, which result respectively from $d \rightarrow \sigma_g^*$ and $d \rightarrow \sigma_u^*$ excitations (the notations σ_g and σ_g^* denote the bonding combinations of the two hydrogen s orbitals which are respectively bonding and antibonding with respect to the M-H bonds, σ_u and σ_u^* being the antibonding combinations of the two hydrogen s orbitals which are respectively bonding and antibonding with respect to the M-H bonds). At somewhat higher energy

Table 5 - Theoretical excitation energies (in cm^{-1}) for $H_2Fe(CO)_4$ from Ref. [14]

Electronic excitation		
$a^1A_1 \rightarrow b^3A_1$	$3d \rightarrow \sigma_g^*$	26290
$a^1A_1 \rightarrow a^3B_2$	$3d \rightarrow \sigma_u^*$	33240
$a^1A_1 \rightarrow a^1B_2$	$3d \rightarrow \sigma_u^*$	39000
$a^1A_1 \rightarrow b^1A_1$	$3d \rightarrow \sigma_g^*$	39760
$a^1A_1 \rightarrow b^3B_2$	$\sigma_u \rightarrow \sigma_g^*$	45560
$a^1A_1 \rightarrow b^1B_2$	$\sigma_u \rightarrow \sigma_g^*$	53560

one finds the states $b^{1,3}B_2$ corresponding to a $\sigma_u \rightarrow \sigma_g^*$ excitation. The states corresponding to the $\sigma_g \rightarrow \sigma_u^*$ and $\sigma_g \rightarrow \sigma_g^*$ excitations should be higher (above 60000 cm^{-1}) since they correspond to the promotion of an electron from an orbital σ_g which is totally bonding relatively to the three-center interaction between iron and hydrogens. The experimental absorption spectrum of $H_2Fe(CO)_4$ shows a shoulder around 37000 cm^{-1} [59],corresponding probably to the excitations to the states a^1B_2 and b^1A_1.

3.7. Excited states of AlPH (P = porphine dianion)

Stricto sensu, the aluminium hydride AlPH is not an organometallic since it lacks a metal-carbon bond. We have included it in this study since it has been used as a model to study the photolytic cleavage of the aluminium-alkyl bond in the aluminoalkylporphyrins[15]. Moreover it provides a nice illustration of the difficulties associated with the calculation of excitation energies for large conjugated systems. The theoretical excitation energies are reported in Table 6 [15]. The two ^1E $\pi \rightarrow \pi^*$

Table 6 - Theoretical excitation energies (in cm^{-1}) for AlPH from Ref.[15]

Electronic excitation		
$^1A_1 \rightarrow 1^3E$	$\pi \rightarrow \pi^*$	20400
$^1A_1 \rightarrow 2^3E$	$\pi \rightarrow \pi^*$	21200
$^1A_1 \rightarrow 1^1E$	$\pi \rightarrow \pi^*$	24600
$^1A_1 \rightarrow 2^1E$	$\sigma \rightarrow \pi^*$	33200
$^1A_1 \rightarrow 3^3E$	$\sigma \rightarrow \pi^*$	35400
$^1A_1 \rightarrow 3^1E$	$\pi \rightarrow \pi^*$	44100
$^1A_1 \rightarrow {}^3A_1$	$\sigma \rightarrow \sigma^*$	58700

states, which correspond to the Q band and to the Soret band, are found at 24 600 and 44 100 cm^{-1}. The 2^1E state at 33 200 cm^{-1} corresponds to a charge transfer $\sigma \rightarrow \pi^*$ excitation (σ denotes the m.o. which is bonding between Al and H). In the electronic spectra of (OEP) $AlCH_3$ and (TPP)$AlCH_3$ [60] (OEP = octaethylporphin dianion, TPP = tetraphenylporphin dianion), the Q band is located at about 16 800 cm^{-1}, thus about one eV below the excitation energy calculated for AlPH. The $\pi \rightarrow \pi^*$ Soret band is found at about 23 500 cm^{-1}, with the corresponding theoretical value for AlPH too high by more than two eV. This is a well-known default of the calculations of $\pi \rightarrow \pi^*$ excitation energies in the porphyrins which has been traced to the lack of σ–π correlation[61-63]. A band at 28 700 cm^{-1} in the experimental spectrum corresponds probably to the transition to the state 2^1E $\sigma \rightarrow \pi^*$ calculated at 33 200 cm^{-1} for AlPH.

4. THE MECHANISM OF PHOTOCHEMICAL REACTIONS.

4.1. General remarks.

When only one photochemical reaction takes place from a reactant, its mechanism may usually be discussed with the help of potential energy curves. On the contrary, potential energy surfaces may be required to understand the mechanism of concurrent photochemical reactions. Potential energy curves may be either bonding or repulsive (also called dissociative) (Fig. 1a) or they may show bumps as the result of an avoided crossing (Fig. 1b). For the sake of convenience, we shall distinguish two types of photochemical reactions :

i) in the first type, the fragment made from one or two leaving ligands is a closed-shell system, such as a molecule of carbon monoxide, or dinitrogen, or dihydrogen, or a hydrocarbon molecule RH or RR', or a silane molecule (note that the corresponding process may be either heterolytic or homolytic) ;

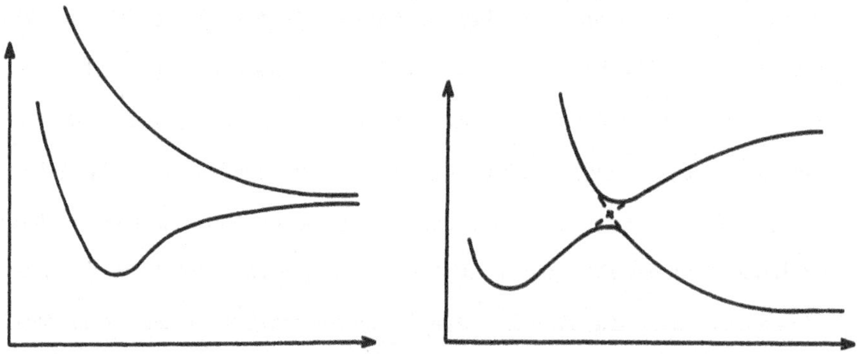

Fig. 1. Potential energy curves : (a) bonding and repulsive ; (b) showing an avoided crossing.

ii) in the second type, the process is homolytic and the two fragments are radicals in a doublet state ; one radical could be either an hydrogen atom or an halogen atom or an alkyl or aryl or silyl group. It could also be an organometallic fragment, such as the product $Mn(CO)_5$ of the photolysis of $Mn_2(CO)_{10}$.

4.2. Photochemical reactions yielding a closed-shell fragment.

4.2.1. Decarbonylation reactions.

The photochemical loss of a carbonyl ligand is a general reaction which is known for a large number of metal carbonyls. The reaction has been reported for the following unsubstituted metal carbonyls[1]

$$M(CO)_6 \xrightarrow{h\nu} M(CO)_5 + CO \quad M = Cr, Mo, W \qquad (1)$$

$$[M(CO)_6]^- \xrightarrow{h\nu} [M(CO)_5]^- + CO \quad M = V, Nb, Ta \qquad (2)$$

$$\mathrm{Fe(CO)_5} \xrightarrow{h\nu} \mathrm{Fe(CO)_4} + \mathrm{CO} \qquad (3)$$

$$\mathrm{Ni(CO)_4} \xrightarrow{h\nu} \mathrm{Ni(CO)_3} + \mathrm{CO} \qquad (4)$$

The details of the mechanism of reaction (1) have been worked out theoretically by Hay[64] and mostly by Burdett *et al.*[65] in order to explain a complicated set of experimental findings obtained from the matrix experiments. Hay pointed out that photodissociation of $Cr(CO)_6$ should yield $Cr(CO)_5$ initially in an excited state since the $^1A_{1g}$ ground state of $Cr(CO)_6$ correlates with the $^1A_1$ ground state of $Cr(CO)_5$, then the potential energy surface of the electronically excited hexacarbonyl must correlate with electronically excited pentacarbonyl. Hay assumed that photodissociation occurs within the singlet manifold and that the excited pentacarbonyl is in the lowest excited state (^1E for the square pyramid). He pointed out that this excited state can eventually yield the $^1A_1$ ground state along certain pathways interconverting the square pyramid (SP) and trigonal bipyramid (TBP) conformations. Burdett *et al.* assumed that the photoactive excited state of $M(CO)_6$ is the $^1T_{2g}$ state (although they also mention a variant of their mechanism which involves the triplet manifold) and analyzed in much detail the pathways interconnecting the ground state $^1A_1$ and the excited state ^1E of the square pyramid conformation of $M(CO)_5$ with the excited state $^1E'$ of the trigonal bipyramid, as shown in Fig. 2. Daniel *et al.* investigated that part of the state correlation diagram connecting $M(CO)_6$ to $M(CO)_5$ assumed to be a square pyramid (this corresponds to the least motion path for the departure of the carbonyl ligand)[40]. They assigned the photoactive excited state of $Cr(CO)_6$ as the $^1T_{1g}$ ligand field state rather than the $^1T_{2g}$ state proposed by Burdett *et al.*. Experimental data seem to support this conclusion, since UV photolysis of $Mo(CO)_6$ occurs upon irradiation at a wavelength of 314 nm (3.95 eV)[65], a value close to the excitation energy of 3.74 eV for the transition $^1A_{1g}$

$\rightarrow {}^1T_{1g}$ but well below the excitation energy of 4.61 eV for the transition ${}^1A_{1g} \rightarrow {}^1T_{2g}$ [1].

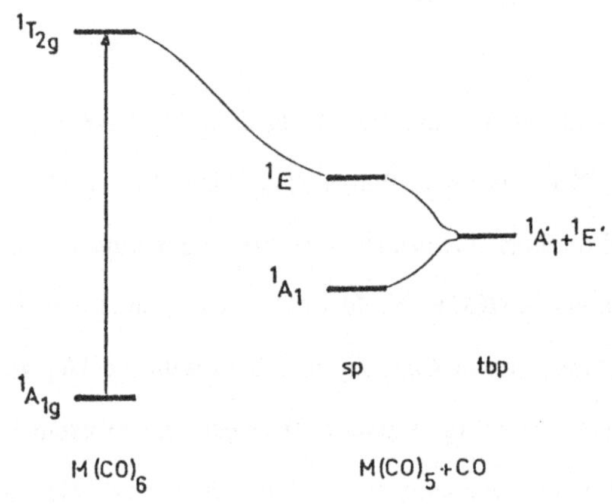

Figure 2 - The state correlation diagram for the photochemical reaction $M(CO)_6 \rightarrow M(CO)_5 + CO$ as proposed in Ref. [65].

The mechanism which has been proposed for the primary step (3) of the photodissociation of $Fe(CO)_5$ is based on state correlation diagrams and potential energy curves[17,40] which show that a single potential energy surface connects, without any barrier, the LF state ${}^3E'$ of $Fe(CO)_5$ to the ground state ${}^3B_2$ of the products $Fe(CO)_4$ + CO. This mechanism is schematically depicted in Fig. 3. Excitation to the LF state ${}^1E'$ is followed by intersystem crossing to the state ${}^3E'$. From there the molecule dissociates to the products of the reaction along the ${}^3B_2$ potential energy surface (with the assumption that C_{2v} symmetry is retained along the reaction path). The conclusion that the LF excited state ${}^3E'$ correlates with the ground states of the products is independant on the assumptions made on the dissociating ligand (equatorial or axial) and on the symmetry retained along the reaction path. This mechanism has been used to interpret the results of laser photolysis experiments on $Fe(CO)_5$ in the gas phase[54].

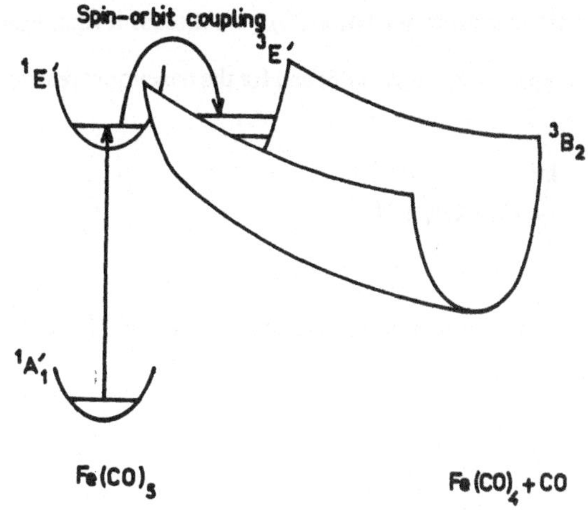

Figure 3 The mechanism proposed for the photochemical dissociation of $Fe(CO)_5$[40].

The photochemical elimination of CO from $Ni(CO)_4$ (reaction (4)) represents another case where the products should be initially in an excited state since the ground state $^1A_1$ of $Ni(CO)_4$ must correlate with the ground state $^1A'_1$ of $Ni(CO)_3$ and the excited states of $Ni(CO)_4$ must correlate with excited states of $Ni(CO)_3$. A luminescence spectrum has been observed upon photolysis of $Ni(CO)_4$ and ascribed to the excited fragment $Ni(CO)_3$ [66]. The photodissociation of $Ni(CO)_4$ has been considered to result from a MLCT excitation into a state ^1E, with the potential energy curve being repulsive[67].

The mechanism of the photochemical loss of a carbonyl ligand has also been studied theoretically for the carbonyl hydrides $HMn(CO)_5$ [13] and $HCo(CO)_4$ [12]

$$HMn(CO)_5 \xrightarrow{h\nu} HMn(CO)_4 + CO \qquad (5)$$

$$HCo(CO)_4 \xrightarrow{h\nu} HCo(CO)_3 + CO \qquad (6)$$

Since the mechanism is simpler for $HCo(CO)_4$, we discuss it first. Potential energy surfaces have been reported for reaction(6) and for the concurrent reaction[4]

$$HCo(CO)_4 \xrightarrow{h\nu} Co(CO)_4 + H \qquad (7)$$

and are shown in Fig. 4 [12]. It was proposed that irradiation at 254 nm (39 400 cm^{-1}),

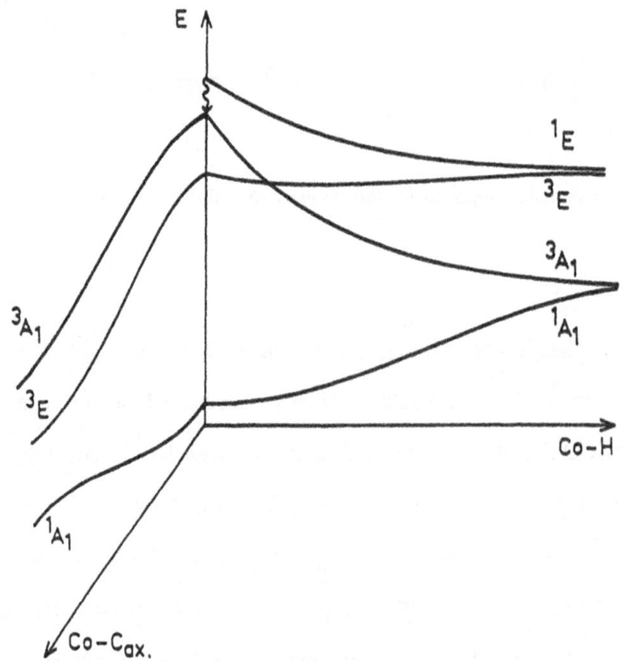

Figure 4 - Potential energy surfaces for the ground and excited states of $HCo(CO)_4$, corresponding to the loss of either the hydrogen atom or the axial carbonyl ligand.

which results in the two photochemical reactions (6) and (7), brings the molecule into the ^1E excited state, this being followed by intersystem crossing to the $^3A_1$ $\sigma \rightarrow \sigma^*$ state. Decarbonylation could take place either along the $^3A_1$ curve (the products being formed in the excited state $^3A_1$) or along the ^3E curve after crossing from the $^3A_1$ to the ^3E surface (the products being formed in the excited state ^3E). Thus chemiluminescence should be observed for the photochemical decarbonylation in the gas phase.

Daniel has reported potential energy surfaces (Fig. 5) for reaction (5) and for the concurrent reaction

$$HMn(CO)_5 \xrightarrow{h\nu} Mn(CO)_5 + H \quad (8)$$

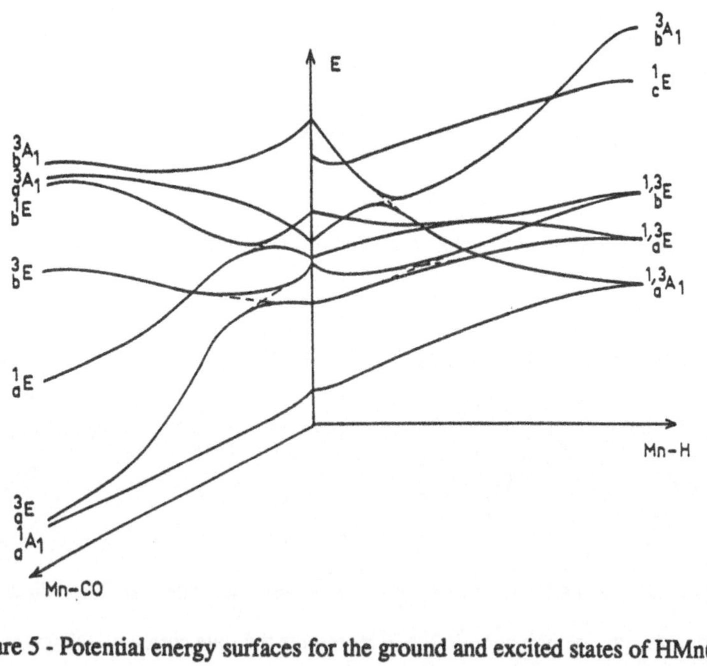

Figure 5 - Potential energy surfaces for the ground and excited states of HMn(CO)5, corresponding to the loss of either the hydrogen atom or the axial carbonyl ligand (reproduced from Ref. [13]).

She proposed that irradiation at 229 nm (43700 cm^{-1}), which results in the decarbonylation reaction (5), brings the molecule into the b^1E state. From there the molecule goes down along the b^1E curve corresponding to Mn-CO elongation until it reaches a potential well corresponding to an avoided crossing. At this point the system evolves to the a^1E state through internal conversion and dissociation to the products CO and $HMn(CO)_4$ will then occur along the a^1E curve [4]. This mechanism produces $HMn(CO)_4$ as a square pyramid with the hydrogen apical, but a subsequent Berry pseudorotation would convert it to a square pyramid with the hydrogen basal (corresponding to the isomer which has been identified experimentally[10]).

Photosubstitution of monosubstituted d^6 metal carbonyls $M(CO)_5L$ may result in CO labilization and this reaction has been found to be highly stereospecific, yielding the cis isomer of the disubstituted product (references may be found in Ref. [20], see also Ref. [68-71]).

$$M(CO)_5\,L + L' \xrightarrow{h\nu} \textit{Cis}\text{ - }M(CO)_4\,LL' + CO \qquad (9)$$

This reaction is the result of a photoelimination followed with a nucleophilic reaction :

$$M(CO)_5\,L \xrightarrow{h\nu} M(CO)_4\,L + CO \qquad (10)$$

$$M(CO)_4\,L + L' \xrightarrow{h\nu} M(CO)_4\,LL' \qquad (11)$$

In that case, SCF and CI calculations were used to generate state correlation diagrams for the primary photoelimination[20]. Fig. 6 represents the state correlation diagram for the photoelimination of the axial carbonyl ligand from $M(CO)_5L$. On the basis of this diagram, the following mechanism was proposed for the photosubstitution of the axial carbonyl :

i) excitation of $M(CO)_5L$ into the ^1E LF state is followed by elimination of the carbonyl ligand with the species $M(CO)_4L$ formed in the excited state ^1E as a square pyramid with the ligand L apical ;

ii) $M(CO)_4L$ evolves along a Berry pseudorotation path first to a trigonal bipyramid, then to a square pyramid with L basal in the $^1A'$ excited state ;

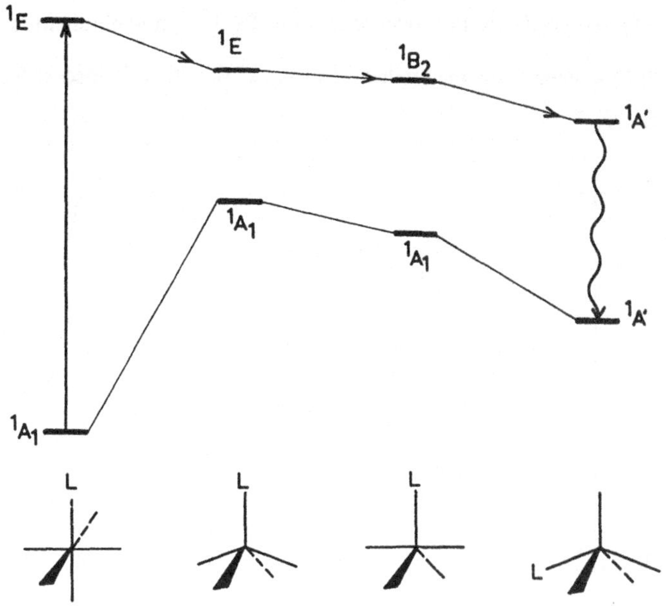

Fig. 6 - State correlation diagram for the photoelimination of the axial carbonyl ligand from $M(CO)_5L$ (for the sake of clarity this ligand is not mentionned in the diagram).

iii) $M(CO)_4L$ being trapped in the potential well corresponding to this excited state $^1A'$ can evolve only through internal conversion to the ground state $^1A'$;

iv) $M(CO)_4L$ is now trapped in the potential well of the ground state $^1A'$, corresponding to a square pyramid with L basal, and as such can react with an incident nucleophile to give $M(CO)_4LL'$ with the *cis* structure (Scheme 1).

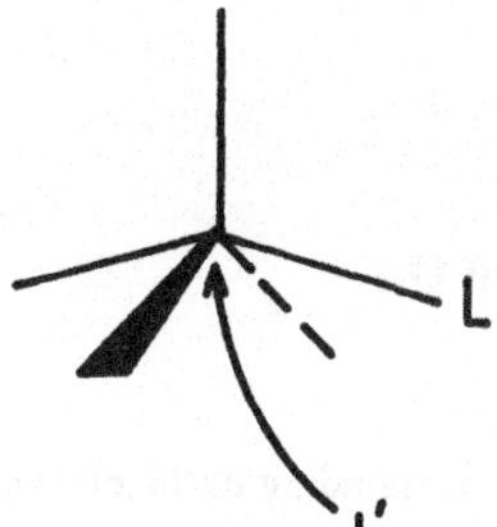

Scheme 1

A similar analysis leads to the same conclusion for the photoelimination of an equatorial ligand, namely that the product $M(CO)_4LL'$ should be formed with the *cis* structure (Fig. 7).

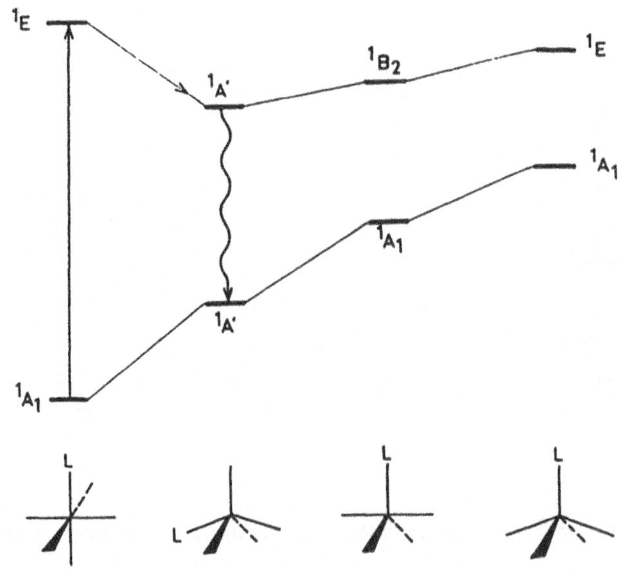

Figure 7 - State correlation diagram for the photoelimination of an equatorial carbonyl ligand from $M(CO)_5L$.

4.2.2. Elimination of molecular hydrogen from a dihydride.

When $H_2Fe(CO)_4$ is irradiated in a low-temperature matrix, the primary process is the loss of dihydrogen[72]

$$H_2Fe(CO)_4 \xrightarrow{h\nu} H_2 + Fe(CO)_4 \quad (12)$$

The potential energy curves corresponding to the elimination of molecular hydrogen under C_{2v} constraint have been calculated by Daniel[14] and are shown in Fig. 8. The

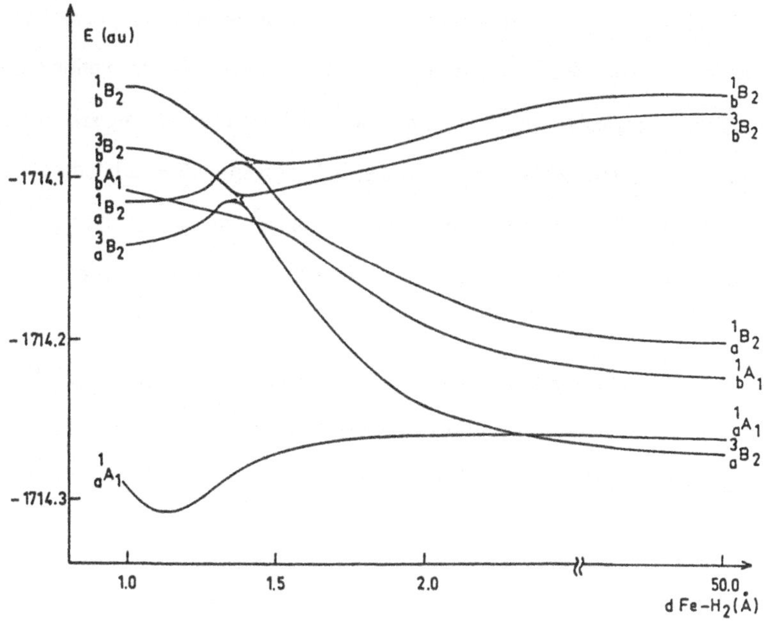

Fig. 8 - Potential energy curves for the ground and excited states of $H_2Fe(CO)_4$, corresponding to the elimination of molecular hydrogen (reproduced from Ref. 14)

curves originating from the states $a^{1,3}B_2$ ($3d \rightarrow \sigma_u^*$) and b^1A_1 ($3d \rightarrow \sigma_g^*$) are dissociative with respect to the elimination of molecular hydrogen. These potential energy curves form the basis for a qualitative understanding of the photochemistry of $H_2Fe(CO)_4$[14]. Irradiation at the experimental wavelength of 254 nm (39 400 cm^{-1}) will bring the molecule into the nearly degenerate states a^1B_2 and b^1A_1. From there the system has the choice between two reactive channels.

i) the dissociation along the b^1A_1 potential energy curve, with formation of the products H_2 and $Fe(CO)_4$, the latter in the excited state b^1A_1,

ii) the population of the a^3B_2 state through intersystem crossing from the a^1B_2 state, followed by elimination of molecular hydrogen along the a^3B_2 curve (with a small energy barrier of the order of 10 kcal/mol), the products being formed in their ground state.

One important feature of this mechanism is the nature of the a^3B_2 curve in its dissociative part, where it corresponds to a $\sigma_u \rightarrow \sigma_g^*$ excitation. The promotion of an electron from a molecular orbital that is hydrogen-hydrogen antibonding and metal-hydrogen bonding to a molecular orbital that is hydrogen-hydrogen bonding and metal-hydrogen antibonding leads to the elimination of H_2. This seems to be a general feature of the photochemistry of the transition metal dihydrides and polyhydrides.

State correlation diagrams have been proposed for the elimination of H_2 from a number of transition metal dihydrides and polyhydrides[73,74]. The elimination of H_2 from $MoCp_2H_2$, probably close to concerted, is well documented[75,76]

$$MoCp_2H_2 \xrightarrow{h\nu} MoCp_2 + H_2 \qquad (13)$$

A state correlation diagram has been proposed, with the assumption that C_{2v} symmetry is retained during the course of the reaction (Fig. 9). The excited state $^3B_2$

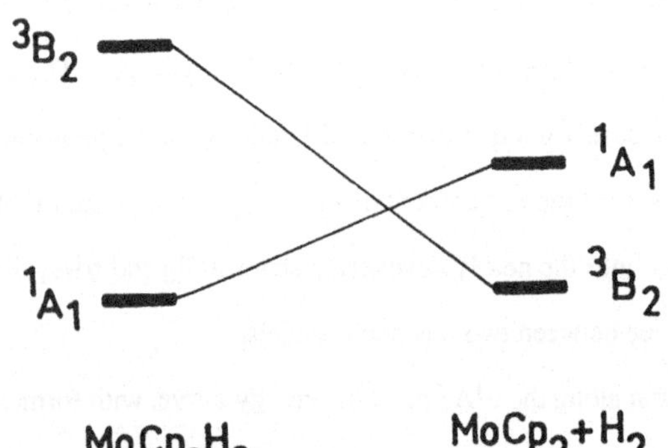

Fig. 9 - State correlation diagram for the elimination of H_2 from $MoCp_2H_2$.

of $MoCp_2H_2$ corresponds to a one-electron excitation from an orbital b_2 that is metal-hydrogen bonding and hydrogen-hydrogen antibonding to an orbital a_1 that is metal-hydrogen antibonding and hydrogen-hydrogen bonding. This state $^3B_2$ could be reached through intersystem crossing after excitation into the corresponding $^1B_2$ state and correlates directly with the ground state of the products.

Irradiation of dihydride complexes of d^6 Ru and Rh leads to elimination of molecular hydrogen[77,78]

$$RuH_2(CO)(PPh_3)_3 \xrightarrow{h\nu} H_2 + Ru(CO)(PPH_3)_3 \quad (14)$$

$$RhH_2Cl(PPh_3)_3 \xrightarrow{h\nu} H_2 + RhCl(PPh_3)_3 \quad (15)$$

A state corelation diagram (Fig. 10) has been derived for the hypothetical reaction

$$MH_2L_4 \xrightarrow{h\nu} H_2 + ML_4 \quad (16)$$

MH_2L_4 being a *Cis* dihydride of a metal d^6, with the assumption that C_{2v} symmetry

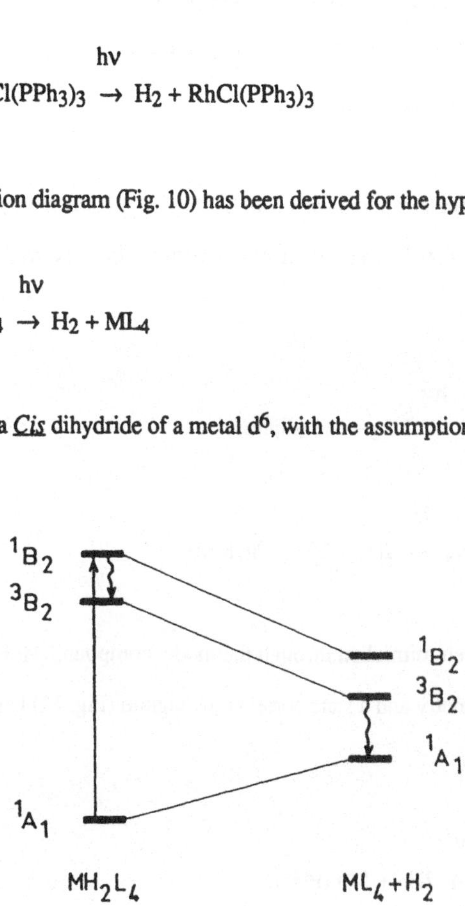

Fig. 10 - State correlation diagram for the elimination of H_2 from ML_4H_2.

is retained along the reaction path. In analogy with $H_2Fe(CO)_4$, the reactant MH_2L_4 has a set of low-lying excited states $^{1,3}B_2$ corresponding to the $\sigma_u \rightarrow \sigma_g^*$ excitation and the corresponding $^{1,3}B_2$ potential energy curves must be dissociative. One may deduce some qualitative features from this diagram :

i) the direct thermal reaction is allowed ;

ii) photochemical elimination of dihydrogen will result from excitation into the $^1B_2$ state, possibly followed by intersystem crossing to the $^3B_2$ state ;

iii) the products should be formed initially into an excited state, $^1B_2$ or $^3B_2$ (chemiluminescence might be observed in the gas phase).

More realistic potential energy curves will probably show avoided crossings between the potential energy curves of different states $^{1,3}B_2$.

Irradiation of $MoH_4(diphos)_2$ and $MoH_4(PPh_2Me)_4$ leads to the elimination of H_2 and probably to the formation of the unstable species $Mo(diphos)_2$ and $Mo(PPh_2Me)_4$ [79]

$$MoH_4(diphos)_2 \xrightarrow{h\nu} 2H_2 + Mo\ (diphos)_2 \quad (17)$$

$$MoH_4(PPh_2Me)_4 \xrightarrow{h\nu} 2H_2 + Mo\ (PPh_2Me)_4 \quad (18)$$

These systems have been mimicked through the model compound $MoH_4(PH_3)_4$ with a structure of D_{2d} symmetry and a state correlation diagram (Fig. 11) has been derived for the reaction

$$MoH_4(PH_3)_4 \xrightarrow{h\nu} 2H_2 + Mo\ (PH_3)_4 \quad (19)$$

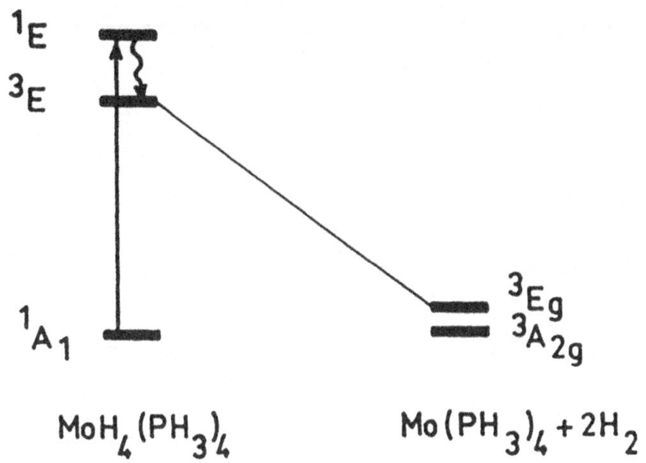

Fig. 11 - State correlation diagram for the elimination of two hydrogen molecules from $MoH_4(PH_3)_4$ (a square-planar structure is assumed for $Mo(PH_3)_4$).

with the assumption that D_{2d} symmetry is retained along the reaction path and that the product $Mo(PH_3)_4$ has a square planar structure (the assumption of a tetrahedral structure would lead to a similar diagram)[73,74]. $MoH_4(PH_3)_4$ has a relatively low-lying set of excited states $^{1,3}E$ corresponding to the one-electron excitation $e \rightarrow a_1$, where e denotes an occupied orbital that is metal-hydrogen bonding and hydrogen-hydrogen antibonding and a_1 denotes an empty orbital that is metal-hydrogen antibonding and hydrogen-hydrogen bonding. The corresponding potential energy curves will be dissociative with respect to the elimination of two hydrogen molecules.

4.2.3. Elimination of molecular hydrogen from dihydrogen complexes.

$Cr(CO)_5(H_2)$, a system with coordinated molecular hydrogen, is destroyed by UV light[80]

$$Cr(CO)_5(H_2) \xrightarrow{h\nu} Cr(CO)_5 + H_2 \qquad (20)$$

A state correlation diagram (Fig. 12) has been derived with the assumption that C_{2v}

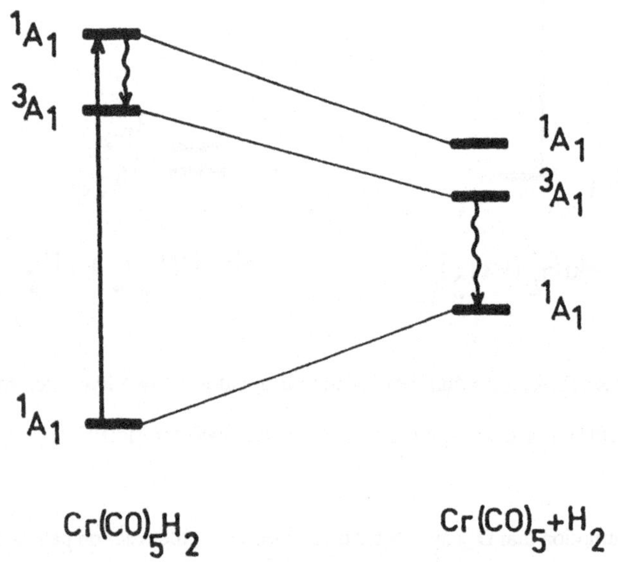

Fig. 12 - State correlation diagram for the dissociation of H_2 from $Cr(CO)_5(H_2)$.

symmetry is retained along the reaction path[73,74]. $Cr(CO)_5(H_2)$ has two low-lying excited states of $^{1,3}A_1$ symmetry corresponding to a one-electron excitation from an occupied orbital that is metal-(H_2) bonding into an empty orbital that is metal-(H_2) antibonding. The corresponding potential energy curves must be dissociative with respect to the elimination of H_2, the products being formed in an excited state. The state correlation diagram for the photoelimination of H_2 from $Ni(CO)_3(H_2)$[81]

$$Ni(CO)_3(H_2) \xrightarrow{h\nu} Ni(CO)_3 + H_2 \qquad (21)$$

should be similar.

4.3. Photochemical reactions yielding radical fragments.

4.3.1. Photolysis of a metal-hydrogen bond.

Photolysis of the cobalt-hydrogen bond of $HCo(CO)_4$ (reaction (7)) occurs upon irradiation at 254 nm[82]. The potential energy curves corresponding to the loss of the hydrogen atom (Fig. 4) form the basis for a qualitative understanding of the mechanism of this photoreaction, the key of this mechanism being the dissociative character of the $^3A_1$ $\sigma \rightarrow \sigma^*$ curve. It has been proposed[12] that irradiation at 254 nm (39 400 cm^{-1}) brings the molecule into the ^1E state (corresponding to the $d_\delta \rightarrow \sigma^*$ excitation). After intersystem crossing to the $^3A_1$ $\sigma \rightarrow \sigma^*$ state, dissociation of the Co-H bond takes place along the $^3A_1$ curve, with formation of the radical products H and $Co(CO)_4$ in their ground states.

Homolysis of the metal-hydrogen bond of $HMn(CO)_5$ (reaction (8)) takes place upon irradiation at 193 nm (51 800 cm^{-1}). The mechanism proposed by Daniel[13] is based on the potential energy curves of Fig. 5 corresponding to the loss of the hydrogen atom. Irradiation brings the molecule into the c^1E state (corresponding to a $d_\pi \rightarrow \pi^*$ excitation). Intersystem crossing at a metal-hydrogen distance of about 1.70 Å will bring the system first into the state b^3A_1 and next through internal conversion into the a^3A_1 $\sigma \rightarrow \sigma^*$ state. From there the molecule will dissociate along this a^3A_1 curve into the radical products H and $Mn(CO)_5$ in their ground state. For both $HCo(CO)_4$ and $HMn(CO)_5$, the $^3A_1$ $\sigma \rightarrow \sigma^*$ dissociative curve is reached through intersystem crossing from near-by singlet states, but the $^1A_1$ $\sigma \rightarrow \sigma^*$ state does not play any role in these mechanisms, since it is a high-lying state computed above 60 000 cm^{-1} for both systems[11,12] (a consequence of its ionic character[9]).

The photolysis of the Al-H bond in AlPH (P = porphine dianion)

$$\mathrm{AlPH} \xrightarrow{h\nu} \mathrm{H} + \mathrm{AlP} \tag{22}$$

has been studied[15] as a model for the photoinduced cleavage of the metal-carbon bond in σ-bonded alkyl and aryl metalloporphyrins[83]

$$\mathrm{MPR} \xrightarrow{h\nu} \mathrm{MP} + \mathrm{R} \tag{23}$$

(R = alkyl or aryl groups, P being a porphin dianion with substituent groups). Potential energy curves for the dissociation of the Al-H bond in AlPH are shown in Fig. 13. The following mechanism has been proposed for the photoinduced cleavage

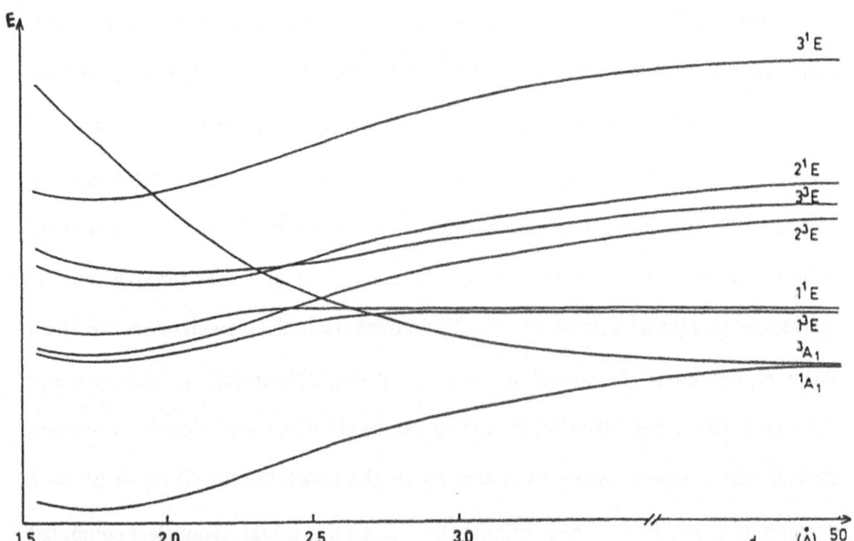

Fig. 13 - Potential energy curves for the dissociation of the Al-H bond in AlPH.

of the Al-H bond in AlPH or of the Al-C bond in the aluminoalkylporphyrins[15]. Irradiation in the Q band or in the Soret band brings the molecule into the 1^1E or 3^1E states. Next intersystem crossing brings the molecule into the lowest excited state 1^3E. Then one possible path corresponds to the molecule moving along this 1^3E curve until it crosses the $^3A_1$ curve (corresponding to the $\sigma \rightarrow \sigma^*$ excitation) around 2.7 Å. Afterwards the molecule dissociates to the products in their ground state along the $^3A_1$ curve. To reach the crossing point at 2.7 Å, the molecule has to go over a barrier of the order of 18 kcal/mole, a fact which probably explains the low quantum yield observed experimentally (between 10^{-2} and 10^{-3}).

4.3.2. Photolysis of a metal-metal bond.

One primary photoreaction of $Mn_2(CO)_{10}$ is the homolytic metal-metal bond cleavage (see Ref. 5 and references therein)

$$Mn_2(CO)10 \xrightarrow{h\nu} 2Mn(CO)_5 \qquad (24)$$

A state correlation diagram[19] (Fig. 14) shows that the excited state $^3B_2$ (correspon-

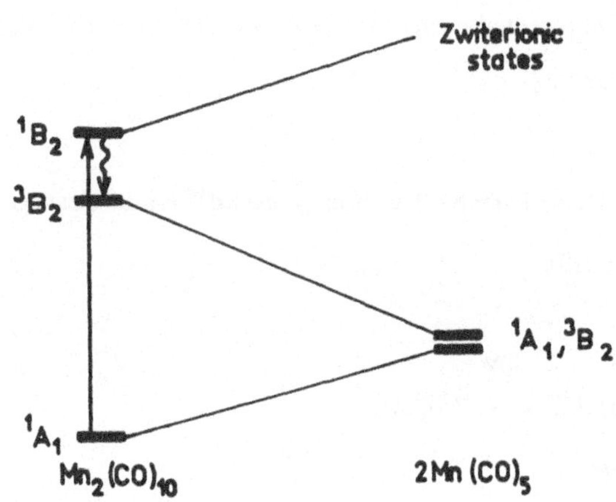

Fig. 14 - State correlation diagram for the dissociation of $Mn_2(CO)_{10}$.

ding to the $\sigma \rightarrow \sigma^*$ excitation, where σ and σ^* denote the molecular orbitals which are respectively metal-metal bonding and antibonding) is connected by a single potential energy surface to the ground state of the products (with the assumption that D_{4d} or C_{4v} symmetry is retained during the course of the reaction). The excited state $^3B_2$ could be reached through excitation to the low-lying state $^1B_2$ followed by intersystem crossing. Presently there are no accurate potential energy curves for the dissociation of $Mn_2(CO)_{10}$, however the state correlation diagram of Fig. 14 may easily be translated into qualitative potential energy curves (Fig. 15).

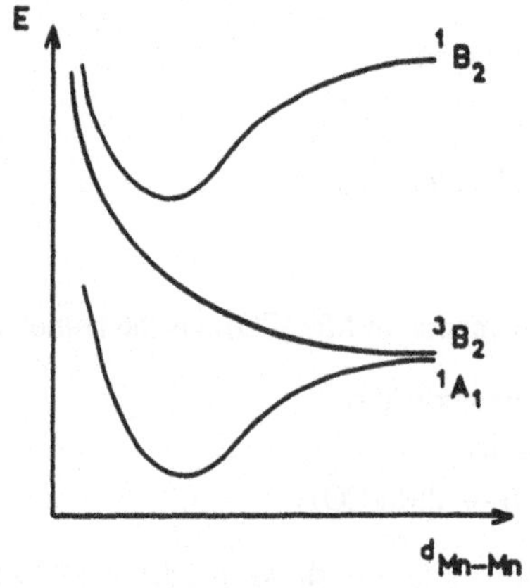

Fig. 15. Qualitative potential energy curves for the dissociation of $Mn_2(CO)_{10}$ into two fragments $Mn(CO)_5$.

Similar state correlation diagrams have been proposed[19] for the cleavage of the metal-metal bond in $Co_2(CO)_8$

$$Co_2(CO)_8 \xrightarrow{h\nu} 2Co(CO)_4 \tag{25}$$

and $[CpMo(CO)_3]_2$

$$[CpMo(CO)_3]_2 \xrightarrow{h\nu} 2\,CpMo(CO)_3 \tag{26}$$

4. CONCLUSION

Through the calculation of potential energy surfaces and the construction of state correlation diagrams, it has been possible to get a better understanding of the mechanism of the photochemical reactions of organometallics. The computation of reasonnably accurate potential energy surfaces for these systems is an expensive and time-consuming process. Fortunately, the mechanisms which have been unveiled must be rather general. In fact, only three different situations in terms of potential energy curves have been found so far, two for the elimination of a closed-shell fragment and one for the homolytic process leading to radical fragments. These situations are depicted in Fig. 16 and correspond respectively to the followings :

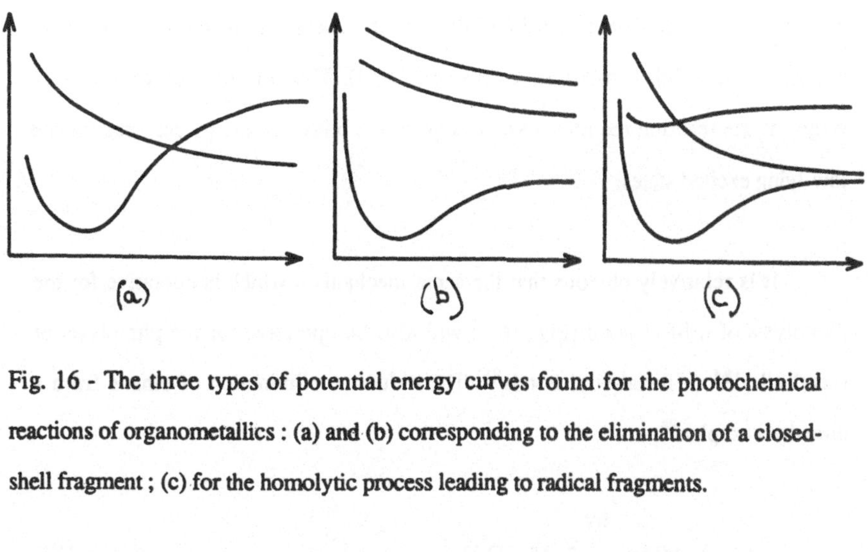

Fig. 16 - The three types of potential energy curves found for the photochemical reactions of organometallics : (a) and (b) corresponding to the elimination of a closed-shell fragment ; (c) for the homolytic process leading to radical fragments.

i) the reactant and the products have different ground states and there is one potential energy curve which connects an excited state of the reactant with the ground state of the products and which is dissociative.

ii) the reactant and the products have the same ground state. There is one set of potential energy curves (corresponding to different multiplicities) which connects some excited states of the reactants to the corresponding excited states of the products and which is dissociative.

iii) for the homolytic process, the potential energy curve corresponding to the triplet state $^3(\sigma^1\ \sigma^{*1})$ is dissociative and leads to the products in their ground state.

One will notice that the potential energy curve leading to the products may be reached only in a late stage of the reaction (when the corresponding excitation energy is too high in the equilibrium region of the reactant). This is the case for the photolysis of the M-H bond in $HMn(CO)_5$ and AlPH and for the photoinduced elimination of molecular hydrogen from $H_2Fe(CO)_4$. This potential energy curve is dissociative since it corresponds to a one-electron excitation from a bonding orbital to the corresponding antibonding orbital. It is reached either through internal conversion at a funnel (this being the case for $HMn(CO)_5$) or through a crossing with a potential energy curve which is associative (case of AlPH). Thus information on the early stages of the reaction requires a knowledge of the potential energy surfaces for the low-lying excited states.

It is relatively obvious that the same mechanism which is operative for the photolysis of a M-H bond (Fig. 16 c) will also be operative for the photolysis of metal-alkyl[84] or metal-silyl bonds[85]. The reductive elimination of ethane from a dimethyl complex[86]

$$L_nMMe_2 \xrightarrow{h\nu} L_nM + C_2H_6 \qquad (27)$$

certainly bears some analogy to the photoelimination of dihydrogen.

Ultimately, changes in the photochemical behavior of related systems may be explained through their potential energy curves. Meyer noticed[2] that the complexes $[(bpy)_2\ M(CO)H]^+$, where M = Ru,Os, are photochemically inert with respect to the homolysis of the metal-hydrogen bond whereas, in the related complexes [(bpy)Re $(CO)_3H$], a relatively efficient Re-H photochemistry does exist, with the MLCT excited state being higher in energy in the latter system. As shown in Fig. 17, raising

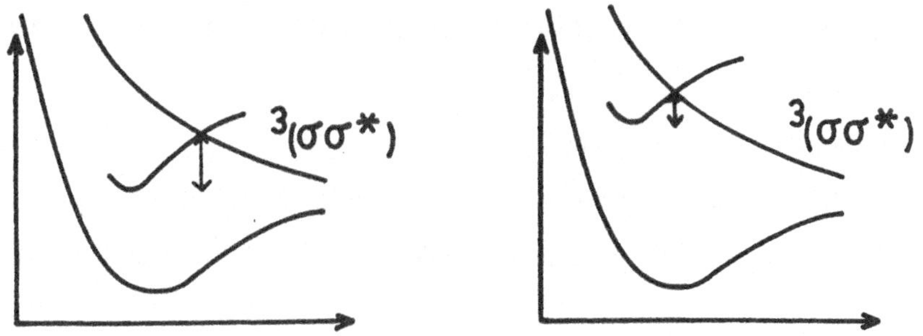

Fig. 17 - The influence of the position of an excited state on the barrier to photolysis of a M-H or M-R bond.

an excited state may lower the energy barrier preventing the crossing to the $^3(\sigma\sigma^*)$ state.

More work will be needed before the mechanism of these photochemical reactions is completely understood. Photochemical studies in the gas phase certainly would be useful, as would be experiments on the wavelength dependence of the photolytic processes. Theoretical evaluations of the probability of nonradiative processes together with estimates of the lifetime of the excited states should provide some information on the efficiency of the possible photoprocesses.

REFERENCES

[1] G.L. Geoffroy and M.S. Wrighton, "Organometallic photochemistry" (Academic Press, New-York, 1979).

[2] E.G. Megehee and T.J. Meyer, Inorg. Chem. 28 (1989) 4084.

[3] K.M. Young, T.M. Miller and M.S. Wrighton, J. Am. Chem. Soc. 112 (1990) 1529.

[4] R.L. Sweany, Inorg. Chem. 21 (1982) 752.

[5] T.J. Meyer and J.V. Caspar, Chem. Rev. 85 (1985) 187 and references therein.

[6] H.K. van Dijk, D.J. Stufkens and A. Oskam, J. Am. Chem. Soc. 111 (1989) 541.

[7] T. Kobayashi, H. Ohtani, H. Noda, S. Teratani, H. Yamazaki and K. Yasufuku, Organometallics 5 (1986) 110.

[8] D.A. Prinslow and V. Vaida, J. Am. Chem. Soc. 109 (1987) 5097.

[9] L. Salem and C. Rowland, Angew. Chem. Intern. Ed. 11 (1972) 92.

[10] S.P. Church, M. Poliakoff, J.A. Timney and J.J. Turner, Inorg. Chem. 22 (1983) 3259.

[11] C. Daniel, unpublished work.

[12] A. Veillard and A. Strich, J. Am. Chem. Soc. 110 (1988) 3793.

[13] C. Daniel, Coord. Chem. Rev. 97 (1990) 141.

[14] C. Daniel, J. Phys. Chem. 95 (1991) 2394.

[15] M.M. Rohmer and A. Veillard, New J. Chem., in press.

[16] N.J. Turro "Modern Molecular Photochemistry", Benjamin, Menlo Park, California, USA, 1978, p. 206 and following.

[17] A. Veillard, Nouv. J. Chim. 5 (1981) 599.

[18] A. Veillard and A. Dedieu, Theoret. Chim. Acta 63 (1983) 339.

[19] A. Veillard and A. Dedieu, Nouv. J. Chem. 7 (1983) 683.

[20] C. Daniel et A. Veillard, Nouv. J. Chem. 10 (1986) 83.

[21] A. J. Rest and J.J. Turner, J. Chem. Soc. Chem. Comm. (1969) 375.

[22] H.K. van Dijk, D.J. Stufkens and A. Oskam, J. Am. Chem. Soc. 111 (1989) 541.

[23] T.J. Meyer, Pure Appl. Chem. 58 (1986) 1193.

[24] D.M. Manuta and A.J. Lees, Inorg. Chem. 25 (1986) 1354.

[25] B.R. James "Homogeneous hydrogenation", J. Wiley, New-York, 1973, p. 7.

[26] R.G. Pearson, Chem. Rev. 85 (1985) 41.

[27] R.H. Crabtree "The organometallic chemistry of the transition metals", J. Wiley, New-York, 1988, p. 55.

[28] T.A. Albright, J.K. Burdett and M.H. Whangbo "Orbital interactions in chemistry", J. Wiley, New-York 1988 and references therein.

[29] A. Marquez and C. Daniel, to be published.

[30] M.J.S. Dewar, E.G. Zocbisch, E.F. Healy and J.J.P. Stewart, J. Am. Chem. Soc. 107 (1985) 3902.

[31] D.R. Salahub and M.C. Zerner, in "The Challenge of d and f electrons", D.R. Salahub and M.C. Zerner eds, ACS Symposium Series 394 (1989) 1.

[32] A. Veillard and J. Demuynck, in "Modern theoretical chemistry.4. Applications of electronic structure theory", H.F. Schaefer ed., Plenum, New-York, 1977, p. 187.

[33] C. Daniel and A. Veillard, in "Quantum chemistry : the challenge of transition metals and coordination chemistry", A. Veillard ed., NATO ASI Series, Reidel, Dordrecht, 1986, p. 363.

[34] A. Veillard, A. Strich, C. Daniel and P.E.M. Siegbahn, Chem. Phys. Let. 141 (1987) 329.

[35] A. Veillard, C. Daniel and M.M. Rohmer, J. Phys. Chem. 94 (1990) 5556.

[36] P.E.M. Siegbahn, Int. J. Quant. Chem. 23 (1983) 1869.

[37] P.E.M. Siegbahn, J. Almlöf, A. Heiberg, B.O. Roos, J. Chem. Phys. 74 (1981) 2384.

[38] C.W. Bauschlicher and S.R. Langhoff, J. Chem. Phys. 86 (1987) 5595.

[39] See for instance J. Matos, B. Roos and P. Malmquist, J. Chem. Phys. 86 (1987) 1458.

[40] C. Daniel, M. Benard, A. Dedieu, R. Wiest and A. Veillard, J. Phys. Chem. 88 (1984) 4805.

[41] R.L. Martin and P.J. Hay, J. Chem. Phys. 75 (1981) 4539.

[42] K.S. Pitzer, Acc. Chem. Res. 12 (1979) 271.

[43] T.J. Barton, R. Grinter, A.J. Thomson, B. Davies and M. Poliakoff, J.C.S. Chem. Commun. (1977) 841.

[44] M. Poliakoff and J.J. Turner, J. Chem. Soc. Dalt. Trans. (1974) 2276.

[45] R.L. Sweany and F.N. Russell, Organometallics 7 (1988) 719.

[46] L.A. Barnes, M. Rosi and C.W. Bauschlicher, J. Chem. Phys. 94 (1991) 2031.

[47] K.E. Lewis, D.M. Golden and G.P. Smith, J. Am. Chem. Soc. 106 (1984) 3905.

[48] F. Ungvary, J. Organomet. Chem. 36 (1972) 363.

[49] L. Versluis, T. Ziegler, E.J. Baerends and W. Ravenek, J. Am. Chem. Soc. 111 (1989) 2018.

[50] N.A. Beach and H.B. Gray, J. Am. Chem. Soc. 90 (1968) 5713.

[51] C. Daniel and A. Strich, unpublished results.

[52] C. Daniel, Thèse de Doctorat d'Etat, Strasbourg, 1985.

[53] J.T. Yardley, B. Gitlin, G. Nathanson and A.M. Rosan, J. Chem. Phys. 74 (1981) 370.

[54] T.A. Seder, A.J. Ouderkirk and E. Weitz, J. Chem. Phys. 85 (1986) 1977.

[55] A. Marquez, C. Daniel and J. Fernandez, submitted for publication.

[56] C. Daniel, to be published.

[57] G.B. Blakney and W.F. Allen, Inorg. Chem. 10 (1971) 2763.

[58] R.L. Sweany, private communication.

[59] R.L. Sweany, in "Transition Metal Hydrides", A. Dedieu ed., Verlag, New-York, in press.

[60] R. Guilard, A. Zrineh, A. Tabard, A. Hendo, B.C. Han, C. Lecomte, M. Souhassou, A. Habbou, M. Ferhat and K.M. Kadish, Inorg. Chem. 29 (1990) 4476.

[61] D.C. Rawlings, E.R. Davidson and M. Gouterman, Int. J. Quantum Chem. 26 (1984) 237, 251.

[62] D.C. Rawlings, M. Gouterman, E.R. Davidson and D. Feller, Int. J. Quantum Chem. 28 (1985) 773, 797, 823.

[63] U. Nagashima, T. Takada and K. Ohno, J. Chem. Phys. 85 (1986) 4524.

[64] P.J. Hay, J. Am. Chem. Soc. 100 (1978) 2411.

[65] J.K. Burdett, J.M. Grzybowski, R.N. Perutz, M. Poliakoff, J.J. Turner and R.F. Turner, Inorg. Chem. 17 (1978) 147.

[66] N. Rösch, M. Kotzian, H. Jörg, H. Schröder, B. Rager and S. Metev, J. Am. Chem. Soc. 108 (1986) 4238.

[67] N. Rösch, H. Jörg and M. Kotzian, J. Chem. Phys. 86 (1987) 4038.

[68] B.H. Weiller and E.R. Grant, J. Am. Chem. Soc. 109 (1987) 1252.

[69] B.H. Weiller and E.R. Grant, J. Phys. Chem. 92 (1988) 1458.

[70] M.F. Gregory, S.A. Jackson, M. Poliakoff and J.J. Turner, J. Chem. Soc. Chem. Commun (1986) 1175.

[71] S.T. Belt, D.W. Ryba and P.C. Ford, Inorg. Chem. 29 (1990) 3633.

[72] R.L. Sweany, J. Am. Chem. Soc. 103 (1981) 2410.

[73] A. Veillard, Chem. Phys. Let. 170 (1990) 441.

[74] C. Daniel and A. Veillard, in "Transition metal hydrides" A. Dedieu ed., Verlag Chemie, New-York, 1991.

[75] G.L. Geoffroy and M.G. Bradley, Inorg. Chem. 17 (1978) 2410.

[76] J. Chetwynd-Talbot, P. Grebenik and R.N. Perutz, Inorg. Chem. 21 (1982) 3647.

[77] G.L. Geoffroy and M.G. Bradley, Inorg. Chem. 16 (1977) 744.

[78] D.A. Wink and P.C. Ford, J. Am. Chem. Soc. 108 (1986) 4838.

[79] R. Pierantozzi and G.L. Geoffroy, Inorg. Chem. 19 (1980) 1821.

[80] R.K. Upmacis, G.E. Gadd, M. Poliakoff, M.B. Simpson, J.J. Turner, R. Whyman and A.F. Simpson, J. Chem. Soc., Chem. Commun. (1985) 27.

[81] R.L. Sweany, M.A. Polito and A. Moroz, Organometallics 8 (1989) 2305.

[82] R.L. Sweany, Inorg. Chem. 19 (1980) 3512.

[83] References may be found in Ref. [15].

[84] H.G. Alt, Angew. Chem. Int. Ed. Engl. 23 (1984) 766.

[85] C.L. Reichel and M.S. Wrighton, Inorg. Chem. 19 (1980) 3858.

[86] A. Becalska and R.H. Hill, J. Am. Chem. Soc. 111 (1989) 4346.

PHOTOCHEMISTRY OF METAL-METAL BONDED CARBONYLS AND ITS RELATIONSHIP TO ELECTRON TRANSFER CHAIN CATALYSIS

D.J. STUFKENS, T. v.d. GRAAF, G.J. STOR AND A. OSKAM
Anorganisch Chemisch Laboratorium
University of Amsterdam
Nieuwe Achtergracht 166
1018 WV Amsterdam
The Netherlands

ABSTRACT. Irradiation of the complexes $(CO)_5MnMn(CO)_3$(α-diimine) into their visible absorption band gives rise to homolysis of the metal-metal bond and release of CO from the $Mn(CO)_3$(α-diimine) moiety. Only the homolysis reaction leads to product formation at room temperature, loss of CO is the only reaction observed at low temperatures. The radicals formed by the homolysis reaction undergo very interesting radical coupling and electron transfer (chain) reactions. The mechanisms of these reactions are discussed and attention is paid to the character of the excited state from which the reactions take place. The corresponding halide complexes $Mn(CO)_3$(bpy)X show a similar formation of $Mn(CO)_3$(bpy) radicals. Preliminary m.o. data of $Mn(CO)_3$(bpy)X (X=Cl,I) and resonance Raman spectra obtained for $Re(CO)_3$(pTol-DAB)Br confirm the presence of a lowest LLCT state for these halide complexes.

1. Introduction

Mechanistic studies in the field of organometallic photochemistry have mainly been confined to transition metal carbonyls. This also holds for the photochemical reactions of metal-metal bonded complexes, for which most mechanistic information has been obtained for the carbonyl dimers $M_2(CO)_{10}$ (M=Mn,Re), $Cp_2Fe_2(CO)_4$ and $Cp_2M_2(CO)_6$ (M=Mo,W) [1-3]. Upon irradiation these complexes undergo homolysis of the metal-metal bond and/or release of CO.

Substitution of two carbonyl ligands by an α-diimine molecule such as 2,2'-bipyridine (bpy) produces a complex with an absorption band in the visible region, which has all the characteristic features (intensity, solvatochromism, resonance Raman spectra) of one or more metal to α-diimine charge transfer (MLCT) transitions [4,5]. Our interest in these α-diimine substituted metal-metal bonded complexes arose from their high photosensitivity, which contrasts with the photostability of most other α-diimine complexes having a lowest MLCT state [6,30]. The stability of such MLCT states is caused by the stronger ionic interaction between the metal and ligand in the MLCT state which normally compensates for the weakening of the metal-ligand covalent bond. Only in a few cases have reactions from a lowest MLCT state been observed although the quantum yields never exceeded the value of 10^{-2} [7-9]. Because of this photostability, complexes such as $Ru(bpy)_3^{2+}$ and $Re(CO)_3$(bpy)Cl have been used as rather stable photosensitizers for electron and energy transfer processes [10,11].

This article describes the primary photoprocess and secondary thermal reactions of the complexes $(CO)_5M'M(CO)_3$(α-diimine) (M,M'=Mn,Re) in which the α-diimine ligand represents one of the molecules depicted in Fig. 1. The results are related to the

E. Kochanski (ed.), Photoprocesses in Transition Metal Complexes, Biosystems and Other Molecules. Experiment and Theory, 217–232.

photochemical behaviour of the mononuclear complexes $M(CO)_3$(α-diimine)X (M=Mn,Re; X=Cl,Br,I).

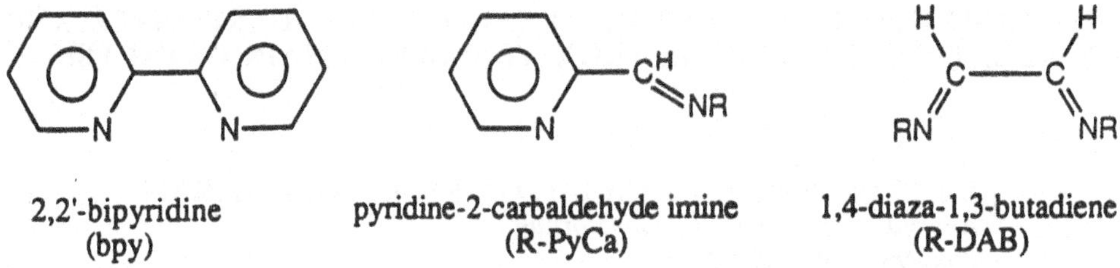

Figure 1. Structures of the α-diimines bpy, R-PyCa and R-DAB in their chelating conformation.

2. Photoreactions of $(CO)_5MnMn(CO)_3$(α-diimine) complexes

The structure of a $(CO)_5MnMn(CO)_3$(R-DAB) complex is shown in Fig. 2 and the absorption spectrum of $(CO)_5MnMn(CO)_3$(iPr-DAB) in *n*-pentane is depicted in Fig. 3. As

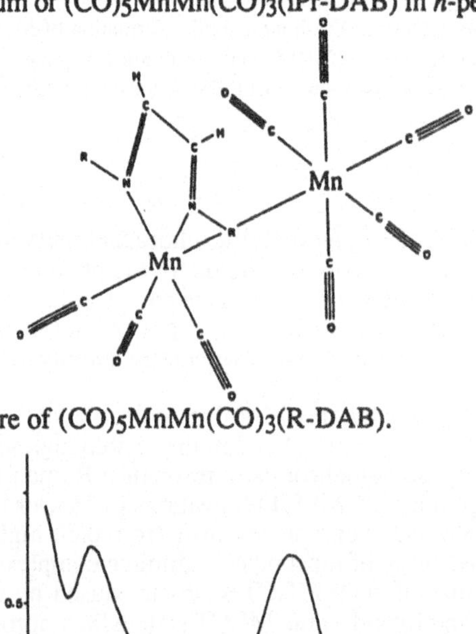

Figure 2. Structure of $(CO)_5MnMn(CO)_3$(R-DAB).

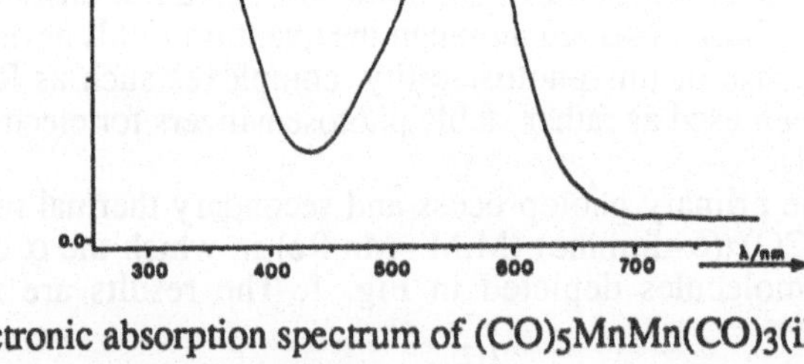

Figure 3. Electronic absorption spectrum of $(CO)_5MnMn(CO)_3$(iPr-DAB) in *n*-pentane.

mentioned in the introduction, the low-energy absorption band of these complexes has a rather high intensity (ε=(8-12)x$10^3M^{-1}cm^{-1}$), is solvatochromic and the resonance Raman spectra closely resemble those obtained by excitation into the visible bands of mononuclear α-diimine complexes such as $M(CO)_4$(α-diimine) (M=Cr,Mo,W) [4]. Based on these observations this low energy band is assigned at least in part to one or more MLCT transitions.
All photoreactions to be described hereafter have been performed by irradiation into this low energy band. An interesting aspect of these complexes is the large variety of reactions taking place in different media and in the presence or absence of Lewis bases.

2.1 PHOTOLYSIS IN NON VISCOUS SOLVENTS

Irradiation of the complexes in toluene or THF at room temperature gave rise to homolysis of the metal-metal bond and formation of the radicals $Mn(CO)_5$ and $Mn(CO)_3$(α-diimine) (reaction 1). In addition, a CO-loss reaction occurred, but the CO-loss product thus obtained reacted back with CO to give the starting complex. This has been established with nanosecond flash photolysis [12].
Both radicals, $Mn(CO)_5$ and $Mn(CO)_3$(α-diimine), dimerised to give $Mn_2(CO)_{10}$ and $Mn_2(CO)_6$(α-diimine)$_2$ respectively (reactions 2 and 3). However, the latter species were partly split into their radicals and the equilibrium of reaction 3 shifted to the left when the α-diimine ligand became either more bulky (e.g. tBu-DAB) or more electron

$$(CO)_5MnMn(CO)_3(\alpha\text{-diimine}) \xrightarrow{h\nu} Mn(CO)_5 + Mn(CO)_3(\alpha\text{-diimine}) \quad (1)$$

$$2Mn(CO)_5 \longrightarrow Mn_2(CO)_{10} \quad (2)$$

$$2\,Mn(CO)_3(\alpha\text{-diimine}) \rightleftharpoons Mn_2(CO)_6(\alpha\text{-diimine})_2 \quad (3)$$

withdrawing (e.g. pTol-DAB). In these cases the radicals could be detected and characterized with ESR without using a spin-trapping agent [13,14]. According to these ESR spectra the unpaired electron is mainly localized at the α-diimine ligand and the $Mn(CO)_3$(α-diimine) radicals are therefore 16-electron radical complexes $^+Mn(CO)_3$(α-diimine$^-$).
The corresponding $Re(CO)_3$(α-diimine) radicals, obtained by irradiation of a $(CO)_5MnRe(CO)_3$(α-diimine), $Ph_3SnRe(CO)_3$(α-diimine) or $Cp(CO)_2FeRe(CO)_3$(α-diimine) complex, did not dimerise. They could therefore more easily be studied with ESR [13] and IR [15] spectroscopy. The reason for this different behaviour of the Mn- and Re-radicals will be discussed hereafter.
If the complexes are irradiated in CCl_4 or $CHCl_3$, the radicals produced in reaction 1 do not dimerise but react instead according to reaction 4 by halogen atom transfer

$$(CO)_5MnMn(CO)_3(\alpha\text{-diimine}) \xrightarrow[CCl_4,\ CHCl_3]{h\nu} Mn(CO)_5Cl + Mn(CO)_3(\alpha\text{-diimine})Cl \quad (4)$$

from the chlorocarbon [16,17]. Upon irradiation of the complexes in CH_2Cl_2, which has a higher C-halogen bond dissociation energy than CCl_4 and $CHCl_3$, the $Mn(CO)_5$ radicals dimerised instead, producing $Mn_2(CO)_{10}$, whereas the $Mn(CO)_3$(α-diimine) radicals still underwent the halogen atom transfer reaction. Apparently, dimerisation of the $Mn(CO)_3$(α-diimine) radical complexes is much slower than that of the $Mn(CO)_5$ radicals. This is not surprising in view of the fact that the $Mn(CO)_5$ radicals are 17e-metal centred radicals and the $Mn(CO)_3$(α-diimine) species are much more stable 16e-ligand centred radicals having a π^* singly occupied molecular orbital (SOMO). Due to the fact that the $Mn(CO)_3$(α-

diimine) radicals are ligand centred radical complexes, their dimerisation will not proceed via a simple metal-metal coupling reaction as in the case of the $Mn(CO)_5$ radicals. Instead, these radicals react with each other by attack of the radical anions at the coordinatively unsaturated and positively charged metal centre of their counterparts. This reaction mechanism, established by irradiation of a $(CO)_5MnMn(CO)_3$(R-DAB) complex at about 180 K, is depicted in Scheme I [14]. The first product formed (A) was characterized by

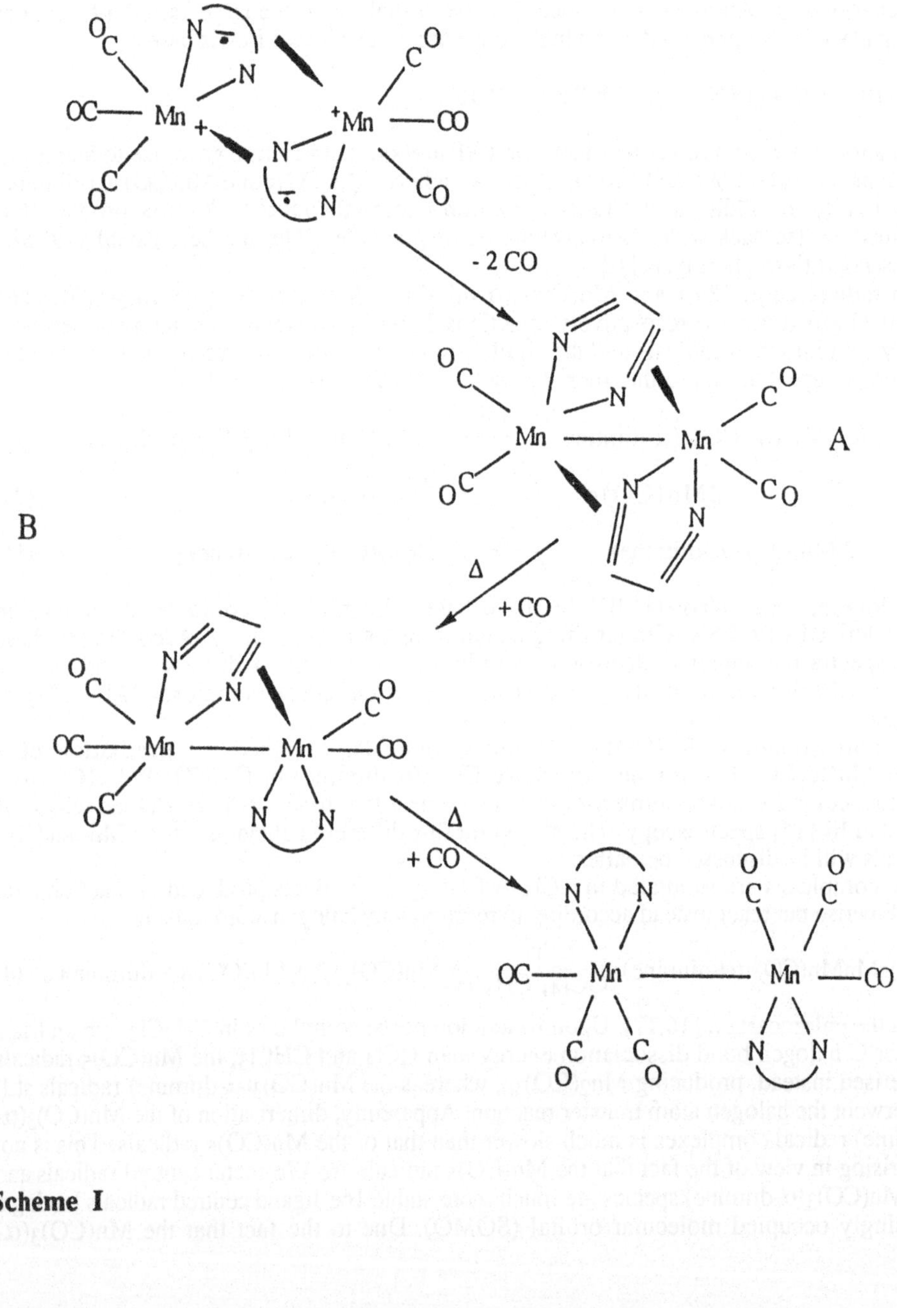

Scheme I

comparing its CO-stretching frequencies with those of the stable analogue $Ru_2(CO)_4$(R-DAB)$_2$ [18,19]. Since the reaction was performed in the IR cell, CO released by formation of A could not escape and reacted back with A to give product B and finally the dimer upon raising the temperature above 200 K. This reaction could of course only be observed for the complexes of the R-DAB and R-PyCa ligands which possess a reactive CN bond that can bind to the metal. As mentioned above dimerisation of the $Mn(CO)_3$(α-diimine) radicals may be prevented by either the electronic or steric influences of the α-diimine ligand. Thus, dimers were formed in the case of the bpy and R-PyCa complexes but not for all R-DAB compounds. For R=iPr reaction 3 still proceeded but for R-DAB complexes having a more electron withdrawing substituent R such as p-Tolyl (pTol) or p-Anisyl (pAn) no appreciable amounts of the dimer were formed. Apparently, the radical is then too stable to dimerise by electron transfer from the π^* orbital of the radical anion to the metal-d orbital of a second radical (Scheme I). This influence of the relative energies of the metal (d) and α-diimine (π^*) orbitals on the dimer formation may also be responsible for the lack of dimerisation in the case of the $Re(CO)_3$(α-diimine) radicals. Due to the much stronger ligand field exerted by Re, electron transfer to the lowest empty d-orbital of this metal from another radical will be much more unfavourable than in the case of the Mn-radicals. Because of this, $Re(CO)_3$(α-diimine) radicals could more easily be detected and investigated spectroscopically [13].

There are in principle two different ways in which the dimerisation of the $Mn(CO)_3$(α-diimine) radicals can be prevented. First, irradiation of the metal-metal bonded complex in a very viscous medium such as paraffin causes the $Mn(CO)_5$ and $Mn(CO)_3$(α-diimine) radicals to stay together in the solvent cage by lack of diffusion. Second, dimer formation can also be avoided by irradiation in a coordinating solvent or by adding a Lewis base to the solution. In both cases the open site at the metal of the 16e-radical will become occupied, thus preventing the dimerisation reaction. In the following two sections the reactions then taking place will be discussed in some detail.

2.2 PHOTOLYSIS IN VISCOUS SOLVENTS

Contrary to the bpy complex, all R-PyCa and R-DAB complexes gave rise to the formation of yet another binuclear photoproduct. The relative yields of these compounds compared to that of the dimer increased by going to more viscous solvents. The products were not formed when CCl_4 was added to the solution. For those R-DAB complexes which did not show dimer formation (e.g. R=pTol or pAn), this new complex was in fact the only product formed. Different products were formed for the R-PyCa and the R-DAB complexes and these could easily be identified and structurally characterized [14]. The products and the proposed mechanism of their formation are schematically depicted in Scheme II. The mechanism is supported by the observation that the reaction is quenched by CCl_4 and that it proceeds more efficiently when the concentration of the $Mn(CO)_5$ radicals is enhanced either by adding $Mn_2(CO)_{10}$ or by photodecomposing the $Mn_2(CO)_{10}$ already present (reaction 2) by shortening the wavelength of excitation. The influence of the viscosity on the product formation is selfevident. It is proposed that the stable product formed in the case of the R-PyCa complexes is an unstable intermediate in the case of the R-DAB complexes.

The relative quantum yield of this reaction with respect to that of reactions 2 and 3 increased upon lowering the temperature of the paraffin solution. In solid paraffin and in a PVC film at room temperature it was even the only reaction taking place.

When a solution of the complexes in 2-MeTHF or toluene was irradiated at still lower temperatures (T<183 K), the above reactions of the radicals could hardly be observed

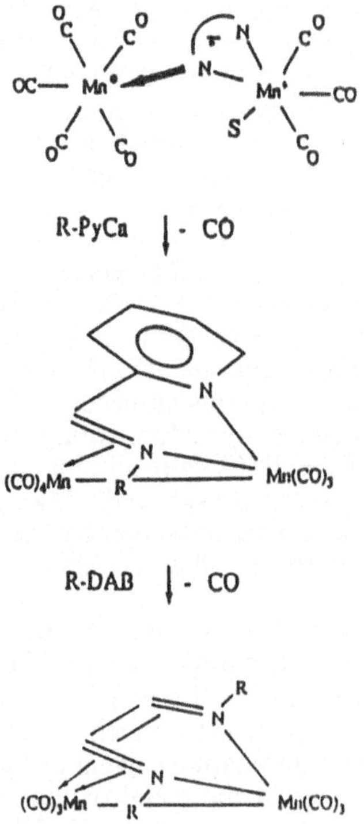

Scheme II.

anymore. Instead a CO-loss reaction 5 occurred with formation of a CO-bridged complex [14].

$$(CO)_5MnMn(CO)_3(\alpha\text{-diimine}) \xrightarrow[T<183\ K]{h\nu} (CO)_4Mn(\mu\text{-}CO)Mn(CO)_2(\alpha\text{-diimine}) + CO \quad (5)$$

This was evident from the appearance of the typical IR bands of free CO (ν=2132 cm^{-1}) and of a semi-bridging carbonyl ($\nu \cong$ 1820 cm^{-1}). The proposed structure of this photoproduct is presented in Fig. 4.

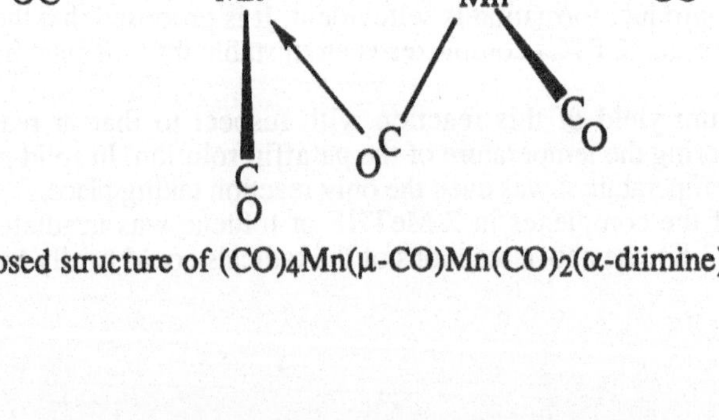

Figure 4. Proposed structure of $(CO)_4Mn(\mu\text{-}CO)Mn(CO)_2(\alpha\text{-diimine})$.

The competition between the formation of CO-loss and homolysis products depended strongly on the temperature and solvent used. Thus, both in 2-MeTHF and toluene, ca. 90% of the complex photoreacted with release of CO at 183 K, whereas homolysis was still the only reaction taking place at this temperature in CH_2Cl_2.

2.3 PHOTOCHEMICAL REACTIONS WITH LEWIS BASES

The $Mn(CO)_3$(α-diimine) radicals are coordinatively unsaturated and can therefore still take up another ligand. This adduct formation with a Lewis base could easily be demonstrated by adding a small amount of $P(OMe)_3$ to a preirradiated solution of $Ph_3SnRe(CO)_3$(tBu-DAB) [13]. This complex had been selected since it produced persistent $Re(CO)_3$(tBu-DAB) radicals that gave rise to an ESR spectrum showing extensive hyperfine splittings. This spectrum then changed immediately into that of a new paramagnetic species with an extra hyperfine splitting of 35.78 G for one ^{31}P nucleus. Because of their reactivity, adduct formation with phosphines could not be studied with ESR for the corresponding Mn-radicals.

Due to the negative charge of the radical anion, the $Mn(CO)_3$(α-diimine) radicals will not easily take up a hard base. This became evident when the complexes were irradiated in neat THF [20]. The same reactions 1-3 then occurred which had been observed before in toluene. When, however, the same reaction in THF was performed at T≤200 K, the complex photodisproportionated into $Mn(CO)_5^-$ and $Mn(CO)_3$(α-diimine)$(THF)^+$. Apparently, the THF molecules can then form a radical adduct $Mn(CO)_3$(α-diimine)(THF) which reduces the parent complex providing $Mn(CO)_3$(α-diimine)$(THF)^+$ and $Mn(CO)_5^-$. The mechanism of this reaction 6 will be discussed hereafter when the photoreactions with phosphines are described.

$$(CO)_5MnMn(CO)_3(\alpha\text{-diimine}) \xrightarrow[\text{THF}]{h\nu} Mn(CO)_3(\alpha\text{-diimine})(THF)^+ + Mn(CO)_5^- \quad (6)$$
$$T \leq 200\ K$$

Irradiation of the complexes in THF at room temperature in the presence of a 50-fold excess of pyridine or NEt_3 gave rise to the formation of both radical coupling and disproportionation products. The relative yields of the disporportionation products increased when a larger excess of the N-donor ligand was used. The radical coupling reaction was completely suppressed when the complexes were irradiated in neat pyridine. Thus, going from THF to the softer base pyridine, the tendency to form adducts increased. Adduct formation with the $Mn(CO)_3$(α-diimine) radicals is most effective for soft bases such as PR_3. This became evident when a mixture of $(CO)_5MnMn(CO)_3$(bpy), pyridine and $P(OMe)_3$ (1:50:50) dissolved in THF, was irradiated [20]. $Mn(CO)_5^-$ and $Mn(CO)_3(bpy)(P(OMe)_3)^+$ were then the only photoproducts formed. This clearly demonstrates that the $Mn(CO)_3$(α-diimine) radicals preferably form adducts with phosphine ligands.

The mechanism of the photodisproportionation in the presence of PR_3 has been studied in some detail [20]. The reactions appeared to be quenched by CCl_4 indicating a radical mechanism. Furthermore, the efficiency of the reaction appeared to be lowest for bulky and for weakly electron donating PR_3 ligands. Thus, solutions of the complexes containing the bulky ligand $P(iBu)_3$ or $P(cHex)_3$ or the very weak donor $P(OPh)_3$ hardly showed a photodisproportionation reaction but reacted instead according to reactions 1-3. From these results and those obtained from the electrochemical measurements on the complexes and the $Mn(CO)_3$(α-diimine)$(P(uBu)_3^+$ cations, a mechanism was derived, that is presented in Scheme III.

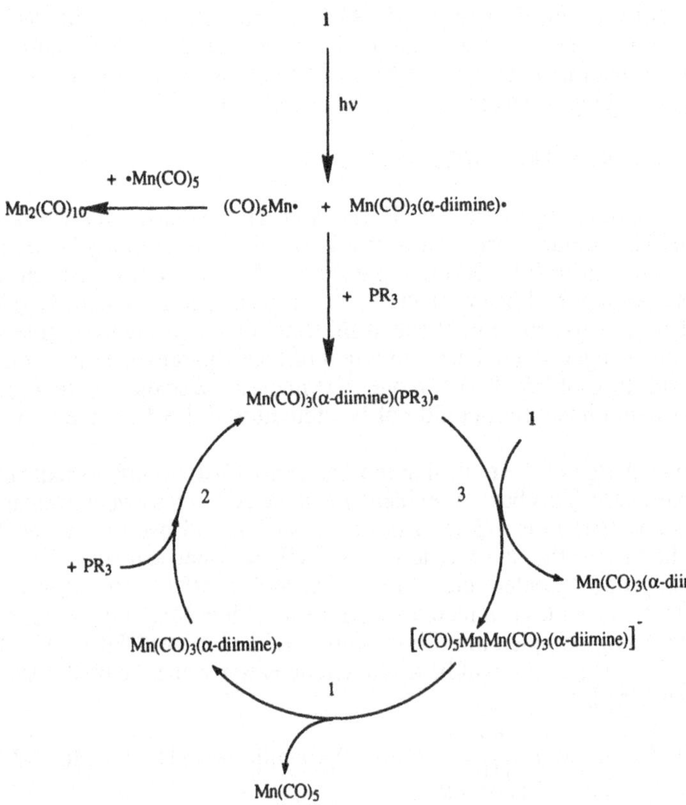

Scheme III

The radicals formed by irradiation of the complex (**1**) take up a PR_3 ligand and the adducts formed then reduce **1** (step 3). The reduced parent complex decomposes into $Mn(CO)_5^-$ and a new $Mn(CO)_3$(α-diimine) radical (step 1) which in turn takes up a PR_3 ligand (step 2) thus closing the catalytic cycle.

The $Mn(CO)_3$(α-diimine)(PR_3) radicals can, however, not only initiate the catalytic disproportionation of the parent complex, they can also reduce other electron acceptors and catalyse substitution reactions of organometallic complexes in an electron transfer chain (ETC) reaction. This was demonstrated for several clusters $M_3(CO)_{12-x}(PR_3)_x$ (M=Fe,Ru,Os; x=0-2) which are known to undergo catalytic substitution of CO in the presence of a reducing agent such as the diphenylketone anion [21,22]. For this purpose the photosubstitution of CO by PR_3 (reaction 7) was studied both in the absence and presence of $(CO)_5MnMn(CO)_3$(bpy).

$$M_3(CO)_{12-x}(PR_3)_x + PR_3 \xrightarrow{h\nu} M_3(CO)_{11-x}(PR_3)_{x+1} + CO \qquad (7)$$

In most cases the rate of the photoreaction was two orders of magnitude larger when the metal-metal bonded complex was present in the solution. Only in the case of the Os-clusters the photosubstitution reaction was not accelerated, which means that the $Mn(CO)_3$(bpy)(PR_3) radicals could not reduce these clusters. In that case only the photocatalytic disproportionation of the complex (Scheme III) was observed. This latter reaction always proceeded to some extent apart from reaction 6 and their relative quantum yields appeared to depend strongly on the reducing power of the $Mn(CO)_3$(α-diimine)(PR_3) radicals. Within the series of complexes used the reducing power was highest for the $Mn(CO)_3$(bpy)(PR_3) radicals and lowest for the corresponding $Mn(CO)_3$(pAn-DAB)(PR_3) species in agreement with the energies of their SOMO's. Thus, for the reaction of $Fe_3(CO)_{12}$ with $P(OMe)_3$ the ratio of the quantum yields of the photosubstitution and the photodisproportionation was about 50 for the bpy complex and ca. 3 for the pAn-DAB compound. In both reactions the $Mn(CO)_3$(α-diimine)(PR_3) radicals start the catalytic chain reaction by reducing either the parent compound or the cluster. In the case of the cluster the reaction will then proceed in the same way as in the presence of the diphenylketone anion as reducing agent [21,22]. The mechanism is presented in Scheme IV for the reaction of a cluster $M_3(CO)_{12}$.

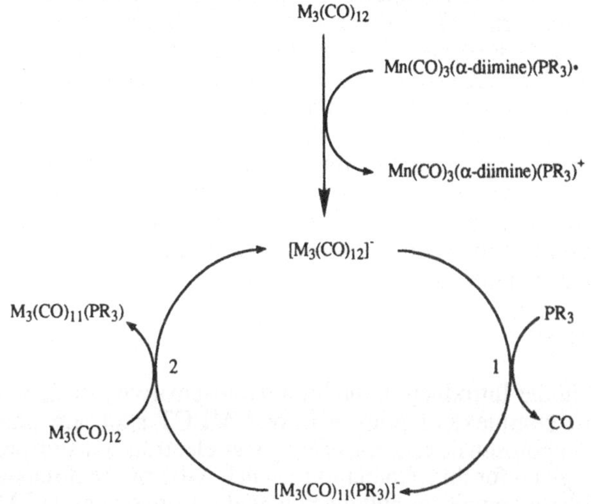

Scheme IV

The catalytic process described in Scheme IV is similar to that induced by the metal centred radicals $CpMo(CO)_3(PR_3)$, $CpFe(CO)_2(PR_3)$, $W(CO)_5(L)^-$, and $Mn(CO)_3(N)_3$ [23]. Although these latter radicals are better reducing agents than the $Mn(CO)_3$(α-diimine)(PR_3) radicals since they have their unpaired electron in a more unfavourable metal orbital, the $Mn(CO)_3$(α-diimine)(PR_3) radicals have the great advantage that they can easily be prepared by exposing the $(CO)_5MnMn(CO)_3$(α-diimine) complex to diffuse daylight.
The disadvantage of the use of these radicals as reducing agent is the presence of a stabilizing PR_3 ligand, which may introduce unwanted sidereactions. Work is in progress to obtain coordinatively saturated α-diimine radicals by intramolecular adduct formation.

3. Photoreactions of other $(CO)_5M'M(CO)_3$(α-diimine) (M,M'=Mn,Re) complexes

It has been shown in the preceding sections that the $(CO)_5MnMn(CO)_3$(α-diimine) complexes undergo both homolysis of the metal-metal bond and release of CO upon irradiation into their visible absorption band. At room temperature only the homolysis reaction leads to the formation of stable photoproducts, in viscous media at low temperatures only the CO-loss reaction is observed. Both reactions are primary photoprocesses as evidenced by nano-second flash photolysis. The other members of the series $(CO)_5M'M(CO)_3$(α-diimne) (M,M'=Mn,Re) behave differently in showing only one of these photoprocesses. Although this has not yet been proven, both reactions probably occur from the same excited state since their quantum yields show the same wavelength independence throughout the absorption band and temperature independence at room temperature [14]. The reaction taking place will then depend on the relative strengths of the M-M' and M-CO bonds in the excited state:

$(CO)_5MnMn(CO)_3$(α-diimine)	Homolysis/Release of CO
$(CO)_5MnRe(CO)_3$(α-diimine)	Homolysis
$(CO)_5ReRe(CO)_3$(α-diimine)	Homolysis
$(CO)_5ReMn(CO)_3$(α-diimine)	Release of CO

Some indication about which reaction might be expected can already be obtained from the various dissociation energies (D) in the ground state. Generally D(Re-Mn)>D(Re-Re)>D(Mn-Mn) [24] and D(Re-CO)>>D(Mn-CO) [25], which means that CO will not easily be released from the $Re(CO)_3$(α-diimine) fragment. The $(CO)_5MnMn(CO)_3$(α-diimine) complexes undergo both homolysis and release of CO in agreement with the weakness of the Mn-Mn and Mn-CO bonds in the ground state. The $(CO)_5ReMn(CO)_3$(α-diimine) complexes have a very strong Re-Mn bond and weak Mn-CO bonds. As a result these complexes only show release of CO. Since the corresponding complexes $(CO)_5MnRe(CO)_3$(α-diimine) do not lose CO from their $Mn(CO)_5$ fragment, it can be concluded that irradiation into the visible absorption band only affects the bonds of the α-diimine fragment of these complexes.

4. Mechanistic aspects

As mentioned already in the Introduction the high photosensitivity of these complexes is remarkable since most complexes having a lowest MLCT state are photostable and therefore often used as photosensitizers for energy and electron transfer processes. Two explanations have been given for this photolability which will now be discussed in the light of the m.o. diagram of these complexes and the potential energy diagram of $Mn_2(CO)_{10}$. The qualitative m.o. diagram presented in Fig. 5, shows the interaction between a $Mn(CO)_5$ radical having its unpaired electron in the d_{z^2} orbital and a $Mn(CO)_3$(α-diimine)

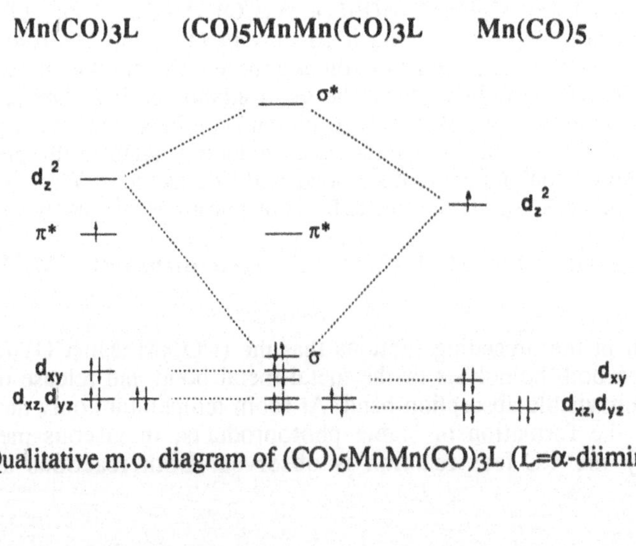

Figure 5. Qualitative m.o. diagram of $(CO)_5MnMn(CO)_3L$ (L=α-diimine).

radical having this electron in the lowest π^* orbital of the α-diimine. Interaction between the d_{z2} orbitals gives rise to the formation of a metal-metal (σ) bond. According to the UV photoelectron spectra the metal-d_π and (metal-metal)-σ orbitals are very close in energy although the σ orbital has the lowest Ionisation Potential (IP) and will therefore be the h.o.m.o. of these complexes [26]. Its IP is largest for the complexes having a Re-Mn bond, in agreement with the high dissociation energy of this bond in $(CO)_5ReMn(CO)_5$ [24].

Electronic transitions are allowed from both the σ and d_π orbitals to the lowest π^* orbital of the α-diimine ligand. However, the characteristic features of the visible absorption band (see Fig. 3) as well as the resonance Raman spectra obtained by excitation into it, can only be interpreted in terms of $d_\pi \rightarrow \pi^*$ excitation [5]. Thus, the changes in MLCT character of these $d_\pi \rightarrow \pi^*$ transitions upon going from one complex to another, are clearly reflected in the resonance Raman spectra and in the solvatochromic effects.

This means that, although the $\sigma \rightarrow \pi^*$ transition may underlie the MLCT band, it hardly contributes to its intensity.

At this moment the reactive excited state responsible for the efficient photochemistry of these complexes has not yet been identified. Morse and Wrighton, who were the first to study the homolysis reactions of several of these complexes in chlorocarbons, concluded from the wavelength independence of their quantum yields that the reactions take place from the $^3\sigma\pi^*$ state [16]. Depopulation of the metal-metal σ bond was assumed to be sufficient to labilize the metal-metal bond. This mechanism is closely analogous to that of the photoreactions of $R_2Zn(\alpha\text{-diimine})$, which produce the radicals R and RZn(α-diimine) upon irradiation into their visible absorption band [27].

Meyer and Caspar presented a different explanation based on the potential energy diagram

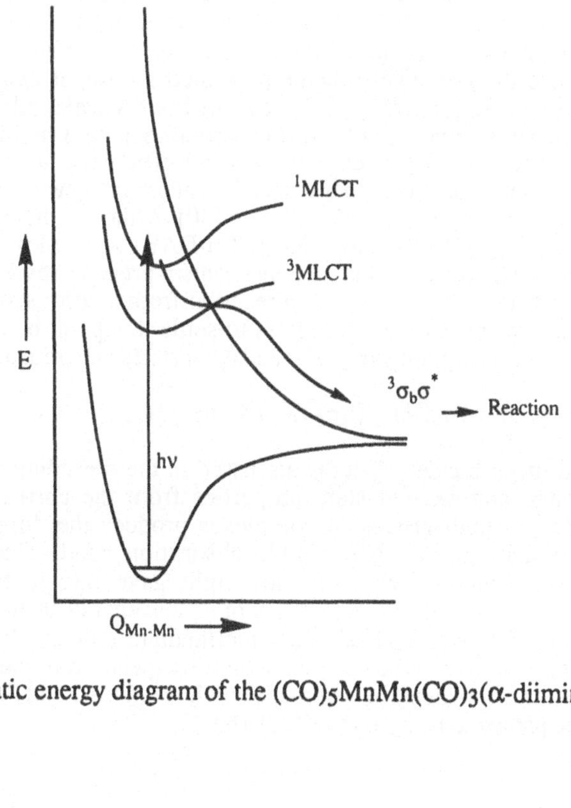

Figure 6. Schematic energy diagram of the $(CO)_5MnMn(CO)_3(\alpha\text{-diimine})$ complexes.

of $Mn_2(CO)_{10}$ [2]. This diagram shows the potential energy curves of the $^1\sigma^2$ ground state and the $\sigma\sigma^*$ excited states, in which σ and σ^* represent the metal-metal bonding and anti-bonding orbital respectively.

According to Meyer and Caspar a similar diagram might be used for the α-diimine substituted complexes. It then contains in addition to the $^1\sigma^2$ ground state and the $^3\sigma\sigma^*$ state, the ^{1}MLCT and ^{3}MLCT states (see Fig. 6). Irradiation into the MLCT band leads to occupation of the MLCT states followed by surface crossing to the repulsive $^3\sigma\sigma^*$ state. The repulsive character of the $^3\sigma\sigma^*$ state then accounts for the high quantum yields of the photoreactions of these complexes. It is not clear from this diagram how irradiation can also lead to release of CO. Recent *ab initio* calculations by Veillard and Strich on the ground and excited states of $HCo(CO)_4$ showed, however, that occupation of the repulsive $^3\sigma\sigma^*$ state of this complex will give rise to both dissociation of the Co-H bond and release of CO [28]. If the metal-metal bonded complexes behave similarly the relative quantum yields of the homolysis and CO-loss reactions will then mainly depend on the relative strengths of the metal-metal and metal-CO bonds in the $^3\sigma\sigma^*$ state.

A similar mechanism has been proposed by Rohmer for the photochemical cleavage of the metal-carbon bond of methylaluminumporphyrin [29]. The *ab initio* calculations performed on this complex showed that the lowest $^3\pi\pi^*$ state of the porphyrin ring has an avoided crossing with the $^3\sigma\sigma^*$ state of the Al-C bond. This means that irradiation into a $\pi\rightarrow\pi^*$ transition eventually leads to homolysis of the Al-C bond. According to the calculations there is a barrier for this photodissociation in agreement with the rather low quantum yield of the reaction.

As mentioned above the visible absorption band of the metal-metal bonded complexes has all the characteristics of a MLCT band. There is no evidence for a $\sigma\rightarrow\pi^*$ transition although such a transition is expected at rather low energy taking into account the low IP of this orbital. Still, the photochemical data are more in support of the interpretation by Morse and Wrighton than of the two-level scheme presented by Meyer and Caspar. For, the quantum yields are very high (>0.5) and they are both wavelength and temperature independent. This means that in case of a surface crossing there is no barrier between the MLCT and $^3\sigma\sigma^*$ states and no increase in crossing efficiency at shorter wavelength excitation. This behaviour is at least rather unusual for those complexes which react from a LF state after surface crossing from a MLCT state [30]. Although these observations and the fact that in the case of the $(CO)_5ReMn(CO)_3$(pTol-DAB) the CO-loss product is already formed within 20 ps [31], do not exclude the mechanism proposed by Meyer and Caspar, they are certainly in better agreement with a reaction from a $^3\sigma\pi^*$ state as proposed by Morse and Wrighton. Work is now in progress to settle this point by a combined time-dependent emission and absorption study of the excited states and primary photoproducts.

5. Photochemistry of $Mn(CO)_3$(bpy)X (X=halide)

Although the metal-metal bonded systems discussed in the preceding sections certainly differ in their ground and excited state properties from the corresponding halides, $Mn(CO)_3$(α-diimine)X, both groups of complexes produce the dimer $Mn_2(CO)_6$(α-diimine)$_2$ upon irradiation into their visible absorption band. Thus, irradiation of $Mn(CO)_3$(bpy)X in toluene with 514.5 nm light gave rise to the formation of $Mn_2(CO)_6(bpy)_2$ as evident from the appearance of its low-energy absorption band and its typical CO-stretching frequencies [32]. The intermediate radical $Mn(CO)_3$(bpy) was trapped by tBu-NO giving rise to the characteristic ESR spectrum of the adduct [13]. The fate of the X radical is not yet clear. In accordance with reactions 1 and 3, reactions 8 and 9 are presented for the photochemistry of $Mn(CO)_3$(bpy)X.

$$\mathrm{Mn(CO)_3(bpy)X} \xrightarrow{h\nu} \mathrm{X + Mn(CO)_3(bpy)} \quad (8)$$
$$\mathrm{2Mn(CO)_3(bpy)} \rightleftharpoons \mathrm{Mn_2(CO)_6(bpy)_2} \quad (9)$$

In order to rationalise this remarkable photochemical behaviour, not observed for the corresponding Re-complexes, we have undertaken a m.o. calculation of $Mn(CO)_3(bpy)X$ (X=Cl,I) [33]. The preliminary results of these calculations show that the h.o.m.o. has mixed metal-halide but mainly $X^-(p_\pi)$ character. The l.u.m.o. is the lowest bpy (π*) orbital and the lowest-energy transition is therefore $X^- \rightarrow$bpy (LLCT). The first MLCT transition of these complexes is expected at somewhat higher energies.
In order to prove the LLCT charcter of the lowest-energy transition we have tried to measure the resonance Raman (rR) spectra of a $Mn(CO)_3(bpy)X$ complex. These attempts failed because of the photolability of the complex. RR spectra of the corresponding photostable $Re(CO)_3(bpy)X$ were also of rather low quality because of the disturbing luminescence of these complexes. However, by using instead the α-diimine para-Tolyl-Diazabutadiene (pTol-DAB) as a ligand, a LLCT transition could be detected in the absorption spectra as a separate transition and identified as such with rR spectroscopy. Figure 7 shows the absorption spectra in CH_2Cl_2 of $Re(CO)_3$(pTol-DAB)Br and of the corresponding triflate complex $Re(CO)_3$(pTol-DAB)(OTf) (OTf^-= trifluoromethane-sulfonate).

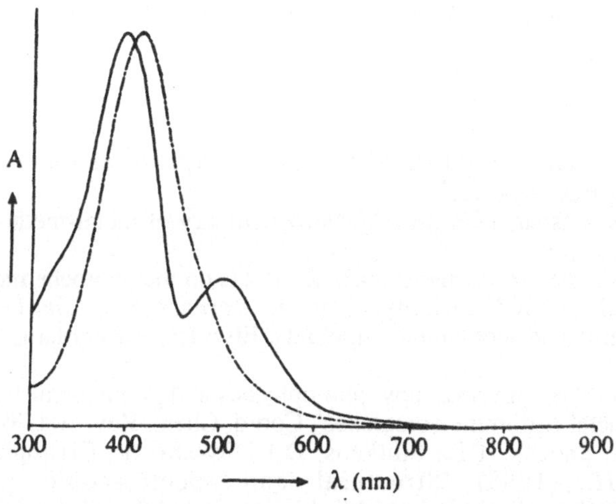

Figure 7. UV-vis spectra in CH_2Cl_2 of $Re(CO)_3$(pTol-DAB)Br (———) and $Re(CO)_3$(pTol-DAB)(OTf) (- - - - -).

The spectrum of the halide complex has two bands at ca. 505 and 400 nm respectively, whereas the OTf^- compound only absorbs at about 400 nm. The Br→pTol-DAB (LLCT) character of the low-energy band of $Re(CO)_3$(pTol-DAB)Br was confirmed by the rR spectra, which showed resonance enhancement of Raman intensity for ν(Re-Br) (190 cm^{-1}) and for ν_s(CN) of the pTol-DAB ligand upon excitation into the 505 nm band. The Raman band belonging to ν(Re-Br) disappeared upon excitation with $\lambda < 514.5$ nm and shifted to 290 cm^{-1} when the Br^- ion was replaced by Cl^-. Upon excitation into the second absorption band, a Raman band belonging to ν_s(CO) was observed next to that of ν_s(CN). These latter rR effects are typical for excitation into a MLCT transition [4]. These results clearly show that the complex $Re(CO)_3$(p-Tol-DAB)Br has close-lying LLCT and MLCT transitions. Replacing pTol-DAB by bpy will cause a shift of the MLCT transitions to lower energy with respect to the LLCT band because of the decrease in metal to ligand π-backbonding. This may even cause the MLCT states to become lowest in energy in the case of the $Re(CO)_3(bpy)X$ complexes in agreement with the assignment from the literature [34]. It has to be realized, however, that in all these $M(CO)_3$(α-diimine)X complexes the LLCT state will still be rather low in energy and therefore may affect the excited state properties of most of these complexes.

6. Summary

Irradiation of $(CO)_5M'M(CO)_3$(α-diimine) (M,M'=Mn,Re) gives rise to homolysis of the metal-metal bond and/or release of CO from the $M(CO)_3$(α-diimine) moiety depending on the relative strengths of the M-M' and M-CO bonds in the reactive excited state. Since the $M(CO)_3$(α-diimine) radicals are coordinatively unsaturated species containing a radical anion, their secondary thermal reactions strongly differ from those of metal-centred radicals such as $Mn(CO)_5$. With Lewis bases the radicals form reducing adducts which can e.g. initiate the catalytic substitution of CO in $Ru_3(CO)_{12}$ by PR_3.
Although the excited state processes leading to these photochemical reactions are not yet known, most experimental data point to a reaction from a $^3σπ^*$ state. Support for this interpretation is provided by the photochemistry of the complexes $Mn(CO)_3$(α-diimine)X (X=halide) which also produce $Mn(CO)_3$(α-diimine) radicals upon irradiation into their lowest energy band.

7. References

1. Geoffroy, G.L. and Wrighton, M.S. (1979) Organometallic Photochemistry, Academic Press, New York.
2. Meyer, T.J. and Caspar, J.V. (1985) 'Photochemistry of metal-metal bonds', Chem. Rev. 85, 187-218.
3. Stufkens, D.J. (1989) 'Steric and electronic effects on the photochemical reactions of metal-metal bonded carbonyls', in I. Bernal (ed.), Stereochemistry of organometallic and inorganic compounds, Elsevier, Amsterdam, Vol. 3, pp 226-300.
4. Stufkens, D.J. (1990) 'Spectroscopy, photophysics and photochemistry of zerovalent transition metal α-diimine complexes', Coord. Chem. Rev. 104, 39-112.
5. Kokkes, M.W., Snoeck, T.L., Stufkens, D.J., Oskam, A., Cristophersen, M. and Stam. C.H. (1985) 'Structural and spectroscopic properties of $[(CO)_5MM'(CO)_3$(R-DAB)] (M,M'=Mn,Re; R-DAB=1,4-diaza-1,3-butadiene) complexes. X-ray structure of $[(CO)_5ReMn(CO)_3$(iPr-DAB)] and infrared and resonance Raman spectra of $[(CO)_5MM'(CO)_3$(R-DAB)]', J. Mol. Struct. 131, 11-29.
6. Meyer, T.J. (1986) 'Photochemistry of metal coordination complexes: metal to ligand charge transfer excited states', Pure and Appl. Chem. 58, 1193-1206.
7. Balk, R.W., Snoeck, T.L., Stufkens, D.J. and Oskam, A. (1980) '(Diimine) carbonyl complexes of chromium, molybdenum, and tungsten: relationship between resonance Raman spectra and photosubstitution quantum yields upon excitation within the lowest metal to diimine charge-transfer band', Inorg. Chem. 19, 3015-3021.
8. Van Dijk, H.K., Stufkens, D.J. and Oskam, A. (1989) 'A mechanistic study of the photochemistry of $Fe(CO)_3$(R-DAB) (R-DAB=1,4-diaza-1,3-butadiene), a unique

group of complexes with two close-lying reactive excited states', J. Am. Chem. Soc. 111, 541-547.

9. Servaas, P.C., Stufkens, D.J. and Oskam, A. (1989) 'Spectroscopy and photochemistry of $Ni(CO)_2$(R-DAB) (R=tBu, 2,6-iPr_2Ph): evidence for two different photoprocesses', Inorg. Chem. 28, 1780-1787.
10. Juris, A., Balzani, V., Barigelletti, F., Campagna, S., Belser, P. and Von Zelewsky, A. (1988) 'Ru(II) polypyridine complexes: photophysics, photochemistry, electrochemistry, and chemiluminescence', Coord. Chem. Rev. 84, 85-277.
11. Meyer, T.J. (1989) 'Chemical approaches to artificial photosynthesis', Acc. Chem. Res. 22, 163-170.
12. Yasufuku, K. and Sakamoto, H., Personal communication.
13. Andréa, R.R., de Lange, W.G.J., van der Graaf, T., Rijkhoff, M., Stufkens, D.J. and Oskam, A. (1988) 'Metal to ligand charge-transfer photochemistry of metal-metal bonded complexes. 5. ESR spectra of stable rhenium-α-diimine and spin-trapped manganese-α-diimine radicals', Organometallics 7, 1100-1106.
14. Van der Graaf, T., Stufkens, D.J., Oskam, A. and Goubitz, K. (1991) 'Metal to ligand charge-transfer photochemistry of metal-metal bonded complexes. 8. Photochemistry of $(CO)_5MnMn(CO)_3$(α-diimine) complexes. Coupling reactions of the radicals formed and x-ray structure of the photoproduct $(CO)_4$Mn(σ-N,σ-N',η^2-CN-iPr-pyca)$Mn(CO)_3$', Inorg. Chem. 30, 599-608.
15. Servaas, P.C., Stor, G.J., Stufkens, D.J. and Oskam, A. (1990) 'Metal to ligand charge-transfer photochemistry of metal-metal bonded complexes. 9. Photochemistry of $CpFe(CO)_2Re(CO)_3L$ (L=4,4'-Me_2-bpy; pyridine-2-carbaldehyde N-isopropyl-imine)', Inorg. Chim. Acta 178, 185-194.
16. Morse, D.L. and Wrighton, M.S. (1976) 'Photochemistry of metal-metal bonded complexes. 5. Cleavage of the M-M bond in $(CO)_5MM(CO)_3L$ by irradiation into a low-lying M→L charge-transfer band', J. Am. Chem. Soc. 98, 3931-3934.
17. Van der Graaf, T., van Rooy, A., Stufkens, D.J. and Oskam, A. (1991) 'Metal to ligand charge-transfer photochemistry of metal-metal-bonded complexes. 11. Halogen atom and electron transfer reactions of $(CO)_5MM'(CO)_3$(α-diimine) (M,M'=Mn,Re) complexes', Inorg. Chim. Acta, in press.
18. Staal, L.H., Polm, L.H., Balk, R.W., van Koten, G., Vrieze, K. and Brouwers, A.M.F. (1980) 'σ^2-N', μ^2-N, η^2-C=N coordination as the activating step in carbon-carbon formation between two α-diimines in dinuclear ruthenium carbonyl complexes. X-ray structure of tetracarbonyl bis[glyoxal bis(isopropyl-imine)]diruthenium', Inorg. Chem. 19, 3343-3351.
19. Polm, L.H., Elsevier, C.J., van Koten, G., Ernsting, J.M., Stufkens, D.J., Vrieze K., Andréa, R.R. and Stam, C.H. (1987) 'Synthesis of [$Ru_2(CO)_5${1,2-bis(μ-alkylamido)-1,2-bis(2-pyridil)ethane}] and [$Ru_2(CO)_4${pyridine-2-carbaldehyde N-alkylimine}$_2$] and the molecular structure of bis(pyridine-2-carbaldehyde N-isopropyl-imine)tetracarbonyldiruthenium. Unprecedented CO-induced carbon-carbon bond formation between two pyridine-2-carbaldimines in dinuclear ruthenium carbonyl complexes', Organometallics 6, 1096-1104.
20. Van der Graaf, T., Hoftra, R.M.J., Schilder, P.G.M., Rijkhoff, M., Stufkens, D.J. and van der Linden, J.G.M. (1991) 'Metal to ligand charge-transfer photochemistry of metal-metal bonded complexes. 10. A photochemical and electrochemical study of the electron transfer reactions of $Mn(CO)_3$(α-diimine) (L)· (L=N-, P-donor) radicals formed by irradiation of $(CO)_5MnMn(CO)_3$(α-diimine) complexes in the presence of L', Organometallics in press.
21. Bruce, M.I., Kehoe, D.C., Matisons, J.G., Nicholson, B.K., Rieger, P.H. and Williams, M.L. (1982) 'Cluster chemistry. Reactions between metal carbonyl

clusters and Lewis bases initiated by radical ions: improved syntheses of substituted derivatives of M_3 and M_4 clusters (M=Fe, Ru, Os or Co)', J. Chem. Soc. Chem. Commun. 442-444.

22. Bruce, M.I., Matisons, J.G. and Nicholson, B.K. (1983), 'Cluster chemistry XVII. Radical ion-initiated syntheses of ruthenium cluster carbonyls containing tertiary phosphines, phosphites, arsines, $SbPh_3$ or isocyanides', J. Organomet. Chem. 247, 321-434.
23. Tyler, D.R. (1988) 'Mechanistic aspects of organometallic radical reactions', Progr. Inorg. Chem. 36, 125-194.
24. Meckstroth, W.K. and Ridge, D.P. (1985) 'Properties of ions and radicals derived from decarbonyldimanganese, decarbonylmanganeserhenium, and decarbonyldirhenium by electron attachment and protonation in the gas phase', J. Am. Chem. Soc. 107, 2281-2285.
25. Connor, J.A. (1977) 'Thermochemical studies of organotransition compounds', Top. Curr. Chem. 71, 71-110.
26. Andréa, R.R., Stufkens, D.J. and Oskam, A. (1985) 'He(I) and He(II) photoelectron spectra of the four metal-metal bonded complexes [$(CO)_5$M-M'$(CO)_3$(1,4-i-Pr_2-1,4-diaza-1,3-butadiene)] (M=M'=Mn or Re; M=Mn, M'=Re; M=Re, M'=Mn)', J. Organomet. Chem. 290, 63-75.
27. Kaupp, M., Stoll, H., Preuss, H., Kaim, W., Stahl, T., van Koten, G., Wissing, E., Smeets, W.J.J. and Spek, A.L. (1991) 'Theoretical and experimental study of diamagnetic and paramagnetic products from thermal and light-induced alkyl transfer between zinc or magnesium dialkyls and 1,4-diaza-1,3-butadiene substrates', J. Am. Chem. Soc. 113, 5606-5618.
28. Veillard, A. and Strich, A. (1988) 'A CASSCF and CCI study of the photochemistry of $HCo(CO)_4$', J. Am. Chem. Soc. 110, 3793-3797.
29. Rohmer, M.-M. (1989) 'Photochemical cleavage of the metal-carbon bond in aluminum porphyrins: insights from ab initio calculations', Chem. Phys. Letters 157, 207-210.
30. Ford, P.C., Wink, D. and DiBenedetto, J. (1983) 'Mechanistic aspects of the photosubstitution and photoisomerization reactions of d^6 metal complexes', Progr. Inorg. Chem. 30, 213-271.
31. Langford, C.H., Personal communication.
32. Stor, G.J., Stufkens, D.J. and Oskam, A., submitted for publication.
33. Stor, G.J., Baerends, E.J., Vernooijs, P. and Stufkens, D.J., unpublished results.
34. Lees, A.J. (1987) 'Luminescence properties of organometallic complexes', Chem. Rev. 87, 711-743.

SUPRAMOLECULAR PHOTOCHEMISTRY: ANTENNA EFFECT IN POLYNUCLEAR METAL COMPLEXES

Vincenzo Balzani
Dipartimento di Chimica "G. Ciamician" dell'Università, 40126 Bologna, Italy

Sebastiano Campagna
Dipartimento di Chimica Inorganica e Struttura Molecolare dell'Università, 98166 Messina, Italy

Gianfranco Denti, Scolastica Serroni
Laboratorio di Chimica Inorganica, Istituto di Chimica Agraria dell'Università, 56100 Pisa, Italy

ABSTRACT. We have synthesized bi-, tri-, tetra-, hexa-, hepta-, and decanuclear metal complexes where the metals (M) are Ru^{2+} and/or Os^{2+}, the bridging ligands (BL) are 2,3-dpp and/or 2,5-dpp, and the terminal ligands (L) are bpy and/or biq (dpp= bis(2-pyridyl)pyrazine; bpy= 2,2'-bipyridine; biq= 2,2'-biquinoline). Luminescence investigations show that the electronic energy generated by light absorption in the various chromophoric units can be channelled towards any desired site of the supramolecular structure by suitable choice and positioning of the M, BL, and L components.

E. Kochanski (ed.), Photoprocesses in Transition Metal Complexes, Biosystems and Other Molecules. Experiment and Theory, 233–252.

1. INTRODUCTION

Electronic energy transfer lies at the heart of important natural phenomena (for example, photosynthesis [1]) as well as of practical applications (for example, spectral sensitization [2]). The possibility of governing the direction of electronic energy transfer in supramolecular arrays may open the way to the design of photochemical molecular devices that can perform a variety of useful functions [3]. Of particular interest is the so called "antenna effect", which consists in an enhanced light-sensitivity obtained by an increase in the overall cross-section for light absorption. To achieve this result, electronic energy should be conveyed from several chromophores to a common component that constitutes the interface toward the use of the absorbed light energy. In the scheme shown in Fig. 1, such a component is a luminophore, but components playing other functions (e.g., that of an electron-transfer photosensitizer as in the natural photosynthetic process [1, 4]) could also be used [3]. Simple antenna devices are the mononuclear lanthanide complexes of polypyridine ligands, where absorption of uv light by the ligands is followed by energy transfer to the metal ion which then gives rise to visible luminescence [5]. Such compounds are promising luminescent labels for biological analyses. For solar energy conversion problems, however, antenna devices capable of absorbing as much visible light as possible are needed. Polynuclear complexes of suitable metals and ligand are good candidates to play this role.

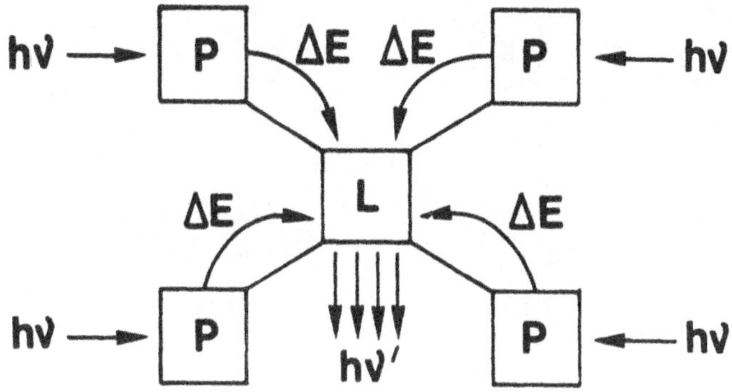

Figure 1. Schematic representation of the "antenna effect".

We are currently engaged in a systematic study of the photochemical, photophysical, and electrochemical properties of strongly light absorbing and luminescent oligometallic compounds [6-17] that can be considered as supramolecular species since they are made of metal-containing building-blocks linked by bridging ligands. In the course of these studies, we have designed and synthesized di-, tri-, tetra-, hexa-, hepta-, and deca-nuclear compounds where the metals M are Ru^{2+} and/or Os^{2+}, the bridging ligands BL are 2,3-dpp and/or 2,5-dpp (dpp= bis(2-pyridyl)pyrazine) and the terminal ligands L are bpy and/or biq (bpy=2,2'-bipyridine; biq=2,2'-biquinoline (Fig. 2). The results obtained show that in these supramolecular systems the electronic

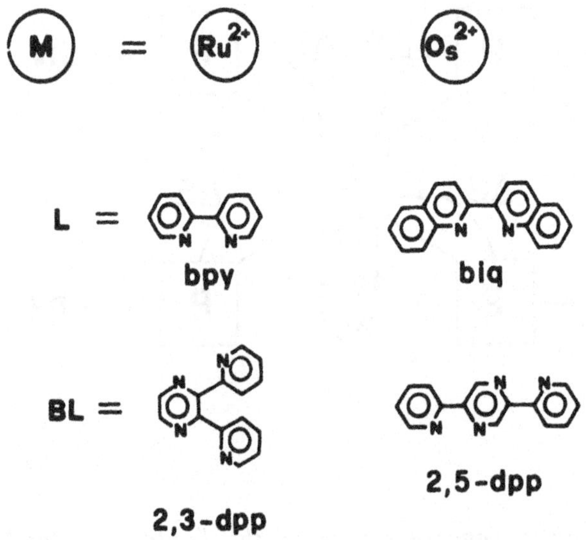

Figure 2. Components of the polynuclear metal complexes.

energy originated upon light absorption in the various chromophoric moieties can be channelled along any desired direction by a suitable choice of the M, BL and L components and of their location in the supramolecular structure.

2. ENERGY ORDERING

Since the pioneering work of Crosby [18], it is known that in Ru(II)- and Os(II)-polypyridine complexes luminescence takes place from the lowest excited level, which corresponds to a formally spin-forbidden metal-to-ligand charge transfer (^{3}MLCT) excited state. As Os(II) is easier to oxidize than Ru(II), the luminescent level of any Os(II)-polypyridine complex lies lower in energy than the luminescent level of the corresponding Ru(II) complex. It is also known that for complexes of the same metal, the energy of the luminescent level can be tuned by changing the type of polypyridine ligand [18,19]. For example, the luminescent level of a $M(bpy)_3{}^{2+}$ complex will always be higher in energy than the luminescent level of the corresponding $M(biq)_3{}^{2+}$ complex because bpy is more difficult to reduce than biq. A third, indirect way to tune the energy of the luminescent levels is that of changing the nature of the ancillary ligands (i.e., of the ligands that are not involved in the lowest energy MLCT transition). For example, when a X ligand of a ML_nX_m complex is replaced by a Y ligand which is a better electron donor that X, the ^{3}MLCT level will move to lower energy because of the increased electron density (and the consequent decrease in the ionization potential) of the metal. Previous work on Ru^{2+} and Os^{2+} complexes has shown that the 2,3-dpp and 2,5-dpp bridging ligands and the bpy and biq terminal ligands can be ordered in the following series as far as the electron-donor power is concerned: bpy>biq>μ-2,3-dpp>μ-2,5-dpp. This means, for example, that coordination of bpy to Ru^{2+} reduces the positive charge of the metal ion more than coordination of μ-2,3-dpp. On the other hand, the energy of the lowest (empty) π* orbital increases in the series μ-2,5-dpp<μ-2,3-dpp<biq<bpy. This means, for example, that in a $Ru(bpy)_2(\mu\text{-}2,3\text{-}dpp)^{2+}$ moiety the Ru ----> (μ-2,3-dpp) excited state lies lower in energy than the Ru ----> bpy excited state.

By using the above guidelines, it is possible to design polynuclear complexes where the component(s) with the lowest energy excited state(s) is (are) located in the desired position(s) of the supramolecular structure. This allows a synthetic control of the direction(s) of energy migration after light absorption.

3. ABSORPTION SPECTRA

A first requirement of an antenna device is strong absorption of light. This is certainly the case of the polynuclear complexes discussed in this paper, as shown, for example, by the spectra displayed in Fig. 3. The very intense bands in the uv region are due to spin-allowed π ----> π^* ligand centered transitions. The intense bands in the visible region are due to spin-allowed MLCT transitions. In species containing different or not equivalent metals and/or ligands, several different MLCT transitions are present. The corresponding bands often overlap, giving rise to very broad absorptions. In the Os(II) moieties, because of the more covalent character of the metal-ligand bonds and the presence of a heavier metal, the formally spin-forbidden MLCT transitions become partially allowed and give rise to noticeable absorptions at low energies.

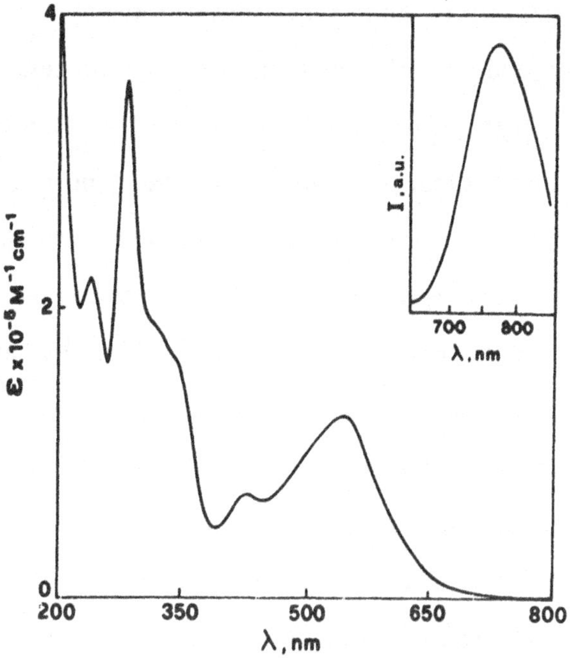

Figure 3. Absorption and (inset) luminescence spectra of the decanuclear complex $Ru\{(BL)Ru[(BL)Ru(bpy)_2]_2\}_3^{20+}$ *in acetonitrile. The luminescence spectrum is uncorrected for photomultiplier response. BL is 2,3-dpp.*

4. ENERGY TRANSFER EFFICIENCY

A second major requirement of an antenna device is a high efficiency of energy transfer between the components. In the Ru(II) and Os(II) complexes excitation in the spin allowed MLCT bands is followed by rapid deactivation to the lowest ^{3}MLCT (luminescent) level of the light-absorbing component. Energy transfer between different components can then follow in competition with intracomponent radiative and radiationless decay. The rate of energy transfer depends on the energetics of the process and on the intercomponent electronic interaction [20]. Electrochemical experiments on polynuclear complexes of the family discussed in this paper show that the metal-metal interaction through a 2,3- or 2,5-dpp bridge is not negligible [8,15]. Therefore, energy transfer between directly connected units can take place by an adiabatic mechanism. Since the reorganizational energy is small, intercomponent energy transfer is expected to be very fast whenever it is exoergonic. As we will see later, the experimental results indicate that the efficiency of energy transfer for exoergonic processes is usually 100% in these systems.

5. ENERGY MIGRATION

We will now examine the patterns for energy migration in species of various nuclearity and we will report examples of compounds where a given pattern is followed. In most cases the relevant compounds have already been prepared and investigated.

From the experimental point of view, energy migration in supramolecular species can be directly demonstrated by the sensitization of the luminescence of the acceptor unit, and its efficiency can be estimated by comparing the absorption and excitation spectra. Indirect evidence of energy migration is also obtained when the donor luminescence, that can be observed in related compounds, is quenched by the presence of the acceptor in the supramolecular array.

5.1. Dinuclear compounds

As shown in Fig. 4, there is obviously only one pattern for energy

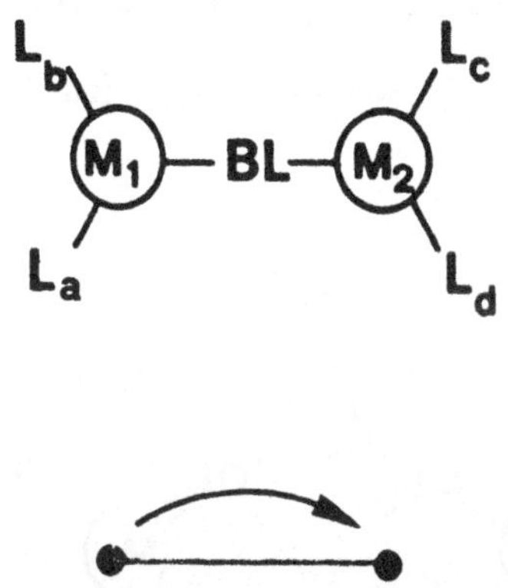

2- I

Figure 4. Energy migration in dinuclear compounds.

migration in dinuclear compounds. An example is given by the compound with $L_a = L_b =$ biq, $L_c = L_d =$ bpy, BL= 2,3-dpp, and M= Ru^{2+}: excitation of the M_1-based chromophore leads to emission from the M_2-based moiety [8]. Corrected excitation spectra show that the process is 100% efficient. In the compound where L= bpy, BL= 2,3-dpp, $M_1 = Ru^{2+}$, and $M_2 = Os^{2+}$, the sensitized luminescence of the M_2-based unit is expected to occur at $\lambda > 900$ nm, a region not covered by our equipment. The lack of luminescence from the M_1-

based moiety (that is observed for the corresponding dinuclear homometallic Ru compound) suggests an efficient energy migration [12,21].

5.2. Trinuclear compounds

In these compounds there are three patterns for energy migration (Fig. 5). Pattern 3-I is expected to operate, for example, when L= bpy, BL= 2,3-dpp, M_1= Os^{2+}, and M_2= M_3= Ru^{2+}. Unfortunately, for

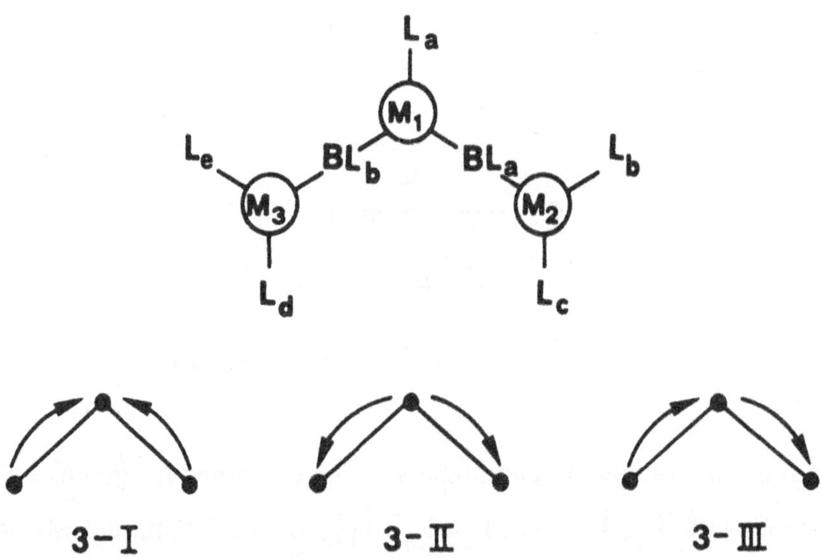

Figure 5. Energy migration patterns in trinuclear compounds.

this complex the luminescence from the M_1-based unit would not be observed in the spectral region covered by the conventional spectrofluorimeters, so that evidence for energy migration would only be indirect (quenching of the emission of the M_2- and M_3-based units). Pattern 3-II is obtained when L= bpy, BL= 2,3-dpp, and M= Ru^{2+} [7]. Corrected excitation spectra show that energy transfer from the M_1-based to the M_2- and M_3-based components is 100% efficient. Pattern 3-III can be predicted, e. g., for L_a= L_b= L_c= bpy, L_d= L_e= biq, BL= 2,3-dpp, M= Ru^{2+}. Attempts to prepare this compound are under way.

5.3. Tetranuclear compounds

The four possible energy migration patterns are schematized in

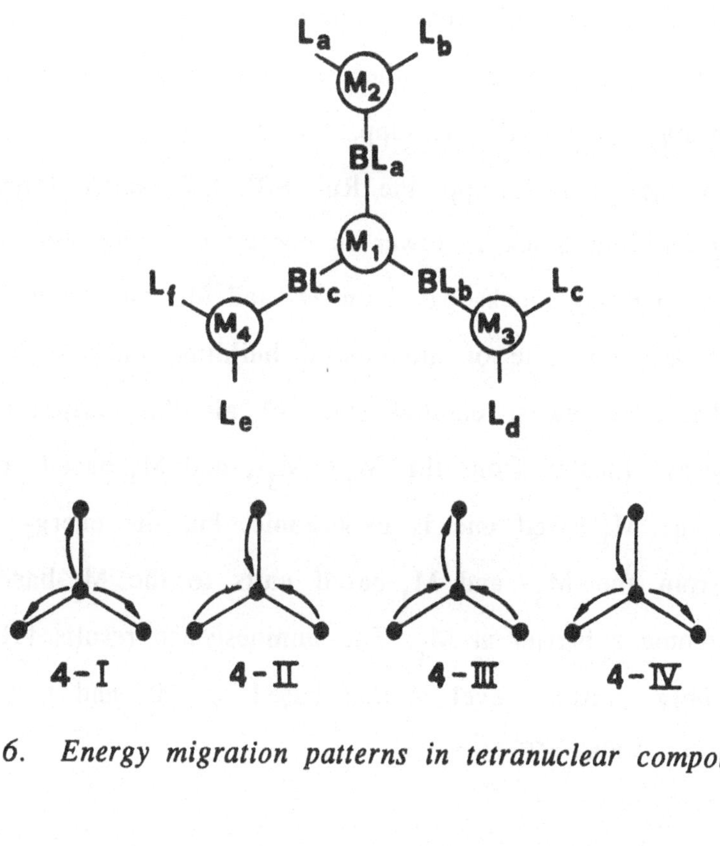

Figure 6. Energy migration patterns in tetranuclear compounds.

Fig. 6. Patterns 4-I obtains for L= bpy, BL= 2,3-dpp, and M= Ru^{2+}. In such a complex, the three peripheral units are equivalent and their lowest excited state lies at lower energy than the lowest excited state of the central unit [6,22]. The luminescence of the central unit is quenched, and that of the peripheral ones is sensitized. The efficiency of energy transfer is difficult to estimate because of strong band overlap in the absorption (and excitation) spectrum. Pattern 4-II is found for L= bpy, BL= 2,3-dpp, M_1= Os^{2+}, M_2= M_3= M_4= Ru^{2+} [6]. The luminescence of the peripheral units is quenched, and the luminescence of the central unit is sensitized with ~100% efficiency.

To channel energy towards a single peripheral unit (pattern 4-III) we have designed and sinthesized a compound with L= bpy, BL_a= 2,5-dpp, BL_b= BL_c= 2,3-dpp, M= Ru^{2+} [10]. Since 2,5-dpp is easier to reduce than 2,3-dpp, the Ru--->BL CT excited state of the M_2-based building block is lower in energy than the Ru--->BL CT state of the building blocks based on M_3 and M_4. On the other hand, the lowest excited state of the central building block is higher in energy than the lowest excited state of all the peripheral ones. Thus, energy transfer from the M_1-, M_3-, and M_4-based building blocks to the M_2-based one is exoergonic, but the energy transfer process from the M_3- and M_4-based units to the M_2-based unit must overcome a barrier at M_1. The luminescence results [10] show that the only emitting level is that based on M_2 and that energy migration is ~100% efficient.

To obtain the migration pattern illustrated by 4-IV we have designed and synthesized a compound with L_a= L_b= biq, L_c= L_d= L_e= L_f= bpy, BL= 2,3-dpp, and M= Ru^{2+} [14]. Since the energy level of the M_1-based building block is (slightly) higher than that of the M_2-based one, the migration process must again overcome an energy barrier. The results obtained with this complex show that energy migration takes place with high efficiency.

5.4. Hexanuclear compounds

The first hexanuclear compound prepared [15], e.g. the one with L= bpy, BL= 2,3-dpp, and M= Ru^{2+}, behaves according to pattern 6-I of Fig. 7. The same pattern is also obtained for L= bpy, BL= 2,3-dpp, M_1= M_2= Ru^{2+}, and M_3= M_4= M_5= M_6= Os^{2+}, although sensitized luminescence cannot be observed for the previously mentioned reasons [17]. Pattern 6-II would require, e. g., a complex with L= bpy, BL= 2,3-dpp, M_1= M_2= Os^{2+}, and M_3= M_4= M_5= M_6= Ru^{2+}; such a complex, however, has not yet been obtained. In order to observe pattern 6-III, we have prepared the compound with L_a= L_b= L_c= L_d= bpy, L_e= L_f= L_g= L_h= biq, BL_b= BL_c= 2,5-dpp, BL_a= BL_d= BL_e= 2,3-dpp, and M= Ru^{2+} [17]. Because of such ligand combination, the lowest energy excited levels are those of the M_3- and M_4-based units, but the M_1-based unit lies at higher energy than the M_5-and M_6- based ones. Experiments on this complex are under way. The nonluminescent compound with L= bpy, BL_a= BL_b= BL_c= 2,3-dpp, BL_d= BL_e= 2,5-dpp, M_1= M_2= M_5= M_6= Ru^{2+}, and M_3 = M_4 = Os^{2+}

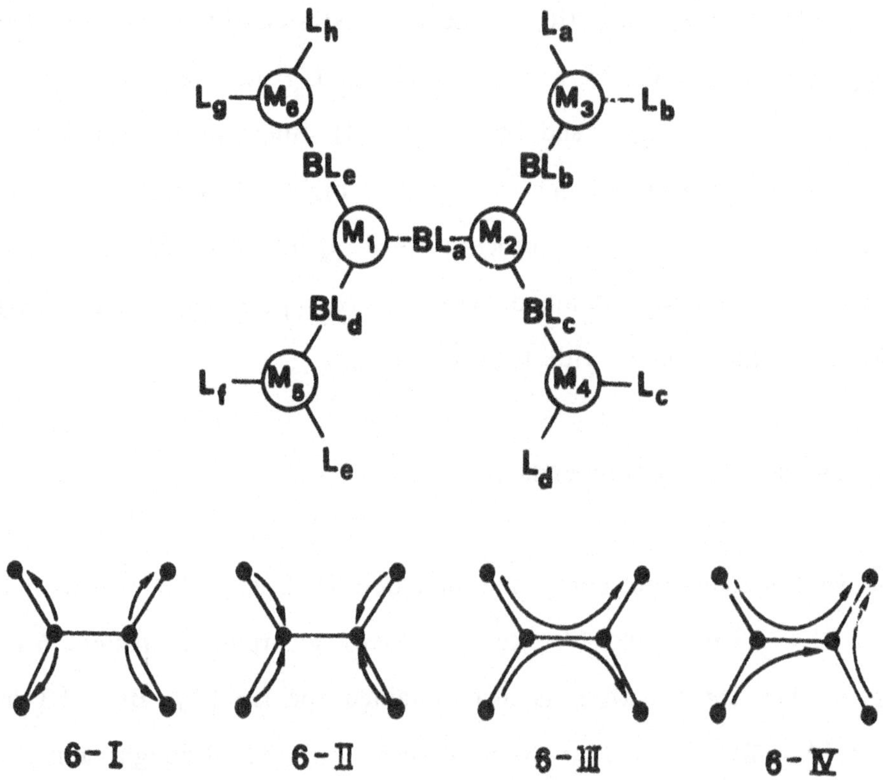

Figure 7. Energy migration patterns in hexanuclear compounds.

should exhibit the same pattern [17]. Since these hexanuclear compounds are prepared from two trinuclear moieties, it is relatively easy to obtain species which behave symmetrically as in patterns 6-I, 6-II, and 6-III. Preparation of a compound capable of displaying pattern 6-IV would require a very specific synthetic procedure.

5.5 Heptanuclear compounds

A heptanuclear compound of the type shown in Fig. 8 has been

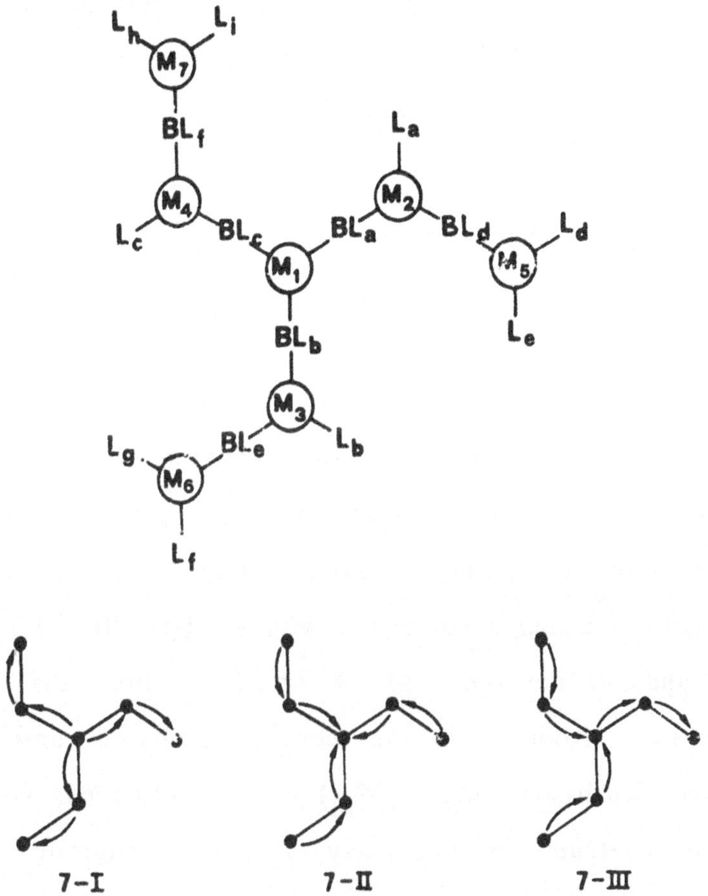

Figure 8. Energy migration patterns in heptanuclear compounds.

prepared [9]. Its composition is L= bpy, BL= 2,3-dpp, and M= Ru^{2+}. In such a compound, the energy migration pattern is of type 7-I. Replacement of Os^{2+} for Ru^{2+} in the central M_1 position would yield pattern 7-II, where the exoergonic energy transfer process from the peripheral units to the central one should exhibit a small activation energy since the intermediate units (that contain the M_2, M_3, and M_4 metal centers) are slightly higher in energy than the peripheral ones. The preparation of a compound capable of dysplaying pattern 7-III is possible in principle, but it would require a very complicated procedure.

5.6. Decanuclear compounds

Several decanuclear compounds of the type shown in Fig. 9 have been prepared very recently [13,16]. The compound with L= bpy, BL= 2,3-dpp, and M= Ru^{2+} behaves according to pattern 10-I [13] with 100% energy transfer efficiency to the peripheral, luminescent units. In order to obtain the reverse energy migration, pattern 10-II, we have prepared a compound with L= bpy, BL= 2,3-dpp, M_1= Os^{2+}, and all the other M= Ru^{2+} [16]. For this compound luminescence from both the central Os-based unit and the peripheral Ru-based (M_5 to M_{10}) units is observed. We are now trying to evaluate the efficiency of energy migration from the peripheral and intermediate units to the central one. Preliminary results show that the energy transfer efficiency from the peripheral units is very low, presumably because their excited state energy is lower than the excited state energy of the intermediate units. Pattern 10-III should be displayed by the compound with L= bpy, BL= 2,3-dpp, M_1= M_5= M_6= M_7= M_8= M_9= M_{10}= Ru^{2+}, and M_2= M_3 = M_4 = Os^{2+}, which is not extremely difficult to synthesize. Less

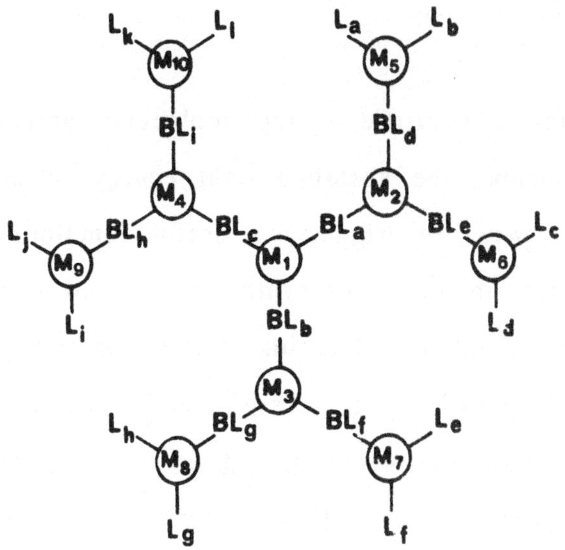

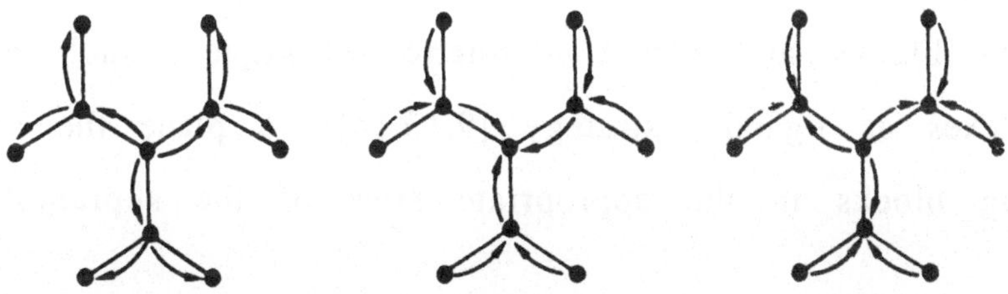

10-I 10-II 10-III

Figure 9. Energy migration patterns in decanuclear compounds.

symmetric energy migration patterns (e.g., energy migration toward a unique peripheral unit) requires complicated synthetic procedures.

6. CONCLUSION

In both natural and artificial supramolecular arrays it is quite important to channel the absorbed light energy towards a specific unit which performs (or triggers) a useful function (for example, charge separation in the photosynthetic reaction center [4]). We have shown that, taking transition metal complexes as building blocks, it is possible to design and synthesize artificial supramolecular systems where the direction of energy migration can be predetermined. The design of such systems requires a deep knowledge of the spectroscopic and excited state properties of the building blocks and clever synthetic techniques (such as the "complexes as ligands" strategy [9,13,16]) to place the various building blocks in the appropriate sites of the supramolecular array.

7. ACKNOWLEDGMENTS

We wish to thank V. Cacciari, G. Gubellini, and L. Minghetti for technical assistance. This work was supported by the Consiglio Nazionale delle Ricerche (Progetto Finalizzato Chimica Fine II) and Ministero dell'Università e della Ricerca Scientifica e Tecnologica.

8. REFERENCES AND NOTES

[1] D.P. Hader and M. Tevini, *General photobiology* (Pergamon, Oxford) 1987.

[2] R. Amadelli, R. Argazzi, C.A. Bignozzi, and F. Scandola, *J. Am. Chem. Soc.* **112**, 7099 (1990).

[3] V. Balzani and F. Scandola, *Supramolecular Photochemistry* (Horwood, Chichester) 1991, Chapter 12.

[4] J. Breton and H. Vermeglio eds., *The Photosynthetic Bacterial Reaction Center. Structure and Dynamics* (Plenum Press, New York) 1988.

[5] L. Prodi, M. Maestri, V. Balzani, J.-M. Lehn, and C. Roth, *Chem. Phys. Lett.*, **180**, 45 (1991), and references therein.

[6] S. Campagna, G. Denti, L. Sabatino, S. Serroni, M. Ciano, and V. Balzani, *J. Chem. Soc., Chem. Commun.* 1500 (1989).

[7] S. Campagna, G. Denti, L. Sabatino, S. Serroni, M. Ciano, and V. Balzani, *Gazz. Chim. Ital.* **119**, 415 (1989).

[8] G. Denti, S. Campagna, L. Sabatino, S. Serroni, M. Ciano, and V. Balzani, *Inorg. Chem.* **29**, 4750 (1990).

[9] G. Denti, S. Campagna, L. Sabatino, S. Serroni, M. Ciano, and V. Balzani, *Inorg. Chim. Acta* **176**, 175 (1990).

[10] G. Denti, S. Serroni, S. Campagna, V. Ricevuto, and V. Balzani, *Inorg. Chim. Acta* **182**, 000 (1991).

[11] G. Denti, S. Serroni, L. Sabatino, M. Ciano, V. Ricevuto, and S. Campagna, *Gazz. Chim. Ital.* **121**, 37 (1991).

[12] G. Denti, S. Campagna, L. Sabatino, S. Serroni, M. Ciano, and V. Balzani, in *Photochemical Conversion and Storage of Solar Energy*, E. Pellizzetti and M. Schiavello eds., Kluwer, Dordrecht, 1991, p. 27.

[13] S. Serroni, G. Denti, S. Campagna, M. Ciano, and V. Balzani, *J. Chem. Soc., Chem. Commun.*, in press.

[14] G. Denti, S. Serroni, S. Campagna, V. Ricevuto, and V. Balzani, *Coord. Chem. Rev.*, in press.

[15] S. Campagna, G. Denti, S. Serroni, M. Ciano, and V. Balzani, *Inorg. Chem.*, in press.

[16] S. Campagna, G. Denti, S. Serroni, M. Ciano, and V. Balzani, manuscript in preparation.

[17] S. Campagna, G. Denti, S. Serroni, M. Ciano, and V. Balzani, work in progress.

[18] G.A. Crosby, *Accounts Chem. Res.* **8**, 231 (1975).

[19] A. Juris, V. Balzani, F. Barigelletti, S. Campagna, P. Belser, and A. von Zelewsky, *Coord. Chem. Rev.* **84**, 85 (1988).

[20] V. Balzani and F. Scandola, *Supramolecular Photochemistry* (Horwood, Chichester) 1991, Chapter 6.

[21] K. Kalyanasundaram and Md.K. Nazeeruddin, *Inorg. Chem.* **29**, 1888 (1990).

[22] W. R. Murphy, K. J. Brewer, G. Gettliffe, and J. D. Petersen, *Inorg. Chem.* **28**, 81 (1989).

PHOTOPHYSICS OF POLYNUCLEAR COMPLEXES. INTERCOMPONENT ENERGY AND ELECTRON TRANSFER PROCESSES.

F. SCANDOLA, C. A. BIGNOZZI, C. CHIORBOLI, M. T. INDELLI, M. A. RAMPI
Dipartimento di Chimica dell'Università, Centro di Fotochimica C.N.R., 44100 Ferrara, Italy

ABSTRACT. Polynuclear complexes add a new dimension to the photochemistry of coordination compounds. Their photophysics is characterized by intercomponent transfer processes: optical electron transfer, photoinduced electron transfer, electronic energy transfer. Examples of such type of processes are discussed in some detail for polynuclear complexes containing Ru(II)-Ru(III), Ru(II)-Ru(II), Ru(II)-Rh(III), and Ru(II)-Cr(III) centers.

1. Introduction.

1.1 FROM MONONUCLEAR TO POLYNUCLEAR COMPLEXES

Compared to typical organic molecules, coordination compounds of transition metals exhibit a remarkable variety of electronically excited state types [1-5]. For most coordination compounds, it is usually possible to assign the various molecular orbitals as predominantly localized on either metal or ligands. As a consequence the excited states of coordination compounds can be classified as metal-centered (MC), ligand-centered (LC), or of charge transfer (CT) type (with the possibility of metal-to-ligand, MLCT, or ligand-to-metal, LMCT, subtypes). A given type of excited state gives rise to typical spectroscopic transitions and (when it occur as the lowest excited state of the system) to a specific type of photophysical and photochemical behavior [1-5]. From a practical point of view, the possibility to play with metal and ligands in a large number of combinations offers to the coordination chemist a remarkable degree of synthetic control on excited state properties.

The variety of excited-state types and behavior exhibited by coordination compounds is clearly related to the "composite" nature of these molecular systems (which are actually often called *complexes*), consisting of covalently bound, but nevertheless distinguishable, metal and ligands. A further, drastic step towards increasing structural complexity can be made by going from mononuclear to *polynuclear complexes*. These are systems in which two (or more) metal complex subunits are connected by one (or more) bridging ligand(s) (Fig 1). Because of this additional complexity, the photophysics of polynuclear complexes is expected to be different from, and more interesting than, that of simple mononuclear species [6]. In this article we illustrate, with a few experimental examples, some of the intriguing aspects of the photophysics of polynuclear complexes. To facilitate the discussion, some general concepts concerning supramolecular systems and intercomponent transfer processes will first be recalled.

E. Kochanski (ed.), Photoprocesses in Transition Metal Complexes, Biosystems and Other Molecules. Experiment and Theory, 253–269.

1.2 POLYNUCLEAR COMPLEXES AS SUPRAMOLECULAR SYSTEMS

With few exceptions, polynuclear complexes can be considered as *supramolecular systems*, with the various metal-containing subunits acting as *molecular components*. The term "supramolecular" was originally coined for multicomponent systems in which the molecular units

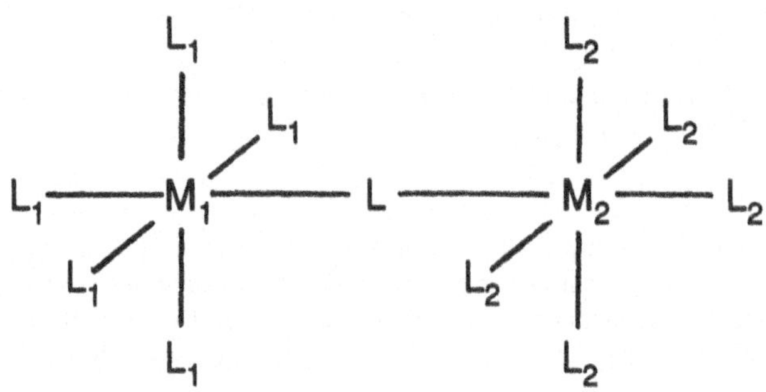

Figure 1. Schematic picture of a binuclear complex

are held together by weak intermolecular forces such as electrostatic interactions, hydrogen bonds, donor-acceptor interactions, host-guest interactions [7]. It seems possible, however, to generalize the definition of supramolecular system to include systems in which molecular components are *covalently* bound, provided that the intercomponent forces are weak enough to leave each molecular component with its essential individual properties [8]. This enlarged definition includes polynuclear complexes, subject to the limitations discussed in the next section.

1.2.1 *Electron Localization.* As stated above, a key requirement for a covalently linked supramolecular system is that the interaction between molecular components is *weak enough* to preserve their characteristic, individual properties. To make this statement more quantitative, the general problem of the degree of electron delocalization in complex chemical systems must be addressed. This has been done in particular detail for a class of complex systems known as mixed-valence compounds, i.e., polynuclear complexes containing the same metal in different oxidation states [9]. The argument can be illustrated using the binuclear complex $(NH_3)_5$Ru-L-Ru$(NH_3)_5^{5+}$ (where L represents a symmetrical bridging ligand) as an example [10]. In a valence-localized description, that is, in terms of integral oxidation states of the metal centers, the overall charge corresponds to a Ru(II)-Ru(III) complex. In a fully delocalized description, on the other hand, a species with both Ru centers in a 2.5 oxidation state would result. The factors determining the localized or delocalized nature of the complex can be easily appreciated following the approach originally developed by Hush [11]. Consider the two valence-localized "electronic isomers" Ru(II)-Ru(III) and Ru(III)-Ru(II). A specific equilibrium geometry corresponds to each of these species, in terms of both *inner* (e.g., Ru-NH_3 distances at both centers) and *outer* (e.g., orientation of solvent molecules around both centers) nuclear degrees of freedom. This is depicted in Fig. 2a using parabolic potential energy curves for the two electronic isomers and a generalized nuclear coordinate involving both inner and outer nuclear displacements. Fig. 2a points out the fact that at the equilibrium geometry of each electronic isomer the other isomer can be considered as an electronically excited state. The energy separation between these two states at the equilibrium geometry is usually called reorganizational energy and is indicated by λ^{el}. At the crossing point both electronic isomers have the same energy and geometry, and can be interconverted nonradiatively without Franck-Condon restrictions. The nuclear configuration of the crossing point corresponds to the classical transition state for electron exchange. Its energy is in this model $\lambda/4$.

If for some reason (for example, very long center-to-center distance, "insulating" character of L) the electronic interaction between the Ru(II) and Ru(III) centers, H_{AB}^{el}, is absolutely negligible, the curves in Fig. 2a accurately represent the system at any geometry along

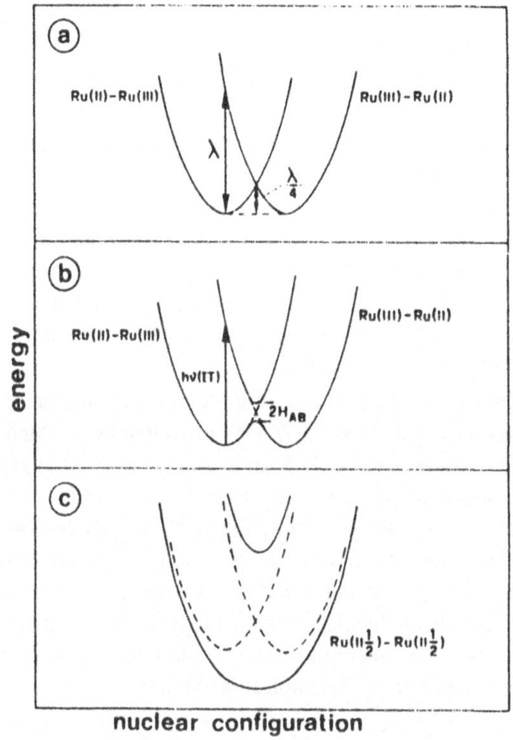

Figure 2. Potential energy curves for mixed-valence compounds with negligible (a), weak (b), and strong (c) electronic coupling. In (b) and (c) the dashed curves represent zero-order states.

the nuclear coordinate. The system is expected to exhibit properties which are a perfect superposition of the properties of isolated $Ru(NH_3)_5L^{3+}$ and $Ru(NH_3)_5L^{2+}$. Furthermore, even if the system acquires sufficient activation energy to reach the intersection region, the probability for electron exchange is negligible. [In the field of mixed-valence chemistry, this is usually called a Class I behavior [12]. An example approaching this type of behavior within the example considered is obtained [10] when L is py-CH_2-CH_2-py (py = 4-pyridyl)].

In most cases, however, some electronic interaction is likely to occur between the Ru(II) and Ru(III) centers, either as a consequence of direct orbital overlap or via some through-bridge mechanism. In such cases, the curves in Fig. 2a should only be regarded as zero-order representations. The electronic interaction has almost no effect on the zero-order curves in the vicinity of the equilibrium geometries, where the difference in energy between the electronic isomers is much larger than H_{AB}^{el}, but causes mixing of the zero-order states (avoided crossing) in the vicinity of the crossing point (Fig. 2b). Systems of this type can still be considered as valence-localized, and will still exhibit the properties of the isolated $Ru(NH_3)_5L^{3+}$ and $Ru(NH_3)_5L^{2+}$ components. However, new properties promoted by the Ru(II)-Ru(III) interaction can also be observed, such as optical transitions (with $h\nu_{max} = \lambda$, often designated as "intervalence transfer" (IT) transitions) or thermally activated electron transfer processes interconverting the two electronic isomers. The barrier to thermal electron transfer is only negligibly smaller than calculated on the basis of the zero-order curves ($\lambda/4$). [This type of behavior is usually called Class II [12]. An example of Class II behavior within the example considered is obtained [10] when L is 4,4'-bipyridine].

If strong electronic coupling is provided by the bridging ligand, the zero-order levels can be substantially perturbed even in the vicinity of their equilibrium geometries. In the limit of very large electronic coupling, when $H_{AB}^{el} \approx \lambda^{el}$, the true first-order curves will show a single minimum at an intermediate geometry (Fig. 2c). In this case, the binuclear complex is better considered a fully delocalized Ru(II1/2)-Ru(II1/2) species, with properties that are mostly unrelated to those of the hypothetical $Ru(NH_3)_5L^{3+}$ or $Ru(NH_3)_5L^{2+}$ components. [This case is commonly indicated as Class III [12]. An example of Class III behavior within the example considered is obtained [10] when L is pyrazine].

The above classification of mixed valence compounds has been illustrated using symmetric redox systems (i.e., systems made of identical subunits in which there is no net driving force for intramolecular electron transfer) but can be easily extended to systems exhibiting redox asymmetry. Clearly, compounds described by Figs 2a and 2b (Class I and II) belong to our operational definition of "supermolecule", while Fig 2c (Class III) describes what is just a "large molecule". The above discussion emphasizes the point that a supermolecule should be amenable to a description in terms of *localized electronic configurations*. This requires that the degree of electronic coupling between the molecular components is small. It is important to recognize that, in this context, "small" is not to be intended in an absolute sense, but with respect to the energy of vibrational trapping of the electron on each molecular component.

1.2.2. *Localization of Excitation Energy*. The conceptual scheme used above to discuss the localization of electrons (oxidation states) can be extended to analyze the localization of excitation energy (electronically excited states) in a supramolecular system. Let us consider the *electronically excited* binuclear complex, e.g., $(NH_3)_5Ru\text{-}L\text{-}Ru(NH_3)_5^{4+}$. Again, three typical situations may be envisaged, depending on the magnitude of center-to-center electronic coupling (see next section for a discussion of the nature of this coupling). In the ideal case of $H_{AB}^{en} = 0$, the excitation will be fully localized on one of the two centers (e.g. $(NH_3)_5{}^*Ru\text{-}L\text{-}Ru(NH_3)_5^{4+}$ or $(NH_3)_5Ru\text{-}L\text{-}{}^*Ru(NH_3)_5^{4+}$), but there would be no possibility for intercomponent energy transfer (by analogy with the electron transfer case, this behavior could be termed Class I). At the other extreme, for very large H_{AB}^{en} (by analogy, Class III), the excitation will be delocalized over the entire binuclear complex (e.g., $^*[(NH_3)_5Ru\text{-}L\text{-}Ru(NH_3)_5^{4+}]$). In the (more common) case where H_{AB}^{en} is non-negligible but smaller than the vibrational trapping energy (by analogy, Class II), the excitation can be still regarded as essentially localized , but now intercomponent energy transfer processes (e.g. $(NH_3)_5{}^*Ru\text{-}L\text{-}Ru(NH_3)_5{}^{4+}$ ---> $(NH_3)_5Ru\text{-}L\text{-}{}^*Ru(NH_3)_5{}^{4+}$ are allowed.

In this extension of the localization argument from the electron to the energy transfer field, the differences between the quantities involved should not be overlooked. Aside from the difference between $H_{AB}{}^{en}$ and $H_{AB}{}^{el}$ (see section 1.3.2), it should be remembered that trapping energies for electronic excitation are expected to involve mainly reorganization of internal vibrational modes (inner-sphere contribution). Reorganization of solvent modes (outer-sphere contribution), which is always an important part of λ in electron transfer, is expected to be negligible for energy transfer unless the ground and excited states have vastly different dipole moments.

1.3 INTERCOMPONENT ELECTRON AND ENERGY TRANSFER PROCESSES.

All of the polynuclear complexes described in this article are supramolecular species in the sense discussed in the previous sections, i.e., they deserve a substantially localized description

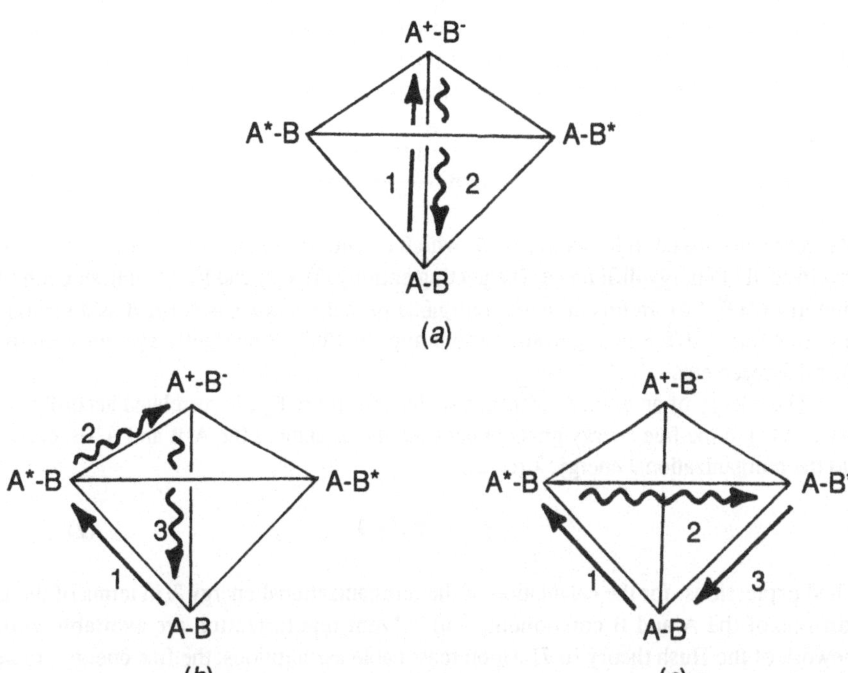

Fig. 3. Schematic representation of the processes involved in (*a*) optical electron transfer, (*b*) photoinduced electron transfer, and (*c*) electronic energy transfer. Full-line, broken-line, and wavy arrows indicate light absorption, emission, and radiationless processes.

both in terms of oxidation states and excitation energy (see later sections). Thus, the spectroscopic, photophysical, and redox properties of their molecular components are similar, albeit somewhat perturbed, to those of the corresponding free species.

To be interesting from a photophysical standpoint, however, a supramolecular species must also exhibit *new* processes, not assignable to the single components. As shown in the previous sections, provided that some electronic coupling is present, a number of intercomponent processes can indeed occur in such systems: *optical electron transfer*, *photoinduced electron transfer*, and *electronic energy transfer*. These processes are schematically shown in Fig. 3. While, strictly speaking, each of these processes is a single step in Fig. 3 (1 in Fig. 3a, 2 in Fig. 3b, and 2 in Fig 3c, respectively), a sequence of steps is involved in typical experimental studies. In particular, charge recombination leading to the ground state (2 in Fig. 3a, 3 in Fig. 3b) usually follows photoinduced electron transfer, and localized emission from the acceptor center (3 in Fig. 3c) is commonly used to detect the occurrence of electronic energy transfer. Figure 3 emphasizes the point that "reactants" and "products" of the various intercomponent transfers are nothing but different electronic states of the supermolecule, which can be interconverted either by optical or by radiationless processes. The factors which determine the probability of such processes (extinction coefficients for the optical process, rate constants for the other processes) are briefly recalled in this section.

1.3.1 *Optical electron transfer.* The absorption spectrum of a supramolecular system can differ substantially from the sum of the spectra of the molecular components. Aside from those small shifts that can be dealt with in terms of perturbation of the spectra of the single components upon bridging, some totally new bands can be present in the spectrum of the supermolecule. These bands correspond to optical electron transfer transitions (often denominated *intervalence transfer* transitions, eq 1). The factors that determine the spectroscopic characteristics of such

$$A\text{-}B \xrightarrow{h\nu} A^{+}\text{-}B^{-} \qquad (1)$$

bands can be discussed in terms of Fig. 4, which is similar to Fig. 2a except for the use of free-energy instead of energy differences (neglecting entropy effects) and for the introduction of redox asymmetry ($\Delta G^0 \approx 0$). In this case, the reorganizational energy λ is defined as a virtual, rather than real, energy difference, pertaining to a hypothetical isoenergetic system with the same nuclear displacements.

The energy of an optical electron transfer transition, E_{op}, is correlated according to Hush theory [10,11] to the free energy gradient between the minima of the A-B and A^+-B^- curves, ΔG^0, and to the reorganizational energy, λ (eq 2).

$$E_{op} = \Delta G^0 + \lambda \qquad (2)$$

Detailed expressions for the calculation of the reorganizational energy λ in terms of the internal distortions of the A and B components and solvent repolarization are available within the framework of the Hush theory [6,7]. Upon reasonable assumptions, the free energy gradient ΔE can be estimated from the standard redox potentials of the isolated components (or of reasonable models thereof). The energy of the optical electron transfer transition can thus be used to obtain an experimental estimate of the reorganizational energy.

The optical electron transfer band is expected to be Gaussian-shaped, with a halfwidth that is directly related to the reorganizational energy by eq 3 [10,11].

$$\Delta\bar{\nu}_{1/2}(\mathrm{cm}^{-1}) = 48.06(E_{op} - \Delta G^0)^{\frac{1}{2}} \qquad (3)$$

The intensity of the optical electron transfer band can be correlated according to Hush [10,11]

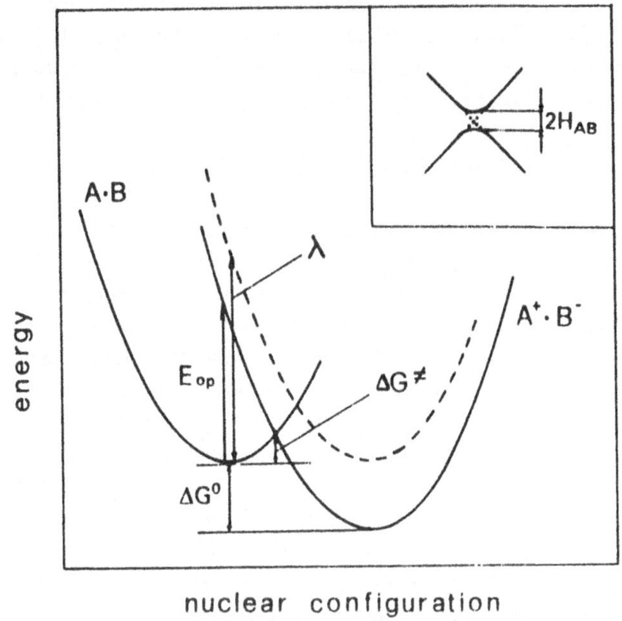

Fig. 4. Energy profiles and relevant parameters for intercomponent electron transfer

to the magnitude of the electronic coupling matrix element H_{AB}. The relationship is given by eq 4, where ε_{max} is the maximum molar absorption coefficient and r (Å) is the intercomponent

$$\varepsilon_{max} = \frac{2380\, r^2}{E_{op}\, \Delta\bar{\nu}_{1/2}} H_{AB}^{\;2} \tag{4}$$

distance (energies and halfwidths, cm^{-1}). We have seen in section 1.2.1 that the magnitude of the intercomponent electronic coupling relative to the reorganizational energy is crucial to determine the localized or delocalized nature of the supermolecule. The above equations show that important information about this point can be obtained, in principle, from the study of the optical electron transfer spectrum of a supermolecule.

Whether or not optical electron transfer transitions are actually observed in the spectrum of a supermolecule depends on several factors. Generally speaking, the observation of such bands requires substantial intercomponent electronic coupling, especially if intense absorption by the molecular components is present in the same spectral region. For example, an $H_{AB} = 200\ cm^{-1}$ is required (assuming common values for the other parameters, i.e., r = 10 Å, $E_{op} = 15000\ cm^{-1}$, $\Delta\nu_{1/2} = 5000\ cm^{-1}$) to have $\varepsilon_{max} = 900\ M^{-1}cm^{-1}$. It will be shown in subsequent sections that in some cases very weak electronic couplings may be sufficient to induce *radiationless* intercomponent processes such as electron or energy transfer. It is evident that in such cases, although the electronic coupling is an important parameter, information about its magnitude cannot be obtained from the spectra, as the optical electron transfer transitions have negligible intensity.

1.3.2 *Kinetics of electron and energy transfer*. Photoinduced electron transfer and electronic energy transfer, despite their clear-cut difference in nature, are amenable to quite similar kinetic models [13-16]. Actually, in the weak interaction (nonadiabatic) limit, a single closed-form expression can be used for the probability of both types of processes, namely [15,17-19]

$$k = (2\pi/h)\, H_{AB}^{\;2}\; \mathrm{FCWD} \qquad (5)$$

where H_{AB} is the electronic coupling term and FCWD is the so-called "Franck-Condon weighted density of states". If one takes the viewpoint that the "reactants" and "products" of energy and electron-transfer processes are but different electronic states of the supermolecule (Fig. 3), both intercomponent processes can be treated as radiationless transitions between electronic states. As a matter of fact, eq 5 is a typical "Golden Rule" expression for radiationless transition probability.

The main difference between the two cases lies in the detailed structure of the electronic factor for electron transfer, $H_{AB}^{\;el}$, and energy transfer, $H_{AB}^{\;en}$. Simply speaking, $H_{AB}^{\;el}$ is a one-electron matrix element involving the HOMO of the (electron) donor center and the LUMO of the (electron) acceptor center. On the other hand, $H_{AB}^{\;en}$ is a two-electron matrix element involving HOMOs and LUMOs on both the (energy) donor and (energy) acceptor centers. Although in a general case $H_{AB}^{\;en}$ may contain both coulombic and exchange terms, the exchange interaction is likely to be dominant for energy transfer processes in involving spectroscopically forbidden states, such as those between coordination compounds. In a supramolecular system, the electronic coupling matrix elements $H_{AB}^{\;el}$ and $H_{AB}^{\;en}$ depend, among other factors, on the intercomponent distance. An exponential decrease with distance (eq 6) is generally expected to occur. In eq 6, r is simply the intercomponent distance if a "through-space" interaction (direct

$$H_{AB} = H_{AB}(0)\exp[-\beta(r - r_0)] \qquad (6)$$

intercomponent orbital overlap) is assumed, but should be considered as a "through-bond" effective distance when the interaction is assumed to be mediated by the intervening bridging groups (e.g., in the so-called "superexchange mechanism"). In eq 6, $H_{AB}(0)$ is the interaction at the minimal intercomponent distance r_0, and β is the attenuation parameter. Because of its two-electron nature, $H_{AB}^{\;en}$ is generally expected to be smaller, in similar conditions, than $H_{AB}^{\;el}$, and to depend more heavily on distance ($\beta^{en} \approx 2\beta^{el}$) [15].

The FCWD term (often referred to as the "nuclear factor" of the rate constant) is a thermally averaged Franck-Condon factor connecting the initial and final states. In particular, it contains a sum of overlap integrals between the nuclear wavefunctions of reactant and product of the same energy. The nuclear wavefunctions include both (inner) vibrational modes and (outer) solvent reorganizational modes. The summation is made over the initial levels of the reactant, suitably weighted for their Boltzmann population. In a general case, the expression of FCWD is very complicated [18], but in the high temperature limit (i.e. when $h\nu < kT$ for all relevant nuclear modes, an approximation which is not too inaccurate for metal complexes at room temperature), it takes the simple form

$$\mathrm{FCWD} = (1/4\pi\lambda RT)^{1/2}\exp[-(\Delta G^0 + \lambda)^2/4\lambda RT] \qquad (7)$$

where ΔG^0 is the thermodynamic driving force of the process. The activation free energy in the exponential term of eq 7 is the same as that predicted by the classical Marcus theory [20] on the basis of parabolic energy curves such as those of Fig. 4. Equation 7 (as well as its more complex general form [18]) predicts three typical kinetic regimes depending on the driving force range: (i) a "normal" regime for small driving forces ($-\lambda < \Delta G^0 < 0$) where the process is thermally activated and is favored by an increase in driving force; (ii) an activationless regime ($-\lambda \approx \Delta G^0$) where no gain in rate can be obtained by changing the driving force; (iii) an "inverted" regime for strongly exergonic reactions ($-\lambda > \Delta G^0$) where the process slows down with increasing driving force. Increasing λ slows down the process in the normal regime, but accelerates it in the inverted regime.

As such, eqs 5 and 6 predict a linear increase of rate constant with H_{AB}^2. This corresponds to considering electronic motion (in classical terms, crossing at the intersection of Fig. 4) as the rate limiting step. In fact, eqs 5 and 6 only hold for relatively small electronic coupling. With increasing H_{AB}, the rate cannot increase indefinitely as the limiting step eventually becomes nuclear motion (in classical terms, climbing the barrier to the crossing point). In this limit, which is called *adiabatic*, the rate constant is classically given by

$$k = \nu_N (1/4\pi\lambda RT)^{1/2} \exp[-(\Delta G^0 + \lambda)^2/4\lambda RT] \qquad (8)$$

where ν represents an effective nuclear frequency [21] for motion along the reaction coordinate. The generally smaller values expected for H_{AB}^{en} relative to H_{AB}^{el} (see above) make the adiabatic regime less likely for energy than for electron transfer.

2. Experimental Examples

2.1 OPTICAL ELECTRON TRANSFER

The system discussed belongs to a series of bi- and trinuclar complexes $X(NH_3)_4Ru\text{-}NC\text{-}Ru(bpy)_2(CN)^{n+}$ and $X(NH_3)_4Ru\text{-}NC\text{-}Ru(bpy)_2\text{-}CN\text{-}Ru(NH_3)_4Y^{m+}$ ($X = NH_3$, py; $Y = NH_3$, py; $m = 4\text{-}6$; $n = 2,3$) [22,23]. The possible combinations of X and Y ligands and of oxidation states of the $Ru(NH_3)_4X$ and/or $Ru(NH_3)_4Y$ subunits give rise to a large number of complexes within this series. Let us consider the species

$$py(NH_3)_4Ru\text{-}NC\text{-}Ru(bpy)_2\text{-}CN\text{-}Ru(NH_3)_5^{5+}$$

in which the oxidation states of the various components are as described below.

In this species, the following electronic states are relevant:

0) ground state: $py(NH_3)_4Ru^{II}\text{-}NC\text{-}Ru^{II}(bpy)_2\text{-}CN\text{-}Ru^{III}(NH_3)_5^{5+}$

1) Ru-->bpy MLCT: $py(NH_3)_4Ru^{II}\text{-}NC\text{-}Ru^{III}(bpy^{\cdot -})(bpy)\text{-}CN\text{-}Ru^{III}(NH_3)_5^{5+}$

2) Ru-->py MLCT: $(py^{\cdot -})(NH_3)_4Ru^{III}\text{-}NC\text{-}Ru^{II}(bpy)_2\text{-}CN\text{-}Ru^{III}(NH_3)_5^{5+}$

3) Ru-->bpy remote MLCT: $py(NH_3)_4Ru^{III}\text{-}NC\text{-}Ru^{II}(bpy^{\cdot -})(bpy)\text{-}CN\text{-}Ru^{III}(NH_3)_5^{5+}$

4) Ru-->Ru IT: $py(NH_3)_4Ru^{II}\text{-}NC\text{-}Ru^{III}(bpy)_2\text{-}CN\text{-}Ru^{II}(NH_3)_5^{5+}$

5) Ru-->Ru remote IT: $py(NH_3)_4Ru^{III}\text{-}NC\text{-}Ru^{II}(bpy)_2\text{-}CN\text{-}Ru^{II}(NH_3)_5^{5+}$

Of these excited states, types 1 and 2 would be present in the isolated components as well, but 3-5 are new states of the intercomponent electron transfer type, characteristic of the polynuclear species as such. Transitions corresponding to the various types of excited states can be easily identified in the absorption spectrum of this complex. The resolution of the absorption spectrum

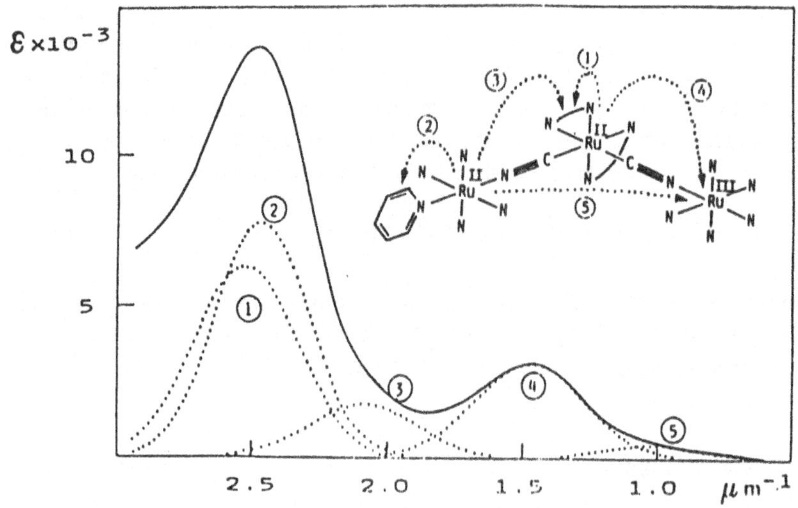

Fig. 5 Resolution of the absorption spectrum of $py(NH_3)_4Ru\text{-}NC\text{-}Ru(bpy)_2\text{-}CN\text{-}Ru(NH_3)_5^{5+}$ into charge transfer transitions.

of the complex into various types of transitions is shown in Fig. 5. By selective oxidation or reduction of the various sites in the molecule, the attribution of the various types of transitions is straightforward [22,23]. Of particular interest from the spectroscopic point of view is the direct observation of *remote* MLCT and of *remote* IT, indicating that, within the limits of an essentially localized description (sections 2.2 and 2.4), sizable electronic coupling between the various sites is present in these systems. For the remote IT transition, the intensity appears to fit a superexchange model for through-bond interaction between the terminal metal centers where the $\text{-NC-Ru(bpy)}_2\text{-CN-}$ fragment is considered as the connector [24].

In this polynuclear complex, no emission can be detected following excitation in the Ru->bpy MLCT absorption band, indicating that efficient pathways are available for quenching of the MLCT state of the $\text{-Ru(bpy)}_2\text{-}$ chromophore. These pathways can be easily identified as two-step electron transfer sequences on the basis of the states detected spectroscopically (1-->3-->0 and 1-->4-->0). Actually, no transients are detected in nanosecond laser experiments, indicating that charge recombination steps (3-->0 and 4-->0) are very fast processes. The reasons for the lack of observable transient electron transfer products in these systems have been discussed in terms of thermodynamic and kinetic factors [22,23]. Briefly, considering the redox properties of the components and reasonable reorganizational energies, all the electron transfer steps are expected to lie in the nearly activationless regime. Furthermore, the strong intercomponent electronic coupling provided by the bridging cyanide, witnessed by the high intensity of the intervalence transfer transition (4 in Fig. 5), is likely to place these electron transfer processes in the adiabatic kinetic limit. Recent ultrafast measurements on a cyano-bridged complex with comparable energetics [25] suggest that an electron transfer step of type 4-->0 may occur in the subpicosecond time scale.

The above discussed system provides an example of the difficulty to correlate optical and radiationless electron transfer. It would obviously be very interesting to use the information obtainable form optical electron transfer (electronic couplings, reorganizational energies, etc., see section 1.3.1) to predict and check kinetic data for the correspondent radiationless electron transfer processes (see section 1.3.2). As shown here, a strong intercomponent coupling makes the optical transitions intense enough to give rise to an experimentally interesting spectroscopy. The same effect, however, contributes to make the radiationless processes so fast as to void the photophysics of any experimental interest. The cross-check of electron transfer spectroscopy and kinetics, allowed in principle by theory, may often prove tantalizing on experimental grounds.

2.2 PHOTOINDUCED ELECTRON TRANSFER

Metal bipyridine complexes are classical mononuclear coordination compounds of photophysical interest [26]. Related bi- and polynuclear species can be designed using a "double bipyridine" ligand, obtained by linking together two bipyridine-type ligands via a short polymethylene chain [27-30]. The example discussed in this section is the Ru(II)-Rh(III) heterodimer [31]

$$(Me_2\text{-phen})_2Ru\text{-}(Me\text{-bpy-}CH_2\text{-}CH_2\text{-bpy-Me})\text{-}Rh(Me_2\text{-phen})_2^{5+}$$

The absorption spectrum of the complex is an exact superposition of the spectra of the isolated mononuclear species, with no new band being present in this case. An approximate energy level diagram for this binuclear complex, based on known excitation energies and redox properties of the components, is shown in Fig. 6. It shows that the lowest excited state of the system is a

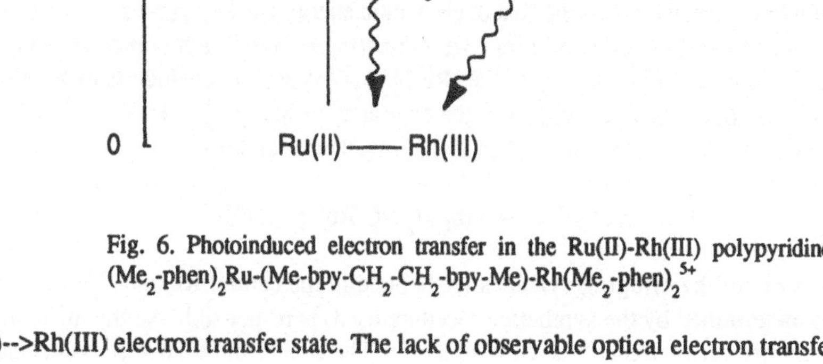

Fig. 6. Photoinduced electron transfer in the Ru(II)-Rh(III) polypyridine heterodimer $(Me_2\text{-phen})_2Ru\text{-}(Me\text{-bpy-}CH_2\text{-}CH_2\text{-bpy-Me})\text{-}Rh(Me_2\text{-phen})_2^{5+}$

Ru(II)-->Rh(III) electron transfer state. The lack of observable optical electron transfer transitions implies that intercomponent interaction is relatively small in this system. On the other hand, photoinduced electron transfer is thermodynamically feasible following excitation of both molecular components.

Upon excitation of the Ru(II) chromophore (which is the only practically feasible one), the typical Ru(II) polypyridine emission is quenched by ca. 90% with respect to that of the free component. The lifetime of the emission, τ, is 5 ns (i.e. ca. 90% shorter than that, τ_0, of the free component). This indicates the occurrence of efficient excited-state electron transfer (Fig. 6), with a rate constant $k = (1/\tau) - (1/\tau_0) = 2x10^8 s^{-1}$ [31]. [A minor component of the emission with somewhat longer lifetime could tentatively be associated with the back electron transfer process (via *Ru(II)-Rh(III)/Ru(III)-Rh(II) excited-state equilibrium), but direct evidence for this hypothesis is not available].

If a comparison is made with the results described in the previous section, the measurable rate obtained in this case for the forward electron transfer step most probably originates from the small driving force ("normal" activated free energy regime) *and* from the small intercomponent electronic coupling (nonadiabatic kinetic regime). Interestingly enough, for a similar Ru(II)-Rh(III) complex recently studied by Furue [32], which has a three-carbon instead of a two-carbon linkage between the components, the electron transfer is almost ten times slower than in this case. This difference is as expected for a nonadiabatic electron transfer process on the basis of the distance dependence of the electronic coupling (section 1.3.2).

A related two-component system, Rh(III)-Q (same type of bridge, Q = diquaternarized bipyridine), has been developed [33] for the study of photochemically induced (eq 9) thermal "charge shift" reaction (eq 10)

$$\text{Rh(III)-Q} + \text{Red} \xrightarrow{h\nu} \text{Rh(II)-Q} + \text{Ox} \qquad (9)$$

$$\text{Rh(II)-Q} \longrightarrow \text{Rh(III)-Q}^{-} \qquad (10)$$

It is possible that fusing the two Ru(II)-Rh(III) and Rh(III)-Q systems may lead to a "triad" suitable for photoinduced charge separation [34].

2.3 ENERGY TRANSFER IN POLYCHROMOPHORIC SYSTEMS.

Systems made up of two or more chromophoric groups that are practically identical in all respects except for small differences in excited-state energy can be denoted as *polychromophoric systems*. In such systems photoinduced electron transfer (which would amount to photodisproportionation) is generally unlikely while, on the other hand, electronic energy transfer processes can be driven by the unbalance in excited-state energies. An interesting class of such compounds is that in which the chromophoric groups are all of the $\text{-Ru(bpy)}_2\text{-}^{2+}$ type, and cyanide is used, both as a terminal and as a bridging ligand, to complete the coordination shell of Ru(II). [35,36]. The example discussed in this section is the trinuclear complex [35]

$$\text{NC-Ru(bpy)}_2\text{-CN-Ru(bpy)}_2\text{-NC-Ru(bpy)}_2\text{-CN}^{2+}$$

The bonding mode of the bridging cyanides (N-bonded to the central Ru and C-bonded to the terminal ones, determined by the synthetic procedure used) is responsible for the differences in excited-state energy between the three chromophores. The energy of the lowest excited state of the central chromophore is estimated to be lower by ca 2000 cm^{-1} than those of the lowest excited states of the terminal chromophores.

In this system, a single emission, attributable to the lowest energy chromophore, is observed (λ_{max} 714 nm). Although no selective population of the various MLCT excited states is possible due to partial overlapping of absorption bands, the exact correspondence between excitation and absorption spectra points towards a very efficient energy-transfer process from the higher-energy chromophores to the lowest energy emitting one. The lack of appreciable risetime in the emission places a lower limit on the rate of intercomponent energy transfer ($k \geq 1 \times 10^9$ s^{-1}. The photophysical behavior of these polychromophoric Ru(II) species is represented in Fig. 7. As

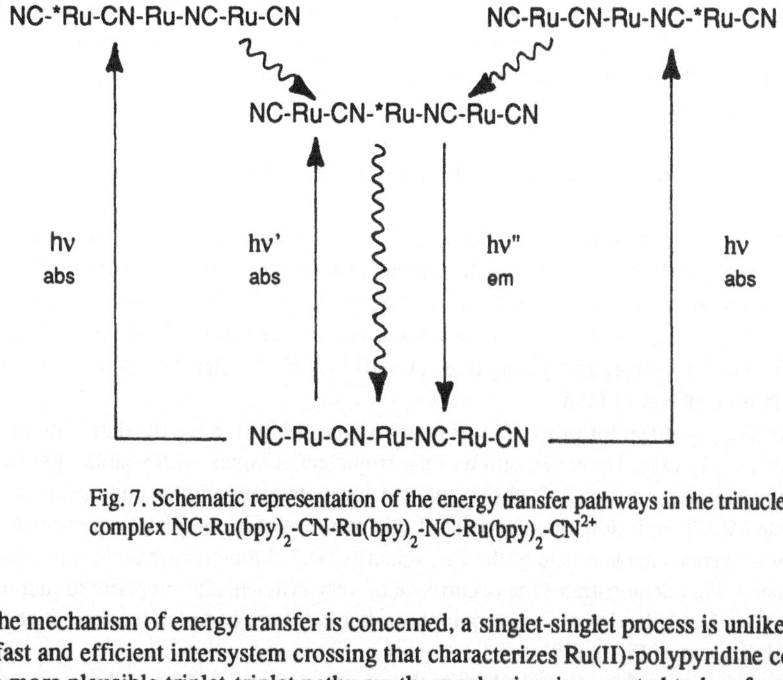

Fig. 7. Schematic representation of the energy transfer pathways in the trinuclear complex NC-Ru(bpy)$_2$-CN-Ru(bpy)$_2$-NC-Ru(bpy)$_2$-CN^{2+}

far as the mechanism of energy transfer is concerned, a singlet-singlet process is unlikely in view of the fast and efficient intersystem crossing that characterizes Ru(II)-polypyridine complexes. For the more plausible triplet-triplet pathway, the mechanism is expected to be of an exchange type. The high rate of energy transfer is understandable as (i) the reorganizational energy is expected to be small for this process and (ii) the electronic coupling provided by cyanide bridges is known to be strong (see section 2.1).

A related trinuclear complex where the two bpy ligands of the central Ru ion bear two carboxylic groups,

$$NC\text{-}Ru(bpy)_2\text{-}CN\text{-}Ru(4,4'(COO)_2\text{-}bpy)_2\text{-}NC\text{-}Ru(bpy)_2\text{-}CN^{2-}$$

has also been studied [37]. This complex is better suited to investigate intramolecular energy transfer since the presence of the carboxylic groups on the ligands of the central chromophoric unit has the effect of further lowering the MLCT levels of this unit. This leads to sizable shifts in MLCT absorption and thus to the possibility to address the individual chromophores with light of different wavelength. Emission from the central chromophore is again observed with constant efficiency, independent on the nature of excited chromophore. This complex has been applied to the problem of the spectral sensitization of wide-band semiconductors. Due to the presence of the carboxylate groups, the complex can be anchored through the central unit to the surface of TiO_2. Here again the incident light energy absorbed by the terminal chromophores is efficiently transferred to the central one, which is then able to inject electrons into the semiconductor, giving rise to observable photocurrents [37]. In this system the terminal units play the so-called "antenna effect", and the trinuclear complex acts as an *antenna-sensitizer molecular device* on the surface of the semiconductor. Promising practical applications of such polynuclear complexes for light energy conversion seem to be at hand [38,39].

2.4 ENERGY TRANSFER IN CHROMOPHORE-LUMINOPHORE COMPLEXES.

This class of polynuclear complexes is ideally suited for the detection and study of intercomponent energy transfer as, contrary to the case of polychromophoric systems discussed above, the molecular components of these systems have very distinct roles. The example discussed in this section is [40]

$$NC\text{-}Cr(cyclam)\text{-}CN\text{-}Ru(bpy)_2\text{-}NC\text{-}Cr(cyclam)\text{-}CN^{4+}$$

The "chromophoric" component is the $-Ru(bpy)_2^{2+}-$ units. Its properties, as seen in the previous section, are: strong MLCT absorption in the visible, relatively long-lived ($\tau = 10^{-7}$-10^{-6} s) MLCT triplet states, broad-band emission in the 600-650 nm, range. The "luminophoric" units are the $Cr(cyclam)(CN)_2^+$ fragments. Their properties are: practically negligible absorption (at 10^{-4} M concentration) in the visible, a very long-lived ($\tau \approx 10^{-4}$ s) MC doublet state, a very typical sharp-band phosphorescence at 715 nm.

The sharply different properties of the two types of unit make the detection of energy transfer particularly easy. Upon excitation of the trinuclear complex with visible light (which is only absorbed by the $-Ru(bpy)_2^{2+}-$ chromophoric component) the following observations are made: (i) the MLCT emission characteristic of this component is completely quenched; (ii) the MC phosphorescence characteristic of the $Cr(cyclam)(CN)_2^+$ chromophoric units is obtained with high efficiency. This demonstrates the occurrence of very efficient chromophore to luminophore energy transfer. The behavior of the chromophore-luminophore complex is schematized on the energy level diagram of Fig. 8 [40].

The energy transfer process, which in this case is definitely of the exchange type, is fast ($k > 10^9$ s^{-1}). Likely, the cyanide bridge is important in providing strong exchange interaction between the Ru(II) and Cr(III) metal centers. An interesting aspect of this (and related) systems is the perturbation of the photophysics of the luminophoric unit by the attached chromophoric component. The emission quantum yields and lifetimes are 5.3×10^{-3} and 260 μs for $NC\text{-}Cr(cyclam)\text{-}CN\text{-}Ru(bpy)_2\text{-}NC\text{-}Cr(cyclam)\text{-}CN^{4+}$, and 3.3×10^{-3} and 310 μs for the isolated $Cr(cyclam)(CN)_2^+$ component. Given the unitary efficiency of population of the emitting state in both systems, this indicates that the radiative rate constant of Cr(III) phosphorescence increases from 10 s^{-1} to 20 s^{-1} upon attachment of the chromophoric unit. This may be due to the increased spin-orbit coupling provided by the presence of the heavier Ru(II) center in the second coordination sphere of Cr(III).

The behavior of this Ru(II)-Cr(III) chromophore-luminophore complex (as well as of some related ones [41]) provides an example of how the properties of a luminophore can be improved by attachment to a suitable chromophoric component (spectral sensitization, antenna

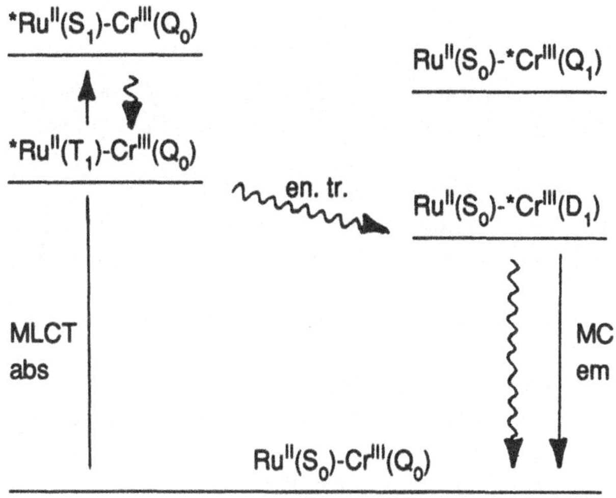

Fig. 8. Energy level diagram and photophysical processes taking place in the chromophore-luminophore complex NC-Cr(cyclam)-CN-Ru(bpy)$_2$-NC-Cr(cyclam)-CN^{4+}

effect, avoidance of quartet photoprocesses). When very specific light absorption and light emission characteristics are required (e.g., in the design of luminescent labels for biochemical applications), the use of chromophore-luminophore systems may represent a convenient strategy. In particular this strategy permits *separate* optimization of absorption and emission properties, a possibility which is precluded in simple molecular species.

Extension of this work to larger systems featuring multistep sequential energy transfer is possible. A polynuclear system containing a Ru(II) chromophore, a Cr(III) luminophore, and an intermediate Cr(III) energy relay has been developed to that purpose [42].

3. Conclusions

The extension from simple mononuclear species to polynuclear complexes represents a new, and still relatively unexplored, dimension in the photochemistry of coordination compounds. As illustrated by the few examples discussed above, the photophysics of polynuclear complexes is generally characterized by facile intercomponent energy and electron transfer processes. The large choice of molecular components and bridging ligands, coupled to the use of rational synthetic strategies, provides a good degree of synthetic control on the intercomponent transfer processes. In suitably designed systems, intercomponent transfer processes may lead to interesting light-induced functions such as, e.g., photoinduced charge separation, antenna effects, and spectral sensitization.

References

[1] Balzani, V. and Carassiti, V. (1970) *Photochemistry of coordination compounds*. Academic

[2] Adamson, A.W. and Fleischauer, P.D. (eds) (1975) *Concepts of inorganic photochemistry*. Wiley

[3] Special issue (1983) *J. Chem. Educ.* **60** 814

[4] Ferraudi, G.J. (1988) *Elements of inorganic photochemistry*. Wiley

[5] Scandola, F. and Balzani, V. (1989). In Serpone, N. and Pelizzetti, E. (eds) *Photocatalysis*. Wiley, p. 9

[6] Scandola, F., Indelli, M. T., Chiorboli, C., and Bignozzi, C. A. (1990) *Top. Curr. Chem.* **158** 73

[7] Lehn, J.-M. (1985) *Science* **227** 849

[8] Balzani, V. and Scandola, F. (1991) *Supramolecular Photochemistry*, Horwood

[9] Brown, D. B. (ed.) (1980) *Mixed Valence Compounds*. Reidel

[10] Creutz, C. (1983) *Prog. Inorg. Chem.* **30** 1

[11] Hush, N. S. (1967) *Prog. Inorg. Chem.* **8** 391

[12] Robin, M. B. and Day, P. (1967) *Adv. Inorg. Chem. Radiochem.* **10** 247

[13] Balzani, V., Bolletta, F., and Scandola, F. (1980) *J. Am. Chem. Soc.* **102** 2152

[14] Scandola, F., Balzani, V. (1983) *J. Chem. Educ.* **60** 814

[15] Closs, G. L., Piotrowiak, P., MacInnis, J. M., and Fleming, G. R. (1988) *J. Am. Chem. Soc.* **110** 2652

[16] Closs, G. L., Johnson, M. D., Miller, J. R., and Piotrowiak, P. (1989) *J. Am. Chem. Soc.* **111** 3751

[17] Ulstrup, J. (1979) *Charge transfer processes in condensed media*. Springer

[18] Jortner, J.(1976) *J. Chem. Phys.* **64**, 4860

[19] Murtaza, Z., Zipp, A. P., Worl, L. A., Graff, D., Jones, W. E. Jr., Bates, W. D., and Meyer, T. J. (1991) *J. Am. Chem. Soc.* **113** 5113

[20] Marcus, R. A. (1964) *Annu. Rev. Phys. Chem* **15** 155

[21] Sutin, N. (1983) *Prog. Inorg. Chem.* **30** 441

[22] Bignozzi, C. A., Roffia, S. and Scandola, F. (1985) *J. Am. Chem. Soc.* **107** 1644;

[23] Bignozzi, C. A., Paradisi, C., Roffia, S., and Scandola, F. (1988) *Inorg. Chem.* **27** 408

[24] Scandola, F. (1989). In: Norris, J. R. Jr. and Meisel, D. (eds) *Photochemical energy conversion*. Elsevier, p. 60

[25] Walker, G. W., Barbara, P. F., Doorn, S. K., Dong, Y., Hupp, J. T. (1991) *J. Phys. Chem.* **95** 5712

[26] Kalyanasundaram, K. (1992) *Photochemistry of Polypyridine and Porphyrin Complexes*. Academic

[27] Sahai, R., Baucom, D.A., and Rillema, D.P. (1986) *Inorg. Chem.* **25** 3843

[28] Furue, M., Kuroda, N., and Nozakura, S. (1986) *Chem. Lett.* 1209

[29] Furue, M., Kinoshita, S., and Kushida, T. (1987) *Chem. Lett.* 2355

[30] Schmehl, R.H., Auerbach, R.A., Wacholtz, W.F., Elliott, C.M., Freitag, R.A., and Merkert, J.W. (1986) *Inorg. Chem.* **25** 2440

[31] Indelli M. T., Bignozzi, C. A., and Scandola, F., manuscript in preparation.

[32] Furue, M., Hirata, M., Kinoshita, S., Kushida, T., and Kamachi, M. (1990) *Chem. Lett.* 2065

[33] Indelli, M. T., Polo, E., Bignozzi, C. A., and Scandola, F. (1991) *J. Phys. Chem.* **95** 3889

[34] Indelli M. T., Bignozzi, C. A., and Scandola, F., work in progress.

[35] Bignozzi, C.A., Roffia, S., Chiorboli, C., Davila, J., Indelli, M.T., and Scandola. F. (1989) *Inorg. Chem.* **28** 4350

[36] Scandola, F., Bignozzi, C.A., Chiorboli, C., Indelli, M.T., and Rampi, M.A. (1990) *Coord. Chem. Rev.* **97** 299

[37] Amadelli, R., Argazzi, R., Bignozzi, C.A., and Scandola, F. (1990) *J. Am. Chem. Soc.* **112** 7099

[38] O'Regan, B. and Graetzel, M. (1991) *Nature* **353** 737

[39] MalMallouk, T. E. (1991) *Nature* **353** 698

[40] Bignozzi, C. A., Bortolini, O., Chiorboli, C., Indelli, M. T., Rampi, M. A., and Scandola, F. (1992) *Inorg. Chem.* **31** 0000

[41] Bignozzi, C.A., Indelli, M.T., and Scandola, F. (1989) *J. Am Chem. Soc.* **111** 5192

[42] Chiorboli, C., Bignozzi, C. A., Indelli, M. T., Rampi, M. A., and Scandola, F. *Coord. Chem. Rev.*, in press.

RIGID ALKANE-BRIDGED DONOR-ACCEPTOR SYSTEMS AS TOOLS FOR THE INVESTIGATION OF SOLVENT-, DISTANCE-, AND CONFORMATION-EFFECTS IN ELECTRON TRANSFER PROCESSES

J.W. VERHOEVEN*
Laboratory of Organic Chemistry, University of Amsterdam, Nieuwe Achtergracht 129, 1018 WS Amsterdam, The Netherlands

M.N. PADDON-ROW
Department of Chemistry, University of New South Wales, P.O. Box 1, Kensington, N.S.W. 2033, Australia

J.M. WARMAN
Interfaculty Reactor Institute, Delft University of Technology, Mekelweg 15, 2629 JB Delft, The Netherlands

ABSTRACT. Several series of D(onor)-bridge-A(cceptor) systems have been developed in which a rigid alkane-bridge maintains a well defined distance and relative orientation of the electron donor-acceptor pair attached. For a given length of the bridge the effect of the solvating power of the medium on the rate of photoinduced intramolecular charge separation was studied as a function of structural variations at the D and/or A sites. The results of these studies are discussed in the context of current models for the thermodynamics and kinetics of electron transfer processes. It is shown that general and simple design criteria can be formulated, which define the properties of a D/A pair that is tuned to realize optimally fast electron transfer, in any solvent, across a bridge of a given length. The influence of the length and configuration of the alkane bridges on the rate of photoinduced charge separation as well as thermal recombination was investigated. Independently the ability of the bridges to mediate electronic coupling was studied in an experimental (PES, ETS) and theoretical investigation of bridged dienes. For the rate of photoinduced charge separation a very strong correlation with the structure dependence of the electronic coupling was observed. This demonstrates the importance of through-sigma-bond interaction as a coupling mechanism for electron transfer, thereby stressing the decisive role of the intervening medium in mediating long-range electron transfer processes. Interestingly, charge recombination appeared to be equally responsive to the length but less responsive to the configuration of the bridge, and not at all responsive to temperature. The latter is attributed to a dominant contribution of nuclear tunneling, typical for electron transfer in the 'inverted region'.

E. Kochanski (ed.), Photoprocesses in Transition Metal Complexes, Biosystems and Other Molecules. Experiment and Theory, 271–298.

1. Results and Discussion

1.1. MINIMIZATION OF SOLVENT POLARITY EFFECTS ON ELECTRON TRANSFER RATES.

While in recent years intramolecular electron transfer in a plethora of D-bridge-A systems has been studied[1], and quite a few will in fact be treated in this paper, a seemingly simple question such as "what is the maximum rate achievable across a given type and length of bridge" is still not easily answered. One of the complications arising in trying to do so relates to the often quite dramatic rate effect of changes in solvent polarity. This makes data sets obtained in different solvents hard to compare and furthermore leads one to question whether the highest rate reported for a given system could perhaps be enhanced in still another solvent than the ones employed.

Evidently the basic cause for solvent polarity effects on charge separation processes must be related to the stabilization of the 'product state' with respect to the 'initial state' in more polar solvents. For photoinduced charge separation this is conveniently quantified[2,3] by eqn (1), which relates the standard Gibbs energy change ('driving force') in a medium with static dielectric constant ε_s to the standard redox potentials of D and A (as measured in acetonitrile, i.e. at $\varepsilon_s = 37$), the E_{00} excitation energy of either D or A (which ever is the lowest), their center to center separation R_c and the average ionic radius r of D^+ and A^-.

$$\Delta G = e[E_{ox}(D)-E_{red}(A)]-E_{oo}-\frac{e^2}{\varepsilon_s R_c}-\frac{e^2}{r}\left(\frac{1}{37}-\frac{1}{\varepsilon_s}\right) \tag{1}$$

Already in an early stage of our investigations on rigidly bridged systems we noted[4] that the solvent dependence of the rate of charge separation may be dramatically different even for systems which are geometrically and structurally closely related thus implying that the solvent dependence of ΔG is very similar. The data (TABLE 1) on **1(5)** and **1(5)a** (Scheme 1), that both incorporate a steroidal bridge which separates D and A by at least five (n=5) sigma bonds and the same methoxybenzene donor but different, although closely related, acceptors, demonstrate this behaviour. While for **1(5)** electron transfer occurs with a rate beyond that measurable by the fluorescence quenching method used to determine it, even in a solvent (n-hexane) where it is calculated to be nearly thermoneutral, the rate for **1(5)a** remains at least three orders of magnitude smaller in a solvent (acetonitrile) where it is calculated to be significantly exergonic. These observations led us to propose[4], that solvent reorganization effects retard the charge separation in **1(5)a** in the relatively polar solvents which are required to make the process exergonic for this compound. Unfortunately, the very high rates for **1(5)** made quantitative comparison of the rate changes with solvent impracticable for the pair **1(5)/1(5)a**.

More recently we studied[5] another pair of rigid D-bridge-A systems (**5(8)/5(8)a**) (Scheme 2) for which kinetic data could be collected in a wide variety of solvents (see TABLE 2). From the data in TABLE 2 it is evident that photoinduced charge separation in **5(8)** should be thermodynamically feasible in virtually all solvents, including satura-

Scheme 1

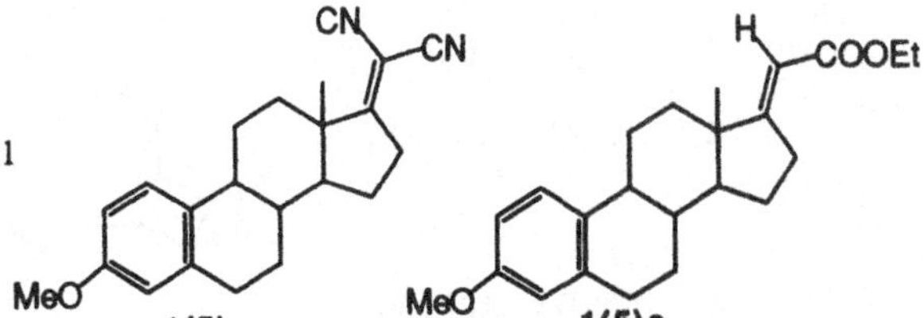

1(5) 1(5)a

TABLE 1. Rates of photoinduced charge separation (k_{et}) observed for **1(5)** and **1(5)a** and the ΔG for this process as calculated via eqn.(1).

solvent	ε_S	**1(5)** ΔG (eV)[a]	**1(5)** k_{et} (s^{-1})	**1(5)a** ΔG (eV)[a]	**1(5)a** k_{et} (s^{-1})
n-hexane	1.90	0.12	$> 10^{11}$	0.63	$< 2x10^7$
di-ethyl ether	4.20	-0.51	$> 10^{11}$	0.00	$< 2x10^7$
acetonitrile	37.50	-0.91	$> 10^{11}$	-0.41	$6.6x10^7$

a) Calculated via eqn.(1) with r= 3.5Å, R_c=7.5Å, $E_{ox}(D)$ = +1.76 V, $E_{00}(D)$= 4.22 eV, $E_{red}(A)$ = -1.7 V for **1(5)** and -2.2 V for **1(5)a**. The latter value is estimated from that for **1(5)**, as measured electrochemically, and the difference (~4000 cm^{-1}) in charge-transfer fluorescence frequencies between **1(5)** and **1(5)a** as determined in various solvents[4].

ted hydrocarbons ($\varepsilon_s \approx 2$), while for **5(8)a** additional solvent stabilization ($\varepsilon_s \geq 3$) is required.

The experimental results (see TABLE 2) are in line with this prediction. Thus for **5(8)** photoinduced charge separation occurs with a high rate in all solvents investigated, but for **5(8)a** it can only be detected if $\varepsilon_S > 3.1$, with the exception of benzene ($\varepsilon_s = 2.28$). The latter is obviously due to specific solvation effects that cannot be accounted for by eqn (1), but the overall data once more underline the remarkable usefulness of this equation for estimating ΔG.

Closer inspection of the data reveals a dramatic difference in solvent dependence of k_{et} for **5(8)** and **5(8)a** in the region where ΔG is negative for both molecules. While k_{et} in **5(8)** shows very minor ε_s dependence over the whole range of solvents, in **5(8)a** it increases by more than an order of magnitude in response to a tenfold increase of ε_s (i.e. from di-isopropyl ether to acetonitrile). Even more strikingly, under conditions of virtually identical ΔG, such as constituted by comparison of **5(8)** in di-n-butyl ether (ΔG = -0.44 eV) and **5(8)a** in tetrahydrofuran (ΔG = -0.45 eV), the rate in **5(8)a** is almost three orders of magnitude lower than that in **5(8)**. While this may seem rather baffling, a simple explanation exists[4,5] in that the solvent reorganization energy (λ_S) incre-

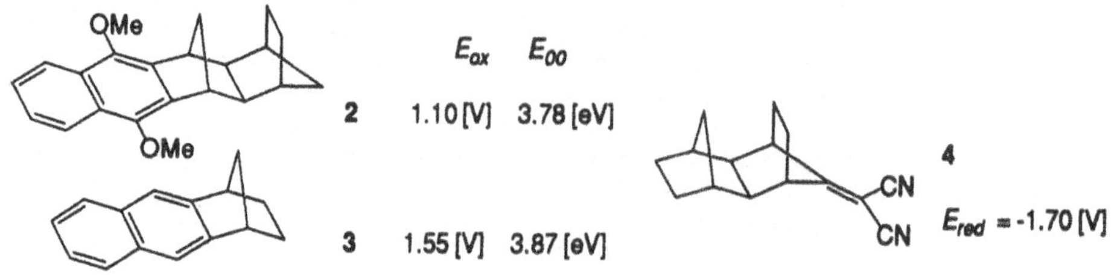

Scheme 2

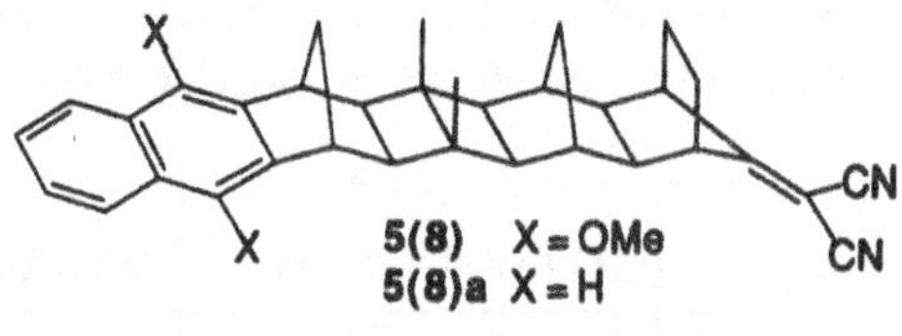

TABLE 2. Measured rate (k_{et}) and calculated driving force (ΔG) of photoinduced charge separation in **5(8)** and **5(8)a**. In the calculations a center to center distance R_c =11.4Å and an average ionic radius r = 4.5Å were employed[6] (see Scheme 2 for the other parameters entering eqn. (1))

			5(8)		**5(8)a**	
solvent	ε_s	n^2	ΔG (eV)	k_{et} ($\times 10^7 s^{-1}$)	ΔG (eV)	k_{et} ($\times 10^7 s^{-1}$)
cyclohexane	2.02	2.03	-0.11	2100	0.25	< 0.1
benzene	2.28	2.25	-0.22	5200	0.14	1.4
di-n-butyl ether	3.10	1.96	-0.44	4700	-0.08	< 0.1
d-isopropyl ether	3.88	1.87	-0.57	-	-0.21	0.5
di-ethyl ether	4.20	1.83	-0.60	4700	-0.25	1.4
ethyl acetate	6.02	1.88	-0.74	4500	-0.38	3.9
tetrahydrofuran	7.58	1.98	-0.81	6700	-0.45	6.5
acetonitrile	37.50	1.81	-1.01	3000	-0.65	11

ases rapidly in more polar solvents, which increases the barrier ($\Delta G^{\ddagger}$) at constant driving force (ΔG). Figure 1 qualitatively illustrates this situation.

Using approximations similar to those used in deriving eqn (1) above, the solvent reorganization energy, λ_s, can be expressed[7] by eqn (2), where n is the solvent refractive index.

$$\lambda_s = e^2\left(\frac{1}{r} - \frac{1}{R_c}\right)\left(\frac{1}{n^2} - \frac{1}{\varepsilon_s}\right) \qquad (2)$$

Quantification of the interdependence between the driving force, the reorganization energy and the barrier to electron transfer is most conveniently derived via the well

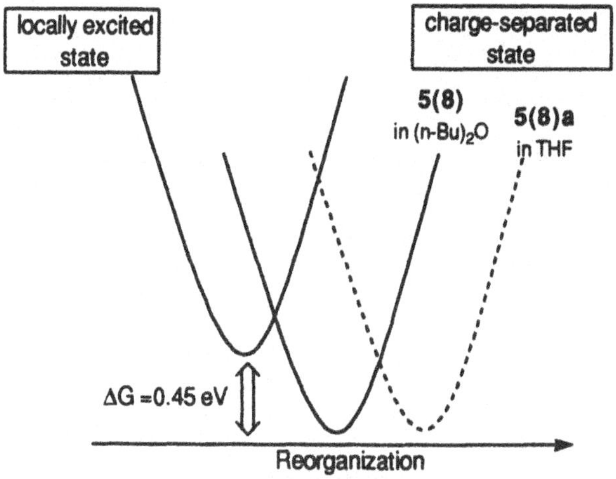

Figure 1. Illustration of the influence of changes in solvent reorganization energy on potential energy surfaces for two charge separation processes with equal exergonicity.

known Marcus expression[8] eqn (3), where λ denotes the sum of the solvent reorganization energy and the internal reorganization energy ($\lambda = \lambda_i + \lambda_s$).

$$\Delta G^{\ddagger} = \frac{(\Delta G + \lambda)^2}{4\lambda} \tag{3}$$

In Fig. 2, the $\Delta G^{\ddagger}$ values for **5(8)** and **5(8)a**, which were calculated from eqns (1-3), are plotted for the solvents experimentally employed (see TABLE 2) and also as a continuous function of ε_s at a fixed value of $n^2 = 2$, which is a reasonable approximation for most (organic) solvents. In calculating the $\Delta G^{\ddagger}$ values, the previously determined[6] λ_i value of 0.6 eV for **5(8)** was used.

In agreement with experiment $\Delta G^{\ddagger}$(**5(8)**) is predicted to be low for all values of ε_s. In contrast, $\Delta G^{\ddagger}$ for **5(8)a** initially falls sharply with increasing ε_s, but it soon reaches a constant value of *ca* 0.14 eV, owing to the compensating effects of changes in λ_s and ΔG. Thus in (very) polar solvents the difference in barrier between **5(8)** and **5(8)a** approaches 0.1 eV implying a ratio k_{et}(**5(8)**)/k_{et}(**5(8)a**) = 50 thereby accounting for the major part of the experimental ratio which amounts to about 300 (see TABLE 2). The latter would imply a barrier difference of 0.15 eV, but it should be borne in mind, that part of the discrepancy may in fact be due to a difference between **5(8)** and **5(8)a** in electronic coupling as a result of different orbital-coefficients and/or -symmetry at the donor site[9].

The intriguing interdependence between the degree of sensitivity of the electron transfer rate to solvent polarity and the extent by which the rate deviates from its optimal value, is more comprehensively visualized in Fig. 3 which plots the barrier for electron

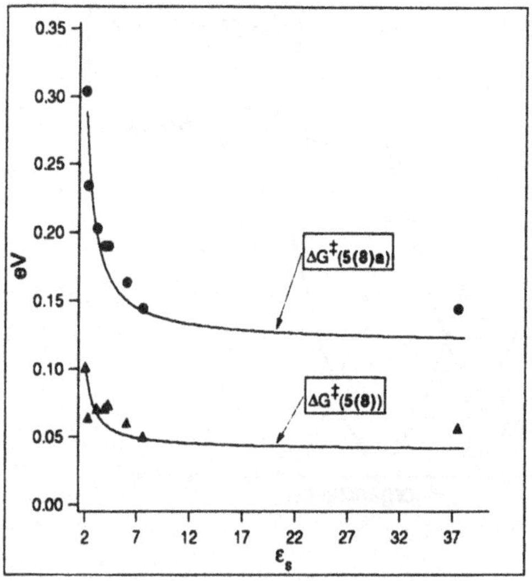

Figure 2. Activation barrier for photoinduced charge separation in **5(8)** (Δ) and **5(8)a** (•) in various solvents as calculated from eqns (1-3). Continuous lines were calculated for $n^2 = 2$.

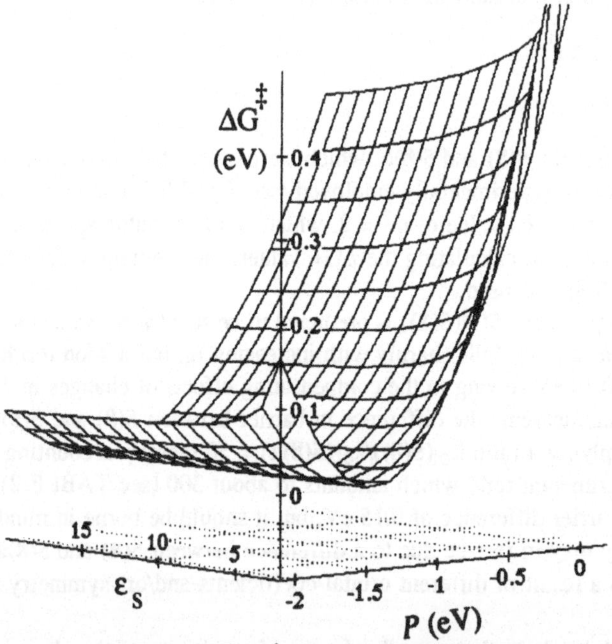

Figure 3. Activation barrier, ΔG‡ (in eV), of photoinduced charge separation for systems with the geometrical properties of **5(8)** and **5(8)a** as a function of the solvent dielectric constant ε_S and the "polar driving force", *P* (in eV), at a fixed value of the refractive index ($n^2 = 2$).

transfer in systems with the geometrical properties of **5(8)** and **5(8)a** as a function of both ε_s and the *Polar Driving Force, P,* defined in eqn (4).

$$P = e[E_{ox}(D)-E_{red}(A)]-E_{oo} \tag{4}$$

The parameter P represents the limiting value of ΔG for charge separation across an infinite distance in polar media and is readily accessible from the independently determined redox potentials (as measured in acetonitrile) and spectroscopic properties (E_{00}) of the separate D and A units. Evidently, the $\Delta G^{\ddagger}$ plots in Fig. 2 represent sections through the three dimensional plot of Fig.3 at two specific values of P (i.e. P = -0.98 eV for **5(8)** and P = - 0.62 eV for **5(8)a**).

The observation that a unique valley, running parallel to the ε_s axis, occurs with $\Delta G^{\ddagger}$ =0 in Fig.3 demonstrates that a single P value (= P_{opt}) can be found at which both optimally fast electron transfer ánd virtual solvent and temperature independence[10] of the rate of this process can be achieved. From the equations given above, together with the condition that $\Delta G^{\ddagger}$=0, eqn (5) can be derived:

$$P_{opt} = -\lambda_i - \frac{e^2}{n^2}\left(\frac{1}{r} - \frac{1}{R_c}\right) + \frac{e^2}{37r} \tag{5}$$

This proves that P_{opt} is independent of the solvent dielectric constant, but depends on the charge separation distance.

On the other hand, since λ_i and the average ionic radius r are mainly determined by the structure of the D and A units employed while n varies only marginally with solvent, this also implies that for any given D/A combination the condition defined by eqn (5) can only be fulfilled at a single charge separation distance R_c, irrespective of the solvent employed. In fact at the distance R_c = 11.4 Å dictated by the structure of compounds **5(8)** and **5(8)a** a P_{opt} = -1.57 eV (see Fig. 3) would be required (assuming the same λ_i and r) to make electron transfer barrierless in <u>all</u> solvents. Clearly, **5(8)** with P = -0.98 eVcomes closer to meeting this requirement than does **5(8)a** (P = -0.62 eV), which is in agreement with our experimental results.

In conclusion eqn (5) provides design criteria for systems intended to display optimally fast electron transfer across a given distance that is virtually independent of both medium effects and temperature.

1.2. DISTANCE DEPENDENCE OF THE RATE OF PHOTOINDUCED CHARGE SEPARATION.

A primary incentive for studying intramolecular electron transfer in rigid D-bridge-A systems has been our wish[11] to reveal the distance dependence of the electronic coupling between D and A as a function of the length and the nature of the intervening bridge. As far as the former is concerned, the data presented in the previous section indicate that variation of the bridge length will not only change the rate of electron transfer by modifying the electronic coupling but also by evoking an unavoidable concomitant

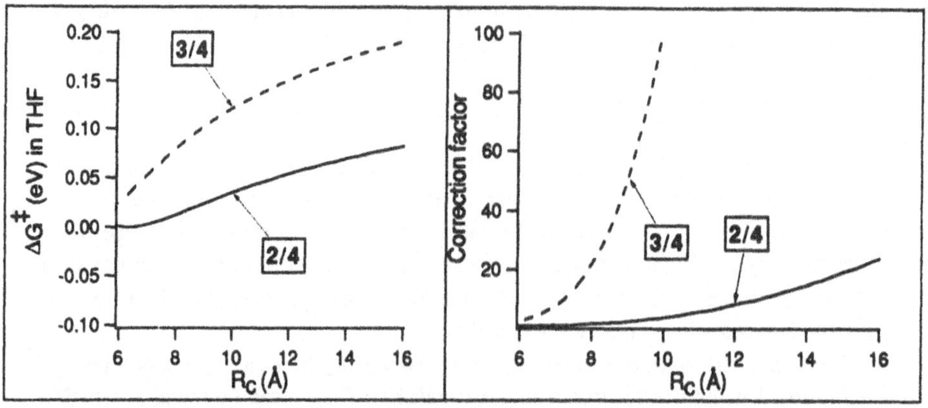

Figure 4. Calculated barrier to photoinduced charge separation ($\Delta G^‡$) as a function of the center to center distance R_c for the D/A pairs **2/4** and **3/4** in tetrahydrofuran (see § 1.1. for parameters and formulae used). The ratio (at 298 K) of the rate under barrier-less conditions ($\Delta G^‡$=0) and for the barrier calculated is displayed as a 'correction factor' in the right hand part of the figure.

change in the height of the barrier $\Delta G^‡$. That this can be a very serious problem is demonstrated in Fig. 4, where we plot the calculated barrier for photoinduced electron transfer between **2/4** and **3/4** , i.e. the D/A pairs incorporated in **5(8)** and **5(8)a** respectively, as a function of the center to center distance R_c, in tetrahydrofuran as a solvent. Although for both pairs electron transfer is exergonic over the R_c range investigated, for **3/4** a significant barrier develops at relatively short distances, which would be enough to depress the rate by more than two orders of magnitude for R_c>10 Å. In principle the calculated $\Delta G^‡$ values could be used to estimate 'correction factors' (see Fig. 4) that, when multiplied to the experimental rate constants, scale these to 'barrierless conditions'. The approximative nature of equations used to calculate $\Delta G^‡$ makes such an approach very tentative however, especially if the resulting correction factors change by (many) orders of magnitude over the R_c range investigated. Thus it would be virtually impossible to extract reliable data on the change of electronic coupling with bridge length from a series of D-bridge-A molecules incorporating a D/A pair such as **3/4**, but with a pair such as **2/4** this appears a worthwhile experiment.

The series of molecules **5(n)** with **n** = 4, 6, 8, 9, 10, 12, and 13, compiled in Scheme 3 comprises one of such series used in our joint research[6,9,11-14] with R_c values ranging from 7 Å for **5(4)** where the bridge separates D and A by an array of four carbon-carbon sigma bonds to about 16 Å in **5(13)**. This series has the unique properties that each member is conformationally fully rigid and that R_c is incremented in relatively small steps for each higher member. In TABLE 3 the rate constants k_{et} of photoinduced charge separation measured[6,14] for **5(n)** in tetrahydrofuran at 298 K are compiled together with the calculated barrier ($\Delta G^‡$) and the 'corrected' optimal value of the rate under barrierless conditions. Furthermore, the center to center (R_c) and edge to edge (R_e) distances between the D and A pi-systems are given.

Scheme 3

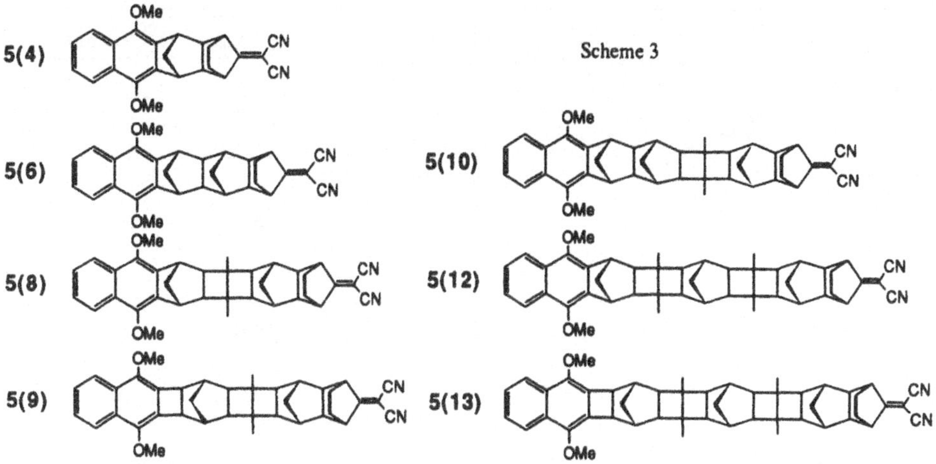

TABLE 3. Donor acceptor distances (edge-to-edge, Re, and center-to-center, R_c) as well as experimental rate constants of photoinduced charge separation (k_{et}, measured in tetrahydrofuran at room temperature[6,14]) for **5(n)**. Furthermore the calculated barrier, $\Delta G^‡$ (see Fig. 4) and the estimated rate constant under barrierless conditions, k_{et}(opt) is indicated.

Compound	R_e (Å)	R_c (Å)	k_{et} ($10^8 s^{-1}$)	$\Delta G^‡$ (eV)	k_{et}(opt) ($10^8 s^{-1}$)
5(4)	4.63	7.1	>5000	0.0035	>5000
5(6)	6.82	9	≥3300	0.0243	≥8250
5(8)	9.4	11.8	670	0.0531	5132
5(9)	10.4	12.5	250	0.059	2400
5(10)	11.47	13.3	120	0.065	1476
5(12)	13.5	14.9	13	0.077	243
5(13)	14.13	15.6	(1.5)[a]	0.0805	(32.8)

a) close to lower limit of detection.

It is generally assumed that the electronic coupling and therefore the rate of electron transfer should show an approximately exponential distance dependence (see also § 1.4). In Fig. 5 we investigate this prediction for the series **5(n)**. As reported earlier[14,15], closely mono-exponential distance dependence is found for the experimental k_{et} values expressing distance either in terms of the number of bonds in the bridge, n, or as the edge-to-edge distance, R_e (see eqn (6)).

$$k_{et} = 10^{13.96}\exp(-0.92n) = 10^{14}\exp(-0.82R_e) \tag{6}$$

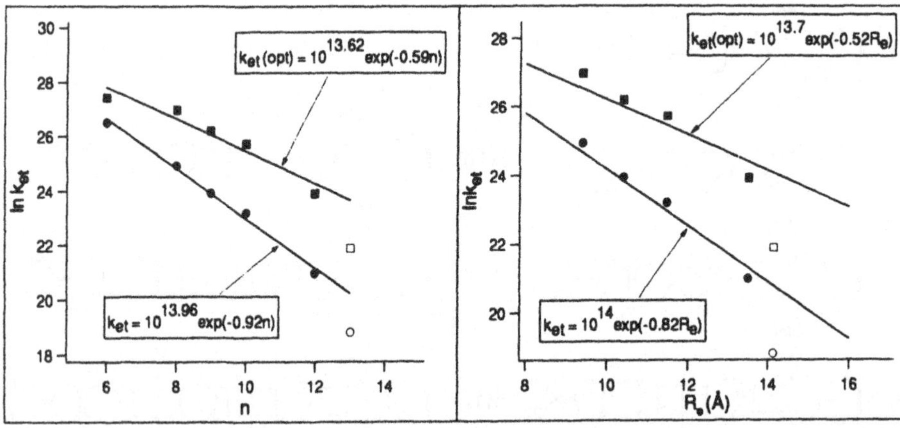

Figure 5. Mono-exponential correlation of experimental (k_{et}) and calculated barrier-less (k_{et}(opt)) rates of photoinduced charge separation in **5(n)** as a function of the bridge-length expressed as number of bonds n, and as closest atom edge-to-edge distance R_e. Solvent tetrahydrofuran at 298 K.

It should be noted, however, that in the regression analysis the data for **5(13)** were omitted. The same test applied to the barrier-corrected rate data k_{et}(opt), yields eqn (7).

$$k_{et}(\mathrm{opt}) = 10^{13.62}\exp(-0.59n) = 10^{13.7}\exp(-0.52R_e) \qquad (7)$$

While this result may suggest that the actual distance dependence of the electronic coupling is much less pronounced than indicated by the experimental rate data, it should be noted that the barrier-correction applied in fact degrades the quality of the exponential fit (see Fig. 5). This demonstrates the extremely problematic nature of such barrier-corrections even for systems, such as **5(n)**, which have been designed to minimize the extent of correction required!

Even without application of a barrier-correction, the distance dependence of intramolecular charge separation in series **5(n)** is rather small as compared to that reported[16] (see eqn (8)) for intermolecular electron transfer under barrierless conditions between donor and acceptor molecules dispersed in a rigid methyl tetrahydrofuran glass.

$$\textit{intermolecular}: k_{et}(\mathrm{opt}) = 10^{13.9}\exp(-1.20R_e) \qquad (8)$$

This has led us to believe that in **5(n)** a major part of the electronic coupling between D and A occurs via the saturated hydrocarbon bridge, i.e. by "through-bond coupling". In the following sections various tests for this hypothesis are provided by (§ 1.3.) investigating the effect of changes in the bridge configuration and (§ 1.4.) correlating the rate of charge separation in **5(n)** with the through-bond electronic coupling between two double bonds connected with the same type of bridges, as measured by photoelectron spectroscopy and electron transmission spectroscopy.

1.3. THE INFLUENCE OF THE BRIDGE CONFIGURATION ON CHARGE SEPARATION IN D-BRIDGE-A SYSTEMS.

From extensive theoretical and experimental studies[17-21] on through-bond interactions it has become evident that an all-*trans* array of sigma bonds provides an optimal coupling path between two functional groups and that introduction of *cisoid* or *gauche*-bonds in such an array severely diminishes its coupling ability. Systems **6(6)** and **6(8)**, see Scheme 4, were designed and synthesized to investigate the effect of such diminished through-bond coupling on the rate of photoinduced charge separation, by comparison to the all-*trans* bridged analogues **5(6)** and **5(8)**. Table 4 compiles relevant data[22,23] from which it is evident that a significant attenuation of k_{et} indeed occurs when the all-*trans* nature of the bridge is lost although this hardly influences the distance between D and A.

For the pair **5(8)/6(8)** data could be collected[23] in a variety of solvents. From these it appears (see TABLE 4) that a much stronger solvent sensitivity of k_{et} applies for **6(8)** than for **5(8)**, although in both cases the barrier is calculated (see § 1.1) to be equally and only very weakly solvent dependent. This remarkable difference is tentatively attributed[23] to a contribution of through-solvent coupling, which may be expected to

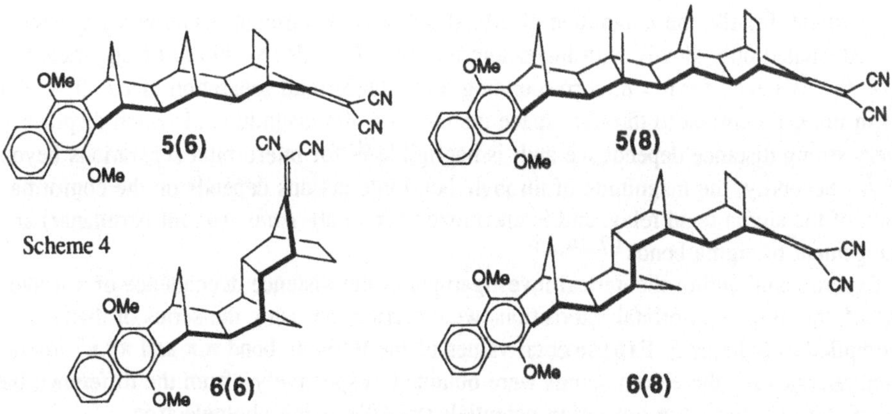

Scheme 4

TABLE 4. Comparison of k_{et} for systems **5(n)** and **6(n)**.

solvent	**5(6)** k_{et} ($10^8 s^{-1}$)	**6(6)** k_{et} ($10^8 s^{-1}$)	ratio **5(6)/6(6)**	**5(8)** k_{et} ($10^8 s^{-1}$)	**6(8)** k_{et} ($10^8 s^{-1}$)	ratio **5(8)/6(8)**
cyclohexane	-	-	-	210	29	7.2
benzene	-	-	-	520	180	2.9
di-ethyl ether	≥ 3300	1250	≥ 2.7	470	72	6.5
ethyl acetate	≥ 3300	770	≥ 4.7	450	89	5.1
tetrahydrofuran	≥ 3300	-	-	670	165	4.1
acetonitrile	-	360	-	300	22	13.6

play a more important role for **6(8)** where the solvent independent, through-bond component is attenuated. Interestingly this leads to a situation where for the apolar but highly polarizable solvent benzene the effect of changing the bridge configuration is minimal and for the polar but unpolarizable solvent acetonitrile it has an effect of more than one order of magnitude.

While these data prove beyond doubt the importance of through-bond coupling in mediating the fast long-range charge separation in **5(n)**, it is thus also evident from the complex solvent dependence that separation of through-bond and through-medium effects is neither a trivial task nor that a simple additivity scheme for such effects is likely to apply.

1.4. THE KOOPMANS' CONNECTION

Hoffmann *et al* .[17,18] introduced the conceptual dissection of orbital interactions into through–bond and through–space types. Through–bond interactions arise from the mutual overlap of the π (or π^*) orbitals of two unsaturated systems with the σ and σ^* orbitals of the relay(s) connecting the two chromophores. The results of Extended Hückel calculations on model diradical systems led Hoffmann *et al.* to make several important predictions about through–bond interactions, of which two are particularly relevant to this article. Firstly, the magnitude of orbital interactions through bonds was predicted to be attenuated only slowly with increasing number of bonds (n), and to be significant (*ca.* 0.2 eV) even for n = 8, corresponding to an interorbital separation of *ca* . 9 Å. This is in marked contrast to through–space interactions, the magnitude of which displays a very strong distance dependence and is negligible[20] for interorbital separations beyond 5 Å. Secondly, the magnitude of through–bond interactions depends on the conformation of the sigma bond relay, and is maximized for an all–*trans* (or antiperiplanar) arrangement of sigma bonds[17,24,25].

Experimental and/or theoretical investigations of the distance dependence of π,π and π^*,π^* through–bond orbital interactions were carried out using the series of dienes compiled in Scheme 5. Experimental values of the through–bond π,π and π^*,π^* interaction energies for these compounds were obtained, respectively, from the difference between the two vertical π ionization potentials (π–ΔIP), using photoelectron spectroscopy[25-29], and from the difference between the two vertical π^* electron affinities (π^*–DEA), using electron transmission spectroscopy[30,31]. The experimental splitting energies are given in TABLE 5. In addition, TABLE 5 lists the calculated π,π and π^*,π^* orbital splitting energies for these dienes, given, respectively, by ΔE_{π} and ΔE_{π^*}. These were obtained using the STO–3G basis set on HF/STO–3G fully optimized structures (within the appropriate symmetry constraints)[32]. The ΔE_{π} and ΔE_{π^*} values are equated, respectively, to the experimental π–ΔIP and π^*–ΔEA values by Koopmans' theorem (KT)[33].

The comparatively large calculated and observed π,π and π^*,π^* splitting energies for these dienes must be due to a through–bond coupling mechanism, involving the hydrocarbon bridge orbitals, since direct through–space interactions between the π (π^*) orbi-

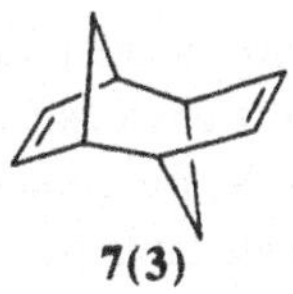
7(3)

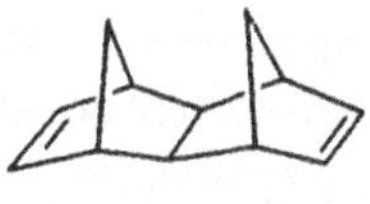
7(4)

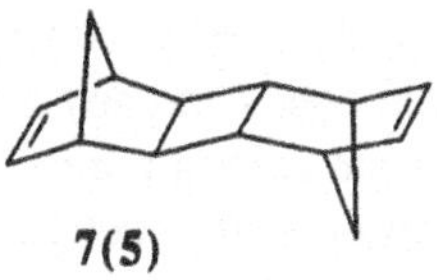
7(5)

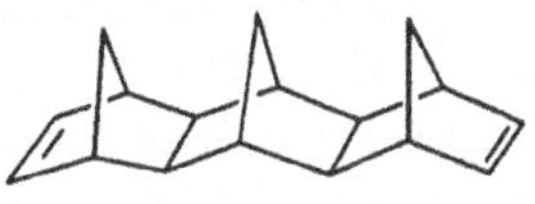
7(6)

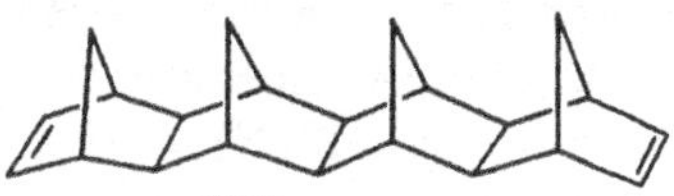
7(8)

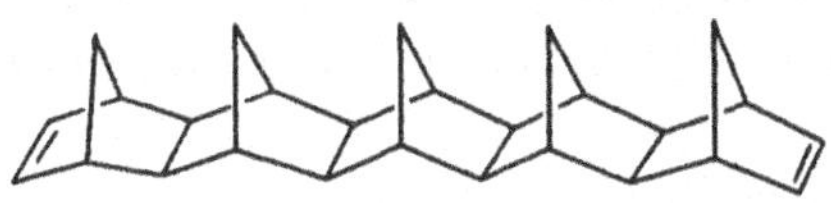
7(10)

Scheme 5

8(6)

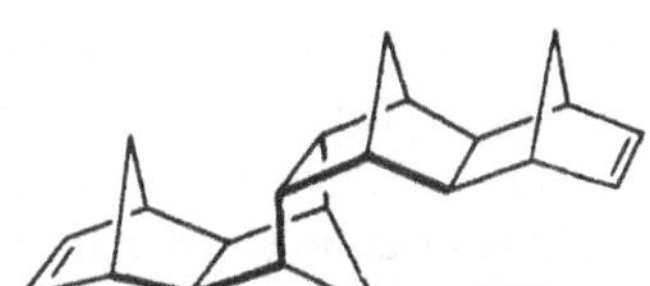
9(8)

TABLE 5. Ionization potential, π-ΔIP (from photo electron spectroscopy), and electron affinity, π^*-ΔEA (from electron transmission spectroscopy), splitting energies (in eV) for the dienes from Scheme 5, together with corresponding KT/STO-3G splitting energies, ΔE_π and ΔE_{π^*}.

Diene	π-ΔIP	ΔE_π	π^*-ΔEA	ΔE_{π^*}
7(3)	1.6	1.787	-	1.177
7(4)	0.87	0.937	0.8	0.841
7(5)	0.43	0.411	0.57	0.406
7(6)	0.32	0.327	0.25	0.233
7(8)	-	0.13	-	0.0653
7(10)	-	0.0601	-	0.019
8(6)	-	0.0577	-	0.0653
9(8)	-	0.0618	-	0.0155
$[7(6):8(6)]^2$		32[a]		12.7[b]
$[7(8):9(8)]^2$		4.4[a]		17.7[b]

a) Square of the ratio of the ΔE_π values for the indicated pair of dienes.
b) Square of the ratio of the ΔE_{π^*} values for the indicated pair of dienes.

tals in these dienes are expected to be negligible, owing to the large spatial separation between the two double bonds and any through-solvent coupling is obviously absent because measurements are carried out in the gasphase. For example, the computed[20] through space π,π splitting energy between two unconnected ethene units placed in parallel planes 7 Å apart, which corresponds to the distance between the two double bonds in **7(6)**, is less than 10^{-4} eV. It is seen that good agreement obtains between the experimental π–ΔIP and π^*–ΔEA values and the respective KT/STO–3G ΔE_π and ΔE_{π^*} values for **7(3)**–**7(6)**. Thus, KT/STO 3G calculations can be used for estimating π–ΔIP and π^*–ΔEA splitting energies for all dienes, including those for which these quantities are expected to be too small to be measured experimentally (<0.15 eV).

It was found that the distance dependence of the KT/STO–3G π,π and π^*,π^* splitting energies for the dienes **7(n)** are well fitted to fairly weak exponential decays[32]:

$$\Delta E_\pi = 5.90 \exp(-0.47n) \quad (9)$$
$$\Delta E_{\pi^*} = 8.22 \exp(-0.60n) \quad (10)$$

where n is the number of sigma bonds in the relay(s) connecting the double bonds. The exponents in these KT based relationships are slightly larger than those obtained using experimental π-ΔIP and π^*–ΔEA data, *i.e.*, –0.45n and –0.58n, respectively.

These results are highly relevant to electron transfer because it can be shown[20] that π,π and π^*,π^* splitting energies for the dienes, obtained either experimentally or via Koopmans' theorem, are related, respectively, to the rates of electron transfer in the corresponding cation radicals and anion radicals of the dienes. Thus consider a gas phase thermal electron transfer process taking place between two identical ethene groups of a cation radical complex, according to (11):

$$E_1^+ + E_2 \rightleftarrows E_1 + E \quad (11)$$

where E_1^+ is the cation radical of ethene 1 (configuration π^1) and E_2 is neutral ethene 2 (configuration π^2).

The reaction coordinate diagram for this process can be constructed from the upper pair of curves in Figure 6. The reaction coordinate represents the geometrical changes of the molecules accompanying electron transfer. For simplicity, these are merely indicated schematically by changes in the double bond length, from a long, stretched bond in the cation radical, to a short bond in the neutral species. The potential energy hypersurface may be analyzed in terms of two intersecting diabatic hypersurfaces. The intersection of the surfaces produces a seam along which the geometries of the two ethene units are identical. The minimum energy point on this seam corresponds to the position on the reaction coordinate, through which electron transfer is most likely to occur. Along the seam of intersection, the two configurations mix most strongly, resulting in an avoided crossing of the surfaces. The magnitude of this avoided crossing is given by 2J, where J is the electron coupling (transfer) integral.

For the case of weak coupling (that is, for small values of J), electron transfer may be

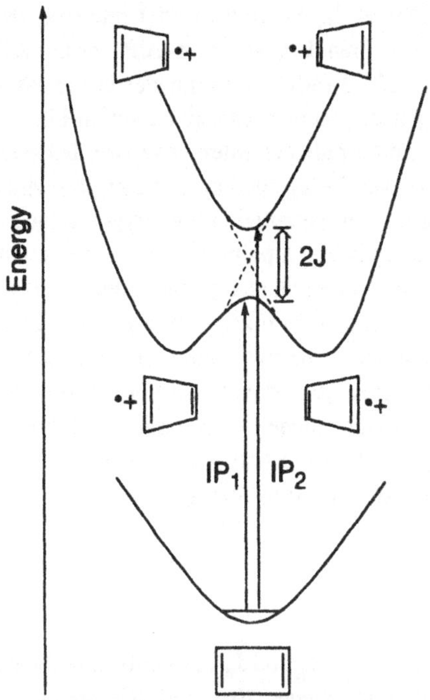

Figure 6. The relationship between the two vertical π-IP's of a symmetrical diene and the splitting, 2J, between the two adiabatic surfaces for the resulting cation radicals.

considered to occur non–adiabatically, and application of the Golden Rule leads to the following expression for the rate constant for electron transfer, k_{et}:

$$k_{et} = \frac{4\pi^2}{h} J^2 \ \mathrm{FCWD} \tag{12}$$

where FCWD is the Franck–Condon weighted density of states[34-36]. In the high temperature limit, FCWD may be represented classically as[34-38]:

$$\mathrm{FCWD} = \frac{1}{\sqrt{4\pi\lambda k_b T}} \exp\left\{\frac{-(\Delta G + \lambda)^2}{4\lambda k_b T}\right\} \tag{13}$$

where ΔG is the standard Gibbs energy of reaction (driving force), and λ is the reorganization energy and contains contributions from solvent as well as from molecular vibrations of the donor–acceptor system (see also § 1.1.). In the high temperature limit, the rate constant, k_{et}, depends on the three variables J, λ, and ΔG (in addition to T of course). Although λ and ΔG are important, it is the dependence of k_{et} on J that largely determines the distance dependence of the electron transfer rate (see also § 1.2.).

Fig. 6 illustrates how the vertical (Franck–Condon) π–ΔIP (or the KT ΔE_{π}) splitting energies provide a direct measure of 2J for electron transfer in the cation radicals of symmetrical dienes. The upper pair of potential energy curves in Fig. 6 represents the adiabatic surfaces for the two π cation radicals. Interconversion between the degenerate charge localized states takes place via the avoided crossing region which presumably has C_{2v} symmetry. The minimum of the potential energy hypersurface of the ground state of the neutral diene lies directly below a point on the avoided crossing seam of the energy hypersurfaces of the two cation radical states that connects all points associated with C_{2v} symmetry. The transitions shown in Fig. 6 represent the (Franck–Condon) vertical π–IP values for the formation of the two cation radical states. The observed splitting energy, π–ΔIP, therefore provides a direct (but not precise) measure of 2J for the cation radical states of that particular diene. An analogous relation exists between π*-ΔEA and 2J for interaction of the anion radical states. Thus, from eqns. (9) and (10), the following distance dependence relationships may be derived:

$$J_{\pi}^{2} = 8.7\exp(-0.94n) \tag{14}$$

$$J_{\pi*}^{2} = 16.9\exp(-1.20n) \tag{15}$$

where $2J_{\pi} = \Delta E_{\pi}$ and $2J_{\pi*} = \Delta E_{\pi*}$. Thus, J_{π} and $J_{\pi*}$ are only weakly attenuated with increasing length of the sigma relay in the dienes **7(n)**, and they are comparatively large (*ca* . 0.01–0.02 eV) even for n=10, corresponding to a separation of about 11Å between the two double bonds.

Thus, our results indicate that, through–bond coupling could facilitate rapid electron transfer over very large distances and since the bridges incorporated in the dienes **7(n)** are closely related those incorporated in the D-bridge-A systems **5(n)** (see Scheme 3) it is of interest to compare the distance dependence of J_{π}^2 and $J_{\pi*}^2$ for the latter systems (see eqns (14) and (15)) with the distance dependence of the rate of photoinduced electron transfer in the former (eqns. (6) or (7)) because, as pointed out above, we may expect that in the series **5(n)** the distance dependence of k_{et} is largely governed by the distance dependence of J^2.

The distance dependence of J_{π}^2 for **7(n)** (eqn(14) matches amazingly well that of k_{et} as measured for **5(n)** in tetrahydrofuran (eqn (6)). It should be pointed out that this may in part be fortuitous, since a much larger discrepancy occurs when barrier-corrected k_{et}(opt) values are applied and furthermore it is not unreasonable to presume[32] that it would be better to use the $J_{\pi*}^2$ values for correlation, since in photoinduced electron transfer the coupling is expected to occur mainly through mixing of D and A π* orbitals with the σ and σ* orbitals of the bridge. As discussed in § 1.2. , however, the procedure required to obtain k_{et}(opt) from k_{et} is quite cumbersome and we therefore feel that the good correlation between the distance dependence of the experimental k_{et} values and either that of J_{π}^2 or of $J_{\pi*}^2$ provides important support for the through-bond nature of the coupling responsible for the rapid electron transfer processes occurring in **5(n)**. In this respect it seems important to note that, as early as 1961, McConnell[39] carried out calculations which suggested that intramolecular electron transfer between two phenyl groups in the anion radicals of the series of α,ω–diphenylalkanes, Phe-$(CH_2)_n$-Phe, could be accelerated by through–bond coupling, but to a much weaker extent than predicted for **5(n)**. The dramatic predicted acceleration of electron transfer by through-bond coupling is clearly seen when the electron transfer rate is recalculated in the absence of any coupling. This can be modelled by replacing the particular diene by two isolated ethene molecules separated by the appropriate distance. Doing this, we find[20] that direct (through–space) electron transfer between two ethene molecules placed 21 Å apart is predicted to be 10^{17} times slower than that mediated by through–bond coupling via an all-*trans* norbornylogous bridge.

An equally striking correlation between the effect of the bridge on ΔE_π and $\Delta E_{\pi*}$ at one hand and on the rate of photoinduced electron transfer at the other, is found by comparison of the influence of the bridge configuration at a constant length, n. In § 1.3. we have shown that substitution of all-*trans* bridges with an effective length of n = 6 and n = 8 bonds by bridges containing one or more *cisoid* bonds reduces k_{et} by about one order of magnitude (see Scheme 4 and TABLE 4).

Inspection of the data for **8(6)** and **9(8)** in TABLE 5 clearly reveals that a significant diminuition of ΔE_π and $\Delta E_{\pi*}$ occurs as compared to the all-*trans* bridged dienes **7(6)** and **7(8)**. The squared ratios, also given in TABLE 5, in fact appear larger than the k_{et} ratios given in TABLE 4, lending some support to the hypothesis (see §1.3.) that the latter are attenuated as a result of solvent effects.

1.5. PROPERTIES OF 'GIANT DIPOLES' FORMED BY PHOTOINDUCED CHARGE SEPARATION IN RIGID D-BRIDGE-A SYSTEMS.

In general more attention has been paid to the study of photoinduced, intramolecular charge separation than to the properties, such as the lifetime and the dipole moment, of the charge separated state resulting from it. This is due, at least in part, to the experimental difficulties involved in monitoring the charge separated states formed, which are often only weakly fluorescent, if at all (see however refs. 4, 9 and 40) for important exceptions). Also, the unequivocal identification and absolute measurement of short-lived radical ion species by their optical absorption is often extremely difficult because of the concurrent bleaching of the absorption bands of the ground state and formation of strongly absorbing local excited singlet and triplet states.

We have applied[12,13,41] mainly the time-resolved microwave conductivity (TRMC) technique[42], which is only sensitive to those states for which a large change in dipole moment compared with the ground state has occurred. The method thereby gives direct information about the kinetics of formation and decay of charge separated states and furthermore can be applied to quantify the degree of charge separation (i.e. the molecular dipole moment) in such states. The method is by its nature limited almost exclusively to the investigation of solutions in nondipolar solvents. Since most of the rigid D-bridge-A systems studied by us readily undergo photoinduced charge separation in such solvents (as these were in fact designed to show a minimal solvent effect on charge separation, see § 1.1. and 1.2.) this limitation did not present a serious problem.

In the sections below we now briefly summarize results obtained mainly by TRMC measurements, of which details have been published elsewhere[13,41], with regard to the dipole moment and the lifetime (i.e. the charge recombination rate) of the 'giant dipole' states created by photoinduced charge separation in the rigid D-bridge-A systems presented in the preceding sections. It should be noted that although TRMC can in principle also be applied to determine charge separation rates, for the present compounds these rates are beyond the time resolution available, which is limited to about 1 ns.

1.5.1. *Dipole Moments of Charge Separated States in Rigid D-bridge-A Systems.* For most of the rigid all-*trans* bridged systems compiled in Scheme 3 photoinduced charge separation occurs with high efficiency even in apolar solvents, thereby enabling TRMC detection of the charge separated state. Figure 7 shows TRMC traces observed for **5(6)** in cyclohexane, benzene and 1,4-dioxane. Quantitative and very fast charge separation (see TABLE 3) occurs but interestingly the charge recombination rate appears to increase dramatically with solvent polarity as will be discussed in more detail in § 1.5.2..

While the fast decay of the charge separated state in 1,4-dioxane prevents quantitative evaluation of the TRMC signal, the dipole moment of the transient could be calculated in both cyclohexane and benzene as a solvent, which gave a uniform value of 36.5±0.5 Debye for **5(6)**. In TABLE 6 we compile dipole moment data for the series **5(n)**.

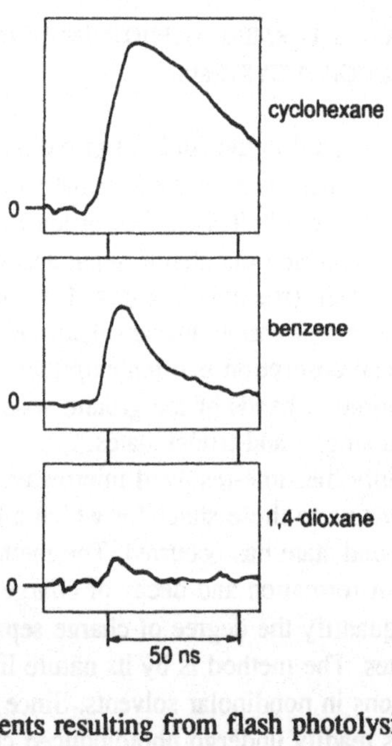

Figure 7. TRMC transients resulting from flash photolysis (8 ns, 9 mJ/cm^2, 308 nm) of ca. 10^{-4} mol/l solutions of **5(6)** in cyclohexane, benzene and 1,4-dioxane. The vertical scales correspond to different sensitivies for the different solvents.

TABLE 6. Dipole moments measured via the TRMC method for the charge separated state created by photoexcitation of some members of the **5(n)** series of rigid D-bridge-A systems (see Scheme 3 and TABLE 3 for structural data)

	dipole moment (Debye) measured in:		
compound	cyclohexane	benzene	1,4-dioxane
5(4)	26	b	b
5(6)	36	37	b
5(8)	63	55	b
5(10)	a	68	65
5(12)	a	77	78

a) No significant charge separation observed.

b) Too short lived.

In Figure 8 we furthermore plot the observed dipole moments as a function of the center-to-center distance, R_C (see TABLE 3), of the D and A groups together with the line indicating the dipole moment expected for full charge separation over a distance R_C. The agreement with the experimental data is seen to be quite satisfying especially for the higher members of the series for which the dipole can be considered as two point charges separated by R_C.

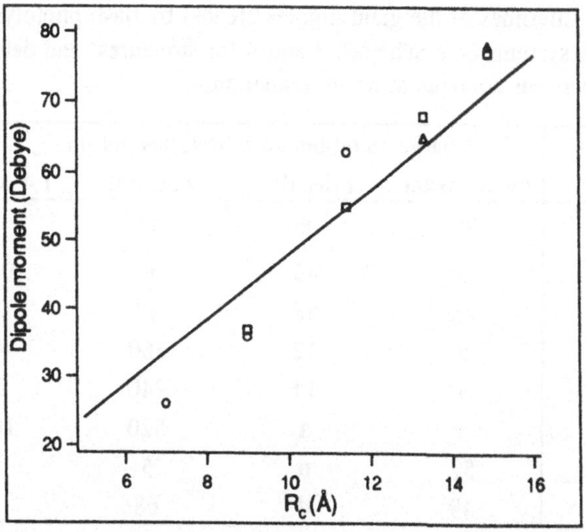

Figure 8. Dipole moments (see TABLE 6) of the charge separated state of compounds **5(n)** as determined from TRMC measurements in cyclohexane (O), benzene (▯), and 1,4-dioxane (Δ). The full line corresponds to the dipole moment for complete charge separation over the center-to-center distance R_C (see TABLE 3)

Thus the dipole moment data obtained from the TRMC measurements provide unequivocal proof for occurrence of fast and complete charge separation in compounds **5(n)** with n values up to and including 12, provided that the solvent polarity is sufficient to make such charge separation thermodynamically feasible. For n = 4-8 the latter condition is fulfilled even in saturated hydrocarbon solvents, while for n =10 and 12 more strongly solvating media are required.

1.5.2. *Distance and Solvent Dependence of Charge Recombination in 'Giant Dipoles'*
As indicated above, TRMC provides a unique tool to measure the recombination lifetime (τ_{cr}) of the 'giant dipoles' created by photoinduced charge separation in D-bridge-A systems. TABLE 7 compiles such lifetime data for most of the rigid systems for which the charge separation kinetics were discussed in § 1.2. and 1.3.

One of the most striking features of the data in TABLE 7 is the sharp variation that the rate of charge recombination ($k_{cr} = 1/\tau_{cr}$) for each of the compounds shows upon a change of solvent. Furthermore this solvent sensitivity appears to be quite different for different compounds thus leading to the, at first glance, unexpected observation of a rate minimum at an intermediate bridge length, e.g. for **5(8)** in *t*-decalin and for **5(12)** in benzene.

It has been observed[43] that the rate increase at greater bridge lengths is always accompanied by the appearance of significant delayed fluorescence component of the donor chromophore, with a lifetime identical to that of the charge separated state. This indicates that an additional decay channel for the charge separated state becomes available

TABLE 7. Lifetimes of the giant dipoles created by flash photolysis of rigid D-bridge-A systems (see Schemes 3 and 4 for structures) and determined by TRMC in various solvents at room temperature.

	Charge recombination lifetimes (ns) in :			
compound	cyclohexane	*t*-decalin	benzene	1,4-dioxane
5(4)	9	8	(1)[c]	b
5(6)	38	45	6	(0.5)[c]
5(8)	28	58	32	2.5
5(10)	a	12	360	43
5(12)	a	11	740	297
5(13)	a	a	520	1050
6(6)	52	d	5	d
6(8)	49	d	68	d

a) No significant charge separation observed
b) Too short lived
c) From in-pulse signal, dipole moment assumed.
d) Not determined

which (re)populates the local donor excited state and thereby enhances the overall rate of charge recombination. This requires that the local excited state and the charge separated state are close in energy, a situation (see § 1.1) augmented by lowering the dielectric constant of the solvent and increasing the length of the bridge. In accordance with this assumption a rate increase with increasing bridge length is not observed in 1,4-dioxane, where even for n = 13 the charge separation appears to be sufficiently exergonic to prevent repopulation of the local donor excited state. It should be stressed that it is the Coulomb term $e^2/\varepsilon_s R_c$ in eqn(1) which governs the change in the energy level of the charge separated state as a function of distance. Evidently this term is quite large and changes rapidly with R_c and ε_s in the low dielectric constant solvents we employ here.

In Figure 9 we give a schematic representation of the relative energy levels for **5(n)** as a function of n in benzene as a solvent and indicate how charge recombination to populate the local excited donor can eventually dominate the overall decay kinetics for n ≥ 12.

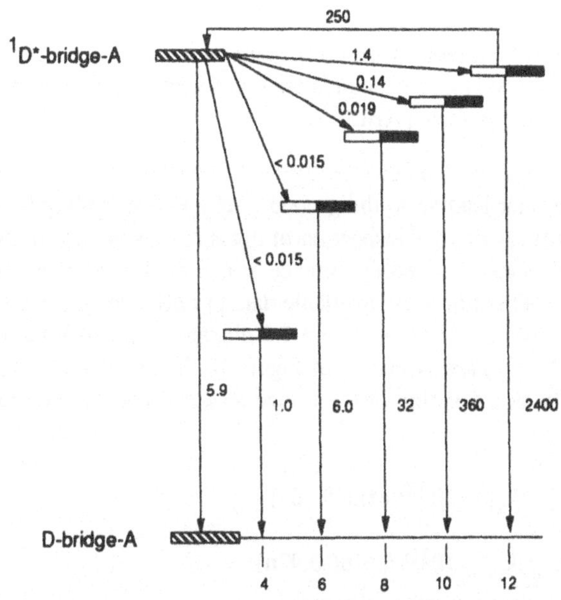

Figure 9. An ergodynamic representation of the changes occurring in the energy level of the charge separated state relative to the ground state (D-bridge-A) and to the lowest local donor excited singlet state (^{1}D*-bridge-A) of **5(n)** as the number of bonds increases from n = 4 to n = 12 together with the resulting changes in the electron transfer kinetics. The diagram refers to the situation in benzene and lifetimes indicated are in nanoseconds.

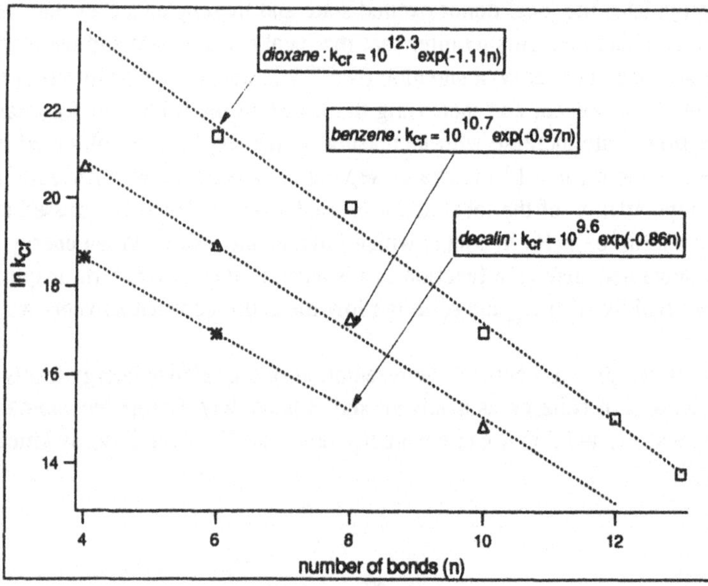

Figure 10. Monoexponential correlation with bridge length, n, of rate constants for charge recombination (k_{cr} in s^{-1}) in compounds **5(n)** as determined by TRMC in various solvents at room temperature (see TABLE 7)

Since it is not an easy task to determine quantitatively the contribution of the two recombination pathways (i.e. that leading to the ground state and that leading to the local excited donor state) we will restrict our inspection of distance dependence to those situations where the latter pathway is of no importance as e.g. evidenced from the absence of delayed fluorescence. This limits the available data to only two in cyclohexane or *t*-decaline (i.e. **5(4)** and **5(6)**) but in benzene and 1,4-dioxane data extending to higher n values (n=10 and n = 13 resp.) are available. In Figure 10 the relevant rate data are plotted semilogarithmically as a function of the bridge length. Good monoexponential fits are found according to:

in *t*-decalin : $k_{cr} = 10^{9.6} \exp(-0.86n)$ (16a)

in benzene : $k_{cr} = 10^{10.7} \exp(-0.97n)$ (16b)

in 1,4-dioxane : $k_{cr} = 10^{12.3} \exp(-1.11n)$ (16c)

Upon comparison with the distance dependence of charge separation rate constants k_{et}, as determined for the same series of compounds (see eqn (6)), it is found that a very similar attenuation of rate with increasing bridge length occurs for both processes. This may appear very gratifying and to indicate that the variation with n of the electronic coupling integrals involved in charge separation and charge recombination is virtually identical, as intuitively expected. It should not be concealed, however, that the interpretation of this similarity in distance dependence is not as straight forward as it may seem. This is because in solvents of low dielectric constant, as indicated in Fig. 9, the energy gap between the charge-separated and ground state, and thereby the driving force for charge reombination, ΔG_{cr}, changes sharply with the bridge length; eqn (17).

$$-\Delta G_{cr} = e[E_{ox}(D) - E_{red}(A)] - \frac{e^2}{\varepsilon_s R_c} - \frac{e^2}{r}\left(\frac{1}{37} - \frac{1}{\varepsilon_s}\right) \qquad (17)$$

A major influence of the length of the bridge on the barrier for charge recombination ($\Delta G_{cr}^{\ddagger}$) would thus be predicted via the Marcus equation, eqn (3).

With the parameters given in § 1.1. and application of eqn (17) we find that in benzene ΔG_{cr} varies from -3.22 eV for **5(4)** to -3.63 eV for **5(10)**. This implies that charge recombination occurs far in the 'inverted region' ($-\Delta G > \lambda$), where an increase of the driving force is predicted to slow down electron transfer. Application of the Marcus equation (3) with $\lambda = \lambda_i = 0.6$ eV gives $\Delta G_{cr}^{\ddagger}$ values of 2.86 eV and 3.82 eV for **5(4)** and **5(10)** respectively. Such barriers are clearly incompatible with both the absolute magnitude and the distance dependence of the experimental k_{cr} values. Thus, assuming a frequency of 10^{13} s^{-1}, the rate of thermally activated charge recombination in **5(4)** would be *ca.* 10^{-35} s^{-1}, whereas the experimental value is *ca.* 10^9 s^{-1} (see TABLE 7). Furthermore a rate decrease of more than sixteen orders of magnitude due to barrier effects alone is predicted between **5(4)** and **5(6)**, while experimentally only a 360fold rate decrease is observed.

These observations once more stress that the simple Marcus equation overestimates inverted region effects, but at the same time indicate that even upon application of more sophisticated treatments[36,44-47] it will not be a simple task to extract unequivocal information about the distance dependence of the electronic coupling from the charge recombination rate data. In fact the similarity between the distance dependence of k_{et} and k_{cr} might be taken as evidence that no inverted region effect on k_{cr} applies at all. This, however, is refuted by the large solvent effect on the preexponential terms in eqns (16). In more polar solvents k_{cr} increases rapidly (more than two orders of magnitude from *t*-decalin to 1,4-dioxane) at constant bridge length, which is consistent with a rate increase when the energy gap between the charge separated and ground state diminishes as a result of solvent stabilization of the former.

1.5.3. *Temperature (in)dependence of the rate of charge recombination in the "inverted region"*. The data presented in the preceding section indicate a very serious deviation of the rates of charge recombination from those predicted when this process is simply considered to be a thermally activated crossing of the Marcus barrier. It therefore appeared of much interest to investigate the temperature dependence of such processes, which clearly occur far into the 'inverted region' especially if an apolar solvent is employed. Unfortunately the TRMC method in its available form did allow us to study only a small temperature range[48]. Especially in the lower members of the **5(n)** series,

TABLE 8. Charge recombination lifetime (τ_{CT}) of **5(6)** as a function of temperature in various saturated hydrocarbon solvents. Lifetimes were determined from the decay of the intramolecular charge transfer fluorescence located around 450 nm in these solvents.

n-hexane		methyl cyclohexane		*t*-decalin	
T (K)	τ_{CT} (ns)	T (K)	τ_{CT} (ns)	T (K)	τ_{CT} (ns)
328	54	328	46	368	50
319	55	318	44	357	50
308	55	308	41	347	49
294	55	292	51	337	48
278	54	276	50	292	43
260	55	258	51	284	43
244	54	240	53	275	45
226	54	222	52	276	46
209	54	204	59	259	49
199	55	186	56	253	50
180	57	167	56	249	53
		152	56	244	53

however, charge recombination is accompanied by a weak but readily detectable charge transfer fluoresence[6,9] (i.e. "exciplex"-type emission), thereby allowing direct monitoring of the decay of the charge separated state also by fluorescence spectroscopy. Of the investigations performed by this technique[43] we only mention here those for **5(6)**, which was studied in three alkane solvents over a wide temperature range, see TABLE 8.

While in the overlapping temperature region agreement between the fluorescence[43] and TRMC[48] data is observed, the former data now even more convincingly show that the rate of charge recombination is virtually temperature independent. These findings are in line with those observed more recently for other inverted region electron transfer processes[49] and indicate that such processes cannot be described as involving the thermally activated crossing of a barrier. Instead, as has been pointed out[43,48,49], a quantum mechanical treatment allowing for a dominant contribution of nuclear tunneling involving relatively high frequency modes appears essential, which makes the rate virtually insensitive to temperature over the experimentally accessible range. A relatively simple 'single-mode' semi-quantum mechanical formulation[47-50] has been proposed to describe the rate under these conditions, but in most cases in a form not readily applicable for media, such as saturated hydrocarbons, with a negligible solvent reorganization energy.

1.5.4. *Bridge Configuration and Charge Recombination.* In the preceding section the question was raised how much of the rate dependence on bridge length for charge recombination can be attributed to a change in the electronic coupling. The major problem being that changing the bridge length is not only expected to change that coupling, but also the driving force (and thereby the Marcus barrier) especially in the low dielectric constant solvents employed for TRMC measurements.

As a possible solution to these problems we also investigated[13] the effect of bridge configuration at (nearly) constant bridge length by comparing the charge recombination kinetics of **6(6)** and **6(8)** with those of **5(6)** and **5(8)**. As evidenced by photo electron spectroscopy and electron transmission spectroscopy data on related dienes (see § 1.4.) as well as charge separation kinetics, the 'kinked' bridges in **6(n)** provide significantly less through-bond coupling than the all-*trans* bridges in **5(n)**.

Inspection of the limited data presently available (see TABLE 7) indicates that also charge recombinations tends to slow down if the bridge looses its 'ideal' all-*trans* configuration. The effect, however, appears to be rather minor. Thus k_{cr} ratios of 1.37 and 1.85 are found for the pairs **5(6)/6(6)** and **5(8)/6(8)** in cyclohexane, while a k_{et} ratio of 7.2 was found for the latter pair in the same solvent (see TABLE 4). In benzene as a solvent, the k_{cr} values for **5(6)** and **6(6)** are virtually identical while a ratio of 2.12 occurs for the pair **5(8)/6(8)**, not too different from the corresponding k_{et} ratio of 2.9. It thus appears that the effect on k_{cr} of the changes in electronic coupling as a result of changes in the bridge configuration is rather strongly swamped by other factors. Whether this relates to 'through-solvent interactions', as especially invoked (§ 1.3.) to rationalize the rather small k_{et} ratio found for **5(8)/6(8)** in benzene as compared to other less polarizable solvents (see TABLE 4) or to minor changes in the driving force or in the high frequency modes thought to be responsible for the nuclear tunneling in 'inverted region electron transfer', cannot be decided from the present data. It would therefore be desirable to collect, by other techniques, k_{cr} values in other, more polar, but little polarizable, solvents. The dramatic overall increase, however, of k_{cr} with polarity presents a serious problem for realization of such experiments with the available pairs of configurationally distinct systems.

2. Concluding Remarks

In the present paper a brief survey has been presented of our continuing research on electron transfer processes in rigidly bridged donor-acceptor systems. This research has enabled us to answer many questions such as those referring to the distance and solvent dependence and to the contribution of long-range electronic interactions via saturated hydrocarbon bridges. It has also enabled us to formulate design criteria for systems intended to display optimally fast and medium independent photoinduced charge separation, which is of obvious importance in applications directed at e.g. photoelectrochemical energy conversion. At the same time, however, many challenges and opportunities for further research remain, especially with regard to the rate and mechanism of charge recombination and the behaviour of systems in which the rigid bridges are substituted by bridges with (limited) flexibility, enabling an interplay between conformational dynamics and electron transfer.

3. Acknowledgements

We wish to acknowledge the contribution of many colleagues and coworkers who contributed to our joint research program on rigidly bridged systems. We are particularly indebted to H. Oevering, J. Kroon, A.M. Oliver, E. Cotsaris, D.C. Craig, S.S. Wong, K.D. Jordan, N.S. Hush, K.J. Smit and M.P. de Haas.

The present investigations were supported in part by the Netherlands Foundation for Chemical Research (SON), with financial aid from the Netherlands Organization for the Advancement of Research (NWO), by the Australian Research Council, by the Australian Federal Government, and by the Netherlands Ministry of Economic Affairs.

4. References.

1. a) M.A. Fox, M. Chanon, (Eds.):"Photoinduced Electron Transfer", Elsevier, Amsterdam, (1988).
   b) D. Gust, T.A. Moore, (Eds.): "Covalently Linked Donor–Acceptor Species for Mimicry of Photosynthetic Electron and Energy Transfer", Tetrahedron Symposium–in–Print No. 39, *Tetrahedron* **45** (1989).
2. D. Rehm, A. Weller, *Ber. Bunsenges. Phys. Chem.* **73** (1969) 834.
3. A. Weller, *Z. Phys. Chem.* **133** (1982) 93.
4. P. Pasman, G.F. Mes, N.W. Koper, J.W. Verhoeven, *J. Am. Chem. Soc.* **107** (1985) 5839.
5. J. Kroon, J.W. Verhoeven, M.N. Paddon-Row, A.M. Oliver, *Angew. Chem.*, (1991) in press.
6. H. Oevering, M.N. Paddon-Row, M. Heppener, A.M. Oliver, E. Cotsaris, J.W. Verhoeven, N.S. Hush, *J. Am. Chem. Soc.* **109** (1987) 3258.
7. N.S. Hush, *Trans. Faraday Soc.* **57** (1961) 557.
8. R.A. Marcus, *J. Chem. Phys.* **43** (1965) 679.
9. H. Oevering, J.W. Verhoeven, M.N. Paddon-Row, J.M. Warman, Tetrahedron **45** (1989) 4751.
10. Under barrierless conditions the solvent reorganization energy and the temperature exert only weak ($\lambda^{-0.5}$ and $T^{-0.5}$) influence on the rate.
11. N.S. Hush, M.N. Paddon-Row, E. Cotsaris, H. Oevering, J.W. Verhoeven, M. Heppener, *Chem. Phys. Lett.* **117** (1985) 8.

12. J.M. Warman, M.P. de Haas, M.N. Paddon-Row, E. Cotsaris, N.S. Hush, H. Oevering, J.W. Verhoeven, *Nature* (London) **320** (1986) 615.
13. J. M. Warman, K.J. Smit, M.P. de Haas, S.A. Jonker, M.N. Paddon-Row, A.M. Oliver, J. Kroon, H. Oevering, J.W. Verhoeven, *J. Phys. Chem.* **95** (1991) 1979.
14. J.W. Verhoeven. J. Kroon, M.N. Paddon-Row, A.M Oliver, in : "Photoconversion Processes for Energy and Chemicals", p. 100-108, D.O. Hall and G. Grassi (Eds.), Elsevier, London (1989).
15. J.W. Verhoeven, *Pure & Appl. Chem.* **62** (1990) 1585.
16. J.R. Miller, J.V. Beitz, R.K. Huddleston, *J. Am. Chem. Soc.* **106** (1984) 5057.
17. R. Hoffmann, A. Imamura, W.J. Hehre, *J. Am. Chem. Soc.* **90** (1968) 1499.
18. R. Hoffmann, *Accounts Chem. Res.* **4** (1971) 1.
19. M.N. Paddon–Row, *Accounts Chem. Res.* **20** (1982) 245.
20. M.N. Paddon–Row, K.D. Jordan in: "Modern Models of Bonding and Delocalization", chap. 3, J. F. Liebman, A. Greenberg (Eds.), VCH Publishers, New York, (1988).
21. J.W. Verhoeven, P. Pasman, *Tetrahedron* **37** (1981) 943.
22. J. Kroon, A.M. Oliver, M.N. Paddon–Row, J.W. Verhoeven, *Recl. Trav. Chim. Pays-Bas* **107** (1988) 509.
23. A.M. Oliver, D.C. Craig, M.N. Paddon–Row, J. Kroon, J.W. Verhoeven, *Chem. Phys. Lett.* **150** (1988) 366.
24. P. Pasman, J.W. Verhoeven, Th. J. De Boer, *Tetrahedron Lett.* (1977) 207.
25. M.N. Paddon–Row, H.K. Patney, R.S. Brown, K.N. Houk, *J. Am. Chem. Soc.* **103** (1981) 5575.
26. F.S. Jørgensen, M.N. Paddon–Row, H.K. Patney, *J. Chem. Soc. Chem. Commun.* (1983) 573.
27. M.N. Paddon–Row, H.K. Patney, B. Peel, G.D. Willett, *J. Chem. Soc. Chem. Commun.* (1984) 564.
28. M.N. Paddon–Row, E. Cotsaris, H.K. Patney, *Tetrahedron* **42** (1986) 2279.
29. W. Grimme, L. Schumachers, R. Gleiter, K. Gubernator, *Angew. Chem. Int. Ed. Engl.* **20** (1981) 168.
30. V. Balaji, K.D. Jordan, P.D. Burrow, M.N. Paddon–Row, H.K. Patney, *J. Am. Chem. Soc.* **104** (1982) 6849.
31. V. Balaji, L. Ng, K.D. Jordan, M.N. Paddon–Row, H.K. Patney, *J. Am. Chem. Soc.* **109** (1987) 6957.
32. M.N. Paddon-Row, J.W. Verhoeven, *New J. Chem.* **15** (1991) 107.
33. T. Koopmans, *Physica* **1** (1934) 104.
34. R.A. Marcus, *J. Chem. Phys.* **24** (1956) 966.
35. R.A. Marcus, *J. Chem. Phys.* **43** (1965) 679.
36. R.A. Marcus, N. Sutin, *Biochim. Biophys. Acta* **811** (1985) 265.
37. N.S. Hush, *Trans. Farad. Soc.* **57** (1961) 155.
38. N.S. Hush, *Coord. Chem. Rev.* **64** (1985) 135.
39. H.M. McConnell, *J. Chem. Phys.* **35** (1961) 508.

40. R.M. Hermant, N.A.C. Bakker, T. Scherer, B. Krijnen, J.W. Verhoeven, *J. Am. Chem. Soc.* **112** (1990) 1214.
41. M.N. Paddon-Row, A.M. Oliver, J.M. Warman, K.J. Smit, M.P. de Haas, H. Oevering, J.W. Verhoeven, *J. Phys. Chem.* **92** (1988) 6958.
42. M.P. de Haas, J.M. Warman, *Chem. Phys.* **73** (1982) 35.
43. H. Oevering, dissertation, University of Amsterdam (1988).
44. T. Kakitani, N. Mataga, *J. Phys. Chem.* **90** (1986) 993.
45. A. Yoshimori, T. Kakitani, Y. Enomoto, N. Mataga, *J. Phys. Chem.* **93** (1989) 8316.
46. A.D. Joran, B.A. Leland, P.M. Felker, A.H. Zewail, J.J. Hopfield, P.B. Dervan, *Nature* (London) **327** (1987) 508.
47. J.R. Miller, J.V. Beitz, R.K. Huddleston, *J. Am. Chem. Soc.* **106** (1984) 5057.
48. K.J. Smit, J.M. Warman, M.P. de Haas, M.N. Paddon-Row, A.M. Oliver, *Chem. Phys. Lett.* **152** (1988) 177.
49. N. Liang, J.R. Miller, G.L. Closs, *J. Am. Chem. Soc* **112** (1990) 5353.
50. I.R. Gould, J.E. Moser, D. Ege, S. Farid, *J. Am. Chem. Soc.* **110** (1988) 1991.

SPECTROSCOPY OF BILIPROTEINS AND ENERGY TRANSFER IN PHOTOSYNTHETIC ANTENNA COMPLEXES

SIEGFRIED SCHNEIDER
Institut für Physikalische und Theoretische Chemie der Universität Erlangen-Nürnberg
Egerlandstr. 3
D 8520 Erlangen
FRG

ABSTRACT. In this contribution the various spectroscopic techniques which have been applied to elucidate the energy transfer in photosynthetic antenna pigments are described and some of the results of a large number of studies are presented. It is not intended to give a complete review on the investigations on different types of photosynthetic organisms but to demonstrate the fundamental principles which are important in the chain of processes (light absorption, energy transfer and charge separation) which is active during the process of converting light energy into chemical binding energy. Since the thermophilic cyanobacterium Mastigocladus laminosus is one which was subject to most intensive studies by all kinds of spectroscopic techniques and the models explaining the energy transfer are well developed, most of the data presented in this contribution will refer to this organism.

1. Introduction

Photosynthetic organisms absorb light and convert the photonic energy into chemical energy by the process of photosynthesis. As a general feature, the photosystems (fig.1) are composed of an intramembrane reaction centre (where this conversion is performed) and an associated antenna of light-harvesting proteins, containing up to over 1000 chromophores [1,2]. The task of this antenna complexes is to increase the cross section for light absorption both with respect to the geometrical cross section as well as with respect to the spectral region where light is absorbed. The energy absorbed in the antenna complexes is transfered in an energetic cascade to the reaction centres, where the chemical events take place.

The light-harvesting antenna allow the photosynthetic organisms to fully utilize the capacity of the electron transport

E. Kochanski (ed.), Photoprocesses in Transition Metal Complexes, Biosystems and Other Molecules. Experiment and Theory, 299–331.

chain and the enzymatic system of carbon dioxide fixation. By variation of the antenna composition the organisms can adapt to different environments and ecological conditions, primarily to light intensity and light quality. Depending on cell structure and organisation two different kinds of light-harvesting systems are formed: (i) membrane-integrated antenna, which represent mainly chlorophyll a/b light-harvesting complexes and (ii) antenna structures outside of the thylakoid or cytoplasmatic membrane [2,3].

Examples of extramembrane antenna systems are the chlorosomes of green bacteria and the phycobilisomes of cyanobacteria and red algae. Phycobilisomes are large supramolecular structures of bright red and blue colored proteins (biliproteins) with an absorption ranging between 450 and 660 nm. Biliproteins therefore fill the gap in the absorption spectrum of chlorophylls which absorb primarily in the blue (430-440 nm) and red region (670 - 680 nm) of the visible spectrum [1-5].

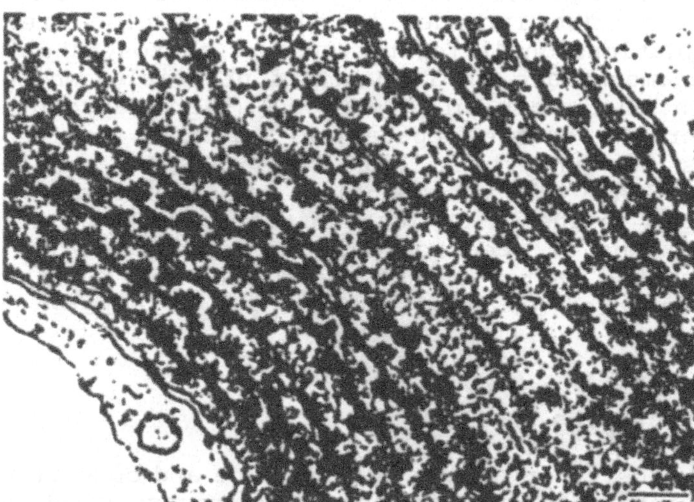

Fig. 1 Enlarged section of the chloroplast of the red alga Rhodella violacea cut perpendicular to the thylakoids with hemi-discoidal phycobilisomes displays in face views as semi-circular aggregates. (From Mörschel and Rhiel [2]). Scale marker indicates 100 nm.

2. Organisation of light-harvesting complexes (phycobilisomes) of photosynthetic algae (cyanobacteria and red algae)

The extramembrane localisation of the phycobilisomes allows to dissociate them from the thylakoid membrane without destroying their function. This point proved very helpful when studying both the structural aspects as well as the energy transfer, since the influences of the thylakoid membrane could be eliminated.

The first successful structural analysis on isolated hemidiscoidal phycobilisomes of the red alga Rhodella violacea and a proposal of a supramolecular model were put forward by

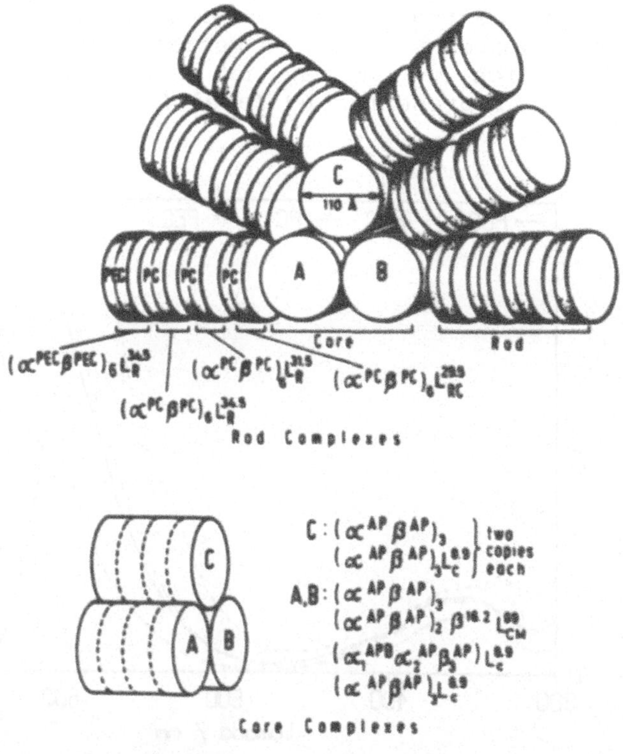

Fig. 2 Phycobilisome model of the cyanobacterium Mastigocladus laminosus. The rods comprise the biliproteins phycoerytrocyanin (PEC) und phycocyanin (PC). The core is made up mainly of allophycocyanin (AP) and one pigment called allophycocyanin B (APB). The assembly is hold together by linker polypeptides (L). From [2].

Wehrmeyer and co-workers. Their investigations showed that the isolated phycobilisomes had a similar size and outline compared to phycobilisomes in situ [2]. They consisted of two morphologically distinct domains, a trigon-shaped core and a periphery built up of short rods (fig.2). The core is normally composed of three ring-shaped cylindrical units of about 11 nm diameter, the two lower cylinders, A and B, represent the original attachment site to the membrane in situ, while the third occupies the grove on top of these. From this core usually 6 peripheral rods extend in a symmetrical, hemi-discoidal array. These rods consist of stacked discs of biliprotein aggregates (trimers) associated face to face, thereby forming double discs (hexamers) with diameters and hight of about 10 nm. Normally 2 - 6 discs make up one rod, the number being dependent on organism and growth condition.

The binding of the phycobilisome to the thylakoid membrane, the binding of the rods to the core and the formation of the rods from the hexameric units is mediated by so-called linker polypeptides which may or may not contain chromophores. In fig. 2 they are represented by label L.

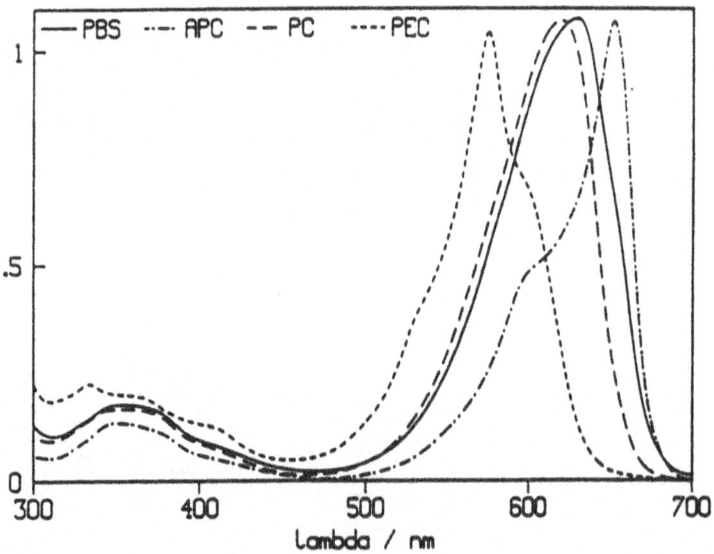

Fig. 3 UV-vis absorption spectra of different biliproteins in the trimeric form. The spectra are normalized; the absorption characteristics of whole phycobilisomes (PBS) vary with the composition of the rods. In the example shown, phycocyanin (PC) is the predominant biliprotein.

The function of phycobilisomes as light-harvesting antenna complexes is guaranteed by the arrangement of the biliproteins according to their spectral properties (fig.3) [4]. The biliprotein complexes, which absorb at shorter wavelength, namely phycoerytrocyanin (PEC), are located at the outer periphery. Allophycocyanin (APC), which absorbs at longest wavelength, is located in the core and next to the reaction centre. In between these two biliprotein complexes one finds phycocyanin (PC) whose absorption maximum lies intermediate between the absorption maxima of the two other complexes. Exitation energy transfer from PEC to PC and APC is therefore downhill, as can be seen from the normalized absorption spectra in fig.3. The absorption characteristics of the whole phycobilisome depend on the composition; in the example shown in fig. 3 it is dominated by the absorption of phycocyanin.

The monomeric units of the biliproteins mentioned above are composed of two polypeptide chains called α and β, respectively (see also fig.2). To each polypeptide chain one or two tetrapyrrole chromophores are covalently bound [5]. It is obvious that the different chromophores shown in fig.4 exhibit different absorption spectra due to the fact that the π-electron systems, are of different size. Moreover, due to the flexible linkage between the four 5-membered rings, the chromophores can adopt various geometries (different configurations and conformations). The theoretical studies [6,7] show

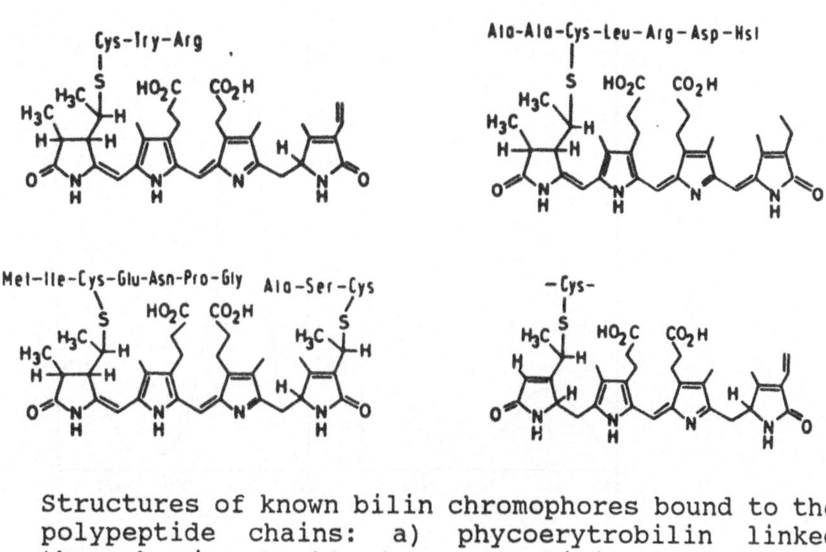

Fig. 4 Structures of known bilin chromophores bound to the polypeptide chains: a) phycoerytrobilin linked through ring A, b) phycoerytrobilin double bound through ring A and D, c) phycocyanobilin linked through ring A and d) phycourobilin linked through ring A. From [2].

that the absorption spectra of these chromophores vary significantly with molecular geometry and it was believed that different biliproteins, which contain exactly the same chromophore, exhibit different absorption spectra because of different chromophore geometry. By comparison with spectra of model compounds, it was for example concluded that the tetrapyrrole chromophores in the native pigment must be present in an extended conformation characterized by an intense absorption about 600 nm (see fig.5); free tetrapyrrole chromophores, on the other hand, are known to adopt the cylic helical structure for which the red absorption maximum is much weaker than the one in the near UV [4]. A large fraction of the work dealing with energy transfer in antenna complexes was therefore devoted to the question of "structure and function" (for a review see for example [8]).

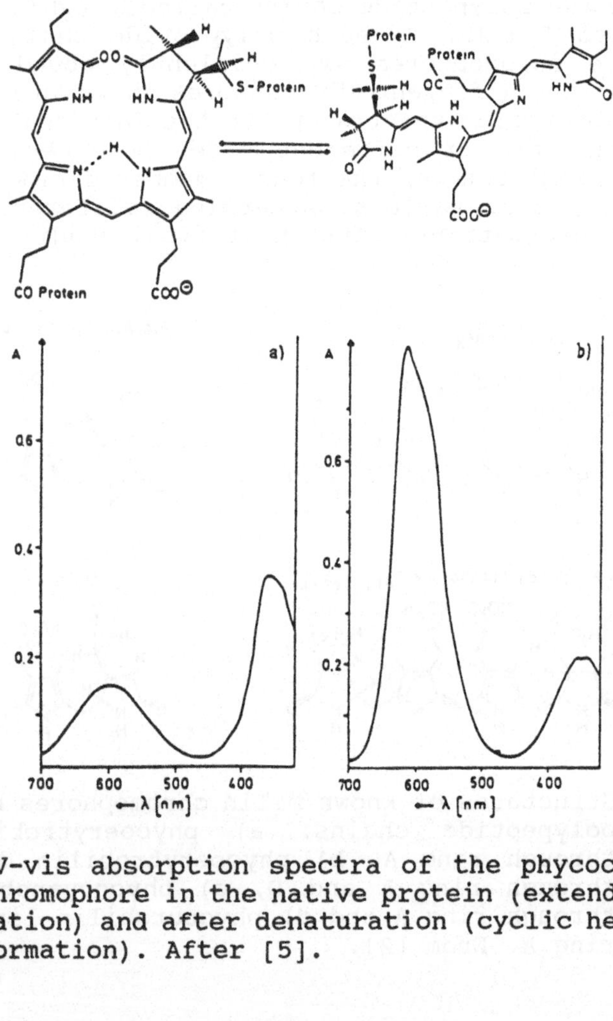

Fig. 5 UV-vis absorption spectra of the phycocyanobilin-chromophore in the native protein (extended conformation) and after denaturation (cyclic helical conformation). After [5].

A major break through in this attempt was the appearance of high resolution X-ray structures of various biliproteins [9,10]. These studies confirmed that the tetrapyrrole chromophores possess an extended geometry and also that the three phycocyanobilin chromophores, which are present in a phycocyanin monomer, exhibit essentially the same geometry. This was surprising, since the absorption spectra of the isolated subunits are significantly different (in the α-subunit one chromophore is bound to the polypeptid chain via a cystein and usually named A84; the β-subunit contains 2 chromophores called B84 and B155 according to their binding sites). The results provided by X-ray spectroscopy therefore confirmed the hypothesis that the fine tuning of the absorption spectra of different biliproteins, containing the same chromophore in a different protein environment, is due to specific interactions between the chromophores and the various amino acids in close neighbourhood to the various chromophores (see below).

An other important result produced by X-ray spectroscopy is the fact that in the crystallized samples the chromophore arrangement in a hexameric unit, that is one double disc in the rods of the phycobilisomes, shows a threefold symmetry axis along the axis of the rods. The existence of such a symmetry implies that the set of rate equations, which describes the excitation energy transfer between the various chromophores, has a largely reduced number of different coupling constants thereby facilitating the solution of that problem.

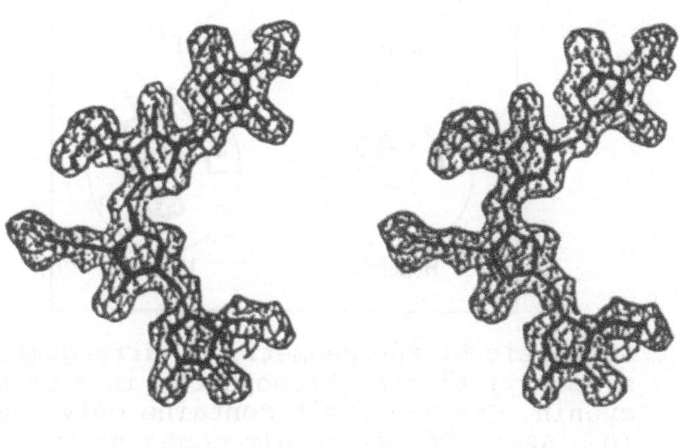

Fig. 6 Electron density map of the phycocyanobilin chromophore bound to the α polypeptide chain (A84). From [10].

3. Problem of micro-heterogeneity of chromophore-protein arrangement

The widely accepted strategy to attack the problem of energy transfer in the complex phycobilisomes was to study first the fluorescence kinetics and energy transfer in the building blocks and the constituent fragments, that is monomers and subunits. As can be seen from the schematic in fig.7, the α-subunit, holds only one chromophore (in fig.7 it is named M, indicating that its absorption maximum is intermediate between those of the two chromophores found in the β- subunit which are called S and F for sensitizing and fluorescing chromophore; for a more comprehensive explanation see below). Assuming a completely uniform sample without any microscopic heterogeneity in the chromophore-protein arrangement, the fluorescence decay of the α-subunit should be mono-exponential, the decay time being equal to the lifetime of the exited state of the A84 chromophore. In case of the β-subunit one expects in the absence of heterogeneity a bi-exponential fluorescence decay with the apparent decay times being connected to the lifetimes of the isolated S and F chromophores and the energy transfer time between those two chromophores. For the monomeric unit,

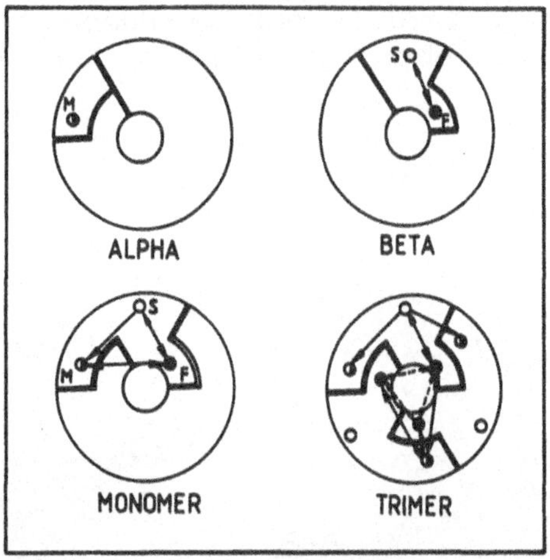

Fig. 7 Schematic of the geometrical arrangement (threefold symmetry) of the chromophores in a trimer of phycocyanin. The α-subunit contains only one chromophore (M = A84). The β-subunit contains two chromophores: the sensitizing (s-chromophore = B155) and the fluorescing (f-chromophore = B84). The number of energy transfer rates increases in the higher aggregates; due to symmetry three rate constants are always equal.

a tri-exponential fluorescence decay is expected with the apparent decay constants being uniquely related to the lifetimes of the three isolated chromophores and the rate constants for energy transfer between these three chromophores as indicated in fig.7. Because of the assumed symmetry, the fluorescence kinetics of the trimer also should be described by three exponentials which are determined by an even larger number of energy transfer rates. It is therefore impossible to derive the rate constants for energy transfer between the various chromophores just on the basis of measured observed fluorescence decay times.

Extensive investigations in distinct laboratories on biliproteins in various states of aggregation, especially in case of the well-studied phycocyanin, revealed that the fluorescence decay curves could not be fitted by multi-exponential decay laws, the number of exponentials being equal to that predicted by the theoretical considerations given above. This finding proved that the chromophore-protein arrangement is not as unique as assumed on the basis of the results of X-ray spectroscopy. Since during the preparation of the isolated biliproteins of different size, extensive chemical treatment including denaturation and renaturation of the biliproteins had to take place, it was believed that the problem of heterogeneity of the samples was related to the preparation procedure. After preparation techniques had developed such that lifetime data and fluorescence yields could be reproduced from one sample to another, the problem of heterogeneity was still pending. Suggestions were made that in contrast to the picture suggested by the X-ray structure, no perfect threefold symmetry should be present in a native biliprotein rod. Instead there should exist different possibilities for the arrangement of chromophore and neighbouring protein [11]. As will be shown below, one source for such a heterogeneity can be the distribution of protons within the functional groups of the various amino acids in the neighbourhood of the chromophores.

4. Spectral hole burning in biliproteins

One experimental evidence for the micro-heterogeneity of the chromophore-protein arrangement is the fact that the technique of spectral hole burning can be applied to biliproteins and phycobilisomes [12,13]. Using the excitation by a narrow bandwidth dye laser at low temperature certain chromophore-protein arrangements can be excited selectively. During the relaxation process a fraction of the excited species converts into a differently absorbing photoproduct, that is a chromophore with a different geometry or the same chromophore in a different protein environment. Such a photoinduced reaction therefore results in the reduction of the absorption cross section at the wavelength of the photolysis laser and shows up as a hole in the absorption spectrum. At elevated temperatures, when the photoproduct returns to its initial state, this photoinduced absorption change disappears and the hole in the absorption spectrum is filled-up again. In fig.8, the absorption spectrum of phycobilisomes of Mastigocladus laminosus at a temperature of 4 K in a saccharose/phosphate buffer solution is shown. The arrows indicate the positions where hole burning was performed. Typical holes are shown as insets, their width is on the order of 0.4 cm^{-1}. It must therefore be concluded

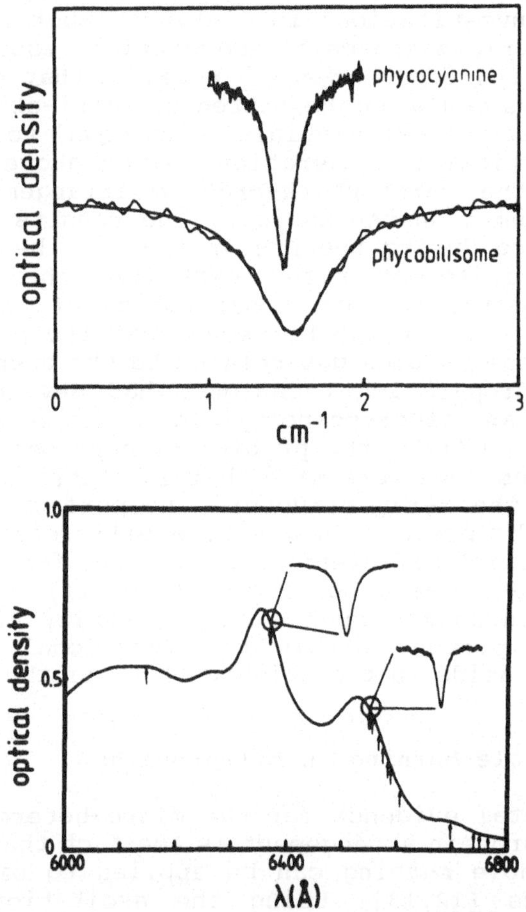

Fig. 8 Hole burning in phycobilisomes of Mastigocladus laminosus. Burning positions are indicated by arrows. Typical hole shapes are shown on an enlarged scale. In the top part, spectral holes in isolated phycocyanin and phycobilisomes generated under identical conditions are compared. For more details see text. From [12].

that the large width of the absorption spectra is due to inhomogeneous broadening and reflects the existence of different chromophore-protein arrangements. It is also interesting to compare the width of the holes obtained in phycobilisomes (PBS) and in isolated phycocyanin (PC) under identical conditions. The hole width is much larger in the whole phycobilisomes than in isolated phycocyanin. This is in accordance with the fact that the lifetime of the excited state of the phycocyanobilin chromophores in phycobilisomes is reduced because of fast energy transfer. From the width of the hole in phycobilisomes the lifetime of phycocyanin can be estimated to about 30 ps.

5. Raman- and CARS-spectroscopy of biliproteins for determination of chromophore geometry

It is clear that it is of utmost importance to determine the structure of the chromophores in the native system or, for smaller aggregates, at least in solution at room temperature. Experimental techniques, which give information on the structure of a compound like NMR-spectroscopy, have not yet been successfully applied to biliproteins, because the signals of the chromophore are burried under the signals of the large protein, which contains about 1000 times as many nuclei of the same type. Information on the chromophore structure therefore relies on techniques which enhance the signal due to the chromophore compared to that of the protein. Among such techniques both (pre-)resonance-enhanced Raman- and CARS-spectroscopy have been applied in the past [14-16]. By use of these techniques, it was possible to show the effect of denaturating reagents and the effect of aggregation on chromophore structure. A detailed interpretation of the differences observed in Raman- and CARS-spectra of biliproteins under different conditions is hampered by the lack of a good normal coordinate analysis. Crude calculations of normal coordinates indicate that the local coordinates are generally heavily mixed wherefore observed vibrational bands can seldom be assigned to certain localized structural elements. Nevertheless it was possible by comparison of a great body of experimental data to deduct some empirical interpretations and rules of thumb. One of these observations is that whenever the chromophore is present in a fully extended conformation, a strong band appears in the Raman spectra around 1650 cm^{-1}. When the chromophore is present in the cyclic helical conformation, two bands are observed around 1620 and 1630 cm^{-1} [15]. Szalontai et.al. [14] could demonstrate that in the Raman spectra of phycocyanin a gradual shift of the strong band at 1653 cm^{-1} is observed upon acidification. This shift to lower wavenumbers can be understood since upon lowering of the pH value the phycocyanin denatures during which process the chromophore changes from an extended to a cyclic helical conformation.

Resonance-enhanced Raman spectroscopy suffers from the overlap of fluorescence and Raman scattered light when the excitation wavelength coincides with the first electronic absorption band. In order to avoid this problem excitation wavelength has been chosen to be in the near UV, that is resonant with higher electronic states or, more recently, in the near IR (using a Nd^{+}-YAG laser). In the latter case the advantage of large resonance enhancement of the signal contributions of the chromophore is lost in part; the advantage is, however, that the samples are less subject to photoinduced deterioration processes.

Coherent anti-Stokes Raman scattering (CARS) has the advantage that the recorded signal is at shorter wavelength than the excitation radiation (anti-Stokes shifted photons). Therefore stray light and fluorescence light do not interfere with the scattered light to be recorded. The disadvantage of the CARS technique is, however, that two laser beams of high intensity have to be focused onto the sample in order to produce in a nonlinear process (third order suszeptibility) the CARS signal. The samples are, therefore, often subject to photodegradation and have to be replaced permanently.

In fig.9, CARS spectra covering the finger print region and the double bond stretching region for biliproteins in various states of aggregation are shown for comparison. The most obvious difference is seen between the spectra of the α- and β-subunit of phycocyanin. Such a large difference is not expected in view of the fact that the X-ray studies claim that the chromophores in both subunits have nearly identical geometry. The electronic structure and, therefore, the force constants describing the movements of the various atoms must be significantly different in the three types of chromophores. The second astonishing feature is the number of observed vibrational bands which is about twice as large in the α-subunit than in the β-subunit, despite of the fact that the α-subunit contains only one chromophore, whereas the β-subunit has two of them. This may be indicative for the existence of more than one chromophore-protein arrangement in the isolated α-subunit. The latter assumption is also in accordance with the above mentioned fact that in contrast to expectation the fluorescence decay of isolated α-subunits is not mono-exponential. The conclusion drawn from this observation is that in the separated subunits micro-heterogeneity can occur based on different chromophore geometries and concomitantly a different interaction with the protein surrounding [17]. The changes observed in the spectra when going from PC trimer to PC hexamer and to whole phycobilisomes are less dramatic. It is interesting to note that in the phycobilisome spectra, a band at 1624 cm^{-1} shows up which is also seen in isolated allophycocyanin. The enhancement of the band at 1235 cm^{-1} relative to that at 1273 cm^{-1} can also be related to the existence of allophycocyanin in the phycobilisomes. It must be noted that despite of the fact that in the phycobilisomes many more different chromophores are present than for example in PC hexamers, the bands are generally narrower in the sample with a higher the state of aggregation. It is believed that this is related to the effect that in higher aggregates the microheterogeneity is reduced in comparison to that of the lower aggregates (perhaps also due to linker proteins).

In X-ray spectroscopy the mercury salt PCMS (p-chloromercury-benzenesulfonate) is applied to produce special scattering centres because it specifically binds to free cystein residues in the amino acid sequence. When solutions of phycocyanin are titrated with this salt it binds to the only one free cystein which is present in the β chain and which is actually located

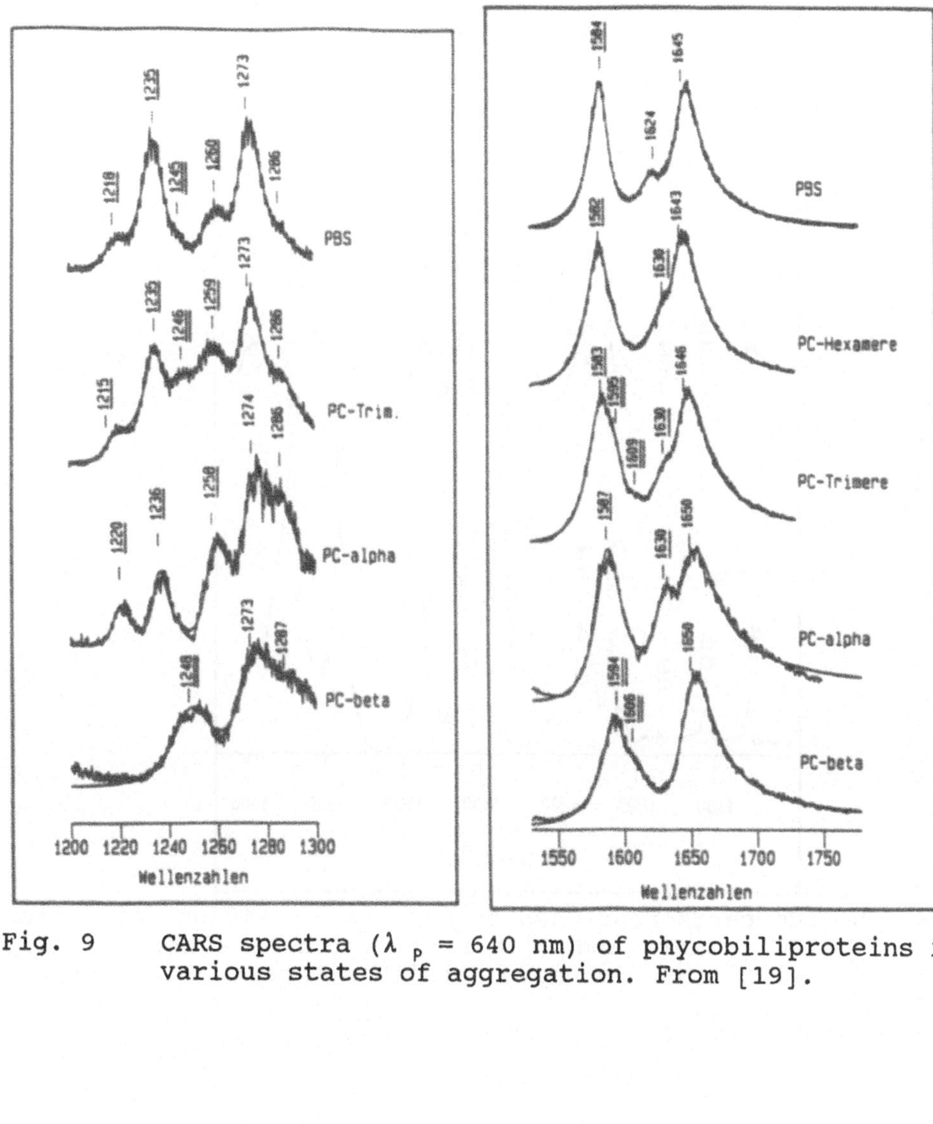

Fig. 9 CARS spectra (λ_p = 640 nm) of phycobiliproteins in various states of aggregation. From [19].

very close to ring D of chromophore B84 [18]. The changes observed in the absorption spectra of the β-subunit and PC trimer can be rationalized by assuming that the presence of the voluminous mercury forces ring D to rotate away from its position in the native pigment. The CARS spectra of the two trimer samples show also significant differences which can be explained by this assumption. The band at 1650 cm^{-1} changes its shape indicating that at least one of the C=C stretching vibrations of the methine groups is changed. Furthermore, the band at 1246 cm^{-1}, which is observed both in the β-subunit as well as in the native trimer spectra, disappears upon PCMS treatment. This indicates that a CH-bending vibration of the methine group is changed. The changes in chromophore geometry and in the electronic excitation spectrum of chromophore B84 manifest themselves also in a modified fluorescence decay kinetics of PCMS-treated phycocyanin trimers (see below) [20].

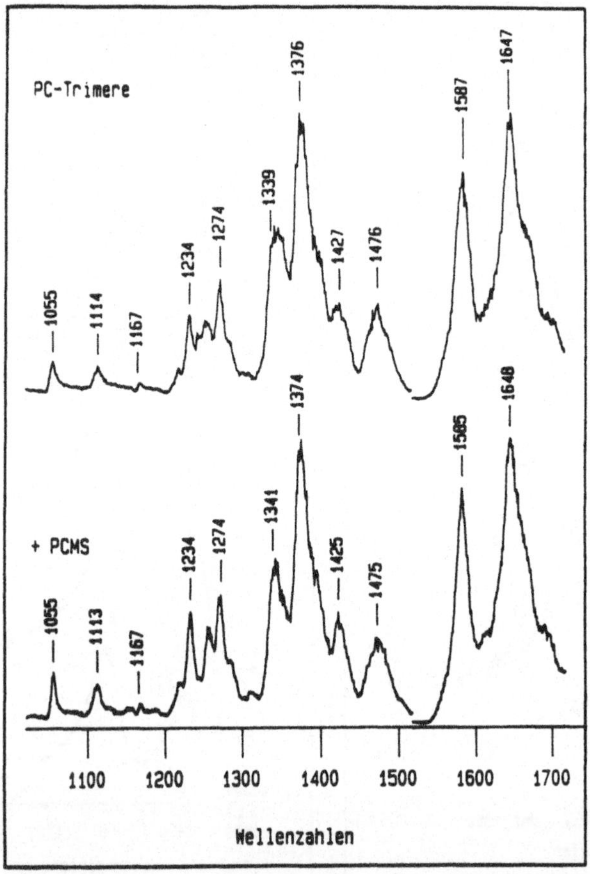

Fig. 10 Comparison of CARS spectra (λ_p = 640 nm) of native PC trimers and PC trimers which have been treated with PCMS. For more details see text. From [19].

6. Fine-tuning of electronic properties by chromophore-protein interaction

Tetrapyrrole chromophores in the extended conformation exhibit a large delocalized π-electron system. Under the influence of electronic charges in the neighbourhood the π-electrons can very easily be polarized with the effect that excitation energies can change dramatically. This fact implies on one hand that the absorption maxima of the same chromophore can shift easily by as much as 100 nm (tuning of the absorption maximum). But it also implies that the excitation energy of the individual chromophores of a trimer is changed and therefore possibly the type of coupling (Förster type or excitonic coupling; see below).

The source of electric charges in the neighbourhood of the chromophores are various amino acids with functional groups like -COOH and -C(NH)(NH_2). Fig.11 presents a sketch of the A84 chromophore and its neighbouring amino acid residues in the relative arrangement suggested by the high resolution X-ray structure [9]. The chromophore adopts the fully extended

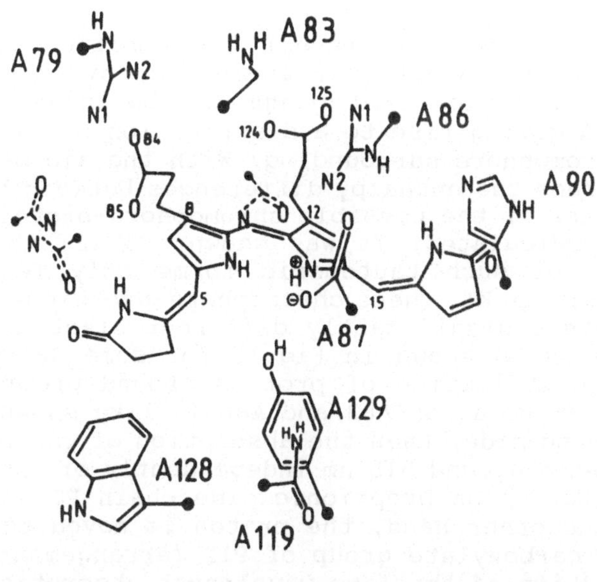

Fig. 11 Sketch of the geometrical arrangement of the chromophore A84 of <u>Mastigocladus laminosus</u> and the amino acids which are located within a sphere of 10 Å around C10 of the chromophore. Note that the position of the hydrogen atoms of functional groups like -COOH and -C(NH)(NH2) is not revealed by X-ray spectroscopy. From [21].

conformation, the configuration can be described as Z-anti, Z-syn, Z-anti. Both ring A (binding site to cystein A84) and ring D (next to histidin A90) are deviating from the "molecular plane" defined by rings B and C. The angle of twist around the single bond C5-C6 is about 60°, that around C14-C15 about 30°. All three chromophores in phycocyanin from <u>Mastigocladus laminosus</u> show one common type of interaction with the protein. They are arched around an aspartate residue (A87 in fig.11). The nitrogens of pyrrole rings B and C (N22 and N23) are within hydrogen bonding distance to one oxygen atom of the carboxylic group of aspartate A87. Therefore, one must assume that a salt bridge is formed which leaves a negative aspartate counter ion and protonates the chromophore on the imin nitrogen N23. Next to the carboxylate group of the propionic acid side chain attached to C8 (henceforth called P8), the arginine A79 is located; near the proprionic side chain connected to C12 (called P12) there is a lysine (A83) and another arginine (A86). These residues are also within hydrogen bonding distance to the carboxylate groups and help to fix the chromophore in its extended geometry. Within a sphere of 10 Å radius around C10 of the chromophore, there are other weakly interacting amino acids like Trp A128, Tyr A129, Asn A119 and His A90.

X-ray analysis does not reveal the position of hydrogen atoms and lone-pair electrons. Therefore, hydrogen atoms of carboxylate or arginine head groups can be located at different positions giving rise to a distribution of partial charges in the chromophore surrounding. With the aid of molecular modeling programs the enthalpy differences between the various tautomeric forms of the possible chromophore-protein arrangements can be calculated. It was shown [21,22] that among a large variety of such tautomeric forms only few ones are isoenergetic within kT. These chromophore-protein arrangements exhibit, however, significantly different electronic excitation energies as is shown in Fig.12 for some selected examples. If the distribution of protons around proprionic side chain P12 and amino acids A83 and A86 is like shown in fig.12 on the right hand side, then the absorption of chromophore A84 should be located around 515 nm independently of the distribution of the protons on proprionic side chain P8 and arginine A79. If, on the other hand, the proton is moved to the other oxygen in the carboxylate group of P12 (arrangement c or d) a bathochromic shift of the long wavelength absorption by about 100 nm is observed. This shift is again very insensitive to the proton distribution around P8/A79. Since in isolated α-subunits the absorption maximum is found around 624 nm, one must conclude that a proton distribution according to arrangement c or d should be the thermodynamically more stable arrangement compared to a or b.

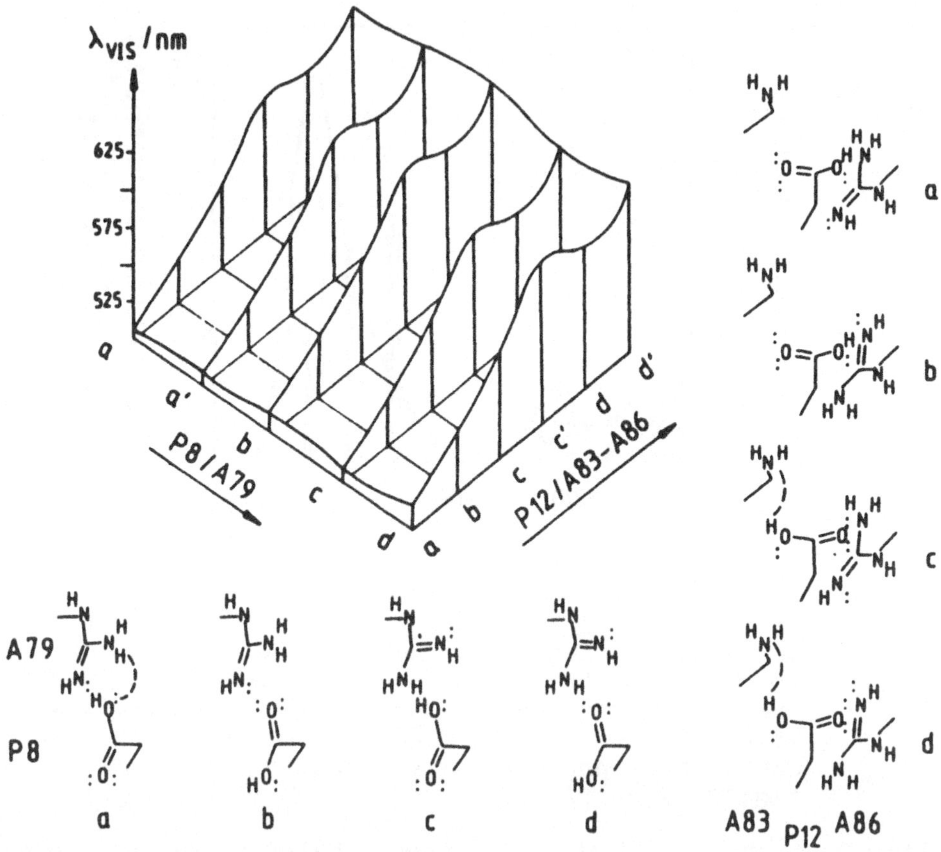

Fig. 12 Schematic of the protein environment around proprionic side chains P8 and P12, respectively, and the various distributions of hydrogen atoms (tautomeric forms of amino acids) considered in quantum-mechanical model calculations. For each combination of the various environments denoted by a-d, the calculated wavelength of the visible absorption band is displayed as vertical bar (primed letters: hydrogen bonded arrangements). From [21].

In figs. 13 and 14 experimental absorption and circular dichroism spectra are compared with theoretical ones. From the 16 possible combinations shown in fig.2, 9 combinations exhibiting the lowest enthalpy values are selected. The statistical weight for each arrangement is taken according to the

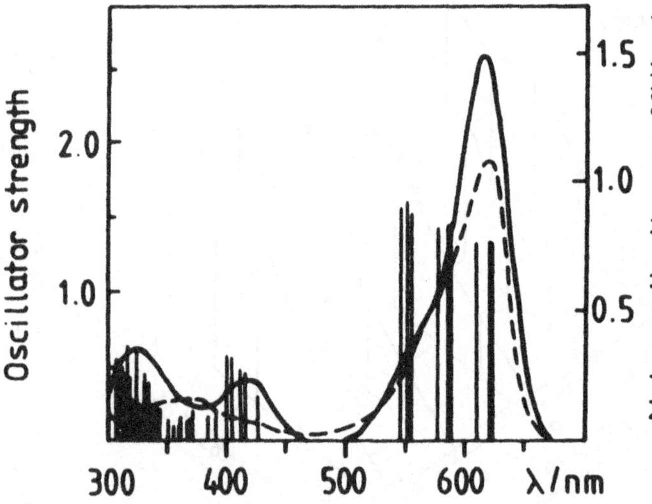

Fig. 13 UV-vis absorption spectrum of the α-subunit of C-PC of Mastigocladus laminosus. Dashed line: experimental spectrum from [17]; solid line: calculated absorption spectrum (details are given in the text). The bars represent the calculated oscillator strengths for the chromophore in the 9 most stable chromophore-protein arrangements displayed in fig.12. From [21].

Boltzmann factor calculated with the enthalpy difference (they are given in square brackets in fig.14). The three combinations with highest statistical weight are P12c/P8b, P12c/P8c and P12c/P8d for which the longest wavelength transition are calculated to be about 600 nm. Proton distributions, which lead to higher excitation energies, are of low statistical weight and may give rise to the short wavelength shoulder observed both in the absorption and circular dichroism spectra.

Similar calculations have been performed for the β-subunit [22] and are under way for the monomeric and trimeric unit. In all cases one finds that the absorption wavelength of the chromophore is varied over a wide range, when the distribution of protons on the functional groups in the neighbourhood of the chromophore is changed. If one therefore assumes as the source of the microheterogeneity the existence of various tautomeric forms in the protein environment, then one can not only explain the large inhomogenous width of the electronic absorption spectrum, but also the features of spectral hole burning. It furthermore explains why fluorencence kinetics

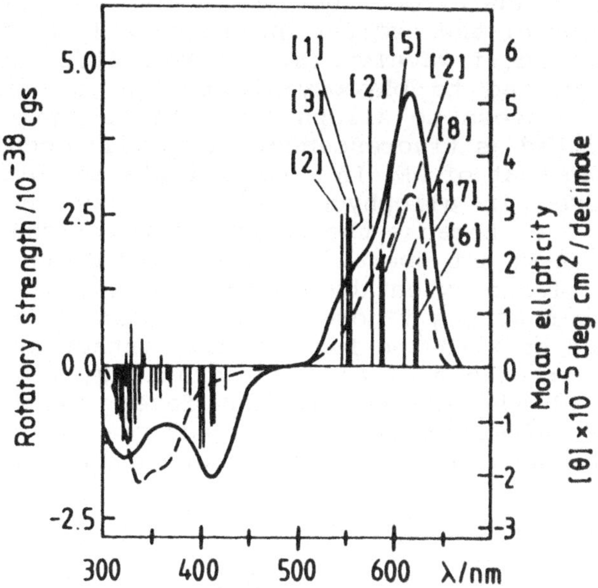

Fig. 14 Circular dichroism spectrum of the α-subunit of C-PC of Mastigocladus laminosus. Dashed line: experimental spectrum from [17]; solid line: calculated circular dichroism spectrum. The bars represent the calculated rotarory strength for the chromophore in the 9 most stable chromophore-protein arrangements of fig.12. From [21].

cannot be explained on the basis of a simple multi-exponential decay law with the lifetimes being independent of excitation and observation wavelength (see below).

7. Picosecond time-resolved absorption and emission spectroscopy

The guide line for doing time-resolved absorption and emission measurements can be understood by inspection of fig.15. As has been mentioned above, the wavelength of the first absorption maximum is decreasing in the order: chlorophyll (reaction centre), allophycocyanin, phycocyanin and phycoerytrocyanin or phycoerytrin. Illumination at short wavelength will therefore preferentially excite the biliproteins which absorb at higher frequency and which are located at the periphery of the phycobilisome (let say PE). According to the simple model most of the excitation energy will be transfered to the chromoprotein which is next in the ladder towards the reaction centre.

Because energy transfer does not occur with 100 % efficiency, a small fraction of the excitation energy will be released as fluorescence at short wavelength. The excitation energy which has been transfered to phycocyanin can be transfered in a radiationless process to allophycocyanin or again, to small extent, be emitted as fluorescence, which will occur at longer wavelength than that of PE. In case of isolated phycobilisomes the chromophores in allophycocyanin act tas the so-called terminal emitters. Because the energy transfer channel to the reaction centre is not operative, the excitation energy must be released to large extent by fluorescence, that is, one can observe a rather strong fluorescence emission around 670 nm. In case of studies with the functionally intact system (whole cells) the fluorescence of allophycocyanin will also be weak because most of the energy is transfered to the reaction centre. In case of the so-called "closed" reaction centres, in which the photoinduced electron separation is prohibited,

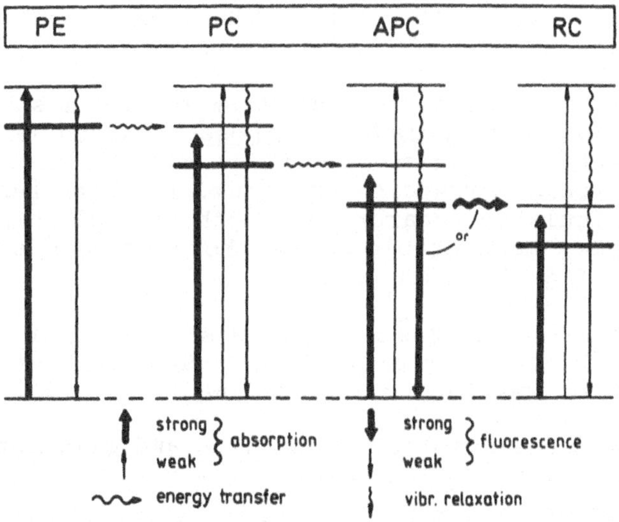

Fig. 15 Schematic of the energy transfer chain in photosynthetic organisms. The wavelength of the absorption maximum increases from phycoerytrin (phycoerytrocyanin) to phycocyanin, allophycocyanin and the chlorophyll antenna in the reaction centre. In case of functionally intact phycobilisomes, the originally absorbed energy is transfered radiationless in a cascade to the reaction centre; the energy transfer process can be studied by monitoring the weak fluorescence of each of the intermediately excited species.

fluorescence of the chlorophyll antenna in the reaction centre can be observed. Since the decay time of the fluorescence represents the inverse of the sum of all deactivation rates, its variation reflects the changes in rates of electron transfer when manipulations are made, which leave the intrinsic relaxation rates unchanged.

Because of this relationship, the so-called pump-probe experiment measuring the time for ground state recovery can also be used to monitor energy transfer in such combined systems. Excitation of PC, for example, by a strong photolysis pulse would deplete the ground state population and therefore result in an increased transmission of a probe pulse, which is delayed in time with respect to the photolysis pulse. When the probe pulse delay is increased, then its delay time dependent transmission monitors the return of the excited molecules to the ground state. If energy transfer takes place, then the rate of molecules returning to the ground state is increased compared to that of an isolated molecule and the apparent decay of the time-resolved transmission curve is faster [23].

In early experiments laser pulses of very high peak power were used to perform such picosecond time-resolved measurements. In these circumstances the density of excited molecules was so high that bimolecular processes between the excited molecules occured (e.g. singlet-singlet annihilation). Decay times measured under these conditions are laser power dependent and have no direct relationship to the processes occuring in the pigments under low level illumination. More recent experiments therefore made use of trains of very weak laser pulses and improved the signal to noise ratio by means of signal averaging.

Another problem connected with this type of experiment is related to the term "photoselection". If one uses linear polarized light for excitation, then from an isotropically distributed sample of absorbers, only those are excited, whose transition dipole moments are parallel to the electric field vector of the exciting radiation. That is to say that the ensemble of excited molecules is no longer isotropic with respect to its orientation in space. As a consequence of this photoselection the fluorescence intensity which is measured with an analyzer being either parallel or orthogonal to the polarization of the exciting light will be different. The intensities are proportional to the number of molecules in the excited states ($N_e(t)$ in fig.16) and to the correlation function P_2 of the transition dipole moments for absorption and emission. This implies that by measuring separately both $I_\parallel$ and $I_\perp$ information about both the lifetime of the excited state and the rotation of the emission dipole moment can be deduced. The isotropic decay function, which equals $I_0 = I_\parallel + 2 \cdot I_\perp$ describes directly the evolution of the number of excited molecules and is measured by setting the analyzer under the so-called "magic angle". The difference D between these two quantities, $D = I_{\parallel} - I_{\perp}$, describes the product of the excited state population and the correlation function. The emission dipole can change its orientation in space by rotation of the whole chromophore. However, the same is true, when the excitation energy is transfered to another, identical molecule with a different orientation in space. This implies that energy transfer between like species is evidenced in the difference of the decay of the terms $I_0 = I_{\parallel} + 2 \cdot I_{\perp}$ and $D = I_{\parallel} - I_{\perp}$. It should be noted that the fluorescence anisotropy R(t), which is the quotient of these two quantities, directly reflects the correlation function of absorbing and emitting oscillator. This expression can, however, be formed meaningfull only, if only one type of emitter is present.

What has been said about polarization of fluorescence is also true for the pump-probe experiment. Usually the probe beam must be polarized under magic angle with respect to the polarized photolysis pulse when the evolution of the excited state population should be monitored.

In fig.17a the isotropic fluorescence decay curves of isolated phycobilisomes of Mastigocladus laminosus recorded for variable observation wavelength with excitation at 600 nm are shown for comparison. The smooth lines represent the fit curves based on three exponentials. The decay times and the corresponding amplitudes of a fit with 3 and 4 exponentials, respectively, are summarized in the graph of fig. 17b. In

$$I_{\parallel} \sim \{5 + 4 \cdot \langle P_2[\vec{E}(0) \cdot \vec{E}(T)] \rangle\} \cdot N_E(T)$$

$$I_{\perp} \sim \{5 - 2 \cdot \langle P_2[\vec{E}(0) \cdot \vec{E}(T)] \rangle\} \cdot N_E(T)$$

$$I_{\parallel} + 2 \cdot I_{\perp} \sim N_E(T)$$

$$I_{\parallel} - I_{\perp} \sim N_E(T) \cdot \langle P_2[\vec{E}(0) \cdot \vec{E}(T)] \rangle$$

$$R = (I_{\parallel} - I_{\perp}) / (I_{\parallel} + 2 \cdot I_{\perp}) = \langle P_2[\vec{E}(0) \cdot \vec{E}(T)] \rangle$$

$\langle \; \rangle$ AVERAGE OVER ALL INITIAL ORIENTATIONS

Fig. 16 Influence of polarization of excitation and emission on fluorescence decay. For more details see text.

accordance with expectation, the fluorescence monitored around 614 nm rises instantaneously and decays very fast (the predominant component has $\tau \sim 20$ ps). The rise of fluorescence monitored above 640 nm is clearly delayed (negative amplitude in fit). Its decay is most consistently described by 3 exponentials with lifetimes of about 100, 200 and 1400 ps, respectively. Fluorescence at this wavelength should predominantly originate from allophycocyanin which is excited via energy transfer from phycocyanin (note that the decay curves are normalized to the same peak height; the fluorescence decreases strongly with increasing recording wavelength). Lifetimes of 1000 - 1500 ps are typical for phycocyanobilin chromophores in the extended conformation, when they are not subject to energy transfer ("isolated" chromophores). The intermediate decay time (100 - 200 ps) must be associated with energy transfer between phycocyanin and allophycocyanin, whereas the fastest decay time (20 ps) reflects energy transfer within phycocyanin. Similar measurements on whole cells show that there the long wavelength fluorescence does not contain a component with $\tau > 360$ ps. This proves that the energy transfer from allophy

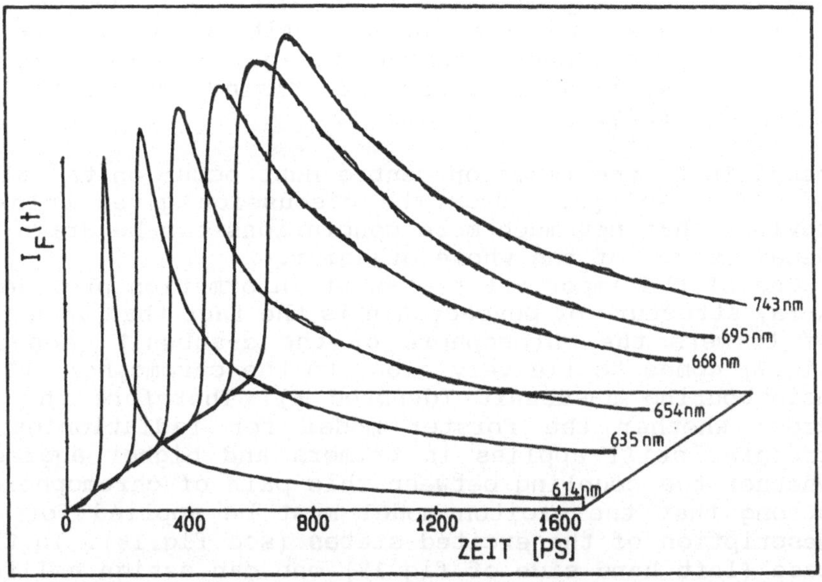

Fig. 17a Variation of fluorescence decay profile of phycobilisomes (<u>Mastigocladus laminosus</u>) with observation wavelength (λ_{ex} = 600 nm). The emission intensity decreases with increasing wavelength; for better comparison the curves are normalized. From [24].

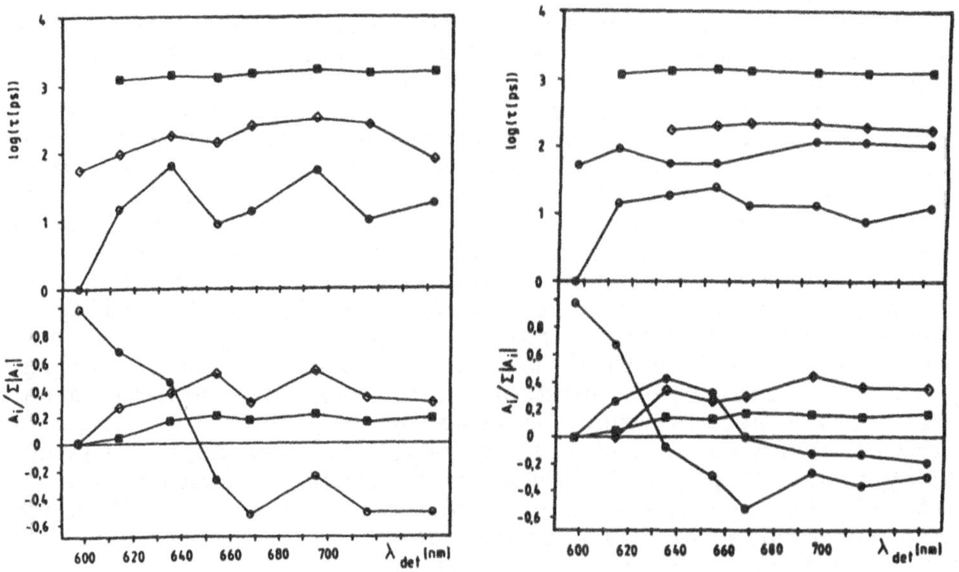

Fig. 17b Summary of decay times and amplitudes derived from a 3- and 4- exponential fit, resp., of the fluorescence decay curves of fig.17a. Note that negative amplitudes represent rising exponentials. From [24].

cocyanin to the reaction centre must occur on the time scale of 300 - 400 ps. From the discussion given below, it is obvious that not much more conclusions can be drawn from the decay curves of the whole organism.

One of the important pieces of information provided by the X-ray structure of phycocyanin is the fact that upon formation of trimers the chromophore of the α-subunit (denoted M in fig.7) comes to lie very close to the chromophore B84 of the neighbouring β-subunit (denoted F). Therefore the question arose whether the Förster model for radiationless energy transfer still applies in trimers and higher aggregates or whether the coupling between this pair of chromophores is so strong that the exciton model must be applied for a proper description of the excited states (see fig.18). In the first case (left hand side of fig.18) one can assign a lifetime to each type of chromophores and in this picture it becomes clear why B155 is called sensitizing (s) and B84 fluorescing chromophore (f). In case of efficient energy transfer excitation of B155 or A84 chromophores leads to fluorescence from the excited state of B84 whereas the fluorescence from the former two

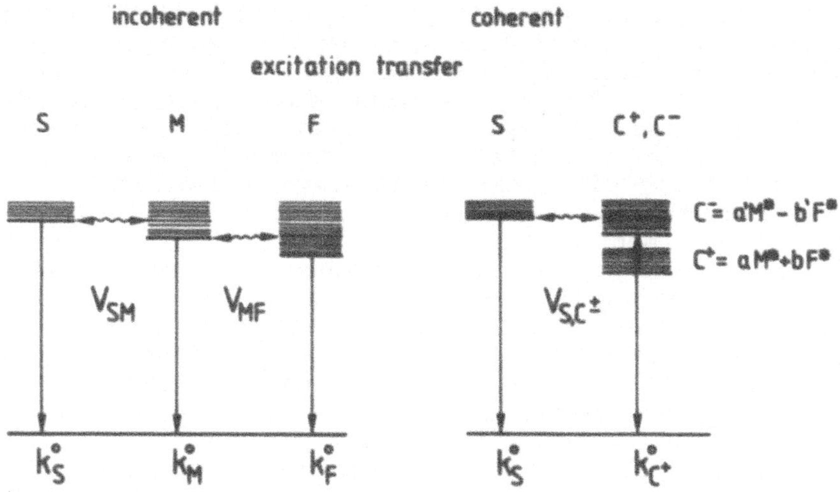

Fig. 18 Comparison of weak and strong coupling limit. Left: Förster type energy transfer (incoherent); right: strong coupling case; formation of delocalized excited states (coherent). For more details see text.

chromophores is extremely weak, but still detectable. In time-resolved measurements one expects three fluorescence decay components, two short-lived ones and one long-lived one. In the case of strong coupling only chromophore B155 can be seen as isolated chromophore whose lifetime is determined by energy transfer to the upper delocalized excited state C^- of the pair of chromophores. Instead of Förster type transfer, a rapid internal conversion process could take place leading to the lowest delocalized excited state termed C^+ in fig.18, whose lifetime is determined by the exact nature of this exited state. In a time-resolved experiment with excitation of chromophore B155 only two decay components should be found if time resolution is not sufficient to resolve a very fast internal conversion process (see however discussion in [25]).

Fig. 19 displays the results of pump-probe experiments performed with phycocyanin trimers. As can be concluded from the signal to noise ratio the maximum bleaching is largest around 620 nm (absorption maximum). The smooth lines represent bi-exponential fits with one lifetime varying between 28 and 97 psec and the second one between 823 and 1220 psec. This variation of the fit parameters with probing wavelength could indicate that actually more than two exponentials are required for a proper reproduction of these curves. A three exponential free fit does not produce a consistent, wavelength-independent set of decay times. The question whether the observed fast component is actually composed of two or more components with similar fast decay times can eventually be decided by pump-probe experiments with higher time resolution or by two color pump-probe experiments.

Additional evidence for the existence of strong coupling between chromophores A84 and B84 of neighbouring monomeric units came from theoretical calculations [26]. According to Förster theory, the rate of energy transfer is governed by the overlap of the fluorescence spectrum of the donor and the absorption spectrum of the acceptor. If one uses the absorption and emission spectra of the isolated subunits, the necessary overlap integrals can be determined under some simplifications. If one assumes furthermore that the transition dipole moments are along the long molecular axis, then the orientation factor and the distance between each pair of chromophores can be determined.

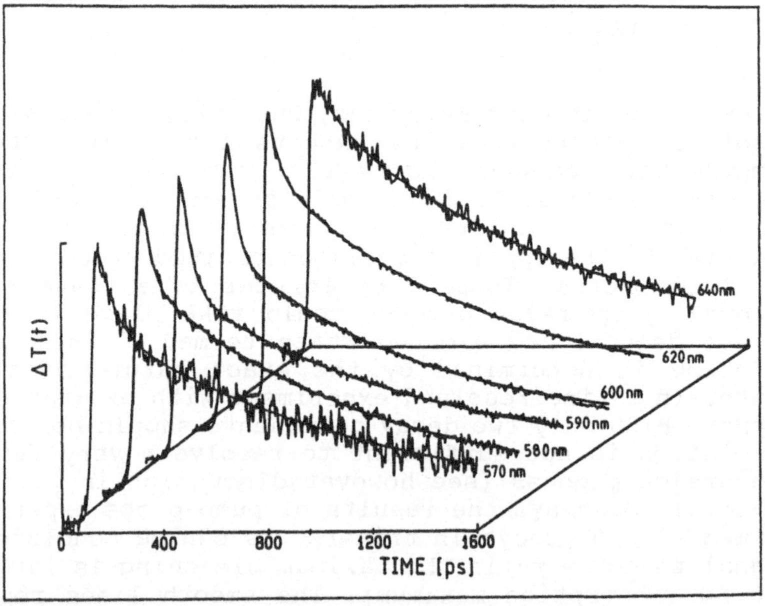

Fig. 19 Result of pump-probe experiment for phycocyanin trimers (Mastigocladus laminosus). For better comparison the curves are normalized to equal maximum transmission change. From [23].

The energy transfer time from B155 is predicted in fair agreement with the experimental observation (several 10 psec) whereas for the transfer between A84 and B84 a subpicosecond time is postulated. Such a short transfer time implies of course a very strong coupling and one must doubt whether the presupposition for applying Förster's formula are still valid. In view of the above exposition on micro-heterongeity one can also doubt that the calculation of spectral overlap on the

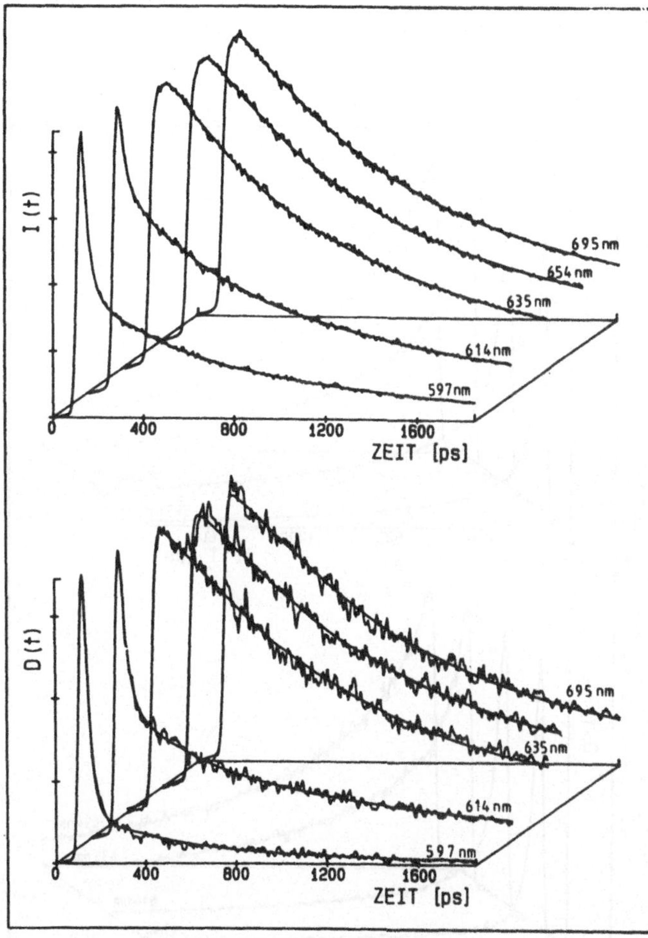

Fig. 20 Fluorescence decay of native, linker-free PC trimers (<u>Mastigocladus laminosus</u>). Top: isotropic decay I_0(magic angle). Botton $D = I_{\parallel} - I_{\perp}$. All curves are normalized to the same peak intensity (λ_{ex} = 580 nm). From [24].

basis of the subunit spectra is correct. Instead of the overlap of the ensemble average, one must average over the coupling of individual chromophore pairs, in which the energetic difference and the coupling strength can be considerably modified with the distribution of the protons of the functional groups of the neighbouring amino acids.

The first hint for the existence of very fast energy transfer in phycocyanin trimers came from polarized fluorescence measurements using a streak camera. In figs. 20 and 21 fluorescence decay measurements using the single photon timing technique are displayed. For comparison fig. 20 shows the

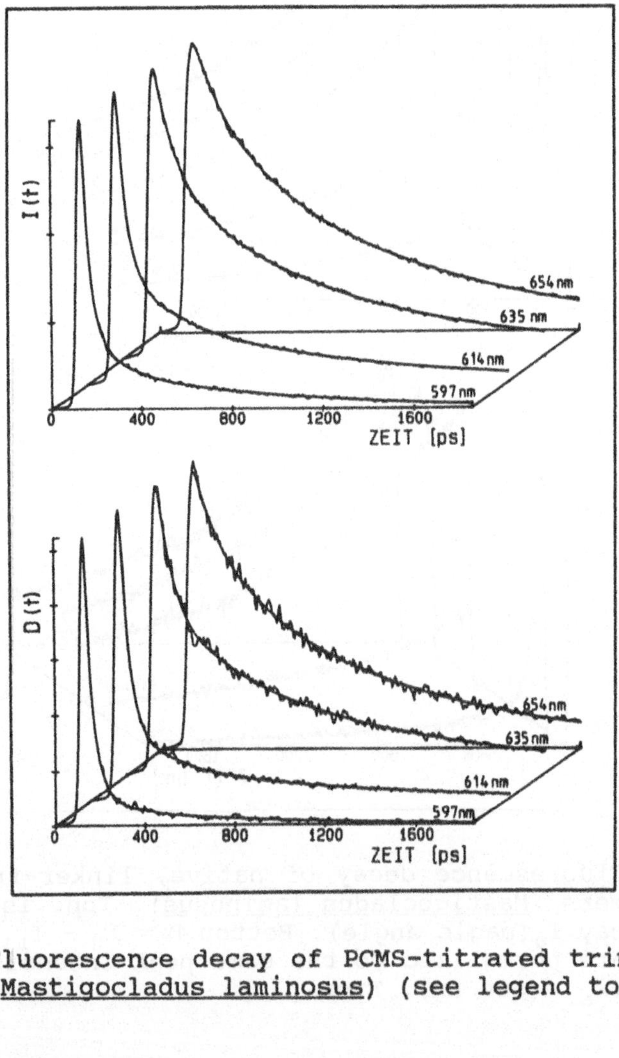

Fig. 21 Fluorescence decay of PCMS-titrated trimers of PC (Mastigocladus laminosus) (see legend to fig.20).

results for linker free, native trimers of phycocyanin whereas fig. 21 refers to PCMS-titrated trimers. The general pattern is similar in both cases inasfar, as upon excitation at 580 nm the fluorescence monitored below 600 nm rises instantaneously. For long wavelength detection the delayed onset is better visible in the native system than in the chemically modified one. The lifetime of the long-lived species is somewhat larger in the native system (1 ns) than in the chemically treated one (0,75 ns). In each case a third component with a lifetime around 200 psec is extracted from the tri-exponential fit. There is general agreement nowadays that this intermediate component is due to chromophore-protein arrangements which are not present in the native phycobilisome. The shortest decay component increases from about 25 psec to 40 psec when going from the native to the PCMS-treated sample. This increase as well as the decrease of the longest lifetime can be rationalised by the fact that the chromophore B84 which is subject to perturbation due to the binding of mercury on cystein 110 (see above) has a lower excitation energy which means faster radiationless deactivation to the ground state and an increased gap for energy transfer from chromophore B155.

There was hope that the long wavelength shift of excitation energy of B84 chromophore upon titration with PCMS would lead to a transition from excitonic coupling to Förster type transfer and thereby help to resolve the question raised above. Since in both systems no significant difference in decay kinetics can be observed, the question must be considered as still pending. This leads to the interesting conclusion that in the energy transfer chain in phycobilisomes both Förster type transfer between pairs of weakly coupled chromophores and coherent transfer between closely lying, strongly coupled pairs of chromophores occurs. What has been said for phycocyanin is namely also true for allophycocyanin [25].

In summary one may state that among the many chromophores found in phycobilisomes there is a clear hierarchy in energy transfer times [23-26]. Energy migration within a trimeric unit occurs on the time scale of tens of picoseconds. Energy transfer between various trimers and hexamers, respectively, occurs on the time scale of a few 100 picoseconds and the time needed for energy transfer from the antenna to the reaction centre is still below 1 nsec.

8. Summary

Nature succeeded in constructing a highly efficient energy collection system by using only one type of chromophores (tetrapyrrole chromophores) and by modifying their properties through interaction with the surrounding protein. The number of different biliproteins and therefore the number of possible intermediates within the energy transfer chain has been increased during the evolution by modification of the protein in which the chromophores are embedded [27], rather than by creating new types of chromophores.

Acknowledgement: That part of the described work which has been performed in the author's laboratorium benefitted much from financial support by Deutsche Forschungsgemeinschaft and Fonds der Chemie.

REFERENCES

[1] Contributions in: H.Scheer and S.Schneider (eds.), Photosynthetic Light-Harvesting Systems, Organization and Function, Walter de Gruyter, Berlin, 1988

[2] E.Mörschel and E.Rhiel, Phycobilisomes and thylakoids: The light-harvesting system of cyanobacteria and red algae, in: J.R.Harris and W.Home (eds.), Electron Microscopy of Proteins, Vol.6: Membranous Structures, Academic Press, London, 1984

[3] H.Zuber
Structure and function of light-harvesting complexes and their polypeptides
Photochem.Photobiol. 42, (1985) 821-844

[4] M.Nies and W.Wehrmeyer
Biliprotein assembly in the hemidiscoidal phycobilisomes of the thermophilic cyanobacterium Mastiglocladus laminosus Cohn. Characterization of dissociation products with special reference to the peripheral phycoerythrocyanin-phycocyanin complexes
Arch.Microbiol. 129 (1981) 374-379

[5] H.Scheer
Phycobiliprotein: Molecular aspects of photosynthetic antenna system. In: F.K.Fong (ed.), Light reaction path of photosynthesis, Springer, Berlin (1986)

[6] G.Wagnière and G.Blauer
Calculations of optical properties of biliverdin in various conformations
J.Am.Chem. Soc. 98 (1976) 7806-7810.

[7] H.Scheer, H.Formanek and S.Schneider
Theoretical studies of biliprotein chromophores and related bile pigments by molecular orbital and Ramachandran type calculations.
Photochem. Photobiol. 36 (1982) 259-272.

[8] A.R.Holzwarth
Structure-function relationships and energy transfer in phycobiliprotein antenna
Physiol.Plant. (1991)

[9] T.Schirmer, W.Bode and R.Huber
Refined three-dimensional structures of two cyanobacterial C-phycocyanins at 2.1 and 2.5 A resolution - a common principle of phycobilin - protein interaction
J.Mol.Biol.196 (1987) 677-695

[10] M.Duerring, G.G.Schmidt and R.Huber
Isolation, crystallization, crystal structure analysis and refinement of constitutive C-phycocyanin from the chromatically adapting cyanobacterium Fremyella diplosiphon at 1.66 Å resolution
J.Mol.Biol. 217 (1991) 577-592

[11] P.Hefferle, W.John, H.Scheer and S.Schneider
Thermal denaturation of monomeric and trimeric phycocyanins studied by static and polarized time-resolved fluorescence spectroscopy
Photochem. and Photobiol. 39, (1984) 221-232

[12] W.Köhler, J.Friedrich, R.Fischer and H.Scheer
Low temperature spectroscopy of cyanobacterial antenna pigments
Ref.1, p.293-306

[13] W.Köhler, J.Friedrich, R.Fischer and H.Scheer
High resolution frequency selective photochemistry of phycobilisomes at cryogenic temperatures
J.Chem.Phys.89 (1988) 871-874

[14] B.Szalontai, V.Csizmadia, Z.Gombos and K.Csatorday
Chromophore conformations in phycocyanin and allophycocyanin as studied by resonance Raman spectroscopy
Ref.1, p.307-316

[15] S.Schneider, F.Baumann, U.Klüter and P.Gege
Resoncance-enhanced CARS spectroscopy of biliproteins
Croat.Chem.Acta. 61 (3) (1988) 505-527

[16] J.Sawatzki, R.Fischer, H.Scheer and F.Siebert
Fourier-transform Raman spectroscopy applied to photobiological systems
Proc.Natl.Acad.Sci.USA 87,(1990) 5903-5906

[17] M.Mimuro, R.Rümbeli, P.Füglistaller and H.Zuber
The microenvironment around the chromophores and its

changes due to the associationn states in C-phycocyanin isolated from the cyanobacterium Mastigocladus laminosus Biochimica et Biophysica acta 851 (1986) 447-456.

[18] S.Siebzehnrübl, R.Fischer and H.Scheer
Chromophore assignment in C-phycocyanin from Mastigocladus laminosus
Z. Naturforsch. 42c (1987) 258-262.

[19] F.Baumann, PhD thesis, Techn.Univ.München (1990)

[20] S.Schneider, P.Geiselhart, F.Baumann, S.Siebzehnrübl, R.Fischer, H.Scheer
Energy transfer in "native" and chemically modified C-phycocyanin trimers and the constituent subunits
Ref.1, p.369-482

[21] C.Scharnagl and S.Schneider
UV-visible absorption and circular dichroism spectra of the subunits of C-phycocyanin. I. Quantitative assessment of the effect of chromophore-protein interaction in the α-subunit
Photochem.Photobiol. B: Biol.3 (1989) 603-614

[22] C.Scharnagl and S.Schneider
UV-visible absorption and circular dichroism spectra of the subunits of C-phycocyanin. II: A quantitative discussion of the chromophore-protein and chromophore-chromophore interaction in the ß subunit
J.Photochem.Photobiol. B: Biol. 8 (1991) 129-157

[23] S.Schneider, P.Geiselhart, S.Siebzehnrübl, R.Fischer and H.Scheer
Energy transfer within PC trimers of Mastigocladus laminosus studied by picosecond time-resolved transient absorption spectroscopy
Z.Naturforsch. 43c (1988) 55-62

[24] P.Geiselhart, PhD thesis, Techn. Univ. München (1988)

[25] A.R.Holzwarth, E.Bittersmann, W.Reuter and W.Wehrmeyer
Studies on chromophore coupling in isolated phycobiliproteins. III. Picosecond excited state kinetics and time-resolved fluorescence spectra of different allophycocyanins from Mastigocladus laminosus
J.Biophys.Soc. 57 (1990) 133-145

[26] K. Sauer and H. Scheer
Excitation transfer in C-phycocyanin. Förster transfer rate and exciton calculation based on new crystal struc-

ture data for C-phycocyanin from Agmanellum quadruplicatum and Mastigocladus laminosus
Biochem.Biophys. Acta, 936 (1988) 157-170

[27] P.Füglistaller, F.Suter and H.Zuber
Linker polypeptides of the phycobilisome from the cyanobacterium Mastigocladus laminosus: amino-acid sequences and relationships
Biol.Chem. Hoppe-Seyler 366 (1985) 993-1001

PHOTOSYNTHESIS : BIOLOGICAL CONVERSION OF LIGHT INTO CHEMICAL ENERGY

P. MATHIS
Département de Biologie Cellulaire et Moléculaire,
Section de Bioénergétique
Bâtiment 532, C.E. Saclay
91191 Gif-sur-Yvette Cedex
France

ABSTRACT. The processes taking place in photosynthetic reaction centers are described with an emphasis on purple bacteria in which the reaction center structure is known well enough to permit detailed studies by photophysical methods and by theoretical treatments of electron transfer. Plant reaction centers and other systems are briefly considered.

1. INTRODUCTION : LIGHT IN BIOLOGY

Within a comprehensive coverage of photoprocesses, starting form basic principles, biological systems have a quite natural place. "Light and living matter" is the title of a famous book (Clayton, 1971) dealing with the way light is absorbed and utilized for several biological reactions. Many progresses have been made since 1971.. The basic photophysics is much better understood, and it also appears that the biological processes which make use of light are much more complex and sophisticated than expected : it is now recognized in fact that such an incredible complexity is a general property of biology.

Before going into a detailed description of photosynthesis, it may be of interest to briefly examine the entire field of photobiology, in order to acquire a superficial view on the importance of light in biology (Smith, 1989). Photobiology can be divided into four domains :

i) Conversion/Storage of energy.

- Photosynthesis is the major process. It relies on photo-induced electron transfer, making use essentially of chlorophyll-like pigments, but also of carotenoids and phycobiliproteins. Photosynthesis occurs in some bacteria and in plant-type eucaryots (including algae).
- Bacteriorhodopsin, a retinal protein, is involved in a specific energy conversion found in some bacteria (Halobacterium sp.). Here, the primary process is H^+ transfer across a membrane, building up an utilizable membrane potential.

ii) Light as a stimulus (Perception/Action)

In this kind of process, light acts as a sensory stimulus : it triggers a molecular modification, which is then amplified in an energy-consuming sequence of events.

- Vision is the best known process in this category. A complex cascade of events is triggered by the absorption of light by a polyene (retinal) bound in a membrane protein (rhodopsin).
- Light-dependent movements, which take place largely in plants and microorganisms. They are quite complex and often poorly understood : phototaxis (kinesis, phobic responses), phototropism, etc. One macroscopic easy-to-observe example is the orientation of flowers (sunflower for instance) with respect to sunlight. The chemical nature of the pigments involved is still a matter of debate (retinals, chlorophylls, flavins, bile pigments ?).
- Photomorphogenesis, in which absorption of light, essentially in plants, triggers various processes such as greening (a plant grown in the dark is usually devoid of chlorophyll ; it contains a precursor, protochlorophyll, which is converted into chlorophyll in a photoreaction), flowering (triggered by a bile pigment, phytochrome, present in very small

E. Kochanski (ed.), Photoprocesses in Transition Metal Complexes, Biosystems and Other Molecules. Experiment and Theory, 333–347.

amounts) or germination (which seems to involve phytochrome and a blue-light receptor, perhaps a flavin).

- Light-dependent rythms, which are found in plants and animals. They are poorly understood. One such rythm is perturbed in the "Jet-lag".

iii) Molecular alterations

While in the previous processes the light receptors are turning on in a cyclic manner, light also induces direct molecular alterations :

- DNA damage, induced by UV radiations. In nucleic acid, bases can be dimerized or oxidized, leading to long-term effects such as mutations or cancers.
- Damage to proteins, lipids and membranes, which is part of the aging of tissues, specially the skin.
- Photosensitizations which are induced by some dyes (naturally occurring like porphyrins, but sometimes present in excessive amounts in some diseases such as porphyria). Light populates the dye triplet state which reacts with oxygen to make the reactive singlet oxygen. This process is used by plants in order to get protection against insects. It is now used in photomedecine and there is much hope in the possibility of curing cancers by phototherapy. It is also used in agriculture (photobleaching herbicides).

This list of examples is not exhaustive. Another well-known photoreaction occurs in the synthesis of vitamin D from its precursors.

iv) Light as analytical tool

Light is now a major tool in cell biology, in particular with the development of fluorescence markers. It is also widely used in biochemistry : fluorescence identification, photo-affinity labelling, etc.

As expected, this long list of photobiological domains goes along with a large number of photosensitive molecules and a large variety of photochemical processes : electron or proton transfer, isomerization, radical reactions, energy transfer, etc.

2. PHOTOSYNTHESIS : GENERAL CONSIDERATIONS

Photosynthesis is the biological process by which plants and some bacteria use the energy of light to build-up their organic molecules. The photochemical reactions are triggered by chlorophylls, which are organized, together with electron carriers (tetrapyrroles, quinones, iron-sulfur centers, ...), into specific, complex proteins named reaction centers. The basic reactions are electron transfers, and the reaction centers can be considered as micro-photocells. Within the context of this book, it may be useful to recall a few of the specific properties of photosynthesis (for reviews, see Amesz, 1987 ; Barber, 1987 ; Govindjee, 1982 ; Hatch and Bardman, 1981 ; Lee, 1991) :

- The process takes place within well-defined structures. This is a general property of biological systems. Sub-cellular organelles (named chloroplasts) or bacteria contain membrane systems which anchor the reactions centers. These are complex proteins with a well-defined rather rigid structure, so that all the active partners have a well-defined relative positioning.
- The photosynthetic apparatus includes two sets of pigments, the functions of which are, respectively, to absorb the light and to transfer excitation energy (the antenna) and to realize electron transfer (the reaction centers). The antenna provides a means for absorbing light very efficiently from below 400 nm to 700 nm (in plants) and up to 1020 nm (in bacteria). It includes several kinds of pigments : chlorophylls, carotenoids, bile pigments. It will not be discussed here (see the chapter by S. Schneider).

- Chlorophylls, which are the essential pigments of photosynthesis (Scheer, 1991) are characterized by a set of unique properties. They have strongly allowed transitions (permitting an efficient light absorption). Their chemical structure includes several groups which allow their non-covalent binding to the protein acting as a scaffolding device : central mamgnesium, several carbonyl groups, and a long hydrophobic flexible tail. These molecules are susceptible of reversible redox reactions (oxidation or reduction) at the potentials required by the biological systems. In addition they are quite stable, at least in their natural environment, and they have a quite rigid planar structure.
- Natural photosynthesis takes place on a very large scale, storing annually around 3×10^{21} J of energy on earth.
- Considering the primary parts of the process, at the level of reaction centers, it is remarquable that the quantum yield of photosynthesis is very high (95 to 100 %, meaning that each absorbed photon results in a stable charge separation), whereas the energetic yield is rather low, from 21 % in bacteria to 50 % in plants. In plants, the overall photosynthetic reaction can be written :

$$6\,CO_2 + 6\,H_2O \xrightarrow{light} C_6H_{12}O_6 + 6\,O_2 \text{ (where } C_6H_{12}O_6 \text{ is an hexose).}$$

This means that electrons are taken from water to reduce carbon dioxyde into a sugar. This process has an overall energetic yield of 8 -9 % under optimum conditions (Bolton and Hall, 1991).

As mentioned earlier, photosynthesis is performed by many classes of organisms, ranging from simple bacteria to higher plants. The process is more complex is those organisms (plants, algae, cyanobacteria) which perform oxygenic photosynthesis, i.e. use water as the ultimate source of electron, and thus evolve O_2. For energetic reasons, they need two types of reaction centers, named PS-1 and PS-2, which operate in series as would do photocells branched in series. These organisms realize the vast majority of the solar energy conversion in the world. Their reaction centers are relatively complex and they will be presented briefly in a second part of this chapter. Most species of bacteria are unable to use water as a source of electrons. Instead, they use reduced sulfur compounds or organic molecules. Each kind of non-oxygenic bacteria includes only one type of reaction center, but altogether they can be classified into four categories : purple, green non-sulfur, green sulfur, and helio-bacteria. Among them, purple bacteria are the best known : their reaction center is (relatively) simple and it will presented here as a starting model for all classes of reaction centers.

3. THE REACTION CENTER OF PURPLE BACTERIA

The way reaction centers work in purple bacteria is illustrated in Fig. 1. Excitation energy is transferred to a pair of bacteriochlorophyll molecules (named special pair or primary donor P), the excitation of which initiates electron transfer to a sequence of redox centers (bacteriopheophytin, quinone Q_A, quinone Q_B). The hole on P is then filled by an electron coming from a reduced cytochrome molecule. The electron path is made cyclic thanks to another protein named b/c complex (which is not part of the reaction center). The cyclic path of electrons serves to pump protons (H^+) through the membrane which hosts the reaction center in a well-defined orientation. The reaction centers can also work in a non-cyclic way in using a source of electrons to reduce a biochemical carrier of electrons (NAD^+) (not shown in Fig. 1).

The existence of reaction centers has been hypothesized well before their biochemical isolation, which was made possible by the use of appropriate detergents, followed by purification steps. Among other interests, this purification led to three classes of approaches :

biochemical analysis, crystallization leading to the elucidation of the 3-D structure, and detailed photophysical studies.

3.1. Biochemical analysis

The reaction center is a complex protein made up of three polypeptides named L, M, H, each of molecular weight about 30 kDa. The amino-acid sequences of these polypeptides is known in detail. Other constituents are four bacteriochlorophylls, two bacteriopheophytins, two quinones with a hydrophobic side chain, and one Fe^{2+} atom. Some reaction centers, such as in Rhodopseudomonas viridis, also have a bound tetraheme cytochrome. How is all this organized ? How does it work ?

3.2. Structure of the reaction center

A lot of structural information was obtained by biochemical methods (for instance from the access of various parts of the polypeptides to proteolytic enzymes or from photoaffinity labelling) and by photophysical methods. For instance, the orientation of the various chromophores relatively to the membrane plane was determined by linear dichroism (Breton and Vermeglio, 198 1). Electron paramagnetic resonance (EPR) was also quite informative in providing approximative distances between redox centers and more importantly in showing conclusively that P is a dimer of bacteriochlorophylls : in P^+, the unpaired electron is delocalized over two bacteriochlorophylls (Norris et al, 1971).

A real breakthrough happened with the crystallization of reaction centers and the elucidation of their 3-D structure at the atomic level by X-Ray crystallography. This was done with the reaction center Rps viridis (Deisenhofer et al, 1985 ; Michel et al, 1986), the structure of which then served to obtain the structure of Rhodobacter sphaeroides (Chang et al, 1986 ; Feher et al, 1989). These studies provided the precise organization of the amino-acid chains of the protein, and the positioning of all the other constituents. There is no room here to detail either the wealth of information contained in the structure or their photophysical implications (see 3.4). Let us already mention, however, that the present level of knowledge is insufficient in at least two respects :

i) the accuracy in atomic positioning is ± 0.2 A at the best. This leaves a large uncertainty, for example in the electronic factors involved for electron transfer. ii) the structure is a "sleeping" one, obtained in the dark, whereas molecular motions and relaxations are very important for electron transfer paths and rates.

3.3. Photophysical studies on reaction centers

Isolated, purified reaction centers revealed to be a fascinating object for sophisticated photophysical techniques. A few aspects can be briefly outlined.

3.3.1. Flash absorption kinetics. In its classical version, with microsecond or nanosecond time resolution, this technique provided the essential features of the successive steps of electron transfer, identifying the redox centers by their difference spectra and determining reaction rates. This tool is still on essential one. The development of picosecond techniques gives access to the real primary step(s), revealing their rapidity and their complexity. Flash absorption is very useful when it is used in conjonction with selective modifications of the reaction center, such as site-directed mutagenesis, revealing the role of a given amino-acid, and such as chemical substitution of redox factors, revealing the importance of their chemical structure on the rate of electron transfer (see e.g. Gunner and Dutton, 1989).

3.3.2. EPR and other magnetic resonance methods. The redox-active species in reaction centers are EPR-silent in their relaxed, neutral state (see, however, 4.2. for the manganese cluster of PS-2). Electron transfer creates radical-anions and cations, and oxidized cytochromes, all of which have caracteristic EPR spectra. Basic informations were obtained on the chemical nature of the redox centers and on the properties of their surrounding. Magnetic resonance techniques have been very much developed and they can provide detailed structural information (see Hoff, 1989).

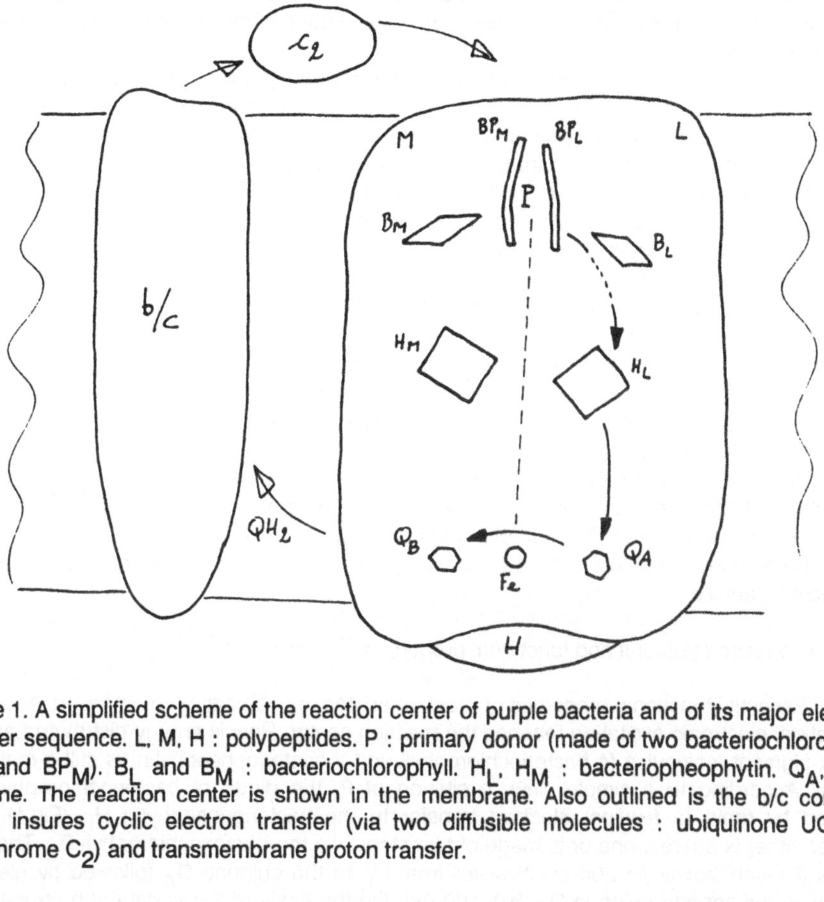

Figure 1. A simplified scheme of the reaction center of purple bacteria and of its major electron transfer sequence. L, M, H : polypeptides. P : primary donor (made of two bacteriochlorophylls BP_L and BP_M). B_L and B_M : bacteriochlorophyll. H_L, H_M : bacteriopheophytin. Q_A, Q_B : quinone. The reaction center is shown in the membrane. Also outlined is the b/c complex which insures cyclic electron transfer (via two diffusible molecules : ubiquinone UQ and cytochrome C_2) and transmembrane proton transfer.

3.3.3.Vibrational spectroscopy. It might seem that detailed studies of vibrational spectroscopy should be hardly feasible in such complicated structures as reaction centers in which the number of vibrational modes is exceedingly large and where essential features may be buried in huge signals due to water, proteins and lipids. Relevant work is now possible by resonance Raman, where one makes use of the wavelength selectivity in scattering excitation, and by differential FT-infra red, using the high sensitivity of the technique to record the spectral differences before and after a light-induced reaction, and comparing the data with those obtained with isolated molecules (see e.g. Lutz and Mäntele, 1991 ; Mattioli et al, 1991 ; Breton et al, 1991 ; Nabedryk et al, 1990). Without detailing the methods and the results, it may be worthwhile to mention why they are important :

i) One of the major problems in the structure of reaction centers is how the cofactors are held by the polypeptides chains in a precise position. X-ray data are very important in that respect, but they are not accurate enough. Potentially, vibrational spectroscopies can determine the chemical groups implicated in binding.

ii) Electron transfer theory predicts that each step of electron transfer is associated with molecular structural changes which contribute to the reorganization energy. It is classically recognized that the vibrational modes involved are those which are different before and after the reaction. These changes implicate the donor and acceptor molecules, and also their "solvent" which, in the case of reaction centers, is made up essentially of the protein. By differential methods, vibrational spectroscopy progresses in understanding these reorganizations, including their kinetic aspects. These informations are practically inaccessible by X-ray crystallography.

iii) The 3-D structure obtained by crystallography requires the availability of good quality crystals. This has been possible in only two kinds of reaction centers. However it is quite illuminating to do comparative studies in the properties of reaction centers from various organisms, including those which are prepared "artificially" by site-directed mutagenesis. Here again vibrational spectroscopies reveal to be extremely useful ; in that respect they behave like many other spectroscopic methods, in which experiments can be performed rather quickly with a small amount of reaction centers.

3.3.4. Absorption and fluorescence methods. These methods are classical in all studies of photophysical and photochemical processes. With respect to reaction centers, it is worth mentioning that some techniques have been greatly refined with the development of instrumentation and the need to answer specific questions (see Mathis, 1990) :

- measurements with polarized light (in absorption or emission), often performed with objects which are oriented physically or by photoselection. Linear polarization is greatly used (Breton and Vermeglio, 1982), but circular dichroism is also helpful ; it can now be performed at the picosecond time scale.
- effects of electric fields are used to probe the charge-transfer character of excited states, and also the geometrical direction of charge displacement in electron transfer.
- hole-burning also contributes to understanding the properties of low-lying excited states in reaction centers.

3.4 How to relate structural and functional properties.

Essential functional properties of reaction centers can be easily summarized. After light absorption, electronic excitation reaches the reaction center. How is energy distributed among the six pigment molecules (4 bacteriochlorophyll and 2 bacteriopheophytin) is still a matter of debate. According to a simple view, excitation of P, the dimer of bacteriochlorophylls, is followed by rapid (a few ps) electron transfer to the bacteriopheophytin H_L (Fig.1). This transfer either is a direct one or is made of two steps, via the bacteriochlorophyll B_L. The next step is a much slower (≈ 200 ps) transfer from H_L to the quinone Q_A followed by electron transfer to the second quinone Q_B (10–100 μs). On the basis of X-ray data, the structure of the reaction center is nearly symmetric (Deisenhofer and Michel, 1989 ; Feher et al. 1989 ; Chang et al. 1986), with an approximate C2 axis running through the dimer P down to a Fe^{2+} atom located between Q_A and Q_B. Electron transfer, however, goes to Q_A through H_L and does not take place through the other side. A few aspects of the structure-function relationships will be discussed here : properties of P ; the primary step of electron transfer ; common properties of electron transfers ; radical-pair mechanism and formation of the triplet state of P ; symmetry/dissymetry of the reaction center ; and a few other properties.

3.4.1. The primary donor P. This species was discovered a long time ago by differential absorption spectroscopy. P has a characteristic absorption maximum in the near infra-red (at 870 or 960 nm in Rhsphaeroides or Rps. viridis, respectively, whence the names P-870 or P-960) which bleaches upon oxidation. The long-wavelength absorption of P was interpreted in the exciton theory as due to two bacteriochlorophylls in interaction. Circular dichroism spectra supported this attribution which then received strong support from the analysis of the EPR spectrum of P^+ by Norris et al (1971) who proposed that P is a "special pair". A rather detailed structure for P^+ was proposed on the basis of ENDOR data (Lubitz et al. 1984). The structure obtained by X–ray diffraction clearly shows that P is made up of two closely spaced bacteriochlorophylls (at an interplane distance of about 3.2 A). The two molecules are not in direct chemical interaction, as also shown by Raman spectroscopy, but they are held in that conformation by the protein. The Mg^{2+} ion of each bacteriochlorophyll of P is pentacoordinated, four ligands being provided by the pyrrole rings and a fifth one by a histidine residue. The complete mode of binding cannot be fully predicted by the X–ray data, but vibrational spectroscopy (resonance Raman and FT–IR) are useful in that respect.

The excited state of P starts the process of photosynthesis and its detailed nature is thus of utmost importance. Stark effect and hole burning experiments conclude to a significant charge–transfer character in P^*, the properties of which are usually analyzed in terms of exciton theory (see Warshel and Parson, 1987). The importance of charge resonance states has been recently emphasized (Thompson et al. 1991).

It is often thought that P is a dimer because that structure offers a low–lying excitonic state and that P is thus an efficient trap for energy. This is certainly not true in general, since in Rps viridis the excited state of P^* (maximum absorption at 960 nm) lies higher in energy than the antenna (maximum at 1014 nm). In photosystem–2, P–680 is nearly isoenergetic with the antenna. Efficient trapping results more probably from dynamic properties, i.e. the very fast electron transfer from P^*. It has also been proposed that internal charge transfer in P^* is important to start electron transfer along the active branch. Recently, however, the chemical nature of P (BChl...BChl) has been modified by site–directed mutagenesis to (BChl...BPheo) or to (BPheo...BChl), where BChl and BPheo are bacterio–chlorophyll or–pheophytin. Both of these modified reaction centers are active. Careful examination permits to conclude that charge transfer character may be detrimental and in any case does not initiate primary electron transfer (see e.g. Mc Dowell et al. 1991). It seems that the reason for the dimeric nature of P is not yet elucidated.

3.4.2. The primary electron transfer step. It is well established that P^* lasts for about 2 ps and that bacteriopheophytin (H_L) becomes reduced in about the same time. This reaction is remarquable in several respects : i) it is very fast, considering the distance between donor and acceptor : 16.5 A center–to–center or 9.5 A edge–to–edge. ii) its temperature dependence shows a progressive acceleration when temperature decreases. In Rps viridis, for instance, the time constant is 0.7 ps at 8 K and 2.7 ps at 294 K (Fleming et al. 1988). iii) the 3–D structure of the reaction center points out to monomeric bacteriochlorophyll (B_L) as an obvious intermediate in electron transfer from P^* to H_L. Several groups, using picosecond spectroscopy, find no evidence for B_L becoming transiently reduced after flash excitation (Martin et al. 1986). Others concluded to a small extent of B_L reduction (Holzapfel et al. 1990). At the present time, two schemes are advocated for the overall reaction :

(a) $P^*B\,H \rightarrow^{3ps} P^+BH^-$. A direct electron transfer would be greatly facilitated by the appropriate positioning of the bacteriochlorophyll B_L. The role of B could take place by superexchange, considering (PBH) as a supermolecule. If the energy of the state (P^+B^-H) is slightly above that of (P^* B H), there will be no chemical reduction of B_L during the reaction.

(b) $P^*BH \rightarrow^{3ps} P^+B^-H \rightarrow^{0.9ps} P^+BH^-$. This sequential mechanism explains a small extent of B^- formation if the second step is faster than the first. Both schemes are not fully exclusive and their respective contributions could be modulated by small perturbations, specially of the energy level of (P^+B^-H) (Bixon et al. 1991). Recently, a mutant was manufactured where H_L is replaced by a bacteriochlorophyll (Kirmaier et al. 1991). It was enough to replace a leucine residue (purely aliphatic, no binding capability) by a histidine (which can coordinate Mg^{2+} of bacteriochlorophyll) to induce the binding of bacteriochlorophyll instead of the Mg^{2+}–less bacteriopheophytin. This work remarkably illustrates how the polypeptide sequence determines the binding of cofactors. Not less remarkably, electron transfer from P^* to the new bacteriochlorophyll takes place nearly as fast as in the wild–type, although the electrochemical properties of both molecules are significantly different.

3.4.3. Properties of the primary radical pair. Triplet state formation. Under normal conditions, H_L^- gets reoxidized by further electron transfer to the quinone Q_A, in a time of about 200 ps. This transfer step cannot take place in several artificial circumstances, such as Q_A pre-reduction or its removal from the reaction center by treatment with a suitable solvent. Under these circumstances, the radical pair (P^+BH^-) lasts for about 10 ns. Its fate and its properties have been studied in detail by photophysical methods (see Budil and Thurnauer, 1991, for a recent review) :

- the radical pair, born as a singlet state, experiences singlet-triplet mixing, induced by hyperfine interactions, which lead it to oscillate between the singlet and triplet manifolds within its lifetime.
- the radical pair eventually recombines. According to its multiplicity, this leads to a singlet state (ground state or P^*) or to a triplet state which remains localized on P (^3P).
- the triplet state ^3P has a distinctive EPR spectrum, showing a preferential population of the To level (these experiments are done in a strong magnetic field of about 3,000 Gauss). The triplet state was also detected by flash absorption and by phosphorescence.

The triplet state revealed to be a nice tool for studying several properties of the reaction center. i) application of a weak magnetic field induces first an increase in ^3P yield, and then a decrease. This effect permits to obtain the singlet-triplet splitting, a parameter which is considered in theoretical treatments of electron transfer from P^* to H (Michel-Beyerle et al. 1988). ii) orientation of the special pair has been probed in great detail, by studying ^3P by EPR in oriented samples (membranes or isolated reaction centers). In a recent development of the method, ^3P was studied in monocrystals and related to the radiocrystallographic data (Norris et al. 1989). Triplet state studies also permitted to study the orientation of P-680 and P-700 in plant reaction centers (Rutherford and Sétif, 1990 , van Mieghem et al. 1991).

It may be worth pointing out that the triplet state of the pigments is normally not populated by intersystem crossing. The primary electron transfer reactions take place from singlet excited states : they are so fast that intersystem crossing has no time to develop. In the antenna a small yield of formation of the triplet state of bacteriochlorophyll (chlorophyll a in plants) takes place. These triplets are rapidly quenched, in a few nanoseconds, by T-T energy transfer to carotenoids. This is an essential protective mechanism which avoids the formation of singlet oxygen.

3.4.4. Effects of temperature. Forward versus back-reactions. In reaction centers, all forward electron transfer steps are much faster than the corresponding reverse reacion (Table 1), insuring a nearly 100 % yield of final state. The reasons for this favorable stituation have to be examined for each case. For example, the ($P^+H^- \rightarrow PH$) back reaction is rather slow because, of the two possible paths, going to P^* or to the ground state, the first one requires a substantial activation energy, while the second has a very large $\Delta G0$. In the case of P^*, it has been proposed that internal charge transfer may increase the rate of radiationless decay, while the rigidity of the macrocycles may decrease its rate, and thus behave as a favorable factor (Mc Dowell et al. 1991 ; Kirmaier et al. 1991).

It is remarkable that temperature has little effect on the intra-reaction center electron transfers. An exception is the transfer from Q_A to Q_B which is dramatically slowed down with decreasing temperature. This reaction is poorly understood. Four reactions display a weak acceleration when temperature is lowered (Table 1). The most straightforward interpretation of this property is that the reactions have a negligible activation energy, i.e. $\Delta G0$ = L (reorganization energy). The small acceleration at low temperature could arise form a contraction or from a more favorable Franck-Condon factor when vibrations are frozen. As pointed out by Gunner and Dutton (1989), the general property that $\Delta G0$ = L would require that the reaction centers are really well engineered. These authors conducted a very important work showing, among other things, that the rate of electron transfer from H_L^- to Q_A remains temperature-independent after rather large changes in $\Delta G0$. They concluded that the temperature independence does not originate in $\Delta G0$ = L, but in the fact that electron transfer is coupled to rather high energy vibrations which are not populated at low temperature. The results demonstrate the role of nuclear tunneling in low temperature reactions.

Table 1. Time constants for electron transfer reactions in the reaction center of Rh.sphaeroides. b.r. : back reaction

	294 K	Low Temperature
$P^* \rightarrow H_L$	4.2 ps	2.3 ps
$H_L \rightarrow Q_A$	230 ps	100 ps
$(P^+ H_L^-)$ b.r.	16 ns	18 ns
$(P^+ Q_A^-)$ b.r.	132 ms	27 ms
$Q_A \rightarrow Q_B$	≈ 100 μs	very long

The case of electron transfer from bound cytochromes to P^+ appears somewhat different. Under conditions where a low-potential heme is the electron donor, the kinetics are slowed between room temperature and ≈ 100 K, displaying an activation energy of 14 kJ. mol^{-1}, and they remain temperature-independent below 100 K. The historical work of De Vault and Chance (1966) was interpreted as a proof of nuclear tunneling at low temperature, but Bixon and Jortner (1989) have recently proposed another interpretation. This field of research is very open and involves several groups of physical chemists and of theoreticians

3.4.5. Symmetric structure/Dissymmetric functioning. Perhaps the most surprising feature in the reaction structure obtained by X-ray crystallography is the occurrence of a C2 symmetry axis which relates the L, M polypeptides as well as the cofactors. The symmetry is not perfect, but it is quite impressive. Among the two possible branches for electron transfer, only one is active, as shown by flash absorption. The inactive branch is at least ten times, but perhaps as much as 200 times less active (Kellog et al., 1989). This situation initiates two types of questions : with the known structure, is it possible to rationalize the dissymmetric functioning ? What is the biological significance (if any) of the existence of an inactive branch, knowing that economy of means is a general law for basic biological functions ?

Several groups have attempted to evaluate quantitatively the effects of the small dissymmetries between the two branches that could influence the rates of electron transfer.

Michel-Beyerle *et al.* (1988) essentially considered the electronic coupling by superexchange which would be much more favorable for B_L than for B_M. Parson *et al.* (1990) gave more importance to the tuning of energy levels of radical-pair states by electrostatic interactions with amino-acid side chains. The essential feature would be a lowering of $(P^+B_L^-)$ compared to $(P^+B_M^-)$ energy. Although those treatments are very worthwhile, it seems that our present knowledge of the structure is not detailed enough to make fully reliable predictions. Site-directed mutagenesis is a complementary approach to the question. A few key amino-acids have been pinpointed on the L or M polypeptide and modified so as to decrease or increase a dissymmetry. For the moment this approach has not yet provided a complete understanding although amazing results, largely negative, were obtained. For instance, two tyrosine residues (M 208 and M 210) are located on the active branch. They were supposed to facilitate electron transfer. The equivalent to M 208 on the inactive branch is a phenylalanine (L 181) ; it has been replaced by a tyrosine, with the consequence of making the active branch primary step faster (2.1 ps instead of 3.5 ps) without improving the inactive one. Robles *et al.* (19) chosed to make reaction centers more symmetric by changing an entire α-helix and leaving the bacteria find functional revertants. This bold approach may pay off in the future.

Why did all purple bacteria develop and keep an inactive branch in a nearly symmetric structure ? It may be speculated that ancestors of bacteria had a simpler reaction center with only one polypeptide. Gene duplication could have occurred, leading to a homo-dimeric structure permitting a better stability. It could then have revealed to be useful to specialize the quinone Q_A (a one-electron carrier which is always bound to its site) and to let Q_B do the job of taking two electrons and two H^+, leave its site and be replaced by another (oxidized) quinone. If only one branch is active (the one which contains Q_A), this specialization could have the advantage of keeping always one quinone ready to accept an electron from the bacteriopheophytin, thus avoiding formation of triplet states in the P^+H^- back reaction. In the present reaction centers, bacteriochlorophyll B_M and bacteriopheophytin H_M may contribute to keep together the L and M polypeptides. Admittedly, most of this is highly speculative...

4. STRUCTURE AND FUNCTION OF PLANT-TYPE REACTION CENTERS

In oxygenic photosynthesis, reaction centers serve to transfer electrons from water to a small iron-sulfur protein named ferredoxin. Oxydation of water results in di-oxygen evolution. A mentioned above, two different reaction centers, named PS-1 and PS-2 have to cooperate for this function. The study of these reaction centers is much less advanced, perhaps for the unique reason that they are much more complex : they include each at least ten different polypeptides. However, it appeared during the last years that all reaction centers are built basically on the same model and work similarly. So the good knowledge acquired on reaction centers from purple bacteria serves as a basis for understanding plant-type ones.

4.1. Photosystem 2

In the PS-2 reaction center, primary reactions take place in a core structure which resembles very much that of purple bacteria (Michel and Deisenhofer, 1988). Two hydrophobic polypeptides, named D_1 and D_2, hold the primary partners : the primary donor P-680, which is presumably a dimer of chlorophyll *a* (see however the discussion by van Mieghem *et al.* 1991), and the series of electron acceptors : a pheophytin *a* and the quinones Q_A and Q_B with an Fe^{2+} ion inbetween. The properties of these acceptors are very similar to those of purple bacteria in many respects : chemical nature of the molecules, electrochemical properties, EPR spectra, kinetics, etc. (see Mathis and Rutherford, 1987). A specially interesting case is offered by inhibitors which bind at the Q_B site in competition with the endogenous quinone. In general the same classes of inhibitors are active on PS-2 and on purple bacteria, in agreement with an important ressemblance between Q_B binding pockets. Significant differences also occur, however, and it is hoped to use them for modelling the Q_B pocket in PS–2 on the basis of the known structure in purple bacteria (see e.g. Sinning et al. 1989).

The donor side of PS–2 is quite unique among photosynthetic reaction centers. Its function is to oxidize water with evolution of molecular oxygen. This reaction requires for the P–680/P–680$^+$ couple a redox potentiel over +1.0 volt, which is highly positive for a biological system. Kinetic experiments showed that P–680$^+$ is re–reduced by a tyrosine residue located on D_1, which in turn oxidizes the active site for water oxidation, including four manganese ions. The structure and function of the water oxidizing catalytic site are of considerable interest and are thus the object of very active research (Babcock et al. 1989 ; Rutherford, 1989). There is a conceptual interest in understanding one basic function required for life, and a potential practical interest in homogeneous catalysis by manganese of water oxidation, in particular with relation to solar energy conversion. The PS–2 site is presently poorly known. Its study brings together the efforts of spectroscopists, who mainly use EPR and EXAFS, of biochemists who study the proteins and the amino–acids involved in the catalytic site, and of chemists who synthesize artificial models, attempting to mimic the properties of the biological system. It seems that no only manganese, but also one amino–acid, a histidine residue, participate in the storage of oxidizing equivalents (Boussac et al. 1990). The structure of the manganese cluster is difficult to resolve. Possible models include cubic structures, or two dimers, or one dimer and two atoms remaining separate. It seems probable that the manganese ions are in the Mn (III) and/or Mn (IV) states. The mechanism of water oxidation is not known.

The crystallization of PS–2 has been attempted, with a very limited success however. A major difficulty results from the complexity : twelve different polypeptides have been shown to be present in PS–2 ... Their roles are not easy to understand : two of them (D_1, D_2) hold the primary partners, two others comprise chlorophyll a molecules and are thus considered as antenna, but all the other polypeptides have an unknown function. In addition to the twelve, three extrinsic polypeptides serve to define a pocket where Mn is bound to the core, and where Ca^{2+} and Cl^- ions are trapped, serving to the function of the catalytic site.

4.2. Photosystem 1.

The PS–1 reaction center receives electron from PS–2 via plastoquinones, a membrane complex of cytochromes b/c and a soluble copper protein named plastocyanim (this insures the "in series" arrangement of plant reaction centers). The PS–1 photoreactions (Mathis and Rutherford, 1987 Golbeck and Bryant, 1991) lead to the reduction of a small iron–sulfur protein named ferredoxin which serves several cellular functions, the most important being to reduce NADP which itself provides the reducing equivalents for the reduction of CO_2 into sugar. The basic modes of organization and of functioning of PS–1 are like other reaction centers, but they differ in two main respects :

i) the primary partners are held by two large polypeptides of about 82 kDa each, which carry a large number of pigment molecules (at least 40 chlorophyll a). In addition PS–1 includes, like PS–2, a large number of small polypeptides, the functions of which are largely unknown. The 3–D structure is far from being solved, although crystals have been obtained.

ii) the electron acceptors have a low redox potential, as required for the reduction of ferredoxin (Em = – 0,42 V). The sequence of acceptors includes one chlorophyll a (equivalent to the pheophytin a in PS–2), one naphtoquinone, one Fe–S center named F_X, and an iron–sulfur protein with two 4 Fe – 4 S clusters named F_A and F_B. As an example, Fx as an Em of about – 0,70 V.

Most of the properties of PS–1 have recently been found in two classes of bacteria : the green sulfur bacteria and a nearly discovered family named heliobacteria. All of these bacteria live in very reducing environments where molecular oxygen is completely absent.

Compared properties of reaction centers may lead to speculations concerning the evolution of photosynthetic organisms (Mathis, 1990 ; Nitschke and Rutherford, 1991).

5. CONCLUSIONS

Reaction centers are fascinating objects for biochemists and photophysicists. In relation with theory, purple bacteria are known well enough to permit very significant treatments. These systems are very complex, however, and thus require appropriate methods (for the theory of photophysics and of electron transfer). A constant feedback takes place between theory and experiments (spectroscopy, biochemistry, X-ray crystallography). It seems that reaction centers, where reactions can be started with a pulse of light, are the best system for confronting theory and experiments related to electron transfer in proteins.

Apart from purple bacteria other reaction centers are too complex and too poorly known for a direct successful theoretical treatment. Many things can be said, however, by reference to the best known case of purple bacteria, and relying on the existence of a few basic common properties of reaction centers. In these cases also, theory is extremely useful for dealing with local problems such as electron transfer in small metalloproteins (plastocyanin or ferredoxin) and with interpretation of spectroscopic data.

REFERENCES

Amesz,J. (1987) "Photosynthesis" (New Comprehensive Biochemistry, Vol. 15) (collective). Elsevier, Amsterdam

Babcock,G.T., Barry,B.A., Debus,R.J., Hoganson,C.W., Atamian,M., Mc Intosh,L., Sithole,I. and Yocum,C.F. (1989) Biochemistry 28, 9557–9565

Barber,J. (1987) "Topics in Photosynthesis. Vol. 8 : The Light Reactions" (collective). Elsevier, Amsterdam

Bixon,M. and Jortner,J. (1989) Chem. Phys. Lett. 159, 17–20

Bixon,M. and Jortner,J. (1989) Photosynth. Res. 22, 29–37

Bixon,M., Jortner,J. and Michel-Beyerle,M.E.(1991) Biochim. Biophys. Acta 1056, 301–315

Bolton,J.R. and Hall,D.O. (1991) Photochem. Photobiol. 53, 545–548

Boussac,A., Zimmermann,J.L., Rutherford,A.W. and Lavergne,J. (1990) Nature 347, 303–306

Boxer,S.G.(1990) Ann. Rev. Biophys. Biophys. Chem. 19, 267–299

Breton,J. and Vermeglio,A. (1982) in "Photosynthesis Vol.1 : Energy Conversion by Plants and Bacteria" (Govindjee, ed.), pp.153–194. Academic Press, New York

Breton,J., Thibodeau,D.L., Berthomieu,C., Mäntele,W., Verméglio,A. and Nabedryk,E. (1991) FEBS Lett. 278, 257–260

Budil,D.E. and Thurnauer,M.C. (1991) Biochim. Biophys. Acta 1057, 1–41

Chang,C.H., Tiede,D. Tang,J., Smith,U., Norris,J. and Schiffer,M. (1986) FEBS Lett. 205, 82–86

Chan,C.K., Chen,L.X.Q., Di Magno,T.J., Hanson,D.K., Nance,S.L., Schiffer,M., Norris,J.R. and Fleming,G.R. (1991) Chem. Phys. Lett. 176, 366–372

Clayton,R.K. (1971) "Light and living matter", Mc Graw Hill, New York

Creighton,S., Hwang,J.K., Warshel,A., Parson,W.W. and Norris,J.R. (1988) Biochemistry 27, 774–781

De Vault,D. and Chance B. (1966) Biophys.J. 6, 825–847

Deisenhofer,J., Epp,O., Miki,K., Huber,R. and Michel,H. (1985) Nature 318, 618–624

Deisenhofer,J. and Michel,H. (1989) EMBO J. 8, 2149–2169

Di Magno,T.J., Bylina,E.J., Angerhofer,A., Youvan,D.C. and Norris,J.R. (1990) Biochemistry 29, 899–907

Feher,G., Allen,J.P.,Okamura,M.Y. and Rees,D.C. (1989) Nature 339, 111–116

Finkele,U., Lauterwasser,C., Zinth,W., Gray,K. and Oesterhelt,D. (1990) Biochemistry 29, 8517–8521

Fleming,G.R., Martin,J.L. and Breton,J. (1988) Nature 333, 190–192

Golbeck,J.H. and Bryant,D.A. (1991) in "Current Topics in Bioenergetics, vol. 16" (C.P. Lee, ed.), pp. 83–177, Academic Press, New-York

Govindjee (1982) "Photosynthesis. Vol.1 : Energy Conversion by Plants and Bacteria" (collective). Academic Press, New York

Gunner,M.R. and Dutton,P.L. (1989) J. Am. Chem. Soc. 111, 3400–3412

Hatch,M.D. and Boardman,N.K. (1981) "The Biochemistry of Plants. Vol. 8 : Photosynthesis" (collective). Academic Press, New York

Hoff,A.J. (1989) "Advanced EPR in Biology and Biochemistry", Elsevier, Amsterdam

Holzapfel,W., Finkele,U., Kaiser,W. Oesterhelt,D., Scheer,H., Stilz,H.V. and Zinth,W. (1990) Proc. Natl. Acad. Sci. USA 87, 5168–5172

Kellog,E.C., Kolaczkowski,S., Wasielewski,M.R. and Tiede,D.M. (1989) Photosynth. Res. 22, 47–59

Kirmaier,C., Gaul,D., De Bey,R., Holten,D. and Schenck,C.C. (1991) Science 251, 922–927

Kirmaier,C., Holten,D., Bylina,E.J. and Youvan,D.C. (1988) Proc. Natl. Acad. Sci. USA 85, 7562–7566

Lee,C.P. (1991) Current topics in Bioenergetics, Vol. 16 (collective). Academic Pres, New-York

Lubitz,W., Lendzian,F., Scheer,H., Gottstein,J., Plato,M. and Möbius,K. (1984) Proc. Natl. Acad. Sci. USA 81, 1401-1405

Lutz,M. and Mäntele,W. (1991) in H. Scheer (ed.), "Chlorophylls", CRC Press, Boca Raton, pp. 855-902

Mäntele,W.G., Wollenweber,A.M., Nabedryk,E. and Breton,J. (1988) Proc. Natl. Acad. Sci. USA 85, 8468-8472

Martin,J.L., Breton,J., Hoff,A.J., Migus,A. and Antonetti,A. (1986) Proc. Natl. Acad. Sci. USA 83, 957-961

Mathis,P. (1990) in "Methods in Plant Biochemistry, Vol. 4" (J. Bowyer, ed.) pp. 231-258. Academic Press, New York

Mathis,P. (1990) Biochim. Biophys. Acta 1018, 163-167

Mathis,P. and Rutherford,A.W. (1987) in "Photosynthesis" (New Comprehensive Biochemistry, Vol. 15) (J. Amesz, ed.), pp. 63-96. Elsevier, Amsterdam

Mattioli,T.A., Hoffmann,A. Robert,B. Schrader,B. and Lutz,M. (1991) Biochemistry 30, 4648-4654

Mc Dowell,L.M., Kirmaier,C. and Holten,D. (1991) J. Phys. Chem. 95, 3379-3383

Mc Lendon,G. (1988) Acc. Chem. Res. 21, 160-167

Michel,H. and Deisenhofer,J. (1988) Biochemistry 27, 1-7

Michel,H., Epp,O. and Deisenhofer,J. (1986) EMBO J. 5, 2445-2451

Michel-Beyerle,M.E., Bixon,M. and Jortner,J. (1988) Chem. Phys. Lett. 151, 188-194

Michel-Beyerle,M.E., Plato,M., Deisenhofer,J., Michel,H., Bixon,M. and Jortner,J. (1988) Biochim. Biophys. Acta 932, 52-70

Nabedryk,E., Bagley,K.A., Thibodeau,D.L., Bauscher,M., Mäntele,W. and Breton,J. (1990) FEBS Lett. 266, 59-62

Nagarajan,V., Parson,W.W., Gaul,D. and Schenck,C. (1990) Proc. Natl. Acad. Sci. USA 87, 7888-7891

Nitschke,W. and Rutherford, A.W. (1991) Trends in Biochem. Sci. 16, 241-245

Norris,J.R., Budil,D.E., Gast,P., Chang,C.H.,El-Kabbani,O. and Schiffer,M. (1989) Proc. Natl. Acad. Sci. USA 86, 4335-4339

Norris,J.R., Uphaus,R.A., Crespin,H.L. and Katz,J.J. (1971) Proc. Natl. Acad. Sci. USA 68, 625-628

Parson,W.W., Chu,Z.T. and Warshel,A. (1991) Biochim. Biophys. Acta 1017, 251-272

Rutherford,A.W. (1989) Trends in Biochem. Sci. 14, 227-232

Rutherford,A.W., and Sétif,P. (1990) Biochim. Biophys. Acta 1019, 128–132

Scheer,H. (1991) "Chlorophylls", CRC Press, Boca Raton

Sinning,I., Michel,H., Mathis,P. and Rutherford,A.W. (1989) FEBS Lett. 256, 192–194

Smith,K.C. (1989) "The Science of Photobiology", Plenum Press, New York

Thibodeau,D.L., Nabedryk,E., Hienerwadel,R., Lenz,F., Mäntele,W. and Breton,J. (1990) Biochim. Biophys. Acta 1020, 253–259

Thompson,M.A., Zerner,M.C. and Fajer,J. (1991) J. Phys. Chem. 95, 5693–5700

Van Mieghem,F.J.E., Satoh,K. and Rutherford,A.W. (1991) Biochim. Biophys. Acta 1058, 379–385

Vos,M.H., Lambry,J.C., Robles,S.J., Youvan,D.C., Breton,J. and Martin,J.L. (1991) Proc. Natl. Acad. Sci. USA (in press)

Warshel,A. and Parson,W.W. (1987) J. Am. Chem. Soc. 109, 6143–6163

Warshel,A., Creighton,S. and Parson,W.W. (1988) J. Phys. Chem. 92, 2696–2701

Wasielewski,M.R., Gaines,G.L., O'Neil,M.P., Svec,W.A. and Niemczyk,M.P. (1990) J. Am. Chem. Soc. 112, 4559–4560

TESTING PRIMARY CHARGE SEPARATION IN PHOTOSYNTHETIC REACTION CENTERS WITH EXTERNAL ELECTRIC FIELDS

A. Ogrodnik and M.E. Michel-Beyerle
Institut für Physikalische und Theoretische Chemie, Technische Universität München, D-8046 Garching

ABSTRACT The mechanism of primary charge separation in photosynthetic reaction centers is yet unclear. The results and shortcomings of directly tracing this process spectroscopically are outlined. Then the expected influence of an electric field on this process is discussed. The method of determining the orientation of the dipole moment of the primary radical pair from the **D**ichroic **E**xcitation spectrum of e**L**ectric **F**ield modulated **Y**ield of the prompt fluorescence (DELFY) will be introduced promising direct access to the identity of the primary electron acceptor. The results of low temperature steady state DELFY experiments on RCs from *Rb.sphaeroides R-26* are presented pointing to single step formation of P^+H^-. The implications of time resolved fluorescence and absorption measurements in an electric field on this result are demonstrated and discussed in the frame of various models.

1. INTRODUCTION

The intention of this article is to outline the present knowledge of the primary charge separation step in photosynthesis. We will give a short summary of the basic results directly obtained from transient optical spectroscopy with the aim to point out the limitations and shortcomings still depriving us from a convincing final view of the primary electron transfer process. In search of a potential way out, the possibility of influencing primary charge separation by an external electric field will be demonstrated. Again, however, crucial drawbacks of the experimental approaches accomplished yet will have to be discussed in detail together with new important consequences for the kinetic scheme of primary processes. Finally promising experimental options for the future will be delineated.

1.1 THE REACTION CENTER

COMPOUNDS: The reaction center (RC) is the smallest photosynthetic entity still fully capable of primary photochemistry [1]. Chromatophores, the next larger unit, in general abundantly contain antenna pigments dominating the characteristics of optical spectra. With the exception of recently found mutants lacking antenna pigments [2] this was a severe obstacle for transient optical spectroscopy of the primary events. The isolation of the RC of the photosynthetic bacterium *Rb. sphaeroides* 20 years ago [3] was a breakthrough opening the door to an enormous wealth of literature on its function and dynamics as an electron transfer system [4-10]. Since we want to pursue the objective of this paper in an exemplary way, we will refer only to RCs of *Rb. sphaeroides*, the most exhaustively investigated system, if not mentioned otherwise. The isolated RC consists of three polypeptides in which six pigment molecules (4 bacteriochlorophylls (B), 2 bacteriopheophytins (H) and two quinones (Q)), the prosthetic groups, are embedded together with a divalent nonheme iron atom [11].

E. Kochanski (ed.), Photoprocesses in Transition Metal Complexes, Biosystems and Other Molecules. Experiment and Theory, 349–373.

STRUCTURE: It took more than a decade before RCs of *R. viridis* could first be crystallized [12] and made accessible for x-ray structure analysis [13,14]. According to these structure data pairs of pigments are each arranged in an approximate C_2 symmetry around an axis perpendicular to the membrane [15]. Near the periplasmic membrane surface two B approach one another very closely forming a dimer P. We will see that P is the primary electron donor. The remaining cofactors form two chains called the A- and B-branch, which lead to the opposite side of the membrane. On each branch the pigments follow in the succession B-monomer, H and quinone, offering themselves as possible intermediary electron acceptors in a sequence of electron transfer steps.

SPECTRAL CHARACTERISTICS: The absorption spectra of B in organic solvent (diethylether/methanol) is characterized by three bands, the Q_y (773nm/772nm), Q_x (577nm/607nm) and the soret band (358nm/365nm) [16]. The Hs have corresponding spectra being blue shifted [17]. These major bands can be found again *in vivo* in the RCs spectrum [1]. Two Bs and two Hs contribute to the Q_y bands at 800 nm and at 760 nm, respectively. The Q_x bands of all four Bs are around 600 nm, while they split for the Hs in 543 nm and 530 nm at the A- and B-branch, respectively. The considerable shifts particularly affecting the Q_y bands partially result from the interaction with specific amino acid residues in the microenvironment of each pigment (hydrogen bonding, ligation, electrochromic effects, geometrical distortion [18]), and partially from pigment-pigment interaction due to their close location in the well defined pigment pockets. On the basis of the x-ray structure data the latter have been treated as excitonic interactions [19-22] which are partially responsible for the redshift of the 860 nm band of the dimer P. In fact, these interactions extend over the whole set of closely spaced pigments inducing not only spectral shifts, but interchanging absorption strength among the bands (intensity borrowing). Further differences in the *in vivo* and *in vitro* spectra relate to the bandwidth, for instance, of Q_y of the B-monomers, being somewhat narrower in the first case thus emphazising the specifity of the protein environment. Furthermore, the "background" absorption in between the bands is by at least a factor of two larger than *in vivo.* After all, the minimum absorption of the unspecific "background" at about 640 nm amounts to about 10% of the largest absorption measured at 800 nm at room temperature. Thus, while the gross features of the absorption spectra can very well be assigned, smaller contributions from the various pigments are spread throughout the spectrum and are difficult to trace. This, together with further difficulties discussed in the next chapter, will impede the thorough interpretation of stationary and transient spectral changes induced by redox changes of the various pigments.

1.2. TRANSIENT ABSORPTION SPECTROSCOPY OF CHARGE SEPARATION

The progress made within the years in detecting fast and ultrafast processes on the basis of the accompanying absorption changes reflects a fantastic development of laser techniques improving spectral flexibility, sensitivity and time resolution. In the following we want to highlight the most essential accomplishments of transient spectroscopy of primary charge separation and dwell on the question of the identity of the primary electron acceptor, an issue which is still under debate.

THE PRIMARY DONOR: The essential spectral features of the primary donor P can be traced back to 1952, when the first reversible light-induced absorption changes in chromatophores (photosynthetic unit including antennas) were detected comprising a bleaching at 860 nm and at 812 nm together with an absorption increase at 790 nm [23,24]. They where closely associated to a light induced ESR signal [25,26,8]. Both optical and ESR signals could also be obtained in redox titrations at potentials above +450 mV [27] indicating the formation of a cation radical. On basis of EPR and ENDOR measurements it was shown later that this cation formed a B-dimer [10,28,29]. It remained unclear, whether this cation radical is a primary reactant emerging immediately on excitation or a secondary oxidation product of the light reaction. This issue provoked the first application of the novel laser-technology to photosynthetic systems demonstrating that bleaching of the 860 nm absorption occurs in less than a microsecond [30]. The full spectral characteristics of the light induced state could be revealed after isolation of the RCs in 1967 [3], since now strong signal distortions due to light scattering from the large particles and due to fluorescence from the antenna were eliminated. In chromatophores the overwhelming absorption of the antenna buries the absorption of the RC cofactors, so that actinic illumination induces only a 2% absorption change at 860 nm as compared to a bleaching of about 88% in the isolated RCs. Additional bleaching at 600 nm and in the Soret band, as well as absorption increase at 435 nm and 1250 nm could now be detected.

INTERMEDIATE ELECTRON ACCEPTORS: First significance of an acceptor anion corresponding to the donor cation was found after long search in ESR experiments [31]. Again RC isolation was the prerequisit for the final identification as a semiquinone radical by ESR [32,33] and optical spectroscopy [34,35]. In case this quinone was prereduced before excitation of the RCs with a nanosecond laser pulse, another, intermediate acceptor could be detected on this time scale. Due to the additional bleachings appearing at 545 nm and 765 nm (the Q_x and Q_y bands of the H), and a broad absorption increase around 665 nm an assignment to H was made [17]. In preparations without blocking of the electron path to the quinone the state P^+H^- was found to transiently exist only on the picosecond time scale [37-39]. The selective bleaching of only one of the Q_x bands of H at 546 (the band at 533 nm remaining unchanged) gave evidence that only one of the pigment branches was active in charge separation. Since loss of the quinone at the B-branch does not impede reduction of the quinone at the A-branch which is bound more tightly it was concluded that charge separation preferentially occurs via this pathway. This notion was assured by the consistency of the orientation of the A-branch H with dichroism of its bleaching [40].

THE PRIMARY ELECTRON ACCEPTOR: The first ps-studies detected bleaching of H within the available temporal resolution. However, the role of the B-monomer molecule as an intermediate electron acceptor has been in discussion for a whole decade [41,42]. When X-ray structural data exhibited the large distance between P and H (17A center to center) which seemed incompatible with electron transfer within few picoseconds, this question became even more pressing. Furthermore the location of the monomer B being close to both P and H became apparent. From that an electron transfer from the excited state P^* to B and subsequently to H in two steps seemed obvious [41-48]

$$P^* \rightarrow P^+B^- \rightarrow P^+H^- \qquad (1)$$

The fact, that intermediate formation of a B anion did not show up in transient measurements does not rule out such a two step pathway, since the transient population of could become arbitrarily small, if the depopulation of the expected intermediate P^+B^- is considerably faster than its population. On the other hand direct electron transfer over the large distance to H in one single step

$$P^* \rightarrow P^+H^- \qquad (2)$$

became understandable from a theoretical point of view, when enhancement of the coupling between the donor and the remote acceptor by a superexchange mechanism mediated by B was invoked [49-60]. In such a case P^+B^- is only populated "virtually" in the sense of quantum mechanical mixing with the initial and final state of the electron transfer step. In order to comply with magnetic field dependent recombination measurements [9,52,61] a relaxation process affecting the electronic coupling between formation and recombination of P^+H^- has to be considered [59]. Both electron pathways may also be utilized simultaneously [62] and may be formulated as extremal cases of more generalized theoretical treatments

[63-67].

In the decade following first detection of P^+H^- as an intermediate, the signal to noise ratio of transient absorption measurements was improved considerably. The spectral range was expanded eventually covering the whole visible and near-infrared region and giving access to all the absorption bands (with exception of the soret region). The energy of the actinic pulse was reduced in order to avoid strange nonlinear effects [68]. Its wavelength was adjusted as to directly excite the lowest energy transition of the special pair. Thus transient signals due to energy transfer amoung the pigments could be ruled out leaving only those originating from processes associated with electron transfer. The application of photoselection techniques allowed further orientational discrimination of the individual pigments involved in the transient states by means of dichroic absorption changes [69]. In a series of comparative measurements performed throughout the temperature range and in different matrices significant effort has been invested to better understand the details of the obtained transient spectra. By monitoring the stimulated emission of P^* the excited state could spectrally be observed in a region with no groundstate absorption from other pigments and gave direct approach to its lifetime [70]. With the new fs-technology time resolution was boosted to better than 150 fs [71]. Thus the lifetime of P^* became well resolvable yielding values between 2.8 ps and 5 ps, depending on preparation conditions, and decreasing with temperature [72-75]. All these measurements failed in detecting a possible intermediate between P^* and the H acceptor since all spectral changes seemed to follow one single time constant [72-77].

SPECTRAL CHARACTERISTICS OF P^+B^-: As mentioned earlier, the absorption bands at 600 nm and 800 nm are characteristic of the B-monomer, while its anion is expected to absorb somewhere around 680 nm [17]. In the following a closer inspection will reveal the particular difficulties of spectrally proving formation of P^+B^-. The band at 600 nm immediately disqualifies itself, since it cannot be discriminated from the Q_x band of the dimer. On one hand P^* and P^+ formation causes significant bleaching of this band. On the other hand these excited states have broad absorptions here and throughout the accessible spectral range.

The most prominent band of the B-monomer at 800 nm coincides with a spectral feature which kept puzzling workers from the very beginning of studying light induced absorption changes. It consists of an "absorption" part at 790 nm and a "bleaching" part at 810 nm. The upper exciton band of the special pair has been suspected to be responsible for the "bleaching" part of the signal. Since the 800nm band has been disclosed with the isolation of the RC the close relation to this band became apparent and opened the way for interpretation as a light induced blue shift. It is not clear which of the indiviual B-monomers contributes to the 800 nm blue shift and to what extent. Furthermore, we do not know how large other contributions to the 800 nm absorption changes are. In polyvinyl alcohol films at low temperatures various bands can be discerned, which coalesce under other conditions. None the less various details remain unclear, e.g. while the expected bleaching of the Q_y band of H seems to be too weak in the P^+H^- difference spectrum a distinct absorptive peak at 765 nm appears in the P^+Q^- spectrum, the origin of which is obscure. Linear dichroism at 812 nm seems to rule out considerable contribution of the upper exciton band of P at this wavelength, in accord with structure based calculations stating that this band should have little oscillator strength. Dichroism at the transition from "bleaching" to the "absorptive" part of the spectrum around 800 nm is complex and not understood.

Witt and Junge have early recognized that the electric field resulting from the separated charges are the physical basis of light induced spectral shifts observed in photosynthetic material [78]. It is very strong and has a fixed orientation within the RC. Thus it might very well give rise to considerable Stark effects which could fully account for the 800 nm blue shift. This notion is supported by the fact, that the blue shift of the B-monomers is more pronounced in the state P^+H^-Q than in the state P^+HQ^- or in the state P^+BHQ since in the first case the electric field on B is largest. Similarly red shifts of the H-bands at 760 nm and the 533 nm have been observed in the state P^+BHQ^- and in the state $PBHQ^-$ [79]. However, some calculations also question such strong electrochromic effects and attribute the signal to spectral rearrangment after loss of excitonic coupling to the bleached special pair ground state [80]. Indeed, the dispersive feature around 800 nm is nonconservative and considerably broader in its absorptive blue wing. Most probably both effects have to be regarded.

Temperature dependent changes of steady state and time resolved absorption spectra give rise to further complications. As an example, temperature dependent changes of the 800 nm blue shift feature (in blocked RCs) were interpreted to reflect the superimposed bleaching of a monomer B due to Boltzmann equilibrium between P^+H^- and P^+B^- which was said to be 0.025 eV higher in energy [81]. However, a measureable thermal admixture of P^+B^- can be ruled out, since according to the magnetic field dependent recombination dynamics of the P^+H^- state P^+B^- must definitively be at least 0.17 eV higher in energy [82-84]. From this we learn that one should be very careful in deducing physical meanings from comparatively small changes in the transient spectra. Recalling the mutual interplay of the excitonically coupled pigments and considering the significant effects of spectral narrowing and shifting in the ground state absorption spectra, a temperature dependence of the transient spectra is not surprising. Finally we are in a complete lack of knowledge on influence of conformational changes of the protein or of nuclear relaxations within the pigments on the absorption properties, their temperature dependence and their time behaviour.

The exact position of the B^- absorption band in the spectral "gap" between 600 and 760 nm is not known. From in vitro spectra, one would expect it to be somewhat red shifted with respect to the broad H^- band around 665 nm. It is clear, however, that both absorptions cannot be discriminated easily from oneanother.

EVIDENCE FOR TRANSIENT P^+B^- FORMATION: The ultimate of all the described accomplishments in ultrafast transient absorption spectroscopy was necessary for the first experimentally reliable signal, which could be indicating the intermediate formation of the state P^+B^- [85-87].

Evidence was based on a 0.9 ps transient with very small amplitude, which is most evident at 785 nm where its amplitude appears to have the opposite sign of the 3.5 ps component. From the transient at this wavelength alone, of course, a specific assignment to the 0.9 ps component cannot be made. Therefore a complete set of transients has been accumulated scanning the spectral range between 500 nm and 1000 nm. Since the 0.9 ps time constant does not emerge in the stimulated emission, it was concluded that the associated kinetic step has to succeed the 3.5 ps step, being the first charge separation step. Thus a global analysis of the whole set of transients with 4 different time constants was performed, accounting for the P^* state (3.5 ps), an unknown intermediate (0.9 ps), the P^+H^- state (220 ps) and the final state P^+Q^- (the lifetime of $\simeq$ 100 ms effectively is infinity in the analysed time window). According to such an analysis, the 0.9 ps component is spectrally characterized by a bleaching at 800 nm without a dispersive shape and a significantly stronger absorption at 665 nm than in the 3.5 ps transient if excitation and probing beam are polarized parallel. Measurements at perpendicular polarisation revealed that the apparent absorption change at 665 nm originates from a dramatic change of dichroism with time. According to these data, the relevant transition moment of the 0.9 ps component has an angle of 36° with respect to the 860 nm transition moment and changes to 68° for the 3.5 ps component. Comparing with crystal structure data these angles are compatible with 29° and 73° expected for B^- and H^- anions respectively.

The main difficulty in dealing with these results and the suggested kinetic scheme originates from the small value of the maximal transient concentration of the unknown intermediate state being less than 20%. Considering the complexity of the transient spectra around 800 nm it is risky to rely soly on such small transient changes. In fact, grossly similar experimental data obtained in this spectral range have called forth an alternative interpretation [88]. In this case the possibility of an additional fast component was ignored. While the spectral dependence of the transient traces at early times seemed to reflect the interplay of the two spectrally independent time constants with changing amplitudes, if processed with the appropriate constraints by Holzapfel et.al. [85] it appeared to exhibit a significant wavelength dependence of the apparent charge separation rate between 740-770 nm and 785-825 nm, if fitted only with a single time constant, which was allowed to change with wavelength, as done by Kirmaier et.al. [88]. Conclusion was made on an inhomogeneous distribution of RCs, in which different charge separation rates have to correlate with different P^+H^- difference spectra.

Evidence for P^+B^- derived from the fast dichroism change observed at 665 nm seemed more promising. It has been pointed out, however, that the excited state absorption of P^* together with its linear dichroism is not known explicitly and may be responsible for the change of dichroism [89]. Furthermore different authors do not agree on the amount of the dichroic change [86,89,90].

With the intention of avoiding difficulties due to the absorption of P^* transient bleaching of H was studied in its Q_x band at 545 nm, in the hope that contributions from P^* can be separated more easily because both signals have opposite sign. Comparison of the transients measured at this wavelength with the stimulated emission signal at 926 nm revealed somewhat different kinetics, the H^- formation appearing to be delayed with respect to P^* decay [89]. This supposed gap between decay of P^* and formation of H^- was filled by assuming a fast intermediate. The claim on intermediate P^+B^- formation is not backed up by direct spectral evidence. Merely the occurrence of an intermediate process, not even necessarily being an electron transfer event, has been proven.

In conclusion we have seen that all investigators agree that spectral dynamics within the first few picoseconds are not consistently single exponential throughout the spectral range. The advocates of P^+B^- formation fail to prove unambiguously the transient existence of P^+B^- while their critics can merely demonstrate the ambiguity of the data, but cannot prove that P^+B^- definitively is not created as a distinct intermediate.

TRANSIENT PROCESSES OTHER THAN ELECTRON TRANSFER: Various processes might be responsible for complex temporal response of the absorption transients. On the investigated timescale excitation energy deposited into the excited state by a coherent pulse has to be regarded as a wavepacket. Its interaction with a coherent probing pulse may give rise to various nonlinear coherent effects, which may appear as damped oscillatory features in the transients. Additionally to these and other purely nonlinear optical effects, electron transfer may have to be considered as a coherent rather than a stochastic process and may deviate from exponential behaviour if dephasing is slow and if coupling to the charge separated state is strong [64,65,91,92] or if motion of the excited state wavepacket along the vibrational coordinates can be assumed to be coherent. The occurrence of such oscillations in the stimulated emission at low temperatures indeed has been claimed [93,89].

Dynamic competition between vibronic relaxation and electron transfer may lead to complex kinetics. According to hole burning studies [94] lifetimes of P^* excited in the zero phonon band match perfectly with the lifetimes measured in stimulated emission from the vibrationally excited P^*. This implies that at low temperature ultrafast electronic relaxation does not occur, and that vibrational thermalisation is fast compared to electron transfer.

Breaking of excitonic coupling due to ground state bleaching of P will induce rearrangment of the excitonic system of the remaining pigments which may be reflected in transient features between 740 and 820 nm. Similarly conformational relaxation of the protein matrix after excitation (induced by the strong charge transfer character of P^* [95-98]) or after electron transfer may lead to additional transients, in particular in sections of the spectra being sensitive to the local electric field. Very recently such transients indeed have been published [99].

A hint on a process influencing the electronic structure of P^* between excitation and electron transfer may be the small cross section of stimulated emission at 920 nm. It is about a factor of 5-10 smaller than that for the P^* bleaching at 860 nm [85]. Since the Einstein coefficients weighted with the level degeneracy are equal for absorption and stimulated emission, the integral over both bands should be equal. According to [85] this is not the case, as long as fortuitious compensation of the emissive signal with an excited state absorption band matching in wavelength is discarded. Thus one would conclude on relaxation influencing the oscillator strength of the transition. Stark effect measurements have revealed a considerable electrical dipole moment of the excited state [95-98]. It is conceivable that such charge transfer contributions can quickly be changed and reoriented during a relaxation process before emission occurs. In such a case the stimulated emission would be expected to change in time correspondingly. Such a change should affect the quantum yield of spontaneous emission as well. From the oscillator strength derived from the integral absorption of the 860 nm band a radiative decay time of 12 ns can be obtained [36]. Together with a time constant of 2.8 ps for charge separation one expects a prompt fluorescence quantum yield of 2.3 10^{-4}, if the electronic structure of P^* remains unchanged. A somewhat larger value of 4 10^{-4} has been obtained and almost the right value for the charge separation rate has been predicted years before the first picosecond measurements [100]. The measured value, however, refered to the total steady state fluorescence. If a considerable amount of the fluorescence originates from other sources with high quantum yield, such as prompt fluorescence of contaminations or minorities with slow charge separation or from delayed fluorescence components (see Chapt. 5), then the true oscillator strength will be correspondingly smaller.

Collecting all the spectral information and all inherent problems of detecting and identifying an ultrafast intermediate with low maximum concentration, it seems hopeless to unambiguously prove or rule out P^+B^- formation after the first charge separation step, exclusively by means of transient optical spectroscopy. Therefore we turn towards a more promising approach, by directly influencing electron transfer with an external electric field.

2. ELECTRON TRANSFER IN AN EXTERNAL ELECTRIC FIELD

2.1. ELECTRIC FIELD EFFECTS ON ELECTRON TRANSFER REACTIONS IN PHOTOSYNTHETIC REACTION CENTERS

The influence of membrane potential on phosphorylation [101] and on proton release [102] has been extensively studied in closed photosynthetic membranes. These observations imply that some of the electron transport steps are electrogenic. Of course they should also be sensitive to external electric fields. Such fields may be applied by imposing a salt [103] or ATP [104,105] gradient over the membranes or by illumination with light [106]. Placing the photosynthetic samples between two electrodes allows a more direct application and control of an electric field. Samples containing RCs may be stacked Langmuir Blodgett monolayers [107,108], membrane suspensions [109,110], black lipid bilayers [111-114] or polyvinyl alcohol (PVA) films [115-121]. In the membrane systems the RCs are oriented, while in PVA films they are distributed randomly, as long as the films are not stretched. Effects of an electric field have been detected on the delayed emission and triplet yield due to recombination of P^+H^- [104-106,121] and on the recombination of P^+Q^- to the ground state [107-115]. Furthermore electric field effects on emission not due to recombination

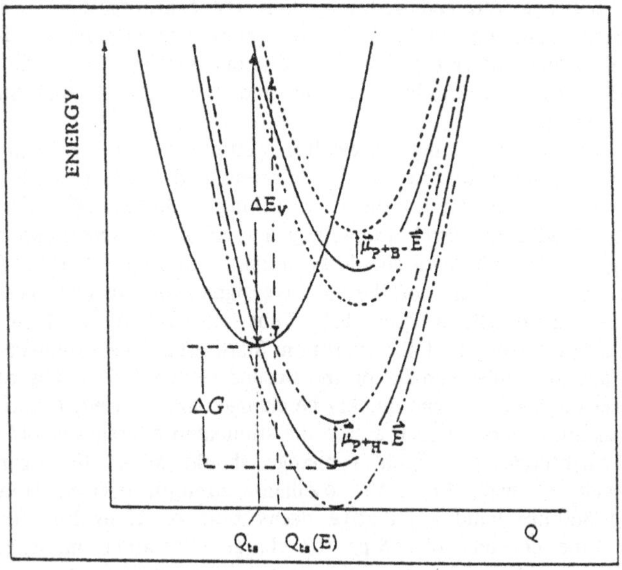

Figure 1: Simplified energy diagramm of an excited neutral state (P^*) prior to charge separation and of the radical pair state formed after electron transfer (P^+H^-) together with a virtually populated radical pair intermediate (P^+B^-) as a function of reaction coordinate (Q) and their changes due to interaction of the radical pair states with an external electric field.

[116-119,123] and on the quantum yield of P^+Q^- [122] and P^+H^- [119,121,123,124] formation, both supposedly being linked with changes in the primary charge separation rate [120] were investigated in detail. In the following the mechanism of influencing an electron transfer rate by an electric field will be pointed out while directly refering to the primary charge separation rate.

2.2. HOW DOES AN ELECTRIC FIELD INFLUENCE PRIMARY CHARGE SEPARATION?

Depending on whether primary charge separation occurs in two steps according to Equ. (1) or in one step enhanced by superexchange according to Equ. (2) the radical pair first formed will be either P^+B^- of P^+H^-, respectively. Due to the large distance of the radical ions, both radical pairs have large electric dipole moments amounting to $|\vec{\mu}_{P^+B^-}|$ = 50 Debye and $|\vec{\mu}_{P^+H^-}|$= 82 Debye. These large dipole moments give rise to significant interaction with an externally applied electric field, as indicated in Fig. 1. For example a field of 1 10^6 V/cm will cause a maximal shift of the free energy of the radical pair states P^+B^- and P^+H^- by ≃ 975 cm^{-1} and 1300 cm^{-1}, respectively, after correcting for the dielectric properties of the sample [96]. Strictly speaking, the free energy difference between the equilibrium configuration of the state prior to the electron transfer process (with dipole moment μ_{P^*}) and after ($\vec{\mu}_{D^+A^-}$) is

$$\Delta G = \Delta G_0 + (\vec{\mu}_{P^*} - \vec{\mu}_{D^+A^-})\vec{E} \tag{3}$$

with ΔG_0 being the free energy difference in absence of a field. $\vec{\mu}_{P^*}$ can be estimated from the difference in dipole moments between ground and excited state $\Delta\vec{\mu}$ as determined from

Stark effect measurements ($\mu_{P^*} \simeq \Delta\mu$ = 8 Debye) [95-98]. As a consequence of changes in ΔG both the energy and the horizonal position of the transition state on the generalized reaction coordinate Q change. The change in free energy is equivalent with a change of the activation energy $E_a(E)=(\Delta G(E)-\lambda)^2/4\lambda$ in the Franck-Condon factor of the electron transfer rate:

$$k = V^2 \frac{1}{\sqrt{4\pi\lambda k_B T}} e^{\frac{-E_a(E)}{k_B T}} \tag{4}$$

The change in nuclear configuration of the transition state $Q_{ts}(E)$ may be reflected in a variation of the electronic coupling matrix element $V(Q_{ts}(E))$, if Born-Oppenheimer approximation is not strikt. In case electron transfer is enhanced due to superexchange coupling, V includes a term:

$$V(E)_{super} = \frac{V_{12}V_{23}}{\Delta E_V}(Q_{ts}(E)) \tag{5}$$

with V_{12} being the coupling between the initial state (P^*) and the mediator state (P^+B^-) and V_{23} being the coupling between the mediator state and the final state (P^+H^-). V_{super} depends on the electric field, since the vertical energy difference between the transition state and the multidimensional potential surface of the mediating state changes due to vertical shifts of the potential surfaces and due to horizonal motion of Q_{ts}. This complicated interplay is treated in more detail in [115].

According to (3) the interaction of the electric field with the vector of the effective electric dipole moment depends on on their mutual orientation, i.e. on the cosine of the angle between these vectors. This angular dependence is the basis for a method to determine the orientation of the primary formed radical pair, as described in the next chapter. This orientation can be compared with x-ray structural data in order to identify the pathway of primary charge separation, since the dipole moments of P^+B^- and P^+H^- differ by 31° in their orientation. By measuring the direction of electron transfer access is made to a completely new quality not suffering from the drawbacks of purely tracing charge separation kinetically by transient absorption. By its very nature this approach is insensitive to the problems of spectroscopic assignment and not affected by short lifetime or small transient concentration of a kinetic intermediate.

In a randomly oriented sample, the orientational dependence of the electric interaction gives rise to a continuous distribution of free energy differences, being reflected in a corresponding distribution of electron transfer rates. The decay characteristics of the intrinsically heterogeneous sample have to be calculated by averaging over all possible orientations. The kinetics will be nonexponential, with slower decay than without field and, except for the case that the rate is activationless and can therefore not be further enhanced, also with faster decay components. Two different approaches have been made to handle the difficulties arising from such complex kinetics when studying the rather slow recombination of P^+Q^-. By fast modulation of the electric field during the electron transfer process, effort has been made to isolate the electric field effect in an experimentally elegant way [107,108]. Alternatively the kinetic trace has been analysed rigorously by numerical fitting [115] making use of cumulant expansion [115,125]. Such analysis necessarily rests on extreme reliability of experimental data with high dynamic range and very good linearity. Thus experiments in vectorially oriented RC preparations are very attractive [111-113].

At first sight the detection of electric field induced changes of the fluorescence quantum yield seem to be an alternative and considerably simpler approach to monitor the change of the very fast primary electron transfer kinetics [116-119]. The fluorescence yield Φ of the primary donor P^* is determined by the rate of radiative decay and the competing dark processes e.g. primary charge separation, internal conversion and others. In RCs where the primary ET provides the dominant decay channel of the excited state P^*, any change of this rate is necessarily accompanied by a corresponding change of the quantum yield of the prompt fluorescence. Even in cases which do not allow the precise determination of this rate kinetically, it should be possible to monitor an averaged rate with high precision via the fluorescence yield. In case steady state fluorescence is utilized as a monitor this approach rests crucially on the assumption that it is purely prompt in nature and not contaminated by significant emission of other source, e.g. delayed fluorescence. As will be shown in Chapter 5, this assumption is not valid in RC preparations.

3. THE DELFY METHOD

A method to determine the 'direction' of the primary charge separation rate is to measure its angular dependence in an external electric field. This brilliant idea was first reported by Lockhart et al. [116] together with an experimental attempt to measure the orientation of the primarily formed dipole moment μ. This first effort to obtain the angle between the primary dipole moment and the transition moment of fluorescence of P^* consisted in measurements of the anisotropy of the fluorescence with respect to an external electric field under the condition of isotropic excitation. This experimental approach suffered from two serious drawbacks: (a) Since the angle under investigation was similar for both radical pairs, conclusions on the ET are not possible [126]. (b) In addition to the relevant prompt fluorescence with ps lifetime, slow fluorescence components could contribute to or even dominate the steady state fluorescence signal. Such additional slow components may also be sensitive to electric fields.

To overcome this indecisive experimental situation according to objection (a), it was suggested to establish the orientation of the primary dipole moment by determining the angles κ of this dipole moment relative to various differently oriented transition moments [127]. According to the X-ray structural data [13-15], the various pigments in the RC supply such transition moments which differ significantly in κ for the two possible states P^+B^- and P^+H^- [117-119]. The new approach consists in the selective excitation of appropriate transitions with polarized light, thereby achieving a defined orientational selection of RCs with respect to the electric field. Thus, measuring the anisotropy of the electric field effect on the fluorescence quantum yield with respect to the polarization of excitation, the angle κ between the selected transition moment of excitation and the dipole moment of the primary radical pair can be determined. Since different transitions with different projection angles κ can be selected by appropriate excitation wavelengths, the steady state Dichroic Excitation spectrum of the eLectric field modulated Fluorescence Yield (DELFY) enables the construction of the vector of the dipole moment of the primary radical pair in the coordinate system of the RC, if the orientation of the transition moments is known.

Polarized excitation of selected transition moments photoselects RCs out of a given isotropic distribution. We assume that energy transfer from any excited cofactor to P occurs exclusively within the same RC. Additional photoselection is achieved by detecting the fluorescence at a defined angle of polarization. These photoselection conditions define an orientational distribution of RCs with corresponding projections of the radical pair dipole moment onto the electric field direction and thus with corresponding changes of the fluorescence yield. The magnitude of the electric field effect with respect to a given condition of photoselection has to be calculated by averaging over all possible orientations of the RCs [127]. In this averaging procedure we only consider fluorescence changes $\Delta\Phi = \Phi(E)-\Phi(0)$ being proportional to $(\mu E)^2$. This is justified for two reasons. (i) For isotropically distributed RCs, the lowest term in the expansion for any field dependence of Φ is quadratic in E, because linear contributions cancel due to mirror symmetry with respect to the electric field, and (ii) a quadratic dependence of $\Delta\Phi$ was indeed observed [116-119], which holds up to external fields of about 8 10^5 V/cm [119].

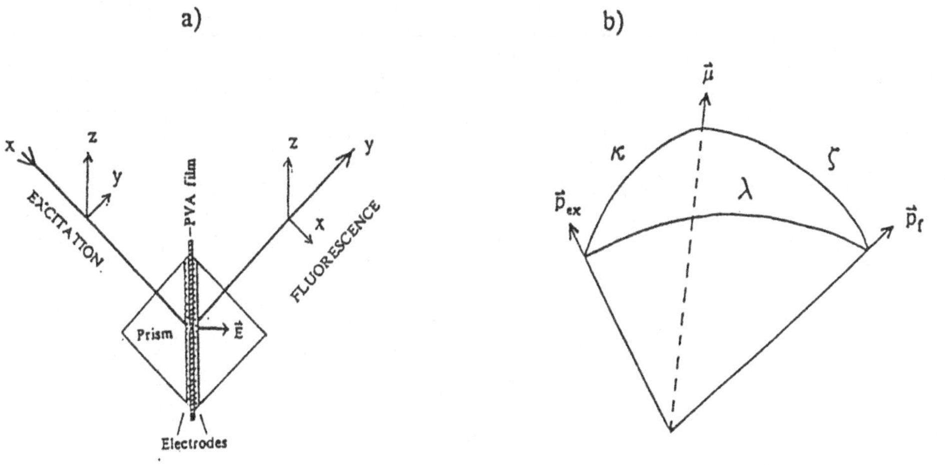

FIGURE 2: (a) Sample configuration: sample sandwiched between two prisms, providing refractive index matching for excitation beam (in x direction) and for emitting fluorescence (in y direction). Both excitation and emission enter the prism perpendicularly and the PVA film under 45°. Excitation may be polarized in the z and y direction and the fluorescence in the z and x direction. The electric field E is applied perpendicularly to the film by two electrodes. (b) Internal configuration of dipole moment $\vec{\mu}$, transition moment of excitation $\vec{p}_{ex}$, and emission $\vec{p}_f$ spanning a tripod with the angles κ, ζ, and λ.

At such fields an electric field induced fluorescence increase of $\simeq$ 60% is observed, whereas saturation of the electric field dependence occurs at fields exceeding 8 10^5 V/cm. For this reason the experiments were carried out at fields $\leq$6 10^5 V/cm (calibrated by a Stark effect on the absorption of $\simeq 10^{-3}$). The possibility of indirect influence of the Stark effect on the detected fluorescence signal has been recognized in [120, 128]. It should show up in spectral shifts, which were not observed [116-119]. This is not surprising, since the effect occurs only under very special conditions. In the context of DELFY measurements the relevant quantity is the *anisotropy* of the electric field effect rather than its magnitude and field dependence. Therefore, uncertainties with respect to the local strength of the electric field do not affect the reliability of the method. However the assumption has to be made that distortion of the internal field due to the anisotropy of the dielectric properties is negligible.

The angle κ between the dipole moment of the primary radical pair and the transition moment of excitation can be deduced from the ratio of field induced fluorescence changes at two distinct photoselection conditions: $\Delta\Phi(yx)/\Delta\Phi(zx)$ [127]. The electric field induced change given in the numerator refers to the geometry given in Fig. 2 with the polarizations of excitation in y-direction and of fluorescence in the x-direction, the one in the denominator is with polarizations of excitation in z-direction and of fluorescence in x-direction. The difference in photoselection conditions can be appreciated by noting that for $\Delta\Phi(yx)$ the electric field vector is in the plane defined by the two directions of

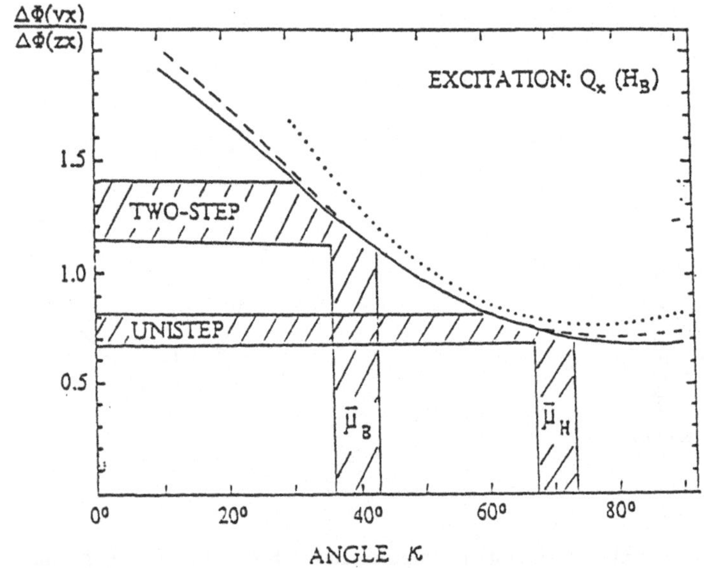

FIGURE 3: The electric field induced excitation anisotropy of the fluorescence yield $\Delta\Phi(yx)/\Delta\Phi(zx)$ as a function of κ with λ=30° (....), λ=50° (- - -) and λ=70° (——). ς is fixed at $\simeq$60° according to theoretical and experimental values for RCs of *Rb. sphaeroides* (see text). As an example, the angles κ between the transition moment of the Q_x band of H_B and the dipole moments μ_B and μ_H are marked [127].

polarization, while for $\Delta\Phi(zx)$ it is out of the plane. Thus the angle between polarization of absorption and the electric field changes from 45° to 90° while the angle between polarization of emission and the field remain unchanged at 45° in both cases. Thus the anisotropy ratio $\Delta\Phi(yx)/\Delta\Phi(zx)$ is most sensitive to the the orientation of the electric dipole moment $\vec{\mu}$ with respect to the transition moment of excitation (angle κ), while it is only slightly sensitive to the angle (ς) between $\vec{\mu}$ and the transition moment of fluorescence. This weak sensitivity holds as long as the angle (λ) between the transition moment of excitation and emission is not too small, since then of course ς and κ are identical and should reveal the same sensitivity. On rotating the polarization of excitation in the above defined way, the polarization vectors of excitation and emission remain orthogonal in both cases with the consequence that the detectability of the fluorecence in absence of an electric field remains unchanged as well, i.e. $\Phi(yx)/\Phi(zx)$=1 for E=0. This restriction dispenses us from normalizing both values of $\Delta\Phi$ and guarantees that $\Delta\Phi(yx)/\Delta\Phi(zx)$ is only weakly dependent on λ. This is illustrated by Fig.3, where the dependence of $\Delta\Phi(yx)/\Delta\Phi(zx)$ on κ is shown for different values of λ, setting $\varsigma\simeq$60°. Due to the small slope of $\Delta\Phi(yx)/\Delta\Phi(zx)$ the discrimination of κ for large values of κ becomes more difficult. Analogously to the determination of κ the value of ς can be obtained from the ratio $\Delta\Phi(yx)/\Delta\Phi(yz)$ when the polarization of emission is rotated [127]. The value of $\varsigma\simeq$60° measured at an excitation wavelength of 870 nm agrees with the one determined in [116].

Finally we want to emphasize two points:

(a) These dichroic ratios can in principle be time resolved.

(b) Analogous dichroic ratios can be defined for transient absorption signals. This holds in particular for the stimulated emission at 920 nm. However the time evolution of these ratios may be complicated.

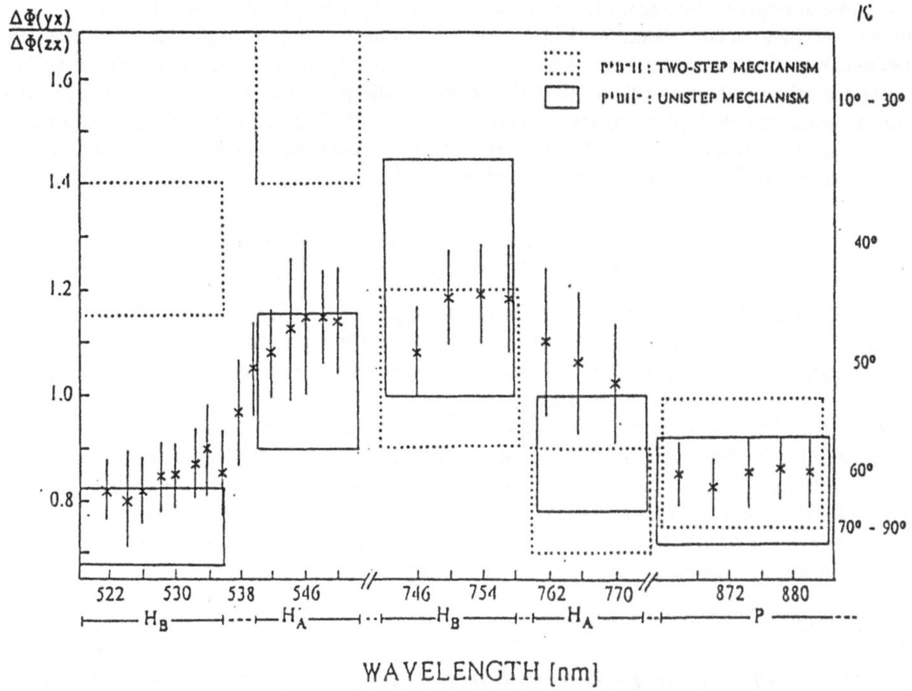

FIGURE 4: The dichroic excitation spectrum of the electric field modulated fluorescence yield in the Q_x and Q_y bands of the bacteriopheophytins and in the special pair band for reaction centers of *Rb. sphaeroides R-26* at 180 K. Left vertical scale: the dichroic ratio $\Delta\Phi(yx)/\Delta\Phi(zx)$, right vertical scale: corresponding angle κ (see Fig. 3). Ranges of calculated values of the dichroic ratio including all orientational uncertainties are marked by solid and dotted boxes for P^+H^- and P^+B^- as the primary radical pair, respectively [118].

4. STEADY STATE DELFY MEASUREMENTS

4.1 EXPERIMENTAL RESULTS

Scrutinizing the optical transitions in the Q_x band of H_B (522-536 nm) and H_A (540-550 nm), in the Q_y band of H_B (746-754 nm) and H_A (758-770 nm) and the special pair transition (866-882 nm) a significant difference of the calculated anisotopy $\Delta\Phi(yx)/\Delta\Phi(zx)$ can be found for the various excitation wavelengths as indicated by the square boxes in Fig.4. Thus the orientation of the effective dipole moment μ can be constructed from the obtained DELFY spectrum. For discrimination of P^+B^- against P^+H^- as the primary radical pair, the DELFY values in the Q_x-band of the H_B at 530 nm should be most useful.

The results of a steady state DELFY measurement obtained at 180K are also shown in Fig. 4. A good correlation of the experimental and theoretical values for the dipole moment P^+H^- can be seen throughout the spectrum. In the two Q_x transitions the deviation from the theoretical values expected for P^+B^- is significant. The Q_y transitions of the bacteriopheophytins are spectrally less clearly resolved than the Q_x transitions. Furthermore in the Q_y region the theoretically expected values of $\Delta\Phi(yx)/\Delta\Phi(zx)$ partially overlap. Nevertheless the experimental values tend to give better agreement with the data of P^+H^-. In the special pair transition the theoretical values for both dipole moments certainly do not allow a discrimination, but again are in good agreement with the experimental results.

From the data accumulated in Fig.4 five different projection angles κ onto the five inspected transition moments have been obtained. Two projection angles κ in general suffice to determine the orientation of the dipole moment of the primary radical pair. A least square fit of all data resulted in an orientational vector for the dipole moment of the primary radical pair with respect to the X-ray structure coordinate system [13] given below. For comparison the dipole vectors μ_B and μ_H are also shown.

$$\mu(\text{fit}) = \begin{pmatrix} 0.14 \\ 0.99 \\ 0.02 \end{pmatrix} \quad \mu_H = \begin{pmatrix} 0.225 \\ 0.974 \\ 0.035 \end{pmatrix} \quad \mu_B = \begin{pmatrix} 0.040 \\ 0.853 \\ 0.520 \end{pmatrix}$$

The deviation of the obtained vector μ(fit) from μ_H is 5.0° while it is 30.6° for μ_B. This result is a convincing evidence that P^+H^- is responsible for the observed electric field effect. The deviation of μ(fit) from μ_B is just as large as the angle between μ_B and μ_H. From the identity of the isotropic excitation spectra of the fluorescence at different wavelengths and of the electric field effect it could be concluded that electric field effects on energy transfer within the RC are neglegible and do not spoil the DELFY spectra [119]. Thus the results of DELFY measurements unambiguously demonstrate that the dipole moment P^+H^- is responsible for the observed electric field effect. It is yet not proven, however, that the steady state fluorescence signal is a good monitor of the primary charge separation rate. Other possible contributions to this signal will be discussed in the following.

4.2 DISCUSSION

POSSIBLE CONTRIBUTION OF DELAYED FLUORESCENCE: By its nature the overall steady state fluorescence signal also contains contributions of delayed fluorescence, if present. Recombination fluorescence according to $P^+H^- \rightarrow P^*H \rightarrow PH + h\nu$ may dominate the obtained electric field effect, if the quantum yield of such a delayed fluorescence $\Phi_{delayed} = k_F \tau_{delayed} \exp(-\Delta G/kT)$ (Equ.1 in [129]) is significant i.e. if the lifetime of P^+H^- is large enough to compensate the reduction of its amplitude due to the Boltzmann factor. The influence of the electric field on the energy level of P^+H^- in properly oriented RCs causes changes of ΔG. This can lead to dramatic changes in the Boltzmann factor and thus of the delayed fluorescence quantum yield. Averaging over all orientations the relative change is given by $(\Phi(E)/\Phi(0))_{delayed} = \langle\exp(-\mu E/kT)\rangle = 0.5\ (kT/\mu E)\ \sinh(\mu E/kT)$. Note that the relative change of this fluorescence component is independent of the energy gap between the fluorescing and the radical pair state and scales with temperature. At an external field of 6 10^5 V/cm a value of $\mu\ E \simeq 960\ cm^{-1}$ (120 meV) is obtained which is much large than k_BT at all temperatures. Thus the exponential behaviour given above together with a significant temperature dependence should become apparent in contrast to the observations [119,120]. Therefore we can conclude, that delayed fluorescence from P^+H^- recombination does not dominate the observed electric field effect and cannot be responsible for its anisotropy.

DOES A MINORITY OF RCs DOMINATE THE ELECTRIC FIELD EFFECT? A fundamental problem of steady state measurements arises, when we consider the possibility of a sample being heterogeneous with respect to the primary charge separation rate (Fig. 7a). In such a case the various subpopulations contribute differently to the magnitude of the fluorescence signal: a subpopulation with larger P^* lifetime produces correspondingly higher fluorescence quantum yield. Thus, we are faced with the possibility that few RCs with a slow charge separation rate dominate the fluorescence signal. This question can only be approached with time resolved measurements.

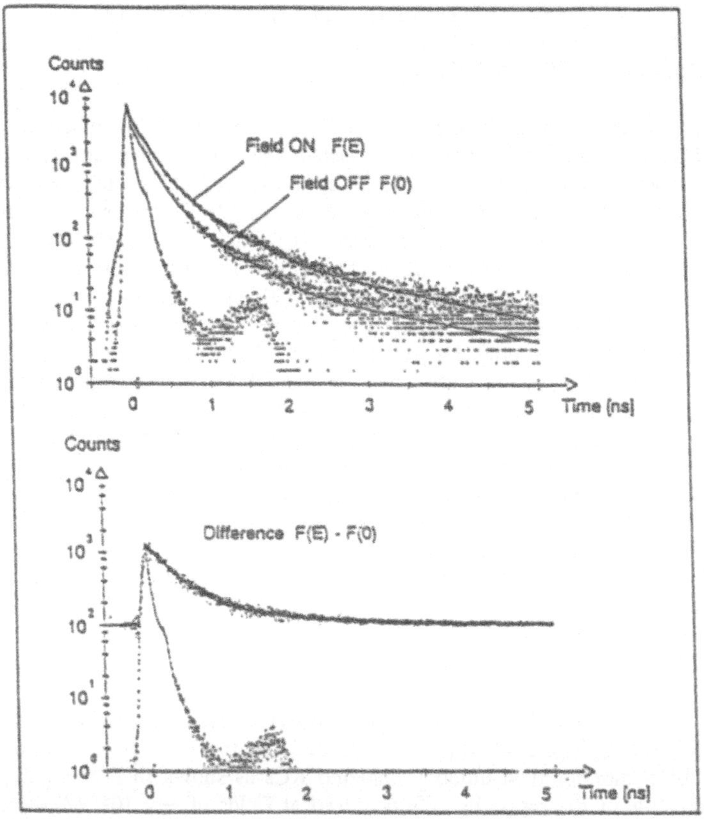

FIGURE 5: (a) Time resolved fluorescence traces F(t,0) in absence and F(t,E) in presence of an electric field of 6 10^5 V/cm detected at 920 nm after 865 nm excitation of reaction centers of *Rb. sphaeroides R-26* at 180 K. (b) Difference F(t,E)-F(t,0) of the fluorescence traces in presence and absence of electric field as given above. To avoid negative numbers not depictable in a logarithmic representation a constant value of 100 was added. Solid lines are 3-exponential fits of the data, with lifetimes τ_i and relative quantum yields Φ_i:

	Lifetime τ_i	Quantum yield Φ_i
F(t,0):	75 ps	0.418
	307 ps	0.501
	2.1 ns	0.081
F(t,E):	77ps	0.480
	332 ps	0.733
	1.9 ns	0.179
F(t,E)-F(t,0):	82 ps	0.172
	362 ps	0.595
	1.67 ns	0.233

In absence of a field the quantum yields are normalized to give $\Sigma\ \Phi_i=1$. In presence of a field they are normalized by the same factor. In this case $\Sigma\ \Phi_i$ represents the modulation $\Phi_{tot}(E)/\Phi_{tot}(0)$ of the total steady state fluorescence which was 1.39 [119].

5. TIME RESOLVED MEASUREMENTS IN AN ELECTRIC FIELD

5.1. TIME RESOLVED FLUORESCENCE

Time resolved fluorescence traces in presence F(t, E) and absence F(t, 0) of an electric field are shown in Fig.5a together with the time constants τ_i and relative quantum yields Φ_i [119].The differences of both traces F(t, E)-F(t, 0) is shown in Fig.5b. Cursory inspection of the data immediately reveals that the main contribution to the fluorescence signal and the main constituent of the electric field effect thereon originates from a component having a lifetime of 300 ps. The contribution of this component to the total fluorescence is $\simeq$ 50%, independent of the electric field. On switching on the field its quantum yield increases by $\simeq$ 46% and the total fluorescence by 39%. The faster component(s) including prompt fluorescence seem to increase only by roughly 15%. Even under the most favourable assumption that prompt fluorescence accounts for the whole of the fastest component in F(t, E)-F(t, 0), its contribution to the total change is not much larger than 17%. The 2ns component appears to change by about 120%, considerably more than the other components, though due to experimental limitations only the major component should be discussed quantitatively. We can summarize making two important statements:

a) The fluorescence has considerable contributions from slow components. Its majority has a lifetime of about 300 ps.
b) All components resolved so far experience electric field effects of similar order of magnitude.

5.2 TRANSIENT ABSORPTION

The influence of an electric field on the light induced transient bleaching of the P band (860 nm) and of the Q_x band of H_A (546 nm) and on the Q_x band absorption of H_B (530 nm) have been monitored in quinone depleted RCs as shown in Fig. 6 [121, 123, 124]. At 546 nm a decrease of the signal by 10% occurred at fields of $\simeq$ 7 10^5 V/cm indicating a corresponding loss of quantum yield of P^+H^- formation ($\Phi_{P^+H^-}$). Similar deficiencies of P^+Q^- formation in an electric field have been observed earlier [122]. While the electric field induced reduction of the bacteriopheophytin signal appears instantaneously, a corresponding reduction of the dimer bleaching is not present at first, but evolves in several hundred ps. This delayed recovery of the groundstate absorption means that the loss channel does not immediately lead to the groundstate. Instead an intermediate state with a lifetime shorter than a nanosecond has to be envisaged. Interestingly enough, partial recovery of the groundstate absorption also occurs in absence of a field [124].

In order to achieve a reduction in $\Phi_{P^+H^-}$ a loss channel has to be effectively competing with charge separation. The electric field would have to alter the branching ratio of P^+H^- formation and the loss channel. According to [120] and judging from the time resolved fluorescence measurements average primary charge separation rate of the majority of RCs does not change more than 50%. Thus, the following problem arises: In order to achieve an average reduction of $\Phi_{P^+H^-}$ by 10%, the loss channel would have to be rather fast on the order of $\simeq$ 1/(10ps). As long as a field induced speeding of the loss channel is not considered the maximum value of $\Phi_{P^+H^-}$ in absence of an electric field would obviously be restricted to values of less than 80%. This value barely complies with measurements of the quantum yield of P^+Q^- [130]. Internal conversion is not likely to be fast enough to account for such losses, since various measurements of the decay of the excited state of the bacteriochlorophyll dimer in systems lacking charge separation showed values of 200-600ps as the fastest time constant [124, 131].

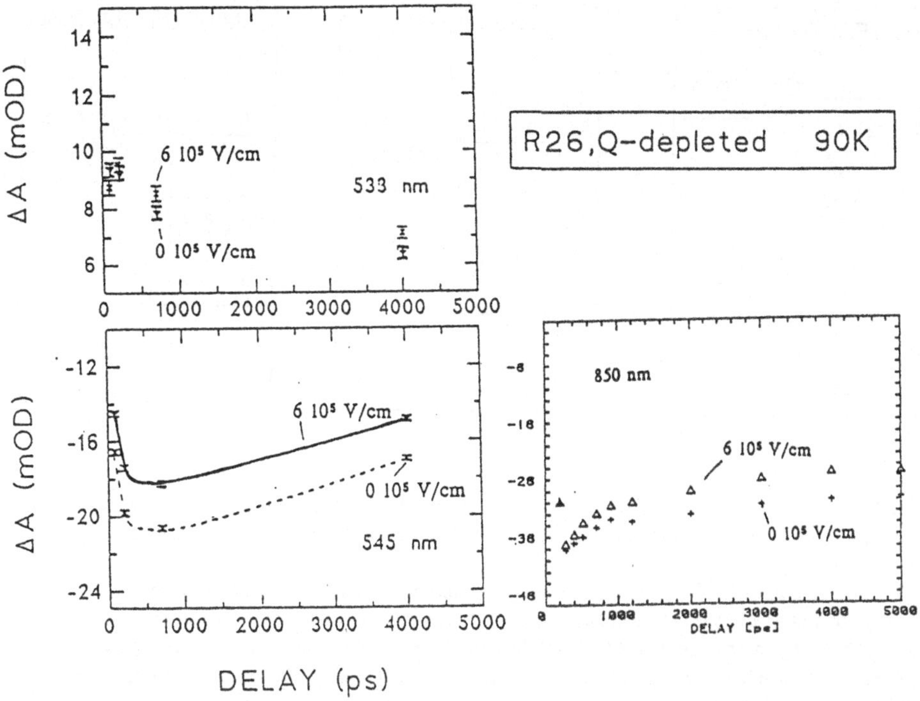

FIGURE 6: Light induced transient absorption changes detected at 546 nm, at 530 nm and at 850 nm in absence and in presence of an electric field of $\simeq$ 6 10^5 V/cm at 80 K in *R. sphaeroides R26* [124].

6. DISCUSSION OF ELECTRIC FIELD EFFECTS

6.1. WHAT IS THE NATURE OF THE SLOWER FLUORESCENCE COMPONENTS?

NONFUNCTIONAL IMPURITIES: The simplest explanation for fluorescence components slower than charge separation is to attribute them to nonfunctional impurity pigments, i.e. to pigments not connected to RCs with intact charge separation. The electric field effect on such pigments, however, is not expected to be larger than a few percent. Such effects may either by a Stark effect on both absorption and emission or a field effect on the relevant quenching process, most probably internal conversion. The lack of large effects on excited bacteriochlorophyll dimers not subdued to charge separation, have been demonstrated experimentally on the D_{LL} mutant and on B800/B850 antenna preparations [123, 132]. In these systems the field effect even turned out to have opposite sign. Thus the contribution from nonfunctional impurities is neglegible. As a result we may state that the electric field effect gives us a very sensitive fingerprint at hand, to discriminate transient spectroscopic components connected to an electron transfer process from those determined by other dynamic processes. This mere fact should prove very useful in investigating other systems as well.

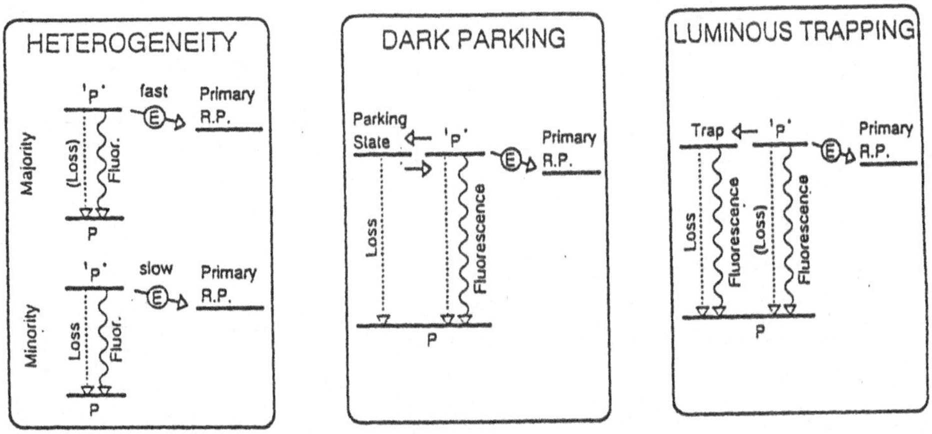

FIGURE 7: Three kinetic schemes illustrating the consequences of an electric field dependence of charge separation on fluoerecence quantum yield and loss of primary radical pair formation.

(a) HETEROGENEITY: Samples are heterogeneous consisting of reaction centers with fast or slow primary charge separation rates: a majority of fast reaction centers governs the signals obtained in transient absorption measurements. A minority of slow reaction centers is exposed to competing loss rates and is responsible for slow fluorescence components with high quantum yields.

(b) DARK PARKING: A parking mechanism (which may also be electric field sensitive) competes with primary charge separation. Delayed repopulation of P^* produces delayed emisson components with amplitudes proportional to the equilibrium constant betwenn P^* and parking state concentration. Losses in the parking state occur at the expense of primary radical pair formation.

(c) LUMINOUS TRAPPING: A trapping state with the ability to fluoresce will directly produce slow fluorescence components with considerably higher quantum yields than in (b), while primary radical pair loss is similarly effective. The slow fluorescence components exhibit the same field dependence as the prompt fluorescence.

DELAYED RECOMBINATION FLUORESCENCE: In light of the time resolved data, we want to resume discussion of the possibility of recombination fluorescence. In RCs in which the reaction $P^+H^- \rightarrow P^+Q^-$ was blocked a fluorescence component somewhat faster than 1 ns and one with about 2 ns lifetime have been detected [36]. They have been assigned as delayed recombination fluorescence of nonrelaxed radical pair states P^+H^- [36, 133-134]. Such components are present in our preparations as well, in which the lifetime of the P^+H^- state is $\simeq$ 100ps and therefore even shorter than the 300 ps component. Thus at least a substantial amount of all these fluorescence components does not origin from P^+H^- recombination. It would be interesting to disentangle the field dependence of the individual components and the temperature dependence on these effects.

As already mentioned, the immediate reduction of the H^- signal in an electric field and the delayed field induced recovery of the groundstate absorption indicates, that part of the excitation energy is stored within a time being shorter than the 40 ps time resolution of the experiment in Fig. 6 into another intermediate state, which decays more slowly. The proposed nonrelaxed radical pair state in principle could be the departure point for such an intermediate. Any kind of loss mechanism emanating from there would though have to be faster than the 40 ps time resolution in order to show up in the absorption signal of H. On the other hand it would have to slow down very quickly with relaxation, since otherwise severe deficiencies in quantum yield of relaxed P^+H^- in blocked RCs would be expected in contrast to experimental results.

RECOMBINATION LOSS FROM P^+B^-: Another departure point for a loss mechanism could be the P^+B^- radical, as proposed in [135]. Provided that this state is formed at all, a loss channel would have to be extremely fast in order to compete with its short lifetime. Recombination to the ground state at such a fast rate would not be compatible with the slow recovery of groundstate absorption, neither could this scheme account for the electric field effects on the slower fluorescence components nor its anisotropy.

DARK PARKING STATE: It seems more likely that such an intermediate is directly formed from P^* in competition with the first charge separation step. A possibility to account for slow fluorescence components is to consider this intermediate as a parking state having essentially no fluorescence quantum yield (dark), but being able to eventually release excitation energy back to the charge separating state P^* (Fig. 7b) or destroying excitation energy via some deactivation route to the ground state. Those excitation quanta escaping the parking state via P^* (contributing to delayed fluorescence) will have the same probability of fluorescing as those which escaped trapping from the beginning (contributing to prompt fluorescence) and will experience the same electric field dependence. Thus the delayed fluorescence yield relative to the prompt fluorescence yield will reflect the relative concentration of RCs being initially captured in the parking state, weighted with the probability of escaping via P^*. The quantum yield of the 300 ps fluorescence component is 50% while that of prompt fluorescence can be considerably less than 42%. This would imply that more than 55% of RCs will be in the parking state and less than 45% would promptly form P^+H^-. Such a possibility is certainly excluded by the transient absorption data. Thus a parking state can only account for part of the slow fluorescence quantum yield.

An obvious candidate for a dark parking channel would be charge separation to the B-branch. In this case both the charge separation process and the parking process are sensitive to the electric field. Since the dipole moments associated with A- and B-branch charge separation are almost orthogonal, the electric field will act on the competing reactions in an antagonistic way. For a certain fraction of RCs the field can slow down electron transfer to the A-branch and simultaneously enhance electron transfer to the B-branch. For this fraction the quantum yield of parking is enhanced by both effects. At the same time the expected increase of the quantum yield of prompt fluorescence due to slowing down at the A-branch may partially be compensated by speeding up at the B-branch. Since the orientation of the dipole moment being effective in DELFY deviates by more than 70° from the dipole moments of both radical pairs on the B-branch, there is no evidence that the latter significantly contribute to the electric field effect on fluorescence. Like for delayed recombination fluorescence from $P^+H_A^-$ one would also expect an exponential dependence of this fluorescence on the electric field. Finally the electric field induced formation of $P^+H_B^-$ is ruled out by transient absorption measurements [123,124,136]. It would be interesting to test whether electric field induced bleaching of the accessory bacteriochlorophyll can be detected.

HETEROGENEOUS CHARGE SEPARATION: As already discussed in Chapter 4 samples might contain a heterogeneous set of RCs with different rates of primary charge separation (Fig. 7a) being reflected by corresponding time constants in the fluorescence decay. According to the very small amplitudes of the slower fluorescence components, only about

3% of the RCs would be responsible for the 300 ps component. Of course these RCs would suffer considerable losses in quantum yield on approaching the internal conversion lifetime. However, the total concentration of such slow RCs would have to be significant in order to account for a 10% change in $\Phi_{P^+H^-}$, which seems to be in contradiction to the fluorescence amplitudes.

LUMINOUS TRAPPING STATE: The delayed fluorescence quantum yield of parking states will always be low, since on reencountering P^* it will suffer fast quenching due to charge separation. A way out is to allow fluorescence to prevail in a trapping state, in which excitation energy escapes very fast quenching (Fig. 7c). Apart from possible differences in oscillator strength, fluorescence yield of this state will increase proportionally to its lifetime, in contrast to the dark parking state. This lifetime may be determined by internal conversion and will lead to recovery of the groundstate. Like in the case of dark parking, the electric field dependence of the charge separation rate together with its anisotropy are reflected in an average increase of the trapping population, an increase of the associated fluorescence quantum yield and an increase of groundstate recovery. The weak temperature dependence of the slow fluorescence components and their respective electric field effects demands that the trap for excitation energy is so deep compared to thermal energy that essentially no reencounter with the initial active state can occur or that the trap is fully entropic.

At very high field strengths deviations from a quadratic dependence of the fluorescence on the electric field were found [119]. The field induced fluorescence increase saturates. Such saturation effects can be understood in both the luminous trapping model as in the dark parking model. The amount of trapped RCs increases if charge separation is slowed in the electric field as long as it is still faster than the trapping rate. If charge separation becomes even slower than trapping more or less all of the RCs affected by the electric field become trapped and their yield saturates. Consequently their contribution to the fluorescence yield saturates as well.

As an example for such a luminous trapping state model we assume that the excited state P^* exists in a form being *active* with respect to charge separation, and in a form being *inactive*, e.g. the P^* state could be in a conformation adverse to charge separation. Such an *inactive* P^* state would indeed be a luminous trapping state. In order to account for the high instant quantum yield of charge separation we further have to assume that after primary excitation essentially *active* P^* states are formed.

7. CONCLUSIONS

The measured steady state DELFY spectrum exposes an electric dipole moment which is parallel in space to that of P^+H^-. The strong influence of this dipole on the fluorescence indicates unistep superexchange charge separation either for the majority or at least for a subset of RCs of our samples. This conclusion is independent of the results of time resolved fluorescence measurements: They reveal that the major contribution to the fluorescence quantum yield and to the electric field effect thereupon originates from fluorescence components with a lifetime of about 300ps. A series of kinetic models could not withstand the additional experimental constraints imposed by fs- and ps-transient absorption measurements and by electric field induced reduction of P^+H^- quantum yield. It is impossible to exclude that a certain fraction of RCs with slow charge separation exists, as reflected by the slow fluorescence components. However, these 'slower' RCs would have to constitute a considerable amount of the total of RCs in order to account for the field effects on P^+H^- formation. More likely, the magnitude of the field effect on P^+H^- reflects losses in the bulk of RCs. The fast charge separation rate in these RCs demands a correspondingly fast loss channel. Since internal conversion is presumably too slow, fast parking or trapping of excitation energy has to be envisaged. In case such a parking/trapping state does not fluoresce an independent source of emission has to exist in order to account for the 300ps fluorescence and its field effect. As an example, electric field induced formation of $P^+B_B^-$ could represent the parking/trapping state while a minority of 'slow' RCs could be the source of slow emission. In case the parking/trapping state does fluoresce, it can account for the observed electric field effect on both the slow fluorescence component and on the P^+H^- quantum yield. Furthermore the results of steady state DELFY measurements would represent the behaviour of the bulk of the RCs implying direct charge separation from P^* to P^+H^- in one step.

There is no doubt, that the possibility of directly influencing primary charge separation with an electric field gives us an attractive access to the mechanism of primary charge separation. Though single step charge separation to P^+H^- has been proven by DELFY measurements at least for a minority of RCs, no conclusion can yet be made for the bulk, in spite of a whole series of experiments in electric fields. The reason for this drawback is the limited time resolution of the present experiments. These measurements did, however, indicate that the presently discussed kinetic scheme of the very fast processes in the RCs, as deduced from fs-transient absorption experiments alone, is incomplete.

Transient absorption and emission measurements with the appropriate time resolution and sensitivity should be able to trace the influence of the electric field on charge separation directly. The electric field effect is an excellent fingerprint for unravelling superimposed signals. Thus it should be useful in deciding the unsettled issues in the kinetic interpretation of transient signals referred to in the introduction. In particular, a time resolved DELFY experiment or a corresponding dichroic absorption experiment in an electric field will be able to establish the mechanism of primary charge separation unambiguously. Since the effect is detected on the P state, it will be detectable, even if the first radical pair formed has an undetectably small lifetime and transient concentration. Additionally, one should try to further trace formation and decay of the proposed parking or trapping states. Investigations on mutants with slow charge separation should reveal such parking or trapping states more clearly, since then larger losses in quantum yield of P^+H^- in presence and absence of an electric field are expected.

8. ACKNOWLEDGEMENT

We are indebted to all our collaborators cited in the text, in particular to U. Eberl, who was deeply involved in all electric field work. Financial support from the Deutsche Forschungsgemeinschaft (Sonderforschungsbereich 143) is gratefully acknowledged.

REFERENCES:

[1] Clayton, R.K. (1980) Photosynthesis: physical mechanisms and chemical patterns, Cambridge University Press, Cambridge.
[2] Robles, .J., Breton, J. and Youvan, D.C. (1990) Science 248, 1402-1405.
[3] Reed, D.W. and Clayton R.K. (1968) Biochem. Biophys. Res. Commun. 30, 471-475.
[4] Okamura, M.Y. Feher, G. and Nelson, N. (1982) in Govindjee (ed.) Energy Conversion by Plants and Bacteria, Academic Press, New York, pp. 195-272.
[5] Parson, W.W. and Ke, B. (1982) in Govindjee (ed.) Energy Conversion by Plants and Bacteria, Academic Press, New York, pp. 331-385.
[6] Parson, W.W. (1987) in: Amesz, J. (ed.) Photosynthesis, Elsevier, Amsterdam.
[7] Kirmaier, C. and Holten, D. (1987) Photosynthesis Res. 13, 225-260.
[8] Hoff, A.J. (1979) Phys. Rep. 54, 75-200.
[9] Hoff, A.J. (1981) Quart. Rev. Biophys. 14, 599-665.
[10] Lubitz, W. (1988) in: Kurreck, H. Kirste, B. and Lubitz W., Electron Nuclear Double Resonance Spectroscopy of Radicals in Solution, VCH-Publishers, Weinheim, pp.279-351.
[11] Straley, S.C. Parson, W.W. Mauzerall, D.C. and Clayton, R.K. (1973) Biochim.

Biophys. Acta 305, 597-609.
[12] Michel, H. (1982) J. Mol. Biol. 158, 567.
[13] Deisenhofer, J. Epp, ., Miki, K., Huber, R. and Michel, H. (1984) J. Mol. Biol. 180, 385-398.
[14] J. Deisenhofer, H. Michel (1989), EMBO J. 8, 2149-2170.
[15] Komiya, H., Yeates, T.O., Rees, D.C., Allen, J.P. and Feher, G. (1988) Proc. Natl. Acad. Sci. USA 85, 9012-9016.
[16] Oelze, J. (1985) Methods Microbiol. 18, 257.
[17] Fajer J. Brune, D.C. Davis, M.S. Forman, A. and Spaulding L.D. (1975) Proc. Nat. Acad. Sci. USA 72, 4956-4960.
[18] Hoff, A.J. and Amesz, J. (1991) in: Scheer, H. (ed.) Chlorophylls, pp.723-738.
[19] Knapp, E.W. Fischer, S.F. Zinth, W. Sander, M. Kaiser, W. Deisenhofer, J. and Michel, H. (1985) Proc. Nat. Acad. Sci. USA 82, 8463-8467.
[20] Parson, W.W. Scherz, A. and Warschel, A. (1985) in: Michel-Beyerle, M.E. (ed.) Antennas and Reaction Centers of Photosynthetic Bacteria, pp. 122-130.
[21] Warshel A. and Parson, W.W. (1987) J. Am. Chem. Soc. 109, 6143-6152.
[22] Scherer, P.O.J. and Fischer, S.F. (1987) Biochim. Biophys. Acta 891, 157-164.
[23] Duysens, L.N.M. (1952) Thesis, Utrecht.
[24] Duysens, L.N.M., Huiskamp, W.J., Vos,J.J. and van der Hart, J.M. (1956) Biochim. Biophys. Acta 19,188-190.
[25] Loach, P.A. and Walsh, K. (1969) Biochemistry 8, 1908-1912.
[26] Bolton, J.R. Clayton and Reed, D.W., (1969) Photochem. Photobiol. 9, 209-218.
[27] Loach, P.A., Androes, G.M., Maksim, A.F. and Calvin (1963) Photochem. Photobiol. 2, 443-454.
[28] Feher, G., Hoff, A.J. Isaacson, R.A., and Ackerson, L.C. (1975) Ann. New York Acad. Sci. 244, 239-259.
[29] Norris, J.R. Scheer,H., Druyan,M.E. and Katz, J.J. (1975) Proc. Nat. Acad. Sci. USA 71, 4897-4900.
[30] Parson, W.W. (1967) Biochim. Biophys. Acta 131, 154-172.
[31] McElroy, J.D., Feher, G. and Mauzerall, D. (1970) Biophys. J. 10, 204a
[32] Loach, P.A. and Hall, R.L. (1972) Proc. Nat. Acad. Sci. USA 69, 786-790.
[33] Feher, G., Okamura,M.Y. and McElroy, J.D. (1972) Biochim. Biophys. Acta 267, 222-226.
[34] Slooten, L. (1972) Biochim. Biophys. Acta 275, 208-218.
[35] Clayton, R.K. and Straley S.C. (1072) Biophys. J. 12, 1221-1234.
[36] Woodbury, N.W. and Parson, W.W. (1984) Biochim. Biophys. Acta 767, 345-361.
[37] Kaufmann K.J. Dutton, P.L. Netzel; T.L. Leigh J.S. and Rentzepis P.M. (1975) Science 188, 1301-1304.
[38] Kaufmann K.J. Petty, K.M. Dutton, P.L. and Rentzepis P.M. (1976) Biochem. Biophys. Res. Comm. 70, 839-845.
[39] Rockley, M.G. indsor, M.W. Cogdell, R.J. and Parson W:W. (1975) Proc. Nat. Acad. Sci. USA 72, 2251-2255.
[40] Zinth, W. Kaiser, W. and Michel, H. (1983) Biochim. Biophys. Acata 723, 128-131.
[41] Shuvalov, V.A. Krakhmaleva L.N. and Klimov V.V. (1976) Biochim. Biophys. Acta 449, 597-601.
[42] Haberkorn, R., Michel-Beyerle, M.E. and Marcus, R.A. (1979) Proc. Nat. Acad. Sci. USA 76, 4185-4188.
[43] Marcus, R.A. (1988) Isr. J. Chem. 28, 205-213.
[44] Marcus, R.A. (1987) Chem. Phys. Lett. 133, 471-477.
[45] Chekalin, S.V., Matveetz, Ya.A., Shkuropatov, A.Ya., Shuvalov, V.A. and Yartzev, A.P. (1987) FEBS Lett. 216, 245-248.
[46] Fischer, S.F. and Scherer, P.O.J. (1987) Chem. Phys. 115, 151-158.

[47] Marcus, R.A. (1988) Chem. Phys. Lett. 146, 13-22.

[48] Creighton, S., Hwang, J.-K., Warshel A., Parson, W.W. and Norris, J. (1988) Biochemistry 27, 774-781.

[49] Fischer, S.F., Nussbaum, I. and Scherer, P.O.J. (1985) in Antennas and Reaction Centers of Photosynthetic Bacteria, (Michel-Beyerle, M.E., ed.), pp. 256-263, Springer, Berlin.

[50] Jortner, J. and Michel-Beyerle, M.E. (1985) in Antennas and Reaction Centers of Photosynthetic Bacteria, (Michel-Beyerle, M.E., ed.), pp. 345-365, Springer, Berlin.

[51] Jortner, J. and Bixon, M. (1987) in Protein Structure Molecular and Electronic Reactivity (Austin, R., Buhks, E., Chance, B., DeVault, D., Dutton, P.L., Frauenfelder, H. and Gol'danskii, V.I., eds.), pp. 277-308, Springer, New York.

[52] Ogrodnik, A., Remy-Richter, N., Michel-Beyerle, M.E. and Feick, R. (1987) Chem. Phys. Lett., 135, 576-581.

[53] Norris, J.R., Budil, D.E., Tiede, D.M., Tang, J., Kolaczkowski, S.V., Chang, C.H. and Schiffer, M. (1987) in Progress in Photosynthetic Research I (Biggens, J., ed.), pp. 363-369, Martinus Nijhoff, Dordrecht.

[54] Michel-Beyerle, M.E., Plato, M., Deisenhofer, J., Michel, H., Bixon, M. and Jortner, J. (1988) Biochim. Biophys. Acta, 932, 52-70.

[55] Bixon, M., Jortner, J., Plato, M. and Michel-Beyerle, M.E. (1988) in The Photosynthetic Bacterial Reaction Center. Structure and Dynamics. (Breton, J. and Vermeglio, A., eds.), pp.399-420, Plenum, New York.

[56] Plato, M., Möbius, K., Michel-Beyerle, M.E., Bixon, M. and Jortner, J. (1988) J. Am. Chem. Soc., 110, 7279-9285.

[57] Michel-Beyerle, M.E., Bixon, M. and Jortner, J. (1988) Chem. Phys. Lett., 151, 188-194.

[58] Bixon, M., Michel-Beyerle, M.E. and Jortner, J. (1988) Isr. J. Chem., 28, 155-168.

[59] Bixon, M., Jortner, J., Michel-Beyerle, M.E. and Ogrodnik, A. (1989) Biochim. Biophys. Acta, 977, 273-286.

[60] Friesner, R.A. and Won, Y. (1989) Biochim. Biophys. Acta, 977, 99-122.

[61] Ogrodnik, A., Volk, M., Letterer, R., Feick, R., and Michel-Beyerle, M.E. (1988) Biochim. Biophys. Acta 936, 361-371.

[62] Bixon, M., Jortner, J. and Michel-Beyerle, M.E. (1991) Biochim. Biophys. Acta 1056, 301-315.

[63] Marcus, R. and Almeida, R. (1990) J. Phys. Chem. 94, 2973-2977.

[64] Almeida, R. and Marcus, R. (1990) J. Phys. Chem. 94, 2978-2985.

[65] Hu, Y. and Mukamel, S. (1989) J. Chem. Phys. 91, 6973-6988.

[66] Sugawara, M., Fujinura, Y., Yeh, C.Y. and Lin, S.H. (1990) J. Photochem. Photobiol. 54, 321-331.

[67] Kharkhats, Y.I. and Ulstrup, J. (1991) Chem. Phys. Lett. 182, 81-87.

[68] Akhmanov, S.A., Borisov, A.Yu., Danielius, R.V., Gadonas, R.A., Kozlowski, V.S., Piskarskas, A.S., Razjivin, A.P. and Shuvalov, V.A. (1980) FEBS Lett. 114, 149-152.

[69] Kirmaier, C. Holten, D. and Parson W.W. (1985) Biochim. Biophys. Acta 810, 49-51.

[70] Parson, W.W. Woodbury, N.W.T., Becker,M., Kirmaier, C. and Holten D. (1985) in: Michel-Beyerle, M.E. (ed.), Antennas and Reaction Centers of Photosynthetic Bacteria, Springer Berlin, pp. 278-285.

[71] Zinth, W., Nuss, M.C. Franz, M.A. Kaiser, W. and Michel, H. (1985) in: Michel-Beyerle, M.E. (ed.), Antennas and Reaction Centers of Photosynthetic Bacteria, Springer Berlin, pp. 286-291.

[72] Woodbury, N.W.T., Becker,M., Middendorf, D. and Parson, W.W. (1985) Biochem. 24, 7516-7521.

[73] Martin, J.-L., Breton, J., Hoff, A.J., Migus, A, and Antonetti, A. (1986) Proc. Nat. Acad. Sci. USA 83, 957-961.

[74] Fleming G.R., Martin, J.-L. and Breton, J. (1988) Nature 333, 190-192.
[75] Breton, J., Martin, J.-L., Fleming, G.R. and Lambry, J.-C. (1988) Biochemistry 27, 8276-8284.
[76] Breton, J., Martin, J.-L., Petrich J., Migus, A. and Antonetti, A. (1986) FEBS Lett. 209, 37-43.
[77] Kirmaier, C., Holten, D. and Parson, W.W. (1985) FEBS Lett. 185, 76-81.
[78] Junge, W. and Witt, H.T. (1968) Z. Naturforsch. 23b, 244-254.
[79] Vermeglio, A. and Clayton, R.K. (1977) Biochim. Biophys. Acta 461, 159-165.
[80] Scherer, P.O.J. (1989) Bull. Soc. Roy. Sci. de Liege, 58e, 247-254.
[81] Shuvalov, V.A. and Parson W.W. (1981) Proc. Nat. Acad. Sci. USA 78, 957-961.
[82] Ogrodnik, A. and Michel-Beyerle, M.E. (1990) in Current Research in Photosynthesis Vol.I (Baltscheffsky, M., ed.), pp. 19-26, Kluwer, Netherlands.
[83] Volk, M., (1991) Thesis, Munich.
[84] Bixon, M., Jortner, J., Michel-Beyerle, M.E. Ogrodnik, A. and Lersch, W. (1987) Chem. Phys. Lett. 140, 626-630.
[85] Holzapfel, W., Finkele, U., Kaiser, W., Oesterhelt, D., Scheer, H., Stilz, H.U. and Zinth, W. (1989) Chem. Phys. Lett. 160, 1-7.
[86] Holzapfel, W., Finkele, U., Kaiser, W., Oesterhelt, D., Scheer, H., Stilz, H.U. and Zinth, W. (1990) Proc. Natl. Acad. Sci. USA 87, 5168-5172.
[87] Lauterwasser, C., Finkele, U., Scheer, H. and Zinth, W. (1991) Chem. Phys. Lett. 183, 471-477.
[88] Kirmaier, Ch. and Holten, D. (1990) Proc. Natl. Acad. Sci. USA 87, 3552-3556.
[89] Chan, C.-K., DiMagno, T.J., Chen, L.X.-Q., Norris, J.R. and Fleming, G.R. (1991) Proc. Nat. Acad. Sci. USA 88, 11202-11206.
[90] Kirmaier, C. and Holten, D. (1991) Biochemistry 30, 609-613.
[91] Friesner, R.A. and Wertheimer, R. (1982) Proc. Nat. Acad. Sci. USA 79, 2138-2142.
[92] Jean, J, Friesner, R.A. and Fleming, G.R. (1991) Ber. Bunsenges.Phys. Chem. 95, 253-258.
[93] Vos, M.H., Lambry, J.-C., Robles, S.J., Youvan, D.C., Breton, J. and Martin, J.-L. (1991) Proc. Natl. Acad. Sci. USA, 88, 8885-8889.
[94] Johnson, S.G. Lee, I.-J. and Small, G.J. (1991) in: Scheer, H. (ed.) Chlorophylls, pp.739-768.
[95] Lösche, M., Feher, G., and Okamura, M.Y. (1987) Proc. Nat. Acad. Sci. USA 84, 7537-7541.
[96] Lockhart, D.J. and Boxer, S.G. (1987) Biochemistry 26, 664-668.
[97] Lockhart, D.J. and Boxer, S.G. (1988) Proc. Nat. Acad. Sci. USA 85, 107-111.
[98] Braun, H.P. Michel-Beyerle, M.E. Breton, J. Buchanan, S. and Michel, H. (1987) FEBS Lett. 221,221-225.
[99] Vos, M.H., Lambry, J.-C., Robles, S.J., Youvan, D.C., Breton, J. and Martin, J.-L. (1992) Proc. Nat. Acad. Sci. USA 89, 613-617.
[100] Zankel, K.L., Reed, D.W. and Clayton, R.K. (1968) Proc. Natl. Acad. Sci. USA 61, 1243-1249.
[101] Witt, H.T. (1979) Biochim. Biophys. Acta, 505, 355-427.
[102] Saygin, Ö. and Witt, H.T. (1984) FEBS Lett. 176, 83-87.
[103] Hardt, H. and Malkin, S. (1973) Photochem. Photobiol. 17, 433-440.
[104] van der Wal, H.H., van Grondelle, R., Kingma, H. and van Bochove, A.C. (1982) FEBS Lett. 145, 155-159.
[105] Borisov, A.Yu., Godik, V.I. Kotova, E.A. and Samulov, V.D. (1980) FEBS Lett. 119, 121-124.
[106] Kotova, E.A., Samulov, V.D., Godik, V.I. and Borisov, A. Yu. (1981) FEBS Lett. 131, 51-54.
[107] Popovic, Z.D., Kovacs, G.J., Vincett, G.S. and Dutton, P.L. (1985), Chem. Phys. Lett.

116, 405-410.
[108] Popovic, Z.D., Kovacs, G.J., Vincett, G.S., Alegria, G. and Dutton, P.L. (1986), Chem. Phys. 110, 227-237.
[109] Arnold, W.A. and Azzi, R. (1971) Photochem. Photobiol. 14, 233-240.
[110] Meiburg, R.F., van Gorkom, H.J. and van Dorssen, R.J. (1983) Biochim. Biophys. Acta 724,352-358.
[111] Gopher, A., Blatt, Y., Okamura, M.Y. Feher, G. and Montal, M. (1983) Biophys. J. 41,121a
[112] Gopher, A., Schonfeld, M., Okamura, M.Y. and Feher, G. (1985) Biophys. J. 48, 311-320.
[113] Feher, G., Arno, T.R. and Okamura, M.Y. (1988) in: The Photosynthetic Bacterial Reaction Center. Structure and Dynamics. (Breton, J. and Vermeglio, A., eds.), pp.271-287, Plenum, New York.
[114] Packham, N., Mueller, P. and Dutton, P.L. (1982) Biophys. J. 37, 465-473.
[115] Franzen, S., Goldstein, R.F. and Boxer, S.G. (1990) J. Phys. Chem. 94, 5135-5149.
[116] Lockhart, D.J., Goldstein, R.F. and Boxer, S.G. (1988) J. Chem. Phys. 89, 1408-1415.
[117] Ogrodnik, A., Eberl, U., Heckmann, R., Kappl, M., Feick, R., Michel-Beyerle, M.E. (1990) J. Phys. Chem. 95, 2036-2041.
[118] Ogrodnik, A., Eberl, U., Heckmann, R., Kappl, M., Feick, R. and Michel-Beyerle, M.E. (1990) in Reaction Centers of Photosynthetic Bacteria (Michel-Beyerle, M.E., ed.), pp. 157-168, Springer Berlin.
[119] Ogrodnik, A., Eberl, U., Keupp, W., Kappl, M. and Michel-Beyerle, M.E. (1992) submitted to Biochim. Biophys. Acta.
[120] Lockhart, D.J., Kirmaier, Ch., Holten, D. and Boxer, S.G. (1990) J. Phys. Chem. 94, 6987-6995.
[121] Volk, M., Häberle, T., Aumeier, G., Ogrodnik, A., Feick, R. and Michel-Beyerle, M.E., to be publ.
[122] Popovic, Z.D., Kovacs, G.J., Vincentt, P.S., Alegria, G. and Dutton, P.L. (1986) Biochim. Biophys. Acta 851, 38-48.
[123] Ogrodnik, A., Eberl, U., Häberle, T., Keupp, W., Langenbacher, T., Siegl, J., Volk, M. and Michel-Beyerle, M.E. (1992) Biophys. J. Abs., in press.
[124] Langenbacher, T., Volk, M., Siegl, J., Eberl, U., Ogrodnik, A., Feick, R. and Michel-Beyerle, M.E., to be publ.
[125] Bixon, M. and Jortner, J. (1988) J.Phys. Chem. 92, 7148-7156.
[126] Ogrodnik, A. and Michel-Beyerle, M.E. (1989) Z. Naturforsch. 44a, 763-764.
[127] Eberl, U., Ogrodnik, A. and Michel-Beyerle, M.E. (1990) Z. Naturforsch., 45a, 763-770.
[128] Boxer, S.G., Lockhart, D.J., Franzen, S. and Hammes, S.H. (1990) in Reaction Centers of Photosynthetic Bacteria (Michel-Beyerle, M.E., ed.), pp. 147-156, Springer Berlin.
[129] Ogrodnik, A. (1990) Biochim. Biophys. Acta 1020,65-71.
[130] Wraight, C. and Clayton, R. (1973) Biochim. Biophys. Acta 333, 246-260.
[131] Breton, J., Martin, J.-L., Lambry, J.-C., Robles, S.J. and Youvan, D.C. (1990) in Reaction Centers of Photosynthetic Bacteria (Michel-Beyerle, M.E., ed.), pp. 293-302, Springer Berlin.
[132] Friese, M. (1991) Diploma thesis, Technical University München.
[133] Logunov, S.L. and Pashchenko, V.Z. (1989) Sov. J. Quantum Electron. 19, 88-91.
[134] Goldstein, R.A. and Boxer, S.G. (1989) Biochim. Biophys. Acta 977, 78-86.
[135] Dutton, P.L. (1988) Biophys. J. 53, 66a: M-PM-F8.
[136] Boxer, S.G., Lockhart, D.J., Hammes, Sh., Mazzola, L., Kirmaier, Ch., Holten, D., Gaul, D. and Schenck, C. (1990) in Current Research in Photosynthesis Vol.I (Baltscheffsky, M., ed.), pp. 113-116, Kluwer, Netherlands.

EXAMINING LONG-DISTANCE BIOLOGICAL ELECTRON TRANSFER WITH SYNTHETIC MODEL SYSTEMS

JONATHAN L. SESSLER*,†, VINCENT L. CAPUANO†, YUJI KUBO†, MARTIN R. JOHNSON†, DARREN J. MAGDA† and ANTHONY H. HARRIMAN‡

Department of Chemistry and Biochemistry† and Center for Fast Kinetics Research‡
The University of Texas
Austin, Texas 78712
USA

ABSTRACT. The use of synthetic model systems in elucidating pathway dependent mechanisms responsible for long-distance electron transfer is discussed. The syntheses of models composed of quinone-substituted, phenyl-linked porphyrin dimers and trimers, porphyrins with *meso*-substituted charge transfer complexes of $Ru(bpy)_3$ (bpy: 2,2'-bipyridyl) and porphyrin aggregates based on multi-point hydrogen bonding are presented along with their steady-state and dynamic photophysical properties.

1. Introduction

The fate of long-distance electron transfer (ET) reactions in proteins is dependent upon several, mutual factors. In order for efficient charge separation to occur over large distances (typically 10-25Å) the electronic coupling between the donor and the acceptor must be enhanced relative to the isolated donor/acceptor pair at such a distance. In terms of the nonadiabatic expression for long-range biological ET (eq 1), the enhanced electronic coupling is reflected in the electronic coupling

$$k_{ET} = \frac{2\pi(H_{AB})^2}{h(4\pi\lambda kT)^{1/2}} \exp\left[\frac{-(\Delta G^0+\lambda)^2}{4\lambda kT}\right] \qquad (1)$$

matrix element, H_{AB}[1]. The degree of electronic coupling is expected to be sensitive not only to the energetics of the donor and acceptor, but also to the nature of the protein environment between and around the redox pair. We have been interested in the factors which govern long-range electronic coupling in biochemical systems and have developed synthetic model systems designed to investigate the enhanced coupling associated with two bridge-mediated pathways:

a) chlorophyll-mediated systems and
b) multi-point hydrogen bonding patterns.

1.1. CHLOROPHYLL-MEDIATED SYSTEMS

1.1.1 *Photosynthetic Reaction Centers.* The initial events occurring within the reaction centers (RC's) of photosynthetic bacteria are long-distance ET reactions which can be classified as chlorophyll-mediated. Clearly, the elucidation of the mechanistic features of this process is a special objective within this class of chemical reactions. Structural information from X-ray diffraction experiments is available for the photosynthetic bacteria *Rhodopseudomonas viridis*[2] and *R. Sphaeroides*.[3] Figure 1 depicts the arrangement of photoactive subunits within the RC. The protein contains four bacteriochlorophylls (BChl), two bacteriopheophytins (BPh; structurally identical with the BChl except that the central magnesium ion is replaced with two hydrogen atoms) and two quinones (Q_A, tightly associated; Q_B, weakly associated). Two of the

E. Kochanski (ed.), Photoprocesses in Transition Metal Complexes, Biosystems and Other Molecules. Experiment and Theory, 375–401.

BChl's are held in close proximity by the protein matrix so that they behave electronically as a dimer ($BChl_2$) often referred to as the "special pair". The remaining vectorial electron transfer pathway consists of a monomeric BChl and a monomeric BPh which form the bridge between the initial electron donor $BChl_2$ and acceptor Q_A.

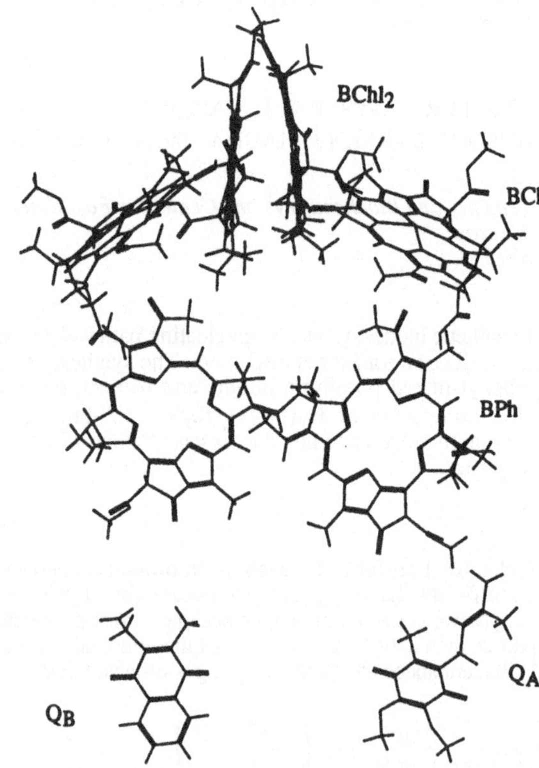

Figure 1. Redox-active chromophore involved in the initial charge separation for the photosynthetic bacteria *Rhodopseudomonas Viridis* taken from reference 2.

In addition to this structural information, a wealth of kinetic data has been obtained from time-resolved spectral experiments performed on native and modified reaction centers.[4] The primary charge separation which occurs within the RC takes place between the initially photoexcited dimer, $BChl_2^*$, and the pheophytin. The reduced pheophytin, BPh^-, appears within 2.8 psec of the initial excitation with a quantum yield approaching one. An ET reaction over a distance of ca. 17Å (center-to-center) in such a short time has no precedent. One area of mechanistic ambiguity which remains involves the role of the monomeric BChl. Several workers have suggested that an enhanced electronic overlap between $BChl_2^*$ and BPh is effected by the presence of the BChl (eq 2),[5] while others contend that the intermediate BChl is a discrete redox

$$BChl_2^*\text{-}BChl\text{-}BPh\text{-}Q_A \rightarrow BChl_2^+\text{-}BChl\text{-}BPh^-\text{-}Q_A \qquad (2)$$

$$BChl_2^*\text{-}BChl\text{-}BPh\text{-}Q_A \rightarrow BChl_2^+\text{-}BChl^-\text{-}BPh\text{-}Q_A \rightarrow BChl_2^+\text{-}BChl\text{-}BPh^-\text{-}Q_A \qquad (3)$$

partner (eq 3).[6] Attempts to identify $BChl^-$ spectrally, in which some recent success has been claimed,[6] have not resolved the controversy as to which mechanism is consistent with the experimental observations. Recently, several synthetic systems have been developed in an effort to resolve this mechanistic controversy as well as to clarify other factors which might govern long-distance biological ET reactions.

1.1.2. *Synthetic Models Related to Photosynthesis*. There are a number of organic and inorganic donor/acceptor systems which have contributed significantly toward the understanding of ET reactions and several reviews have appeared in the literature.[7] Given the scope of this discussion, however, only those synthetic systems incorporating the porphyrin macrocycle will be presented as this is not intended to be an exhaustive review. Representative examples of porphyrin-based model compounds designed to separate charge and/or investigate factors effecting ET rates are shown in Figure 2. The synthetic routes taken in the syntheses of compounds **I-III** are flexible enough to allow for the systematic variation of the factors controlling the ET rate such as distance, orientation and driving force. One critical characteristic that these three systems have in common is rigidity. Flexibility in a system which has been designed to investigate intramolecular interactions is fatal to the premise on which molecular modelling is built and, in the opinion of the authors, is something that should be strictly avoided.

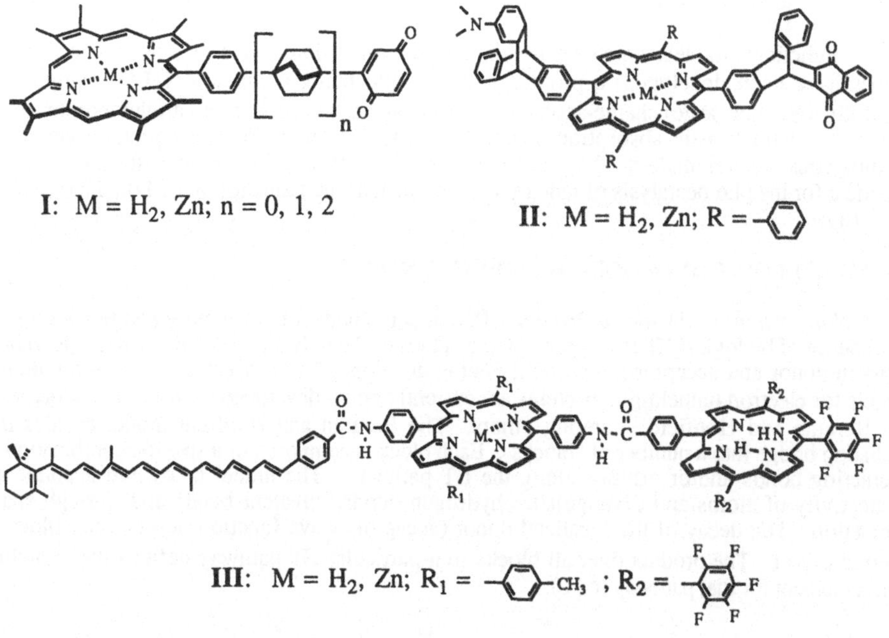

Figure 2. Covalently linked porphyrin-quinone systems designed for the investigation of photoinduced electron transfer. See references 8, 10 and 11.

The rigid bicyclo[2.2.2]octyl spacers of compounds I allow for the examination of distance effects upon reaction velocity without the concern for conformational motions.[8] In the simplest sense, the effect of distance on reaction rate is expected to decrease exponentially with the two-center one-electron integral between parallel aromatic molecules (eq 4). The constant α is

$$V(R) = V_0 e^{-\alpha R} \quad (4)$$

dependent upon the ET pathway and for propagation across a bicyclo[2.2.2]octyl spacer was determined by porphyrin fluorescence decay as $\alpha \geq 0.7$ Å^{-1}.[8] Alternatively, the effect of driving force on reaction rate can be examined at a fixed distance by systematically altering the exothermicity of the reaction. This has been done for a number of systems such as I and II in which a variety of substituted quinones were used to control ΔG. The Marcus expression (eq 5),

$$k = A \exp\left[\frac{-(\Delta G + \lambda)^2}{4\lambda k_B T}\right] \quad (5)$$

relating driving force to ET rate, predicts a maximum velocity occurring at $\Delta G = \lambda$.[9] The quantity λ is the total reorganization energy, including contributions from both inner-sphere vibrational reorganization and outer-sphere solvent reorganization. As predicted by eq 5, the ET rate was observed to decrease at $\Delta G > \lambda$ for systems **I** and **II** with λ = 1.0-1.3 eV and λ = 0.9 eV respectively.[8,10]

Functional models which act as solar energy conversion devices and not as mechanistic probes have been developed, a prototypical example being compound **III**.[11] Long-lived (typically μsec time scale) charge-separated states have been observed for carotenoporphyrins in which the characteristic absorption spectrum of the carotenoid radical cation serves as an unambiguous spectral marker. The generation of sufficiently long-lived charge-separated species is critical for the photocatalysis of redox processes such as the reduction of carbon dioxide ($CO_2 + e^- \rightarrow CO$).

1.2. MULTI-POINT HYDROGEN BONDING SYSTEMS

1.2.1. *Noncovalent Pathways in Proteins*. Practical methods for calculating and predicting the dependence of biological ET rates based on the details of the polypeptide environment intervening between donor and acceptor are currently being developed. For instance, detailed theoretical models for electron tunneling in proteins are currently being developed by Kuki and Wolynes[12] and Beratan and Hopfield,[13] among others. The Beratan and Hopfield model divides the mediating bridge into a number of "blocks". Each block is composed of a specific combination of interacting bonds and/or orbitals along the ET pathway. The model takes into account the connectivity of atoms and distinguishes hydrogen bonds, covalent bonds and through space interactions. The decay of the localized donor (acceptor) wave function across each block is referred to as ε. The product over all blocks in a particular ET pathway defines the tunneling matrix element for that pathway (eq 6).

$$t_{DA} = \text{prefactor} \prod_{i=1}^{N} \varepsilon_i \quad (6)$$

The prefactor is characteristic of the initial interaction between the donor (acceptor) and the first block of the pathway. The determination of ε for wave function propagation through hydrogen bonding networks is currently based on rough approximations to covalent systems[14] and a detailed analysis of this mediation mechanism is clearly needed. The parameters necessary to calculate the mediating effects of hydrogen bonding in an ET pathway may come from studying appropriate model systems.

1.2.2. *Model Systems Incorporation Hydrogen Bonding Arrays.* Advances in molecular recognition and self-assembly over the past decade have led to the development of sophisticated molecular aggregates.[15] The design of donor/acceptor systems linked *via* complementary multi-point hydrogen bonding arrays, however, has become practical only since 1990. At present, several groups are working to develop multichromophore structures incorporating molecular recognition features and electron/energy transfer properties.[15-17] However, with the exception of those systems developed in our laboratory (see Results section 2.1.2), the only known systems involving the porphyrin macrocycle are shown in Figure 3. Nonetheless, what makes this approach attractive is that it is quite amenable to electronic and steric modification of the chromophores, a sometimes formidable task when covalent attachments are required.

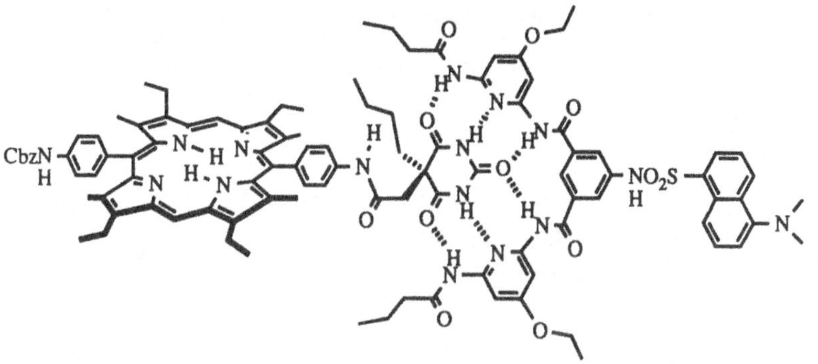

IV

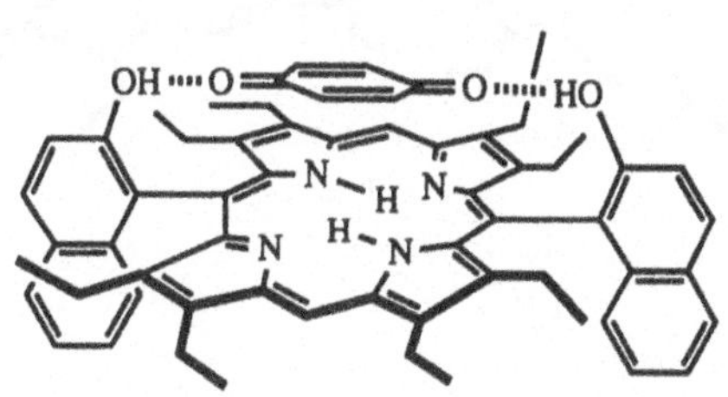

V

Figure 3. Model systems with donor/acceptor assembly *via* hydrogen bonding interactions. See references 16 and 17.

System **IV** relies on the complementarity between barbituate derivatives and two 2,6-diaminopyridine units for the aggregation of porphyrin and dansyl chromophores.[16] The synthetic methodology is flexible enough to allow for other fluorescent or redox components such as porphyrin, napthyl and ferrocene derivatives to replace the dansyl group. Energy transfer over ~23 Å was assessed by using time-resolved fluorescence spectroscopy and indicated efficient fluorescence quenching of the dansyl group by the aggregated porphyrin. The porphyrin-quinone system **V** is complexed not only by two-point hydrogen bonding, but also by the electrostatic or charge transfer interactions between the porphyrin and the quinone.[17] Porphyrin fluorescence in the aggregate is completely quenched by the complexed quinone as determined by fluorescence spectroscopy. Systems such as **IV** and **V** represent a new, general strategy for the construction of biologically significant structures in which ET reactions occurring through hydrogen bonding patterns can be investigated.

2. Results

The construction of elaborated donor/acceptor systems incorporating the porphyrin macrocycle necessarily involves a substantial amount of organic synthesis. We present here a brief description of our previously reported syntheses of systems composed of phenyl-linked, quinone-substituted porphyrin dimers[18] and cytosine-cytosine aggregated porphyrin dimers.[19] In addition, we will discuss in detail our latest synthetic efforts which have produced phenyl-linked, quinone-substituted porphyrin trimers,[20] porphyrin-bridged charge-transfer complexes of Ru, cytosine-cytosine aggregated porphyrin-quinone complexes and cytosine-guanine porphyrin-quinone complexes.[21] Steady-state and dynamic photophysical results will be presented in section 2.2.

2.1. SYNTHESIS

2.1.1. *Model Systems Related to Chlorophyll-Mediated Electron Transfer.* Our initial attempts at modelling events occurring within reaction centers of photosynthetic bacteria were aimed at compounds **1** and **2**.[18] They are composed of 1,3- ("gable") and 1,4- ("flat") phenyl-linked,

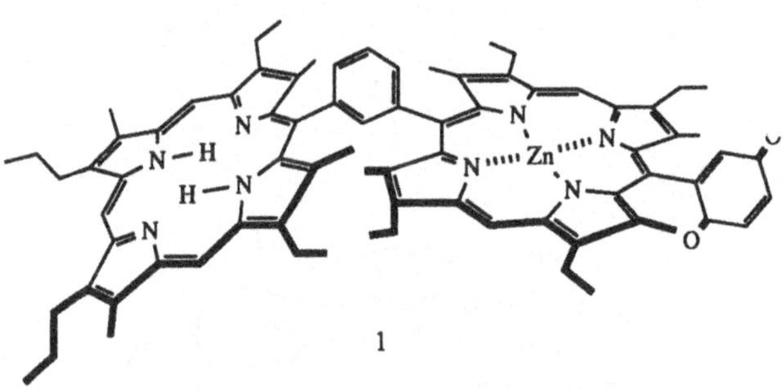

1

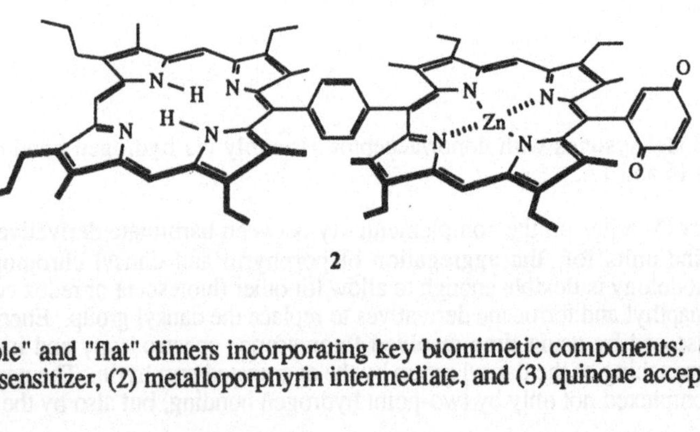

2

Figure 4. "Gable" and "flat" dimers incorporating key biomimetic components: (1) free-base porphyrin photosensitizer, (2) metalloporphyrin intermediate, and (3) quinone acceptor.

quinone-substituted ocataalkyl porphyrin dimers. Scheme I outlines the final steps in the synthesis of compound **1**. The porphyrin-forming reaction between tetra(2'-unsubstituted)pyrromethane **3** and di(2'-formyl)pyrromethanes **4** and **5** gave the dimethoxy-substituted porphyrin dimer **6** in 7-9% yield after careful purification by column chromatography. Deprotection of the hydroquinone dimethyl ether with BBr_3 and subsequent oxidation with DDQ afforded the free base quinone-substituted dimer. Titration of the free-base hydroquinone with one equivalent of zinc acetate, prior to DDQ oxidation afforded predominantly the monometallated dimer with the regiochemistry shown in **1**.

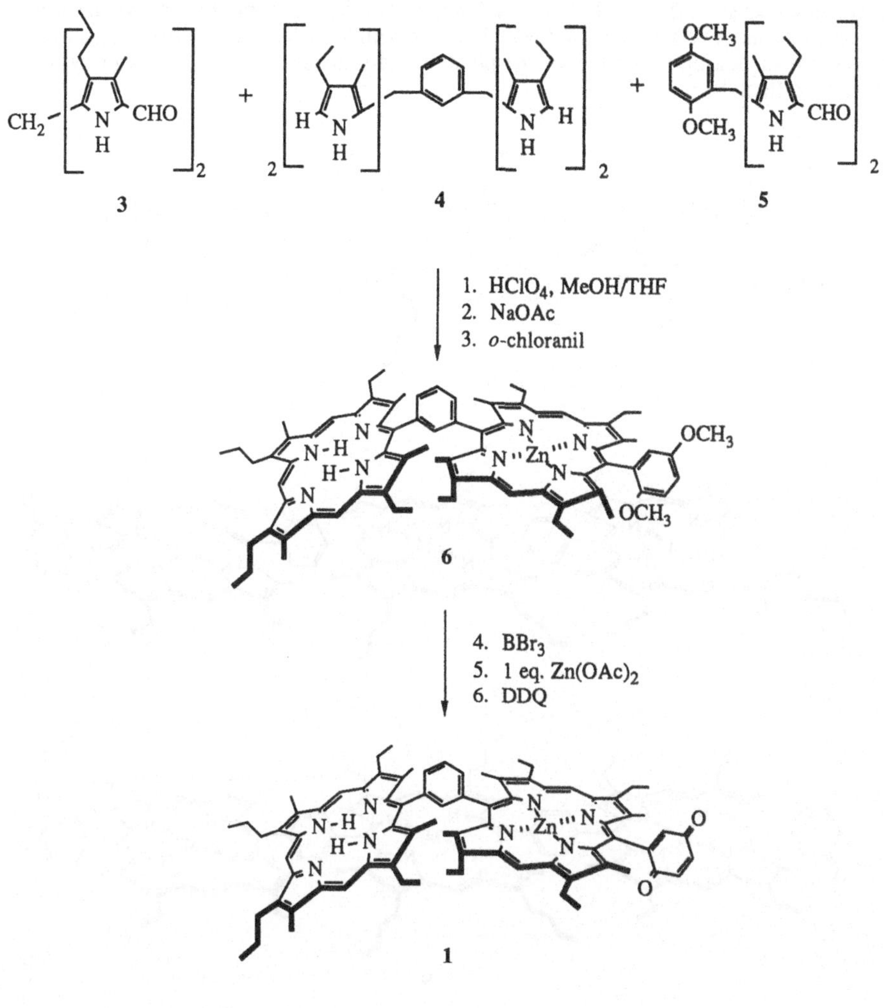

Scheme I. Final synthetic sequence and monometallation strategy for the 1,3- ("gable") phenyl-linked porphyrin dimer **1**.

An important feature of the dimeric models is that they are conformationally rigid and structurally well defined (X-ray diffraction structures of certain bis-copper derivatives have been obtained in both the 1,3-and 1,4-phenyl linked series). The edge-to-edge porphyrin separation is 5.01 and 5.86 Å in the gable and flat dimers, respectively. The distant (distal) porphyrin to quinone edge-to-edge distance is estimated to be 12.4 and 14.0 Å in the gable and flat systems, respectively. In all complexes, the edge-to-edge distance between the quinone and the adjacent porphyrin is 1.5 Å. These structural parameters will prove valuable when analyzing the distance and orientation effects on the ET rate (see Discussion).

In order to investigate the mediating properties of the porphyrin macrocycle in ET reactions we sought to develop a series of compounds in which the only variable was an incremental change in the number of bridging macrocycles. Toward this end we have recently completed a study of compounds **7-9** which, along with the "flat" dimer system **2**, allows us to compare the mediating properties of the phenylporphyrin spacer to that of other systems (eg. bicyclo[2.2.2]octyl).

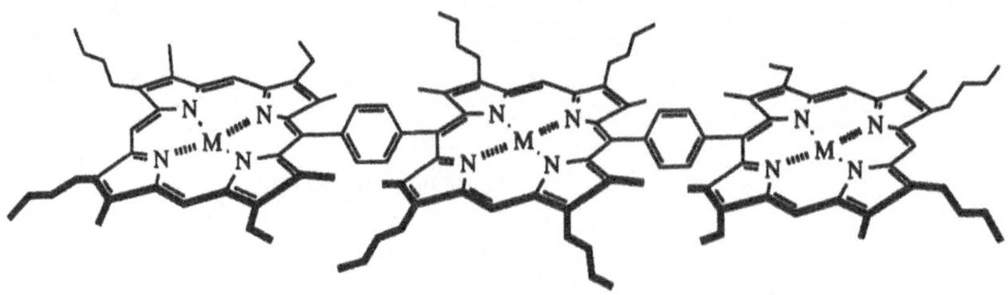

7

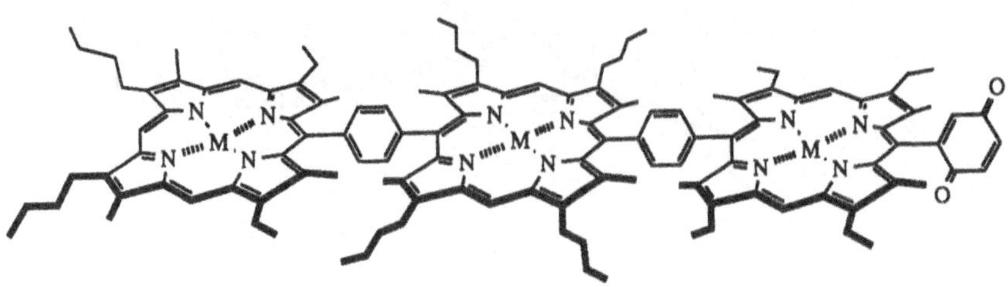

8

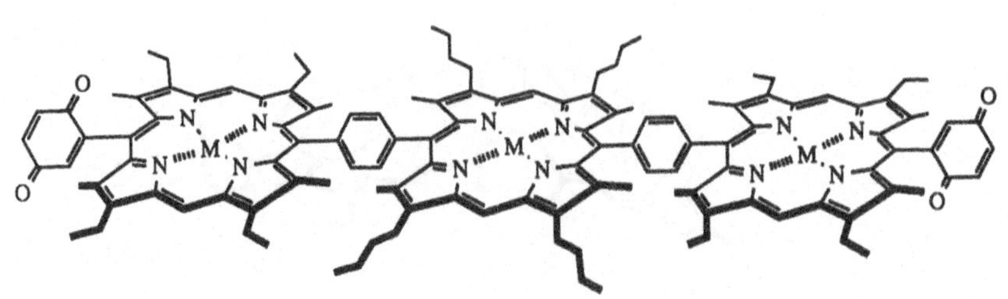

9

Figure 5. Quinone-substituted, phenyl-linked trimeric porphyrins (M = H_2, Zn, Ni).

The synthesis of the phenyl-linked trimeric porphyrins relies on the availability of several key, functionalized porphyrin monomers. The key porphyrin aldehydes **15** and **17** were prepared by methods which we[20] and others[22] have recently reported and is shown in scheme II. Specifically, dipyrrylmethane **12** was prepared by the acid-catalyzed condensation of 4-(hydroxymethyl)benzaldehyde **10** (introduced in the form of its diethyl acetal) with ethyl 3-ethyl-4-methyl-2-pyrrolecarboxylate **11** (75% yield) and was saponified and decarboxylated in boiling ethylene glycol to give **13** in 95% yield). Compound **13** was then condensed with the known di(formylpyrryl)methanes **3** and **5** under conditions of acid catalysis in THF/MeOH (2/1) to give the hydroxymethylphenyl-substituted porphyrin monomers **14** and **16**, the direct precursors to **15** and **17**, respectively. Finally, oxidation with pyridinium dichromate and subsequent recrystallization from $CHCl_3$/hexanes completes the preparation of the intermediates (**15** and **17**) in good yield.

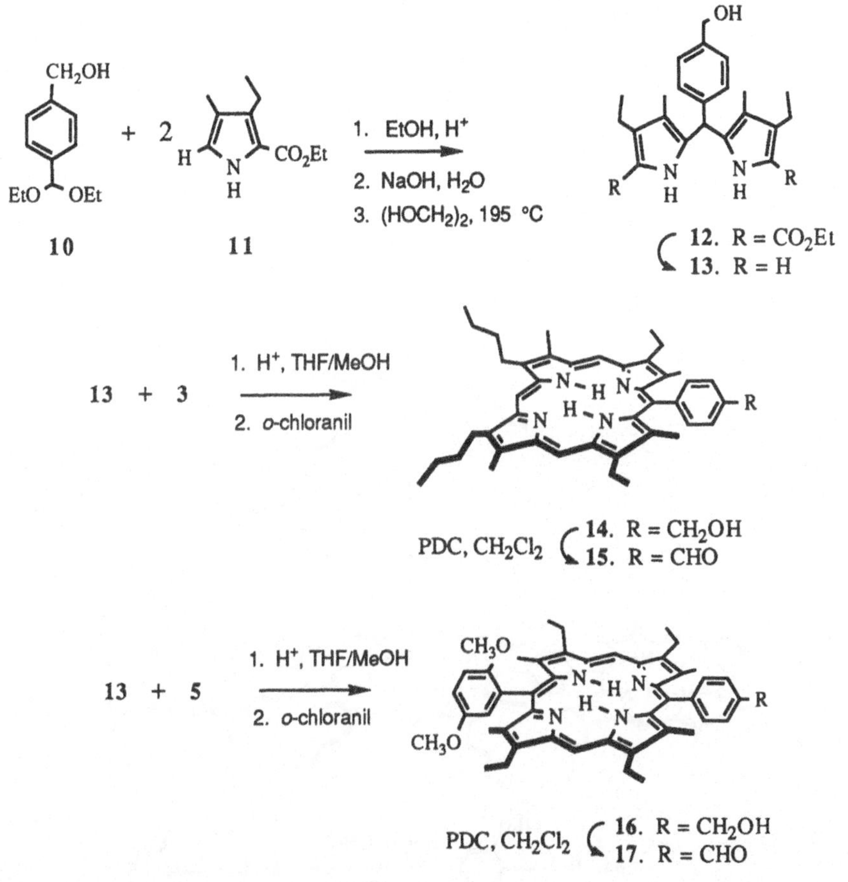

Scheme II. Syntheses of functionalized porphyrin monomers and precursor dipyrrylmethanes.

The syntheses of the trimeric porphyrins **7-9** are based on the condensation between the α-unsubstituted dipyrrylmethane **18** and the functionalized porphyrins **15** and **17** using a modification of a procedure used to form simpler, symmetric porphyrin trimers.[22] The central porphyrin is constructed in the final bond-forming sequence. Using the conditions optimized by

Lindsey[23] for porphyrinogen-based tetraphenylporphyrin synthesis (reactant concentrations of 10^{-2} M in CH_2Cl_2 with trifluoroacetic acid (10^{-3} M) as the acid catalyst), we have been able to isolate pure trimeric porphyrins in >80% yield. This fact is not surprising when one considers that the use of dipyrrylmethanes and aryl aldehydes requires the formation of half the number of C-C bonds as the Lindsey procedure, which involves monomeric pyrroles. The pathway to the monoquinone trimer **8** is outlined in Scheme III. Thus, under the reaction conditions described, a 1:1:1 mixture of precursors **15, 17,** and **18** gave compounds **7, 8a** and **9a** (the hydroquinone methyl ether corresponding to bis-quinone **9**) in roughly a 1:2:1 ratio which could be separated by careful chromatography on silica gel. Compound **8a,** obtained in 23% yield, was then treated with BI_3 in CH_2Cl_2 followed by DDQ to arrive at the monoquinone trimer **8**. Trimers **7** and **9a** were obtained independently by using compound **18** with the appropriate porphyrinaldehyde, **15** or **17**, respectively.

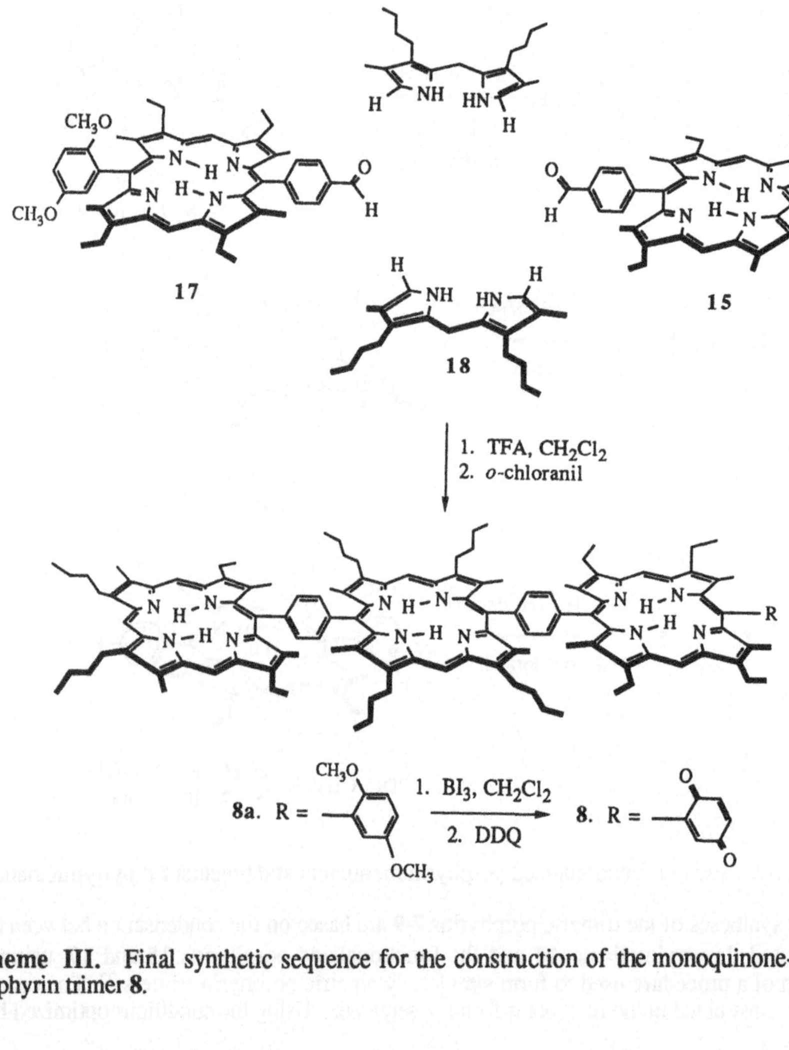

Scheme III. Final synthetic sequence for the construction of the monoquinone-substituted porphyrin trimer **8**.

In order to vary the photo- and electrochemical properties of the macrocycles relative to one another, we are currently examining the metal binding selectivities of compounds **7**- **9** with a variety of metal cations (eg. Zn^{2+}, Ni^{2+}, Cu^{2+}). One approach toward selective metallation involves the titration of one or two equivalents of the metal carrier into a solution of the trimeric porphyrin. For example, titration of 2 equivalents of $NiCl_2$ (solution in methanol) into a 5% MeOH/$CHCl_3$ solution of **7** gives, predominantly the bis-Ni species in which metallation occurs at the outer porphyrins. The selectivity in the formation of this regioisomer, denoted **NiH$_2$Ni-7**, is probably a result of the greater steric accessibility of the outer porphyrins relative to the inner porphyrin. In addition to high resolution mass spectral data, the regiochemistry of the metal binding is supported by ^{1}H NMR spectroscopy. As shown in the figure below, examination of the *meso* and bridging phenyl proton resonances for compounds **7**, **NiH$_2$Ni-7** and the fully metallated **Ni$_3$-7** reveals the regiochemical selectivity of the metal binding.

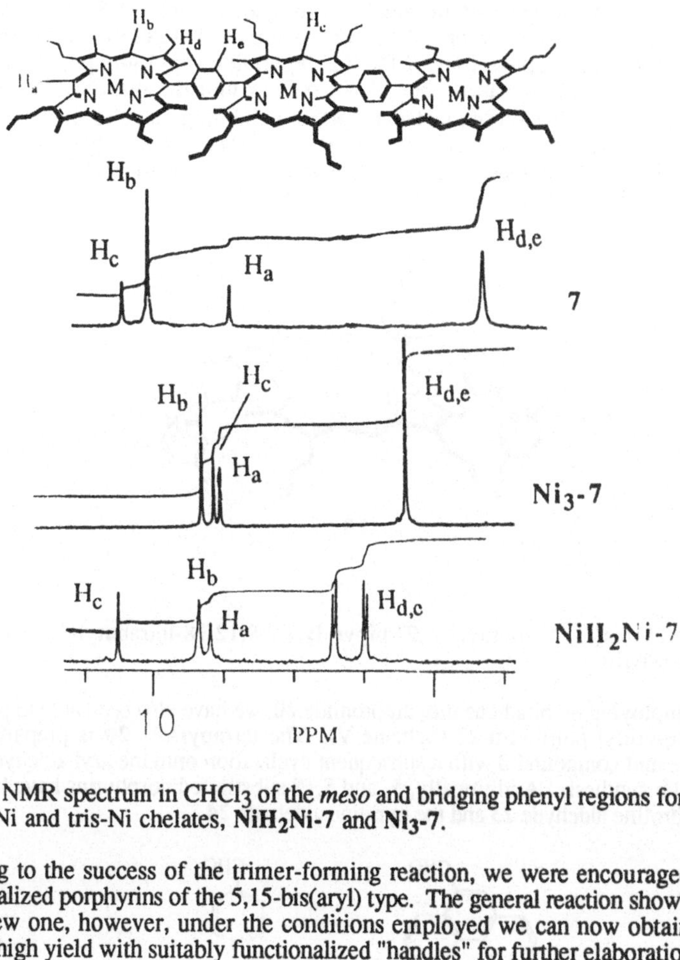

Figure 6. ^{1}H NMR spectrum in $CHCl_3$ of the *meso* and bridging phenyl regions for compound **7** and its bis-Ni and tris-Ni chelates, **NiH$_2$Ni-7** and **Ni$_3$-7**.

Owing to the success of the trimer-forming reaction, we were encouraged to prepare other functionalized porphyrins of the 5,15-bis(aryl) type. The general reaction shown in Scheme IV is not a new one, however, under the conditions employed we can now obtain porphyrin monomers in high yield with suitably functionalized "handles" for further elaboration. Standard methods for the preparation of functionalized porphyrins employ variations on tetraphenylporphyrin syntheses and are plagued by the difficulties associated with the separation of the desired porphyrin from the many unwanted side-products. Through the use of 2,2'-unsubstituted dipyrrylmethanes, the target porphyrin can be obtained typically in > 80% yield.

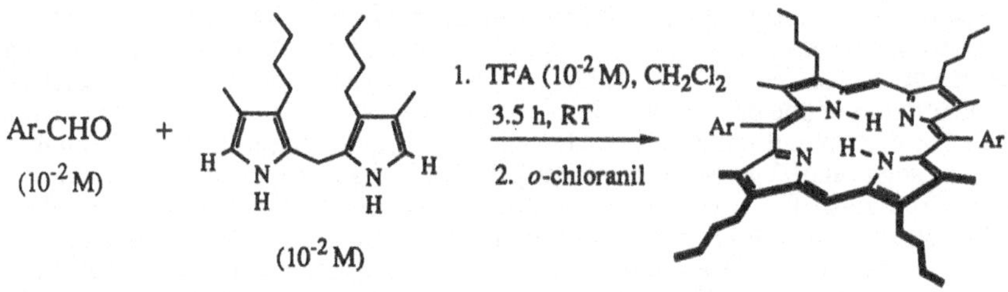

Scheme IV. Reaction conditions for the preparation of 5,15-bis(aryl)porphyrins.

One particular 5,15-bis(aryl)porphyrin in which we have been interested is porphyrin **19** shown below. With the appropriate aldehyde (prepared from the SeO_2 oxidation of 4,4'-dimethyl-2,2'-bipyridine) this compound has been synthesized as per the conditions described in Scheme IV. The ligating properties of the 2,2'-bipyridine (bpy) group make this porphyrin unique in that metal cations can be bound not only within the macrocycle, but also to the bipyridine ligand(s).[24] In addition, metal complexes exhibiting metal-to-ligand charge transfer (MLCT) can be constructed in which the charge transfer complex is held in a rigid manner orthogonal to the porphyrin plane. This may allow us to examine the ability of the porphyrin to bridge, electronically, redox active complexes such as $Ru(bpy)_3$.

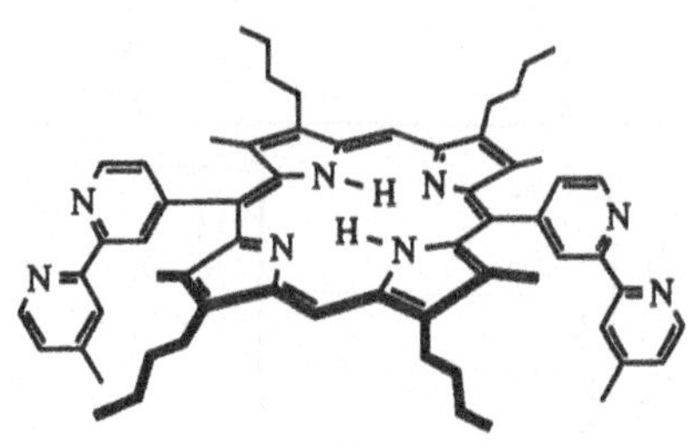

19

Figure 7. 5,15-bis(4'-(4''-methyl)-2',2''-bipyridyl)-2,8,12,18-tetrabutyl-3,7,13,17-tetramethylporphyrin.

By employing *ac*-biladiene dihydrobromide **20**, we have also been able to prepare the 5-substituted bipyridyl porphyrin **22** (Scheme V). The tetrapyrrole **20** is prepared from 3,4-diethylpyrrole and compound **3** with a subsequent cyclization onto the aryl aldehyde **21** serving to complete the synthesis. Additionally, 5- and 5,15-substituted porphyrins have been obtained with phenanthroline aldehyde **23** and the pyridine aldehyde **24**.

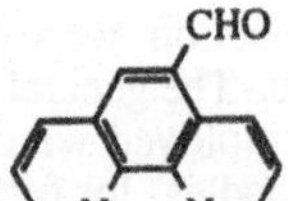

23

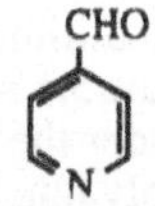

24

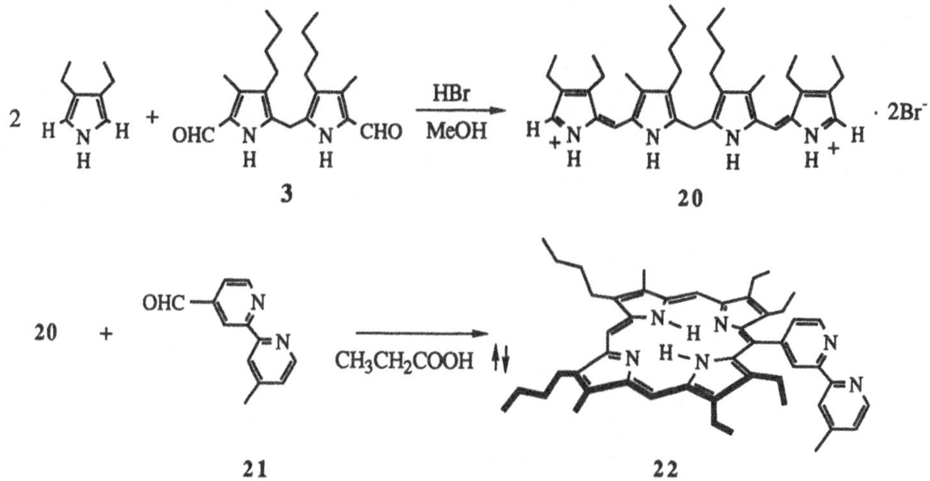

Scheme V. Synthesis of 5-substituted bipyridyl porphyrin **22** via *ac*-biladiene dihydrobromide **20**.

Reaction of 5,15-bis(bpy)porphyrin **19** with an excess of *cis*-Ru(bpy)$_2$Cl$_2$ in refluxing EtOH resulted in the formation of Ru(bpy)$_3$ complexes in which metal centers are chelated in both the 5- and 15- positions of the porphyrin **23** (Scheme VI). The reaction was also successful with **Zn-19**. Alternatively, the porphyrin could be metallated subsequent to complexation with Ru. The presence of both the MLCT band of the charge transfer complex and the chromophores corresponding to the porphyrin macrocycle were evident by UV-VIS absorption spectroscopy (see steady state photophysical results, section 2.2.1).

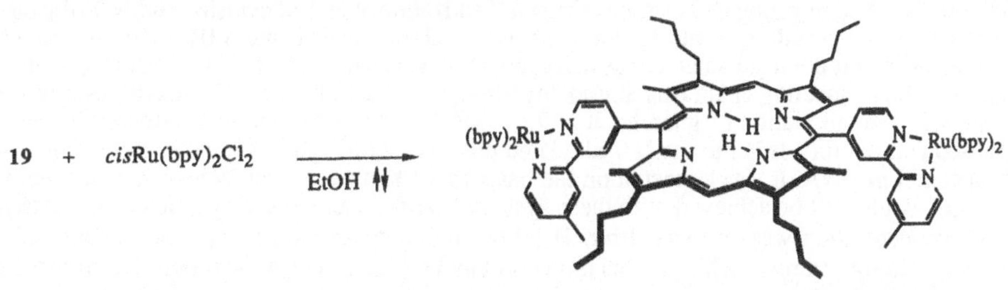

Scheme VI. Formation of *meso*-substituted porphyrin with MLCT complexes of Ru(bpy)$_3$.

2.1.2. *Model Systems Based on Hydrogen Bonding Aggregates.* Our approach toward the construction of model systems in which the ET pathway involves a hydrogen bonding pattern is based on the self-association of nucleic acid bases.[25] This design is outlined in Scheme VII whereby a nucleic acid base has been covalently bound to a porphyrin or a quinone. Upon mixing in aprotic solvents, self-association of the nucleic acid bases occurs *via* multi-point hydrogen bonding to generate a non-covalent, photo-redox ensemble. Excitation leads to energy and electron transfer across the hydrogen bonded zinc-free base porphyrin dimer and the porphyrin-quinone dimer, respectively. Porphyrin-cytosine, porphyrin-guanine and quinone-cytosine conjugates have been synthesized in our laboratory and this section outlines the details of their preparation. The photochemical reactivity of these aggregates has been investigated and is discussed in section 2.2.2.

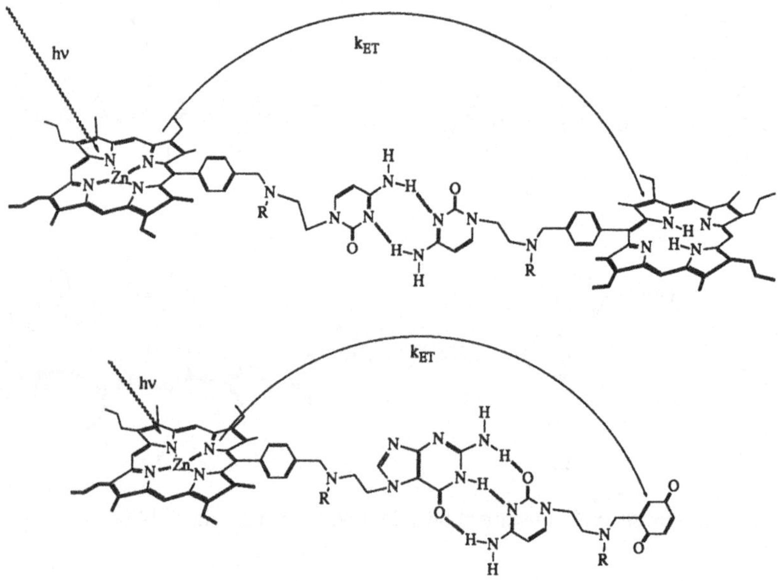

Scheme VII. Structures of the models and schematic representation of energy and electron transfer across hydrogen bonded supramolecular aggregates. M = H_2, Zn; R = polyether solubilizing group.

The inherent molecular recognition features of the DNA bases guanine and cytosine were chosen as the aggregating units. Aminoethyl derivatization at N-1 of cytosine and N-7 of guanine was achieved according to the synthetic sequence shown in Scheme VIII. Due to the poor solubility in non-polar solvents, it was necessary to attach a polyether solubilizing group at the N-atom in the connecting chain (the aminoethyl group). A key element of the synthesis was the regioselective alkylation of guanine at N-7. Alkylation of guanosine with ethylene oxide and subsequent hydrolysis led to the N-7 alkylated product. Although this regioisomer differs from that found in DNA, it was expected on the basis of CPK models that Watson-Crick type base pairing would still be achieved with these systems. Self-association of the nucleic acid bases at high concentration was confirmed by ^{1}H NMR studies made with compounds **24** and **25** in $CDCl_3$ solution, from which an association constant of 1220 ± 200 M^{-1} was derived for cytosine guanine base pairing.

The formation of cytosine-porphyrin, guanine-porphyrin and cytosine-quinone conjugates relies on the reactivity of the appended aminoethyl group on the base with an appropriate electrophile located on the porphyrin macrocycle. We have found that the derivatized porphyrins used in the porphyrin trimer synthesis can also be incorporated into a useful strategy for the formation of these systems as well. Accordingly, the base-assisted reductive amination of the porphyrin aldehyde **15** with the cytosine derivative **25**, followed by trityl group deprotection, results in the formation of the cytosine porphyrin **29** (Scheme IX). Similarly, reaction of the guanine derivative **27** with porphyrin **15** gives the guanine-porphyrin conjugate **30** under identical reaction conditions. Liberation of the free guanine derivative is then achieved by removing the trityl protecting group by treatment with NaOMe in MeOH. The cytosine quinone system **34** is prepared in an analogous manner and relies on two successive reductive

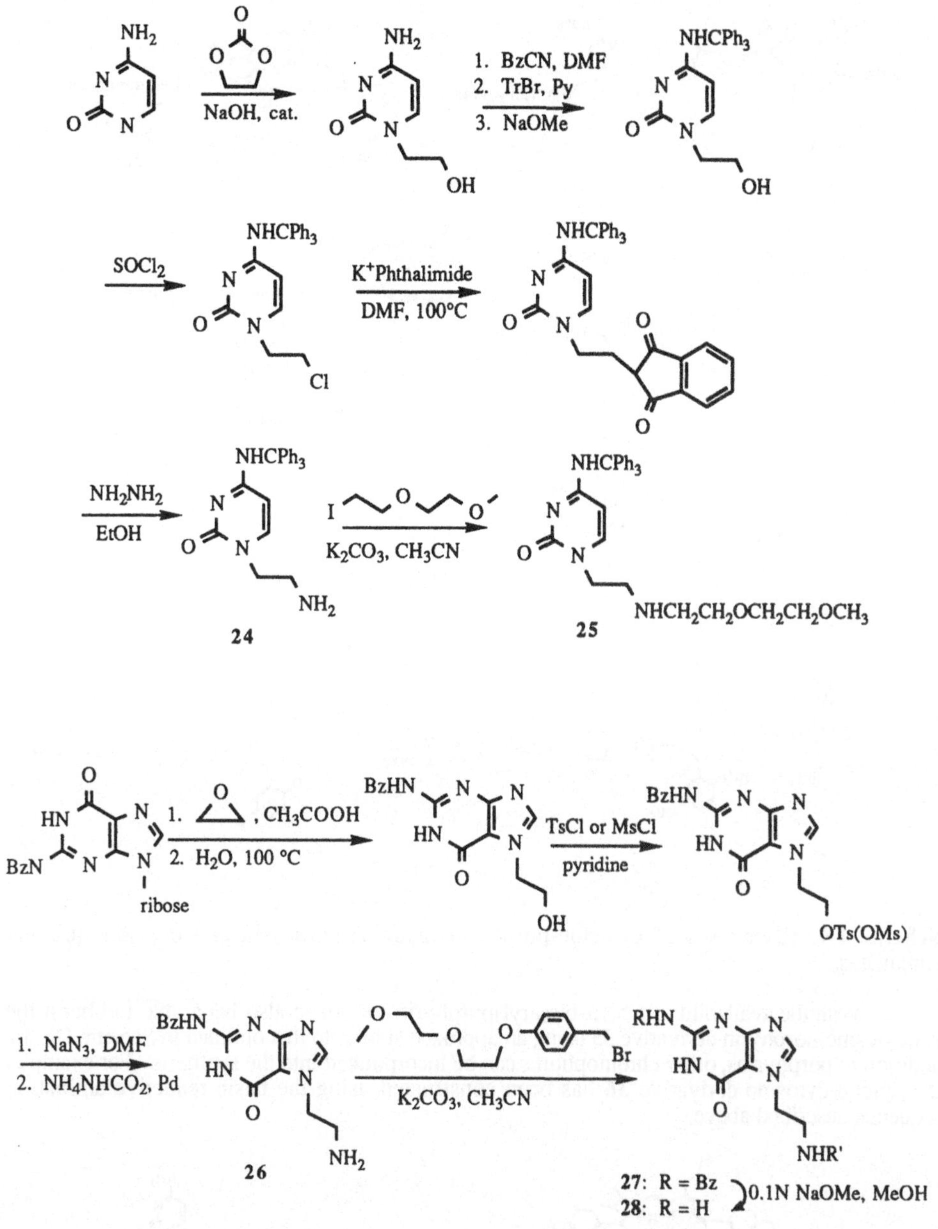

Scheme VIII. Synthetic sequences for the preparation of aminoethyl-cytosine and aminoethyl-guanine derivatives.

aminations including polyether-aldehyde **32** and 2,5-dibenzyloxybenzaldehyde **33**, respectively. All new compounds give satisfactory high-resolution mass spectra, ^{1}H NMR and ^{13}C NMR spectra.

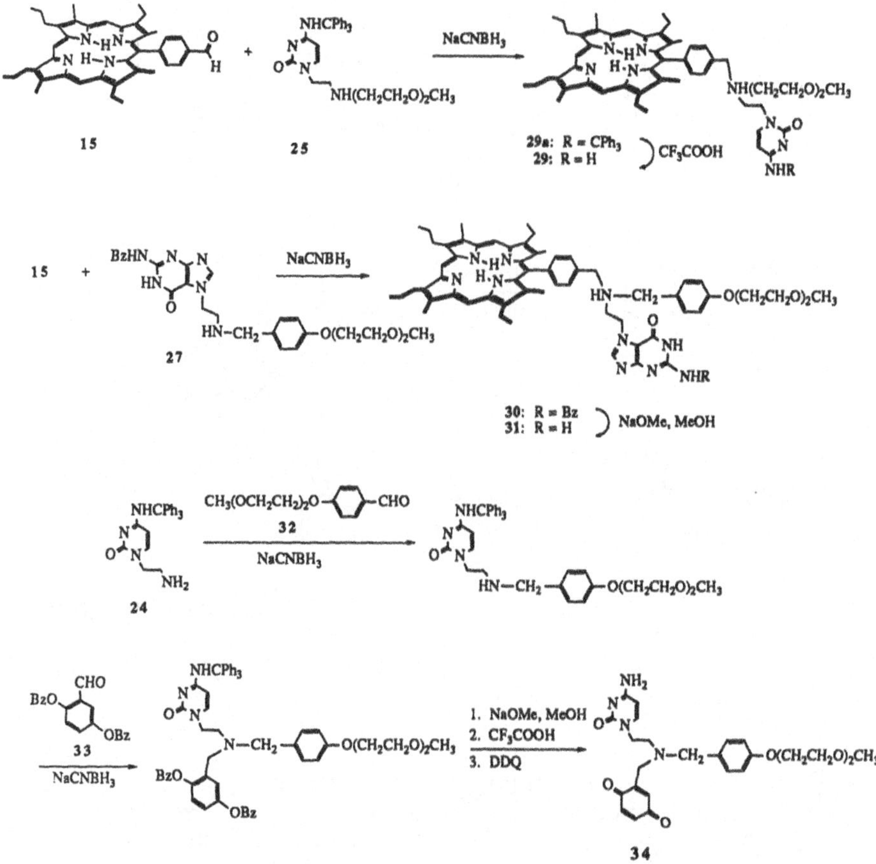

Scheme IX. Syntheses of cytosine-porphyrin, guanine-porphyrin and cytosine-quinone conjugates.

With the availability of 5,15-bis(aryl)porphyrins we have also been able to obtain the bis(cytosine)porphyrin derivative **35** using an approach similar to that outlined in Scheme IX. In addition to porphyrins, other chromophores can be incorporated into the synthesis. For example, the pyrene-cytosine derivative **36** has been constructed using the basic reductive amination sequence described above.

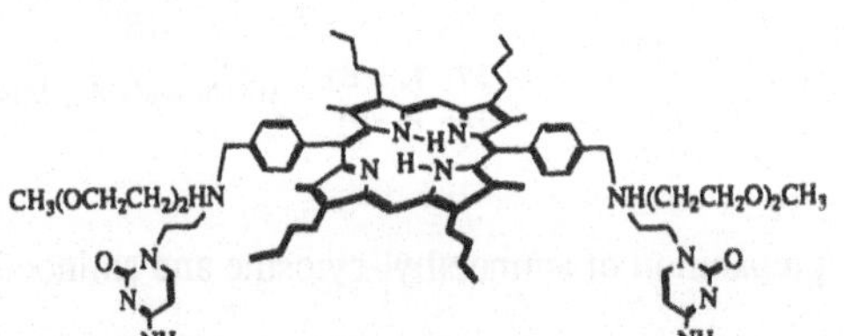

35

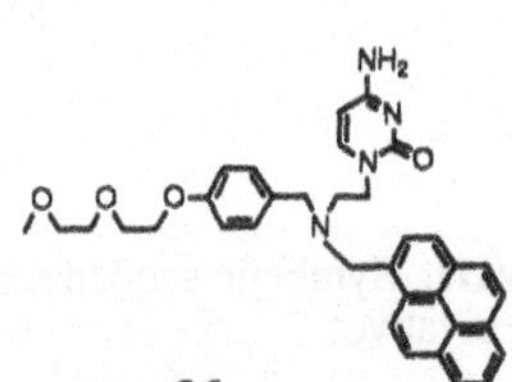

36

Figure 8. Bis(cytosine)-porphyrin **35** and pyrene-cytosine conjugate **36** prepared using an approach similar to that outlined in Scheme IX.

Without the solubilizing polyether group, the primary aminoethyl sidearm could be alkylated with two equivalents of bromomethyl porphyrin **37** (prepared by the treatment of **14** with HBr/HOAc) to give cytosine-bis(porphyrin) **38** (Scheme X). Zinc complexes of all porphyrin-nucleobase derivatives were prepared by conventional methods.

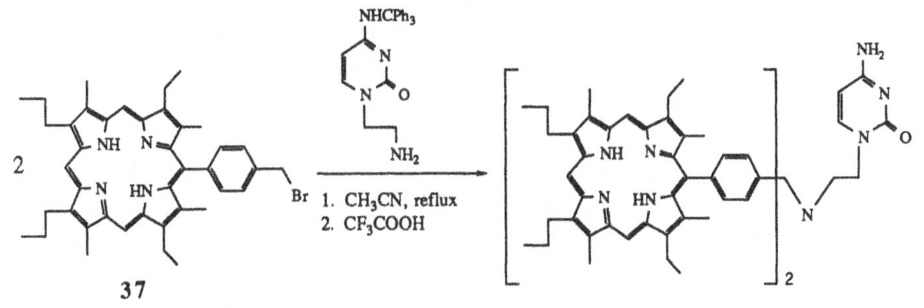

Scheme X. Synthesis of cytosine-bis(porphyrin) derivative **38**.

2.2. STEADY-STATE AND DYNAMIC PHOTOCHEMISTRY

In this section we present the results of static and dynamic spectral experiments performed on the model compounds described above. The techniques employed are the conventional methods now standard in the study of photoinduced electron transfer and include steady state absorption and emission spectroscopy, time-correlated fluorescence spectroscopy operating in the single photon counting mode, and laser flash photolysis. Results for quinone-substituted dimers **1** and **2** and cytosine-cytosine porphyrin aggregates **29** and **35** have been described previously in the literature and will be mentioned only briefly in this section.

2.2.1. *Steady-State Absorption and Emission Spectroscopy.* Ground state absorption features characteristic of free-base (H_2)porphyrin macrocycles include a strong near-UV "Soret" transition ($S_2 \rightarrow S_0$) along with four "Q bands" ($S_1 \rightarrow S_0$) typically near 505, 535, 570, and 625 nm. The ground state spectral properties of phenyl-linked porphyrins differ dramatically from those of isolated porphyrin monomers. In other words, the three porphyrin chromophores in compounds **7**, **Ni$_3$-7** and **NiH$_2$Ni-7** do not act as three independent light absorbing entities as evidenced by their corresponding absorption spectra shown in Figure 9. Rather, as has been observed for other phenyl-linked porphyrin systems,[26] these compounds show evidence of optical coupling as revealed by split, broadened, and/or shifted Soret bands. This optical coupling has been ascribed to excitonic interactions and a detailed analysis of the coupling parameters in the gable and flat dimers has been performed.[27] The extent of coupling becomes more pronounced upon complexation with metal cations such as Zn^{2+} and Ni^{2+}, with selectively complexed trimers such as **NiH$_2$Ni-7** showing coupling which is intermediate between the metal-free trimer **7** and the fully-metallated trimer **Ni$_3$-7**. This kind of electronic coupling in a supramolecular system is similar to that which occurs within the reaction centers of photosynthetic bacteria although the coupling mechanisms operating are probably different. Reaction center chromophores do not behave as isolated macrocycles and often exhibit absorption properties characteristic of strongly coupled chlorophyll species. It is this coupling between macrocycles, or orbital exchange, which makes available pathways for electron transfer *via* what is commonly referred to as superexchange.

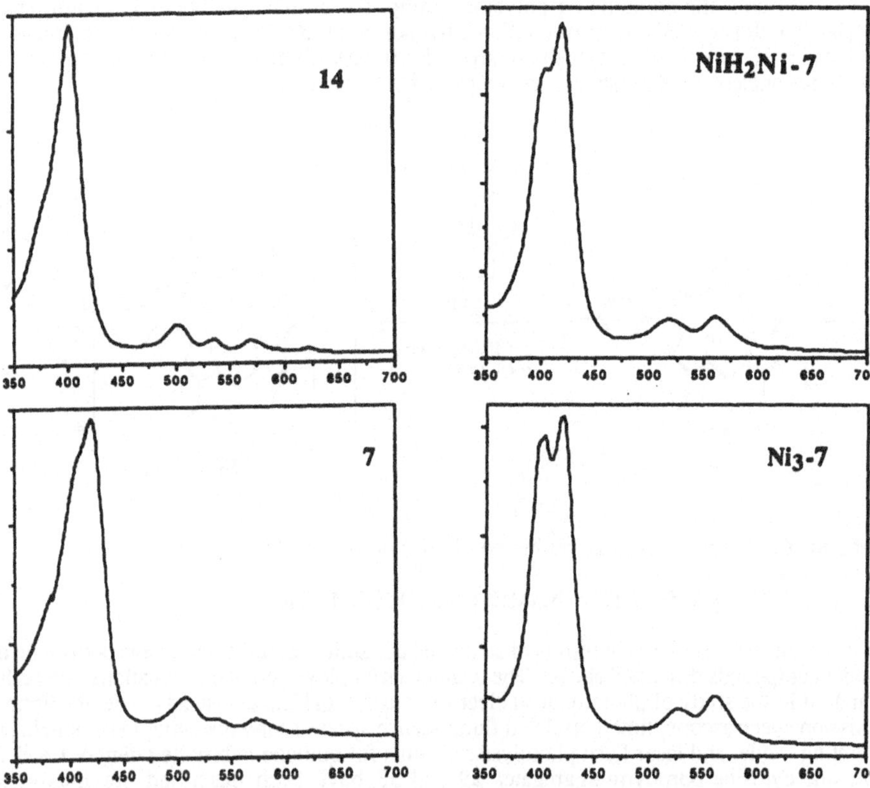

Figure 9. The UV-visible spectra of compounds 14, 7, Ni_3-7, and NiH_2Ni-7.

The UV-visible spectra for porphyrin-nucleobase conjugates are essentially identical to those for monomeric porphyrins (eg. **14**) as are those for 5- and 5,15-bipyridyl porphyrins **19** and **22**. The absorption spectrum for 5,15-bis(Ru(bpy)$_3$)-porphyrin **23** is shown below. In addition to the characteristic free-base porphyrin bands, the spectrum reveals the presence of the metal-to-ligand charge transfer band of the complex at ~455 nm. A nearly identical spectrum was observed by Hamilton for a Ru(bpy)$_2$ complex "strapped" above the plane of a porphyrin macrocycle.[24]

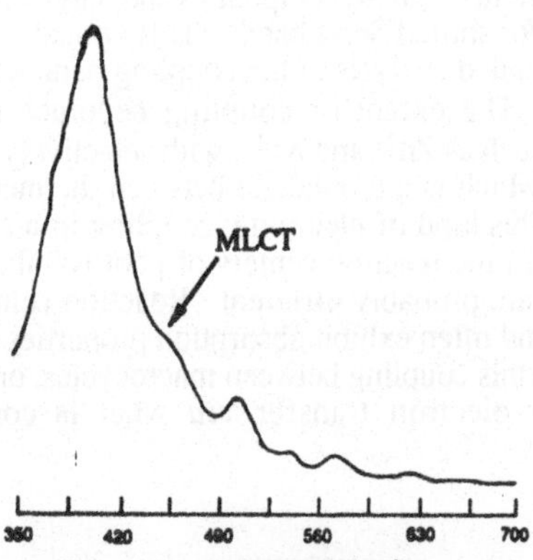

Figure 10. UV-visible spectrum of 5,15-bis(Ru(bpy)$_3$porphyrin^{4+} · $4PF_6^-$ (23).

In linked porphyrin-quinone systems the fate of the photoexcited porphyrin is governed by the competing processes of radiative return to the ground state and non-radiative electron transfer to the quinone (Scheme XI). This situation makes the examination of steady-state

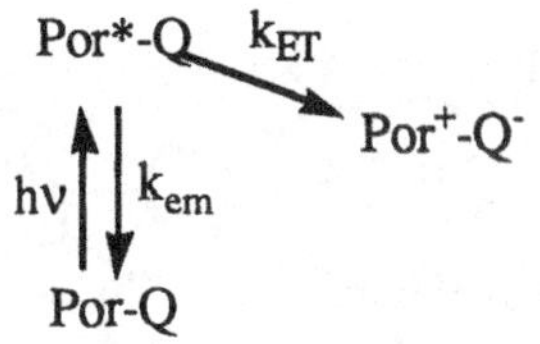

Scheme XI. Reaction sequence illustrating the competing events of radiative decay to ground state (k_{em}) and electron transfer to quinone (k_{ET}) in porphyrin-quinone systems.

porphyrin fluorescence quenching valuable in terms of calculating rates for the electron transfer event. It is possible to carry out a kinetic analysis by comparing the relative quantum yield of fluorescence for systems undergoing the reaction sequence of Scheme XI with those in which the electron transfer reaction does not compete.[28] For example, trimeric porphyrins **7**, **8**, and **9** are all capable of photoexcitation and radiative return to ground state. However, fluorescence quenching *via* electron transfer to a quinone acceptor is possible for **8** and **9** but not for **7**. Figure 11 shows the steady state fluorescence spectra for equimolar concentrations of compounds **7**, **8**, and **9**. Fluorescence quenching in compounds **8** and **9** is attributed to electron transfer from the photoexcited porphyrins to the appended quinone(s). The increased fluorescence intensity of compound **8**, relative to compound **9**, is attributed to emission from the porphyrin farthest from the quinone in this unsymmetrical system.

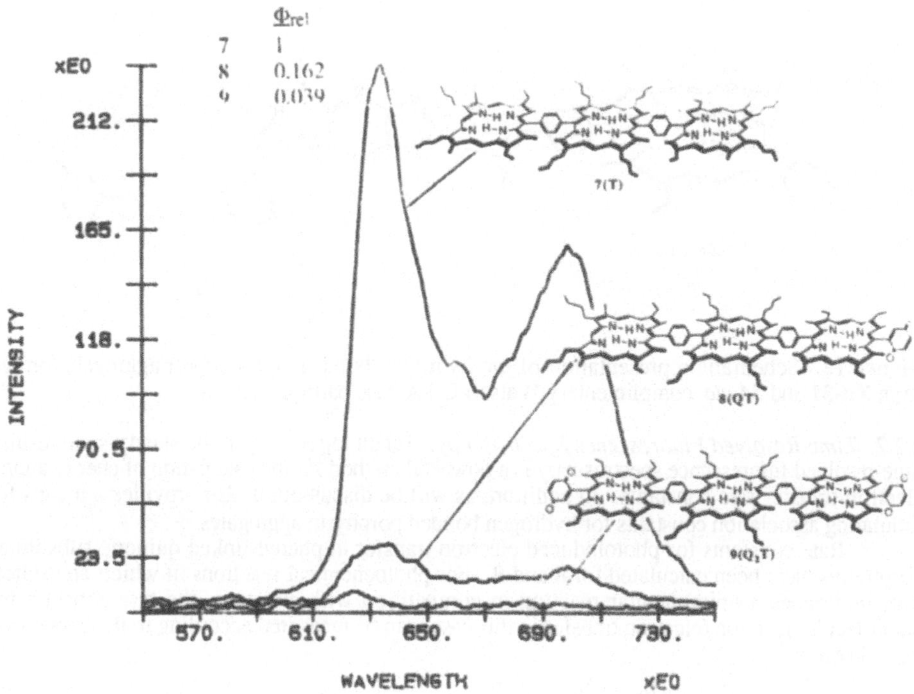

Figure 11. Steady state fluorescence spectra of compounds **7**, **8**, and **9** ($\lambda_{excit.}$ = 410 nm).

In addition, the ca. 4% relative fluorescence of compound **9** is significantly greater than that for an equimolar amount of bis(quinone) dimer **39**, in which no detectable fluorescence is observed. This relatively higher intensity is thus attributed to the central porphyrin in **9**.

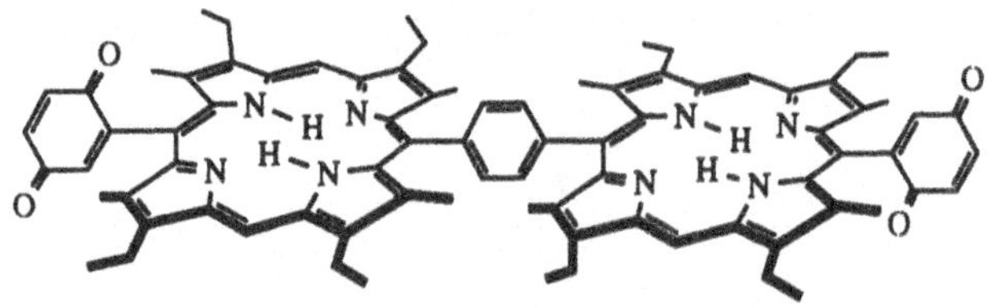

39

Fluorescence quenching of the guanine-porphyrin conjugate **Zn-31** by the addition of the cytosine-quinone conjugate **34** can also be followed by steady state emission spectroscopy. In CH_2Cl_2, fluorescence from the porphyrin subunit in **Zn-31** was quenched upon the addition of high concentrations of **34**. The fluorescence quantum yield (excitation $\lambda = 530$ nm) decreasing with increasing concentration of **34** until a limiting value, corresponding to ca. 35% quenching, was reached. Replacing **34** with 2-methyl-1,4-benzoquinone failed to effect fluorescence quenching at the relevant concentration (< 0.02 M). Additionally, **34** failed to quench the fluorescence of zinc octaethylporphyrin at concentrations below 0.02 M. These steady state results are consistent with the model shown in Figure 12 below in which aggregation *via* multi-point hydrogen bonding leads to the donor-acceptor pair capable of "intramolecular" electron transfer. Further support for this model has been obtained through time-resolved fluorescence experiments. These are described in the following section.

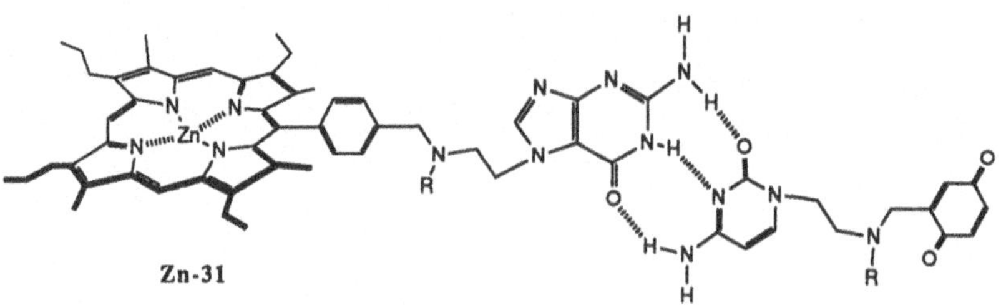

Zn-31

34

Figure 12. Schematic representation of the hydrogen bond donor-acceptor aggregate formed from **Zn-31** and **34** *via* complimentary Watson-Crick base pairing.

2.2.2. *Time-Resolved Fluorescence Spectroscopy*. Taken together with the steady-state results, time-resolved fluorescence spectroscopy is a powerful method for the calculation of energy and/or electron transfer rate constants. In addition, as will be discussed, it also provides a means for estimating association constants for hydrogen bonded porphyrin aggregates.[28]

Rate constants for photoinduced electron transfer in phenyl-linked quinone-substituted porphyrins have been calculated for **8** and **9**. For photochemical reactions in which an excited state undergoes a unimolecular reaction in competition with emission, the rate constant for unimolecular reaction (electron transfer in this case) can be measured according to the expression (eq. 7) below.

$$\frac{I_0}{I} = \frac{\phi_0}{\phi} = k_{ET}\tau_0 + 1 , \qquad (7)$$

Here, ϕ and ϕ_0 are the fluorescence quantum yields for the quenched (e.g. **8** and **9**) and reference (e.g. **7**) molecules, respectively, and τ_0 represents the fluorescence lifetime of the porphyrin chromophore in the reference compound. These fluorescence lifetimes are available through time-resolved fluorescence techniques. The kinetic trace for the decay of the excited state of trimer **7** is shown in Figure 13.

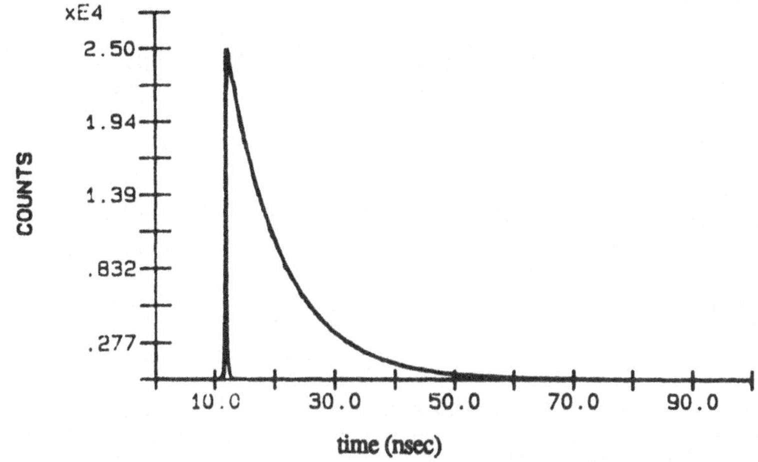

Figure 13. Fluorescence decay of **7** at 298K in benzene with monoexponential analysis.

From the measured lifetime of τ_0 = 9.05 nsec for **7** and the relative fluorescence intensities given in Figure 11, the calculated electron transfer rate constants are 5.73 x 10^8 s^{-1} and 2.74 x 10^9 s^{-1} for the quinone-substituted trimers **8** and **9**, respectively.

Base association constants (K_B) for nucleobase aggregated porphyrins were determined by monitoring fluorescence emanating from one nucleobase-porphyrin conjugate as a function of the concentration of the other added nucleobase moiety. As an example, we describe the association between guanine-porphyrin conjugate **Zn-31** and cytosine-quinone conjugate **34**. Fluorescence intensity (I_f) versus time (t) profiles were analyzed as the sum of two exponentials

$$I_f(t) = A_1\exp(-t/\tau_1) + A_2\exp(-t/\tau_2) \qquad (8)$$

where τ_1 is the fluorescence lifetime of **Zn-31** and τ_2 is the fluorescence lifetime shortened by electron transfer to the added (and aggregated) cytosine-quinone **34**. The fractional amplitude of the shorter lived component (A_2) provides a measure of the relative concentration of the hydrogen-bonded base-derived donor-acceptor pair. The derived lifetimes (see Table 1) remained constant but the relative contribution of the shorter-lived component increased with increasing concentration of **34**, the longer lived component contributing only ca. 5% to the initial emission intensity. The association constant is then calculated as

$$K = A_2[\mathbf{34}]/[(1 - 2A_2)[\mathbf{34}]]^2 \qquad (9)$$

and is estimated to be 1290 ± 230 M^{-1}. In addition to the three-point hydrogen bond association between cytosine and guanine, a two-point hydrogen bond self-association of cytosine or guanine is also possible. Self-association constants, for aggregates such as **Zn29-29**, are determined using the same technique employing eq. 10.

$$K = 2A_2[\mathbf{29}]/[(1 - 4A_2)[\mathbf{29}]]^2 \qquad (10)$$

The derived values of K for cytosine-cytosine and guanine-guanine self association as well as cytosine-guanine association are listed in Table 1. The trend in these values confirm that association affinities increase in the order guanine-guanine < cytosine-cytosine < cytosine-guanine.

Table 1. Time-resolved fluorescence data and derived association constants (K) for nucleobase aggregated porphyrins.

mixture	τ_1(nsec)	τ_2(nsec)	K(M^{-1})	Bases
Zn38-38	1.42	0.61	40	C-C
Zn29-38	1.60	0.77	48	C-C
Zn38-29	1.37	0.80	41	C-C
Zn29-29	1.58	1.16	51	C-C
Zn31-31	1.41	0.82	24	G-G
Zn31-29	1.47	0.87	225	G-C
Zn29-31	1.52	0.86	190	G-C
Zn31-38	1.44	0.63	200	G-C
$Zn_2$38-31	1.51	0.65	215	G-C
Zn31-34	1.50	0.94	1290	G-C

The electron transfer rate constant for charge separation in aggregate **Zn-31···34** can be calculated as $k_{ET} = 1/\tau_2 - 1/\tau_1$ and is estimated according to the data above to be $(4.2 \pm 0.7) \times 10^8\ s^{-1}$ at 298K in CH_2Cl_2.

2.2.3. *Laser Flash Photolysis*. The nature of the transient species generated upon photoexcitation can be identified by their characteristic absorption spectra on the picosecond and/or nanosecond time scale subsequent to the laser flash. This is a very powerful tool for examining photoinduced electron transfer reactions and, taken with the fluorescence experiments described previously, can often serve to clarify the kinetics associated with charge separation and recombination. In earlier, now published, work we used this method to define the excited state kinetics of systems **1** and **2**.[18b] More recently we have applied this technique to the study of phenyl-linked porphyrins **7-9**. Absorption spectra taken at ca. 6 psec intervals after excitation (λ = 532 nm) for quinone-substituted trimer **8** are shown in Figure 14 below. The insert reveals the kinetic trace at 460 nm showing the growth and subsequent decay of this transient species identified as the excited singlet state on the basis of its absorption spectrum. From these data a calculated excited singlet state lifetime of 56 psec was obtained. These data match very well with the fluorescence lifetime determined from the dynamic fluorescence experiments described above.

We have recently initiated flash photolysis experiments which monitor the triplet state of porphyrins in the hydrogen bonded ensembles described above. Here, we have been examining the absorption spectra on the microsecond timescale. Concentration-dependent behavior analogous to that occuring in fluorescence lifetime experiments was observed giving support to the binding model proposed earlier for nucleobase aggregation (see Fig. 12). The distinction between unimolecular and bimolecular processes indicates that intramolecular quenching of the porphyrin excited triplet state in **Zn-31···34** takes place presumably *via* electron transfer through the hydrogen bonded ensemble. Thus, in addition to obtaining information concerning the postulated ET reaction, this technique provides further evidence *inter alia* for the proposed molecular recognition *via* base-pairing. As such, it serves to compement both the ^{1}H NMR and fluorescence results.

3. Conclusions

The compounds developed for this investigation along with the photophysical techniques employed in their study are tools for examining long distance intramolecular electron transfer reactions. An analysis of the reactions described above is intended to provide a further understanding of electron transfer and, more specifically, of the mediating properties of the environment between the redox pair. In most general terms we have been examining two bridge-mediated pathways (covalently linked and hydrogen-bound)and wish to discuss here the significance of our results.

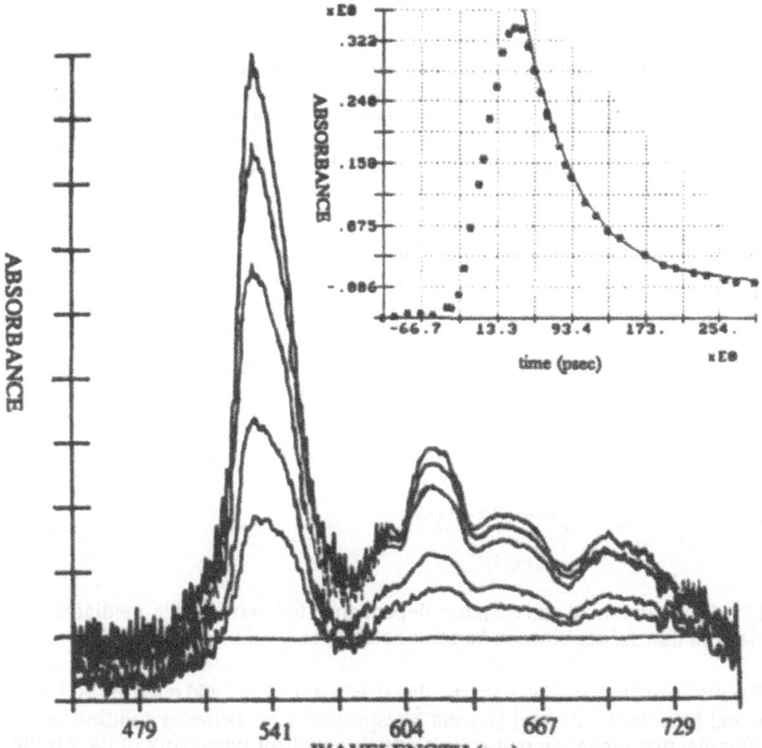

Figure 14. Absorption spectra taken at ca. 6 psec intervals after photoexcitation at 532 nm for 8 in benzene solution. Inset is the kinetic trace at 460 nm).

The first area in which our results are considered significant is in the area of distance effects. Here, theory suggests that the dependence of the electron transfer rate on distance should be a strong function of the medium in the space between the donor and the acceptor. As discussed earlier, the rate is expected to decrease exponentially with distance along with the two-center one-electron integral between parallel aromatic molecules (eq. 4).

$$V = V_0 \exp(-\alpha R) \qquad (4)$$

where α reflects differences in the ET properties of different "mediating" media. The availability of X-ray diffraction structures (and therefore accurate distance parameters) make the phenyl-linked porphyrin series suited to this type of analysis. The compounds in the series (**2**, **8**, **9**, and **39**) allow for the determination of the bridge-mediating ability of *covalently-linked phenylporphyrin* spacers shown below.

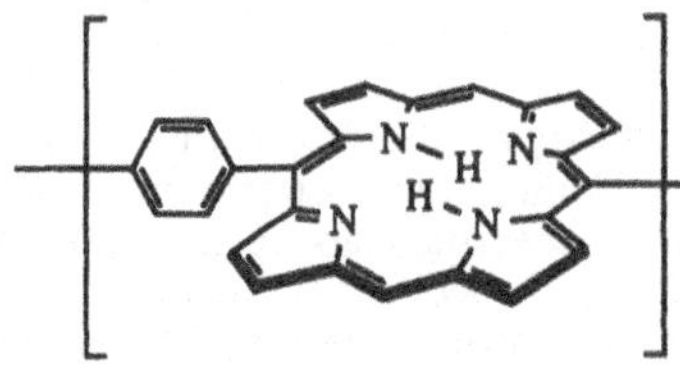

n = 0, 1, 2

The appropriate distances are 19.25 Å and 32.02 Å (porphyrin to quinone, center-to-center) for n =1 (**9** and $\mathbf{H_2 2}$) and n = 2 (**8**) respectively, as estimated from the X-ray crystal structure data for the bis-copper derivative of the 1,4-phenyl-linked dimer. The current absence of X-ray structural information makes an extrapolation to the trimer series necessary. However, such an extrapolation seems reasonable given the rigidity of the systems. These data are plotted in Figure 15. Also shown for comparison are those for the bicyclo[2.2.2]octyl spacer series.[8]

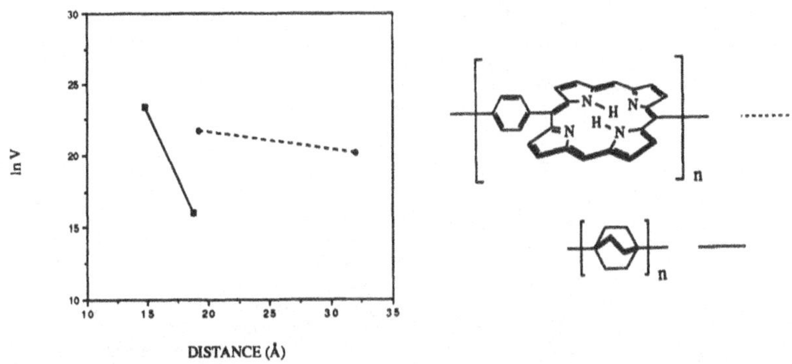

Figure 15. Comparison of the distance dependence for two "bridge-mediated" porphyrin to quinone electron transfer reactions in benzene solution at 298K.

The distance dependence was calculated as $\alpha = 0.12$ Å^{-1} and $\alpha = 1.86$ Å^{-1} for the phenyl porphyrin and bicyclo[2.2.2]octyl spacers,[8] respectively, in benzene solution at 298K. Both systems offer electron tunneling pathways along the covalent framework of their bridge although there is a greater dependence on distance for the saturated bicyclo[2.2.2]octyl spacer. Qualitatively, the reduced distance dependence, or greater mediating ability, of the phenyl porphyrin spacer is due to the overlap of π systems between porphyrins and ultimately between porphyrin and quinone. The bridging phenyl groups should also open a channel for complete conjugation across the system. Any deviation from orthogonality between the porphyrin macrocycle and the phenyl ring will allow for overlap between π systems creating a globally delocalized "supramolecular" molecule. Although the coupling is strong, the very existence of a distance dependence indicates that there is an element of isolation among porphyrin subunits.

Since any covalent attachment deviates substantially from systems of biological interest, we have, as discussed at length earlier, turned to the use of non-covalent ensembles based on multi-point hydrogen bonding. Here our results are less clear in terms of fundamental theory. We have, however, established a synthetic route toward the construction of self-assembled photon antennae and have observed electron and energy transfer in these systems. Moreover, the degree of association for nucleobase aggregation has been quantitated by a number of methods and good evidence has been obtained in support of the contention that these systems exhibit molecular recognition *via* base-pairing. Now, variations in donor-acceptor distance, driving force, etc. should allow us to determine the extent, if any, to which hydrogen bonding networks could play a role in regulating (i.e. facilitating) long range biological electron transfer processes.

4. Acknowledgment

This work was supported by the National Science Foundation (PYI 1986), the Camille and Henry Dreyfus Foundation (Teacher-Scholar 1988), the Sloan Foundation (Sloan Fellowship 1989), the Robert A, Welch Foundation (F-1018) the National Institutes of Health (GM 41657) and the Texas Advanced Research Program (no. 3658-016).

5. References

1. Marcus, R. A. in *Light-Induced Charge Separation in Biology and Chemistry*; Chance, B.; DeVault, D.; Frauenfelder, H.; Marcus, R. A.; Schrieffer, J. R.; Sutin, N., Eds.; Verlag Chemie, Berlin, **1979**, pp. 109-127.
2. (a) Deisenhofer, J.; Epp, O.; Miki, K.; Huber, R.; Michel, H. *J. Mol. Biol.* **1984**, *180*, 385-398. (b) Deisenhofer, J.; Epp, O.; Miki, K.; Huber, R.; Michel, H. *Nature*, **1985**, *318*, 618-624. (c) Deisenhofer, J.; Michel, H. *Angew. Chem. Int. Ed. Engl.* **1989**, *28*, 829-847.
3. (a) Chang, C.-H.; Schiffer, M.; Tiede, D.; Smith, U.; Norris, J. *J. Mol. Biol.* **1985**, *186*, 201-203. (b) Allen, J. P.; Feher, G.; Yeates, T. O.; Komiya, H.; Rees, D. C. *Proc. Natl. Acad. Sci. USA* **1987**, *84*, 5730-5734. (c) Allen, J. P.; Feher, G.; Yeates, T. O.; Komiya, H.; Rees, D. C. *Proc. Natl. Acad. Sci. USA* **1987**, *84*, 6162-6166.
4. (a) Woodbury, N. W.; Becker, M.; Middendorf, D.; Parson, W. W. *Biochemistry* **1985**, *24*, 7516-7521. (b) Martin, J.-L.; Breton, J.; Hoff, A. J.; Migus, A.; Antonetti, A. *Proc. Natl. Acad. Sci. USA* **1986**, *83*, 957-961. (c) Wasielewski, M. R.; Tiede, D. M. *FEBS Lett.* **1986**, *204*, 368-372. (d) Fleming, G. R.; Martin, J. L.; Breton, J. *Nature*, **1988**, *333*, 190-192. (e) Kirmaier, C.; Holten, D. *FEBS Lett.* **1988**, *239*, 211-218. (f) Boxer, S. G.; Lockhart, D. J.; Middenhorf, T. R. *Chem. Phys. Lett.* **1986**, *123*, 476-482. (g) Lockhart, D. J.; Boxer, S. G. *Biochemistry*, **1987**, *26*, 664-668. (h) Lockhart, D. J.; Boxer, S. G. *Chem. Phys. Lett.* **1988**, *144*, 243-250. (i) Lockhart, D. J.; Boxer, S. G. *Proc. Natl. Acad. Sci. USA* **1988**, *85*, 107-111. (j) Boxer, S.G.; Goldstein, R.A.; Lockhart, D.J.; Middendorf, T.R.; Takiff, L. *J. Phys. Chem.* **1989**, *93*, 8280-8294. (k) Tang, D.; Jankowiak, R.; Gillie, J. K.; Small, G. J.; Tiede, D. M. *J. Phys. Chem.* **1988**, *92*, 4012-4015. (l) Johnson, S. G.; Tang, D.; Jankowiak, R.; Hayes, J. M.; Small, G. J.; Tiede, D. M. *J. Phys. Chem.* **1989**, *93*, 5953-5957.
5. (a) Bixon, M.; Jortner, J. *J. Phys. Chem.* **1986**, *90*, 3795-3800. (b) Michel-Beyerle, M.E.; Plato, M.; Deisenhofer, J.; Michl, H; Bixon, M; Jortner, J. *Biochem Biophys. Acta* **1988**, *932*, 52-70. (c) Ogrodnik, A.; Remy-Richter, N.; Michel-Beyerle, M. E.; Feick, R. *Chem. Phys. Lett.* **1987**, *135*, 576-581. (d) Bixon, M.; Jortner, J.; Michel-Beyerle, M. E.; Ogrodnik, A.; Lersch, W. *Chem. Phys. Lett.* **1987**, *140*, 626-630. (e) Michel-Beyerle, M. E.; Bixon, M.; Jortner, J. *Chem. Phys. Lett.* **1988**, *151*, 188-194. (f) Bixon, M.; Jortner, J. *J. Phys. Chem.* **1988**, *92*, 7148-7156. (g) Plato, M.; Möbius, K.; Michel-Beyerle, M. E.; Bixon, M.; Jortner, J. *J. Am. Chem. Soc.* **1988**, *110*, 7279-7285. (h) Bixon, M.; Jortner, J. *Chem. Phys. Lett.* **1989**, *159*, 17-20. (i) Bixon, M.; Jortner, J.; Michel-Beyerle, M. E.; Ogrodnik, A. *Biochim. Biophys. Acta* **1989**, *977*, 273-286. (j) Won, Y.; Friesner, R. A. *Proc. Natl. Acad. Sci. U.S.A.* **1987**, *84*, 5511-5515. (k) Won, Y.; Friesner, R. A. *Biochim. Biophys. Acta* **1988**, *935*, 9-18. (l) Friesner, R. A.; Won, Y. *Biochim. Biophys. Acta* **1989**, *977*, 99-122. (m) Hu, Y.; Mukamel, S. *Chem. Phys. Lett.* **1989**, *160*, 410-416. (n) Feher, G.; Allen, J. P.; Okamura, M. Y.; Rees, D. C. *Nature* **1989**, *339*, 111-116.
6. (a) Holzapfel, W.; Finkele, U.; Kaiser, W.; Oesterhelt, D.; Scheer, H.; Stilz, H. U.; Zinth, W. *Chem. Phys. Lett.* **1989**, *160*, 1-7. (b) Holzapfel, W.; Finkele, U.; Kaiser,

W.; Oesterhelt, D.; Scheer, H.; Stilz, H. U.; Zinth, W. *Proc. Natl. Acad. Sci. U.S.A.* **1990**, *87*, 5168-5172.

7. For recent reviews see: (a) Wasielewski, M. R. *Photochem. Photobiol.* **1988**, *47*, 923-929. (b) Gust, D.; Moore, T. A. *Science* **1989**, *244*, 35-41. (c) Connolly, J. S.; Bolton, J. R. in "Photoinduced Electron Transfers, Part D," Fox, M. A.; Channon, M., Eds. Elsevier, Amsterdam, 1988; Ch. 6.2; pp. 303-393. For reviews of earlier work see: (a) Fendler, J. H. *J. Phys. Chem.* **1985**, *89*, 2730-2740. (b) Boxer, S. G. *Biochim. Biophys. Acta* **1983**, *726*, 265-292. See also *Tetrahedron* **1989**, *45(15)*, a special "Symposium in Print" issue devoted to the topic of covalently linked donor-acceptor photosynthetic model systems, Gust, D.; Moore, T. A., Eds.

8. (a) Joran, A. D.; Leland, B. A.; Geller, G. G.; Hopfield, J. J.; Dervan, P. B. *J. Am. Chem. Soc.* **1984**, *106*, 6090-6092. (b) Leland, B. A.; Joran, A. D.; Felker, P. M.; Hopfield, J. J.; Zewail, A. H.; Dervan, P. B. *J. Phys. Chem.* **1985**, *89*, 5571-5573. (c) Joran, A. D.; Leland, B. A.; Felker, P. M.; Zewail, A. H.; Hopfield, J. J.; Dervan, P. B. *Nature (London)* **1987**, *327*, 508-511. (d) Joran, A. D.; Dissertation, California Institute of Technology, **1986**. (e) Leland, B. A.; Dissertation, California Institute of Technology, **1987**.

9. Marcus, R. A. *J. Chem. Phys.* **1956**, *24*, 966-977. (b) Marcus, R. A. *Discuss. Faraday Soc.* **1960**, *29*, 21-33. (c) Marcus, R. A.; Sutin, N. *Biochim. Biophys. Acta.* **1985**, *811*, 256-322.

10. Wasielewski, M. R.; Niemczyk, M. P.; Svec, W. A.; Pewitt, E. B. *J. Am. Chem. Soc.* **1985**, *107*, 5562-5563.

11. (a) Moore, T. A.; Gust, D.; Mathis, P.; Mialocq, J.-C.; Chachaty, C.; Bensasson, R. V.; Land, E. J.; Doizi, D.; Liddell, P. A.; Lehman, W. R.; Nemeth, G. A.; Moore, A. L. *Nature (London)* **1984**, *307*, 630-632; (b) Liddell, P. A.; Barrett, D.; Makings, L. R.; Pessiki, P. J.; Gust, D.; Moore, T. A. *J. Am. Chem. Soc.* **1986**, *108*, 5350-5352; (c) Gust, D.; Moore, T. A.; Moore, A. L.; Barrett, D.; Harding, L. O.; Makings, L. R.; Liddell, P. A.; De Schryver, F. C.; Van der Auweraer, M.; Bensasson, R. V.; Rougee, M. *J. Am. Chem. Soc.* **1988**, *110*, 321-323.

12. Kuki, A.; Wolynes, P. G. *Science* **1987**, *236*, 1647-1652.

13. (a) Beratan, D. N.; Onuchic, J. N.; Hopfield, J. J. *J. Chem. Phys.* **1985**, *83*, 5325-5329. (b) Beratan, D. N.; Onuchic, J. N.; Hopfield, J. J. *J. Chem. Phys.* **1987**, *86*, 4488-4498. (c) Beratan, D. N.; Onuchic, J. N. *Photosynth. Res.* **1989**, *22*, 173-186. (d) Onuchic, J. N.; Beratan, D. N. *J. Chem. Phys.* **1990**, *92*, 722-733.

14. To a first approximation, a hydrogen bond is approximately equivalent to two covalent bonds in terms of its mediating effects on wave function propagation. See refernce 13(d) for a discussion.

15. Dugas, H. *Bioorganic Chemistry* Springer-Verlag, New York, **1989**.

16. Tecilla, P.; Dixon, R. P.; Slobodkin, G.; Alavi, D. S.; Waldeck, D. H.; Hamilton, A. D. *J. Am. Chem. Soc.* **1990**, *112*, 9408.

17. Aoyama, Y.; Asakawa, M.; Matsui, Y.; Ogoshi, H. *J. Am. Chem. Soc.* **1991**, *113*, 6233-6240.

18. (a) Sessler, J. L; Johnson, M. R.; Creager, S. E.; Fettinger, J. C.; Ibers, J. A. *J. Am. Chem. Soc.* **1990**, *112*, 9310-9329. (b) Rodriguez, J.; Kirmaier, C.; Johnson, M. R.; Friesner, R. A.; Holten, D.; Sessler, J. L. *J. Am. Chem. Soc.* **1991**, *113*, 1652-1659.

19. (a) Harriman, A. H.; Magda, D. J.; Sessler, J. L. *J. Chem. Soc., Chem. Commin.* **1991**(5), 345-348. (b) Harriman, A. H.; Magda, D. J.; Sessler, J. L. *J. Phys. Chem.* **1991**, *95*, 1530-1532.

20. Sessler, J. L.; Capuano, V. L. *Angew. Chem.* **1990**, *102*, 1162-1165; *Angew. Chem. Int. Ed. Engl.* **1990**, *29*, 1134-1137.

21. Harriman, A. H.; Kubo, Y.; Sessler, J. L. *J. Am. Chem. Soc.*, submited for publication.

22. Abdalmuhdi, I.; Chang, C. K. *J. Org. Chem.* **1985**, *50*, 411-413, and references cited therin.

23. (a) Lindsey, J. S.; Wagner, R. W. *J. Org. Chem.* **1989**, *54*, 828-836. (b) Lindsey, J. S.; Schreiman, I. C.; Hsu, H. C.; Kearney, P. C.; Marguerettaz, A. M. *J. Org. Chem.* **1987**, *52*, 827-836.
24. For an example of a $Ru(bpy)_3$ covalently linked across the face of a porphyrin see: Hamilton, A. D.; Rubin, H. -D.; Bocarsly, A. B. *J. Am. Chem. Soc.* **1984**, *106*, 7255-7257.
25. Pitha, J.; Jones, R. N.; Pithova, P. *Can. J. Chem.* **1966**, *44*, 1045-1050. (b) Newmark, R. A.; Cantor, C. R. *J. Am. Chem. Soc.* **1968**, *90*, 5010-5017. (c) Kyoguko, Y.; Lord, R. C.; Rich, A. *Biochim. Biophys. Acta.* **1969**, *179*, 10-18. (d) Petersen, S. B.; Led, J. J. *J. Am. Chem. Soc.* **1981**, *103*, 5308-5313. (e) Williams, L. D.; Chawla, B.; Shaw, B. R. *Biopolymers* **1987**, *26*, 591-603. (f) Williams, L. D.; Williams, N. G.; Shaw, B. R. *J. Am. Chem. Soc.* **1990**, *112*, 829-833.
26. (a) Tabushi, I.; Sasaki, T. *Tetrahedron Lett.* **1982**, *23*, 1913-1916. (b) Tabushi, I.; Kugimiya, S.-I.; Kinnaird, M. G.; Sasaki, T. *J. Am. Chem. Soc.* **1983**, *107*, 4192-4199. (c) Osuka, A.; Ida, K.; Maruyama, K. *Chemistry Letters* **1989**, 741-744. (d) Osuka, A.; Furuta, H.; Maruyama, K. *Chemistry Letters* **1986**, 479-482. (e) Osuka, A.; Maruyama, K. *Chemistry Letters* **1987**, 825-828;
27. Won, Y.; Friesner, R.A.; Johnson, M.R.; Sessler, J.L.; *Photosynthesis Research* **1989**, 22, 201-210.
28. Turro, N. J. *Modern Molecular Photochemistry*; Benjamin/Cummings, 1978; 246-248 and references cited therin.

TUTORIAL WORK

A MISCELLANY OF AB INITIO QUANTUM CHEMISTRY THE HELIUM ATOM STUDIED BY VARIOUS METHODS

Alain Strich
Laboratoire de Chimie Quantique
UPR 139 du CNRS
Université Louis Pasteur
4 Rue Blaise Pascal
F-67000 Strasbourg (France)

I Introduction

The aim of the following Problem is not to make the participants of this Institute experts of *ab initio* computational quantum chemistry (AICQC), but rather give them a "flavour" of this science.
Roughly speaking, there are three ways of practicing AICQC:

1. Purchase a widely distributed quantum chemistry package from the Quantum Chemistry Program Exchange[1] or the CPC Program Library[2] and use it as a black box to produce a huge amount of numbers that will be exploited more or less intelligently.
2. Use a quantum chemistry package written by others, but try to know a little bit about what is being done in the black box.
3. Write one's own package.

Hypothesis 3 supposes that one is already an expert of quantum chemistry, and I am not addressing this kind of audience. The non-expert has the choice between Solutions 1 and 2. As you can imagine, I will favor Choice 2 and the different questions of the Problem will give you some "feeling" of what is going on in the black box, what are the advantages, the drawbacks, the limitations, the basic assumptions of the various methods of AICQC, which you will probably have heard of in the lectures by S. Peyerimhoff, B. O. Roos, L. Salem, and A. Veillard.
We will not make use of any computer, but the problem will be the occasion to approach many of the facets of contemporary quantum chemistry.

E. Kochanski (ed.), Photoprocesses in Transition Metal Complexes, Biosystems and Other Molecules. Experiment and Theory, 403–419.

II Problem: A Two-Electron System, the Helium Atom, Studied by Various Methods.

All the necessary data and formulae for solving the Problem can be found in Section III.
Throughout the Problem, atomic units will be used.

1. ϕ_1 and ϕ_2 are respectively the 1s and 2s atomic orbitals of the hydrogenlike ion He^+.

$$\phi_1(r) = \frac{Z^{\frac{3}{2}}}{\sqrt{\pi}} \exp -Zr \qquad (II-1)$$

$$\phi_2(r) = \frac{Z^{\frac{3}{2}}}{4\sqrt{2\pi}} (2 - Zr) \exp -\frac{Zr}{2} \qquad (II-2)$$

r is the nucleus-electron distance (in bohrs), Z is the charge of the helium nucleus (Z=2). It is recommended to carry out the calculations with Z, and replace it by 2 only when numerical values need to be evaluated.
Check that ϕ_1 and ϕ_2 form an orthonormal basis, namely:

$$< \phi_i | \phi_j > = \delta_{ij} \qquad (II-3)$$

Why are the two functions said to possess "spherical symmetry"?

2. Draw $\phi_i(r)$ and the radial density $D_i(r)$ as a function of r. $D_i(r)$ is defined as:

$$D_i(r) = 4\pi r^2 \phi_i^2(r) \qquad (II-4)$$

What is the physical interpretation of $D_i(r)$? What feature distinguishes $D_2(r)$ from $D_1(r)$?

3. The atomic radius of an orbital ϕ_i is defined as the following expectation value:

$$< r >_i = < \phi_i(r) | r | \phi_i(r) > \qquad (II-5)$$

Calculate $< r >_1$ and $< r >_2$.

4. Write down the expression of the Hamiltonian of He^+. Check that ϕ_1 and ϕ_2 are eigenfunctions of this operator. What are the corresponding

eigenvalues? What can be said about the resolution of the Schroedinger equation?

5. Write down the expression of the Hamiltonian of the neutral helium atom. How does it differ from the previous one? What are the electronic configuration and the spectroscopic notation for the ground state of the helium atom?

From now on, our approach to the solution of the Schroedinger equation will be based on approximation methods. There are two types of such methods: variation and perturbation. We shall restrict ourselves to variation methods, but we have to mention that perturbation methods are also widely used in AICQC (e.g. the Moeller-Plesset MPn method, n=2,3,4,...being the order of the perturbation).

6. We wish to describe the ground state with a Slater determinant wavefunction. Which of the functions ϕ_1 or ϕ_2 will be used to set up the determinant? Write down the determinant in abridged notation. Then, by developing it, show that it can be decomposed into a product of a spatial wavefunction and of a spin wavefunction, each of them being normalized. Give the expressions of these two functions.

7. A useful relationship of the algebra of one-electron spin-operators is the following:

$$\hat{S}^2 = \hat{S}_+\hat{S}_- + \hat{S}_z(\hat{S}_z - 1) \qquad (II-6)$$

In this expression $\hat{S}_+$ and $\hat{S}_-$ are the ladder operators, respectively step-up and step-down, defined as:

$$\hat{S}_+ = \hat{S}_x + i\hat{S}_y \qquad (II-7)$$

$$\hat{S}_- = \hat{S}_x - i\hat{S}_y \qquad (II-8)$$

Demonstrate (II-6), using the commutation relationship:

$$[\hat{S}_x, \hat{S}_y] = i\hat{S}_z \qquad (II-9)$$

Recall the action of the spin operators $\hat{S}_z$, $\hat{S}_+$, $\hat{S}_-$, $\hat{S}^2$ on the one-electron spin functions α and β. Formula (II-6) also holds for many-electron spin operators.

In the case of a two-electron system like helium, all the 2-electron operators on the r.h.s. of (II-6) obey the equality:

$$\hat{S}_\mu(1,2) = \hat{S}_\mu(1) + \hat{S}_\mu(2) \qquad (\mu = + ; - ; z) \qquad (II-10)$$

whereas

$$\hat{S}^2(1,2) \neq \hat{S}^2(1) + \hat{S}^2(2) \qquad (II-11)$$

Using these relationships, show that the 2-electron spin function obtained at question 6 is an eigenfunction of the 2-electron spin-operator $\hat{S}^2$. What is the corresponding eigenvalue?

8. Calculate the expectation value of the energy using the hamiltonian of question 5 and the wavefunction of question 6.
Express this energy as a function of the following integrals:

the one-electron Hamiltonian integral (define operator $\hat{h}(1)$)

$$h_{ii} = \langle \phi_i(1) | \hat{h}(1) | \phi_i(1) \rangle = \int \phi_i^*(1)\, \hat{h}(1) \phi_i(1)\, dv_1 \qquad (II-12)$$

the Coulomb integral

$$J_{ij} = \langle \phi_i(1)\phi_j(2) | \frac{1}{r_{12}} | \phi_i(1)\phi_j(2) \rangle = \int\int \phi_i^*(1)\phi_j^*(2) \frac{1}{r_{12}} \phi_i(1)\phi_j(2)\, dv_1\, dv_2 \qquad (II-13)$$

the exchange integral

$$K_{ij} = \langle \phi_i(1)\phi_j(2) | \frac{1}{r_{12}} | \phi_j(1)\phi_i(2) \rangle = \int\int \phi_i^*(1)\phi_j^*(2) \frac{1}{r_{12}} \phi_j(1)\phi_i(2)\, dv_1\, dv_2 \qquad (II-14)$$

Show that the expression obtained is a special case of the general energy expression for a closed-shell 2N-electron system described by a Slater determinant built up on N spatial orbitals:

$$E = V_{nucl} + \sum_{i=1}^{N} (2h_{ii} + J_{ii}) + 2\sum_{i=1}^{N-1} \sum_{j=i+1}^{N} (2J_{ij} - K_{ij}) \qquad (II-15)$$

where V_{nucl} is the nuclear repulsion energy.

Using (II-15), write down the expression of the energy for the beryllium atom and the water molecule.

9. Calculate h_{11} as a function of Z. It is possible (although lengthy) to show that the value of J_{11} is:

$$J_{11} = \frac{5}{8} Z \qquad (II-16)$$

Rewrite the expression of the energy found in question 8 as a function of Z.
Calculate the energy of the helium atom. Compare this value to the best variational calculation[3], yielding an essentially exact energy (to 10^{-9} Hartree) of -2.903724375 E_h. Convert 10^{-9} Hartree into cm^{-1}.

From the ionization energy of helium, equal to 24.5876 eV, calculate the "experimental total energy" of the atom. What is the origin of the discrepancy with the best value?

10.a) For approximate atomic or molecular calculations, Slater has introduced a basis function commonly referred to as the Slater type function (STF, or STO, O standing for Orbital) $\chi(r)$ defined as:

$$\chi(r) = Nr^{n-1} \exp - \zeta r \qquad (II-17)$$

where N is a normalization constant, n a principal quantum number whose value is 1 for a 1s type STF, 2 for a 2s or 2p type STF, etc...ζ is called the exponent of the STF.
Show that a 1s type STF is equivalent to function ϕ_1 defined in (II-1), in which Z has been replaced by ζ.

b) In the LCAO approach -a more exact term would be LCBF, BF standing for basis function- an atomic (or molecular) orbital ϕ is expanded in terms of M basis functions χ_p according to:

$$\phi = \sum_{p=1}^{M} C_p \chi_p \qquad (II-18)$$

The expansion coefficients C_p are the LCAO coefficients. If only one basis function makes the major contribution in the expansion (II-18) of an atomic orbital, the basis set is said to be of single zeta quality. If two functions contribute, then we have a double zeta basis set, etc...
How many -and which- basis functions (STF's) would be needed to describe in a double zeta manner an atom of the Second Period (B $\rightarrow$ Ne), of the Third Period (Al $\rightarrow$ Ar), or of the First Transition Metal Series?

11. We come back to our helium atom which we now describe using a single STF. What quality of basis set do we have? Using the results of Questions 9 and 10, express the energy as a function of ζ.
In the variational method, the expectation value of the energy is minimized.Which value of ζ makes the energy minimal? Explain why one could have expected a value of ζ comprised between 1 and 2.
Calculate the value of the energy and compare it to the values of Question 9.

12.a) We now use a double zeta basis set of STF's. Clementi and Roetti[4] have found that the best exponents for the two STF's χ_1 and χ_2 are:

$$\zeta_1 = 1.45363 \qquad \zeta_2 = 2.91093 \qquad \text{(II} - 19)$$

When we carry out a Hartree-Fock-Roothaan self-consistent field (SCF) calculation with this basis, we have to do with an iterative process. At each iteration, the energy and the orbitals (i.e. the C_p coefficients of (II-18) are evaluated until both converge.
The present calculation has been carried out with SCF2EL [5], and the convergence is the following:

Iteration Number	Total Energy (E_h)	Max Difference in LCAO coeff.
1	-2.776626155	0.20
2	-2.861501100	0.85E-02
3	-2.861672185	0.42E-03
4	-2.861672597	0.21E-04
5	-2.861672598	0.10E-05
6	-2.861672598	0.50E-07

At convergence, the atomic orbitals (the "vectors") are the following:

$$\psi_1 = 0.8438\,\chi_1 + 0.1807\,\chi_2 \qquad \text{(II} - 20)$$

$$\psi_2 = 1.6240\,\chi_1 - 1.8212\,\chi_2 \qquad \text{(II} - 21)$$

Name these two atomic orbitals, whose energies are respectively:

$$\varepsilon_1 = -\,0.917935\ E_h \qquad \varepsilon_2 = 2.820957\ E_h \qquad \text{(II} - 22)$$

Compare the total energy with those obtained previously.
If we increased the basis set to infinity, the energy would reach a limit, the Hartree-Fock limit. In the case of helium, the corresponding value is -2.8616800 E_h [6]. Clementi and Roetti[4], with a basis set of 5 STF's have reached the value of -2.8616799 E_h.
What matrices are involved in the SCF process? What is their size?

b) To set up the Fock matrix, two-electron integrals are needed. They are defined as:

$$(pq|rs) = \int\int \chi_p(1)\chi_q(1)\frac{1}{r_{12}}\chi_r(2)\chi_s(2)\, dv_1\, dv_2 \qquad (II-23)$$

(The * have been omitted, since STF's χ_t are real functions).
List all the different two-electron integrals, and establish the formula that gives their number.

c) Suppose we wish to carry out a double zeta calculation on a model of a ferroporphyrin, with a chemical formula $FeN_4C_{20}H_{12}$. What would be the size of the matrices, and how many 2-electron integrals would be needed? Actually, a certain percentage of them will vanish. Why? Same question for the recently characterized Buckminster fullerene C_{60}.

d) Suppose a linear molecule A-B placed on the z-axis. If χ_1 to χ_4 are respectively the 2s, $2p_x$, $2p_y$, $2p_z$ STF's of atom A and χ_5 to χ_8 those of atom B, list a few 2-electron integrals that will vanish by symmetry.

13. Now we wish to go beyond the single determinant approximation, in other words to carry out a configuration interaction (CI) calculation.
a) Set up all the possible two-electron determinants D_i built up on the two functions ϕ_1 and ϕ_2 defined in (II-1) and (II-2).

b) Check on a few of them that they form an orthonormal basis, i.e.

$$< D_i | D_j > = \delta_{ij} \qquad (II-24)$$

c) Are all of them eigenfunctions of $\hat{S}^2$? By separating the spatial wavefunction from the spin wavefunction, check what component of what spin function they represent.

d) Construct the appropriate combinations of determinants to go from the basis of determinants to a basis of configuration state functions (CSF's) which are eigenfunctions of $\hat{S}^2$.

e) Using (II-6), check the action of $\hat{S}^2$ on the $M_S=0$ component of the triplet.

f) In (II-25), we write down a CI wavefunction as a linear combination of two CSF's, Ψ_1 representing the SCF ground state, Ψ_2 a singly excited configuration.

$$\Psi_{CI} = a_1\Psi_1 + a_2\Psi_2 \qquad (II-25)$$

a_1 and a_2 are the CI coefficients to be determined variationally. Find Ψ_1 and Ψ_2 among the CSF's constructed in question 13d.

g) Set up the secular determinant.

h) Calculate the Hamiltonian matrix elements H_{ij}, using the following values for the integrals:

$$J_{12} = \frac{17}{81} Z \quad K_{12} = \frac{16}{729} Z$$
$$< \phi_1(1)\phi_1(2) | \frac{1}{r_{12}} | \phi_1(1)\phi_2(2) > = 0.179 \ E_h \qquad \text{(II} - 26)$$

What is the physical meaning of matrix elements H_{11} and H_{22}?
Why is the $^3S(1s^12s^1)$ lower in energy than the first excited $^1S(1s^12s^1)$?

i) Solve the secular equation and discuss the energies obtained. Calculate the correlation energy which has been recovered.

j) What are the expressions of the CI wavefunctions?

k) Using the results of Question 12a, suggest a better CI calculation.

l) Group in a Table all the energies which have been obtained by the different methods and discuss their relative values.

14. It is nowadays possible to carry out CI calculations with more than one billion determinants! Such a calculation has been reported recently for the magnesium atom[7].

In transition metal chemistry, many small systems are still a challenge to AICQC. A classical example is the dimer Cr_2, formally sextuply bonded[8]. Use Weyl's formula (III-3) to determine the number of CSF's in a full CI correlating all the valence electrons in all the valence orbitals.

III Data and Formulae

The atomic units form a natural system in atomic and molecular quantum mechanics. In this system, m_e, e, $\hbar$ and $\kappa_0 (= 4\pi\varepsilon_0)$ all take unit values.

Other Units:

Nature	Name	Symbol	Value
Length	bohr	a_0	0.52918×10^{-10} m
Energy	Hartree	E_h	27.2116 eV
			219475 cm^{-1}

The following integral

$$\int_0^{\infty} r^n e^{-ar}\, dr = \frac{n!}{a^{n+1}} \qquad \text{(III – 1)}$$

will be useful in many calculations.

In spherical coordinates (r, ϑ, φ), the Laplacian operator has the following expression:

$$\Delta = \frac{\partial^2}{\partial r^2} + \frac{2}{r}\frac{\partial}{\partial r} - \frac{1}{r^2}\hat{L}^2 \qquad \text{(III – 2)}$$

where $\hat{L}^2$ is an operator involving only derivatives with respect to ϑ and φ.

The number μ of CSF's for a system characterized by a spin quantum number S, and by N electrons that can be distributed among n orbitals, is given by Weyl's formula:

$$\mu = \frac{2S+1}{n+1}\binom{n+1}{\frac{N}{2}-S}\binom{n+1}{\frac{N}{2}+S+1} \qquad \text{(III – 3)}$$

where

$$\binom{p}{q} = \frac{p(p-1)(p-2)\ldots(p-q+1)}{q!} \qquad \text{(III – 4)}$$

IV Conclusion and Remarks

The foregoing Problem will have been the occasion to encounter a large number of aspects of AICQC, namely:

- an orthonormal basis (1)*

* Numbers in parentheses refer to the numbering of questions in Section II.

- the nodes in orbitals (2)
- the calculation of properties (3) '
- the expression of a quantum-mechanical operator (4,5)
- the eigenvalues and eigenfunctions of an operator (4,7,13)
- exact solutions (4)
- the Slater determinant (6)
- spatial vs. spin wavefunction (6)
- algebra of one- and many-electron spin-operators (7,13)
- typical integrals of quantum chemistry: one-electron, Coulomb and exchange (8)
- the expectation value of the energy for a Slater determinant wavefunction (8)
- integral evaluation (9)
- the physical meaning of the total energy (9)
- the STF (10)
- the LCAO approximation (10)
- the quality of a basis set (10)
- the non-linear variation method (11)
- the linear variation method (12,13)
- the screening effect (11)
- the principle of the Hartree-Fock-Roothaan SCF method, the matrices and integrals it involves (12)
- the Hartree-Fock limit (12)
- the huge numbers (of integrals/ configurations) which are reached for moderately-sized systems (12, 14)
- the construction of multi-determinantal spin eigenfunctions (13)
- the orthogonality of determinants (13)
- the principle of the CI method (13)
- the correlation energy (13)
- the important role of the one-particle basis (13)

Let me make a few final remarks:

- The non-quantum chemist reader may have inferred from the Problem that AICQC molecular calculations are usually carried out with STF basis sets. This is not true: actually, more than 99.9% of the calculations use Gaussian basis sets, but for the non-expert, this is just "technical", and it does not affect the conclusions we have reached.

- We have been able to separate the wavefunctions into a spatial part and a spin part. This is no longer possible for systems with more than two electrons.

- The main interest of this Summer School lies in excited states, and we have essentially discussed ground states, but the basic methods for treating these two kinds of states are the same ones.

V Answers to the Problem

1. Use

$$< \phi_i | \phi_j > = \int_0^{\infty} 4\pi r^2 \phi_i \phi_j dr$$

and integral (III-1).

2.

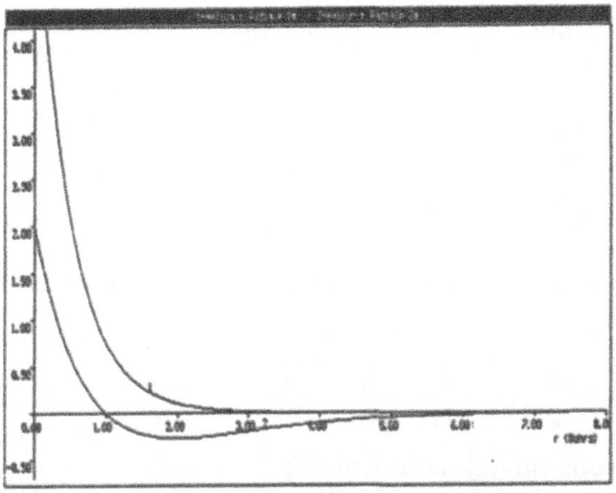

Fig. 1. Plot of $\phi_1(r)$ and $\phi_2(r)$

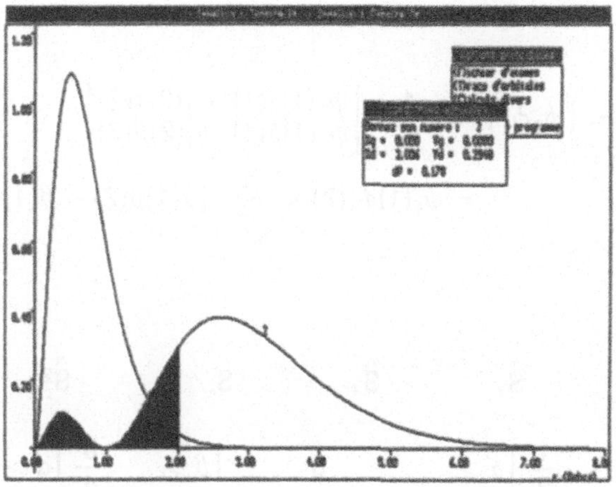

Fig. 2. Plot of $D_1(r)$ and $D_2(r)$

These curves are drawn by RADAO[9], a personal computer software for radial orbitals and radial densities. As ϕ_1 and ϕ_2 are normalized, the area under each of the curves of Fig. 2 is equal to 1. The shaded area under the 2s curve represents the probability of finding a 2s electron of He^+ inside a sphere of radius equal to 2 bohrs. The program calculates a value of 0.178 for this probability.

For the 2s atomic orbital, notice the presence of a node (Figs. 1 and 2) at r= 1.00 bohr.

3. $< r >_1 = 0.75$ bohr $\qquad < r >_2 = 3$ bohrs.

4. $\hat{H} = -\frac{1}{2}\Delta - \frac{Z}{r}$

Eigenvalues : -2 E_h ; -0.5 E_h

ϕ_1 and ϕ_2 are exact solutions of the Schroedinger equation.

5. $\hat{H} = -\frac{1}{2}\Delta_1 - \frac{1}{2}\Delta_2 - \frac{Z}{r_1} - \frac{Z}{r_2} + \frac{1}{r_{12}}$

The two electrons have labels 1 and 2.

He $1s^2$ ^1S

6.

$$D = |\phi_1\overline{\phi_1}| = \frac{1}{\sqrt{2!}}\begin{vmatrix} \phi_1(1)\alpha(1) & \phi_1(2)\alpha(2) \\ \phi_1(1)\beta(1) & \phi_1(2)\beta(2) \end{vmatrix}$$
$$= \phi_1(1)\phi_1(2) \times \frac{1}{\sqrt{2}}[\alpha(1)\beta(2) - \beta(1)\alpha(2)]$$

7.

	$\hat{S}_z$	$\hat{S}_+$	$\hat{S}_-$	$\hat{S}^2$
$\|\alpha>$	$\frac{1}{2}\|\alpha>$	0	$\|\beta>$	$\frac{3}{4}\|\alpha>$
$\|\beta>$	$-\frac{1}{2}\|\beta>$	$\|\alpha>$	0	$\frac{3}{4}\|\beta>$

Eigenvalue : 0

8. $\hat{h}(1) = -\frac{1}{2}\Delta_1 - \frac{Z}{r_1}$

$< E > = 2h_{11} + J_{11}$

Be: 4-electrons system (N=2), with 1s and 2s atomic orbitals occupied.

$$E = 2(h_{1s1s} + h_{2s2s}) + J_{1s1s} + J_{2s2s} + 4J_{1s2s} - 2K_{1s2s}$$

H_2O : 10-electrons system (N=5), with MO's $\phi_1 \rightarrow \phi_5$ occupied:

$$\begin{aligned} E = {} & \frac{1}{R_{HH}} + \frac{16}{R_{OH}} + 2(h_{11} + h_{22} + h_{33} + h_{44} + h_{55}) \\ & + J_{11} + J_{22} + J_{33} + J_{44} + J_{55} \\ & + 4J_{12} - 2K_{12} + 4J_{13} - 2K_{13} + 4J_{14} - 2K_{14} + 4J_{15} - 2K_{15} + 4J_{23} - 2K_{23} \\ & + 4J_{24} - 2K_{24} + 4J_{25} - 2K_{25} + 4J_{34} - 2K_{34} + 4J_{35} - 2K_{35} + 4J_{45} - 2K_{45} \end{aligned}$$

9. $h_{11} = -\frac{Z^2}{2}$ $\quad E = -Z^2 + \frac{5}{8}Z = -2.750000\ E_h$

"Experimental Total Energy " = -2.903571 E_h
Origin of the discrepancy: nuclear motion not taken into account.

10. b) Basis Sets for	B → Ne	4s2p
	Al → Ar	6s4p
	1st Transition Metals Series	8s4p2d

11. $E = \zeta^2 - \frac{27}{8}\zeta$

ζ_{opt} = 1.6875. This is a screening effect.
E= -2.847656 E_h

12. a) ψ_1 and ψ_2 are respectively the 1s and 2s atomic orbitals.
Given the geometry and the basis set, the overlap matrix and the one-electron hamiltonian matrix are evaluated, as well as the two-electron integrals. At each iteration, the density matrix is set up from the LCAO coefficients, and the Fock matrix from the density and one-electron Hamiltonian matrices and the two-electron integrals. All matrices are square symmetric and have the size of the basis set.

b) Due to the definition (II-23) of (pq | rs), the following equalities hold:

$$(pq|rs) = (pq|sr) = (qp|rs) = (qp|sr)$$
$$= (rs|pq) = (sr|pq) = (rs|qp) = (sr|qp)$$

So, only the six following two-electron integrals (listed in "canonical order") need to be evaluated:

(11|11); (21|11); (21|21); (22|11); (22|21); (22|22)

M= size of basis $\quad n_1$ = number of one-electron integrals

n_2 = number of two-electron integrals

$$n_1 = \frac{M(M+1)}{2} \qquad n_2 = \frac{n_1(n_1+1)}{2}$$

c)

	Size of Matrices	Number of 2-el. Integrals
$FeN_4C_{20}H_{12}$	294	0.94 billion
C_{60}	600	16.3 billions

d) An integral (pq|rs) will vanish by symmetry if the direct product of the irreducible representations $X_p \otimes X_q \otimes X_r \otimes X_s$ does not contain the totally symmetric representation of the point group (with function m being a basis of the irreducible representation X_m).

Examples of integrals that will vanish by symmetry are:

(43|21); (87|65); (62|31); (41|32) etc...

In the (43|21) case, we have (in the C_{2v} point group): $A_1 \otimes B_2 \otimes B_1 \otimes A_1 = A_2 \neq A_1$

13. a)

$$D_1 = |\phi_1 \overline{\phi_1}| \quad D_2 = |\phi_2 \overline{\phi_2}| \quad D_3 = |\phi_1 \phi_2|$$
$$D_4 = |\overline{\phi_1} \overline{\phi_2}| \quad D_5 = |\phi_1 \overline{\phi_2}| \quad D_6 = |\overline{\phi_1} \phi_2|$$

b) Use (II-3) and $< \alpha(\mu) | \beta(\mu) > = 0$.

c)

	D_1	D_2	D_3	D_4	D_5	D_6
$\hat{S}_z$ eigenvalue : M_S	0	0	1	-1	0	0
Eigenfunction of $\hat{S}^2$	Yes	Yes	Yes	Yes	No	No
$\hat{S}^2$ eigenvalue : $S(S+1)$	0	0	2	2	-	-

d) The first four CSF's are determinants D_1 to D_4. The last two are the following:

$$|S; M_S> = |1; 0> = \frac{1}{\sqrt{2}}(D_5 + D_6)$$

$$|S; M_S> = |0; 0> = \frac{1}{\sqrt{2}}(D_5 - D_6)$$

e) $\hat{S}^2 \,|1;0> \; = 2\,|1;0>$

f) $\Psi_1 = D_1 \qquad \Psi_2 = |0;0>$

g) Linear Variation Method:

$$\begin{vmatrix} H_{11} - E & H_{12} \\ H_{21} & H_{22} - E \end{vmatrix} = 0$$

$H_{ij} = H_{ji} = <\Psi_i|\hat{H}|\Psi_j>$, $\hat{H}$ being the hamiltonian of question 5.

h) $H_{11} = -2.75 \quad H_{12} = 0.25314 \quad H_{22} = -2.03635$

H_{11} and H_{22} are respectively the single-configuration energies of the ground state (GS) ^1S and of the first excited (E1) ^1S.
The ^3S is lower by $2K_{12}$ than the E1 ^1S

i) $E_{GS} = -2.83067\ E_h$
$E_{E1} = -1.95568\ E_h$
Absolute value of correlation energy recovered for GS $= 0.08067\ E_h$.
But notice that E_{GS} is worse than the uncorrelated energy obtained at question 11 with an optimal basis function, hence the important role of the choice of the one-particle basis.

j)

$$\Psi_{GS} = 0.953\Psi_1 - 0.303\Psi_2$$
$$\Psi_{E6} = 0.303\Psi_1 + 0.953\Psi_2$$

k) Use a CI expansion with determinants built-up on ψ_1 and ψ_2 of equations (II-20) and (II-21).

14. 226512 CSF's.

VI Bibliography

A. General

An excellent, thorough and rigorous introduction to quantum mechanics and quantum chemistry may be found in:

1. Levine, I.N., *Quantum Chemistry, Fourth Edition,* Prentice Hall, Englewood Cliffs, NJ, **1991**.

More advanced treatments will be found in
2. Szabo, A.; Ostlund, N.S., *Modern Quantum Chemistry Introduction to Advanced Electronic Structure Theory*, First Edition Revised, McGraw-Hill, New York, **1989**.

3. Mc Weeny, R., *Methods of Molecular Quantum Mechanics, Second Edition,* Acad. Press, London, **1989**.

A very concise, non-mathematical, applications-oriented introduction (110 pages) to AICQC is:
4. Hinchcliffe, A., *Computational Quantum Chemistry,* Wiley, Chichester, **1988**.

An article entitled *An Experimental Chemist's Guide to ab Initio Quantum Chemistry,* introducing the non-specialist to the jargon and methods of AICQC, has been published recently:
5. Simons, J. J. Phys. Chem., **1991**, 95, 1017.

Modern methods of AICQC are developed at length in:
6. Lawley, K.P., ed., *Ab Initio Methods in Quantum Chemistry,* Adv. Chem. Phys. , vols. 67 and 69, **1987**.

A huge number of numerical results (mainly in the field of organic chemistry) derived from AICQC calculations will be found in :
7. Hehre, W.J.; Radom, L.; Schleyer, P.v. R.; Pople, J.A., *Ab Initio Molecular Orbital Theory,* Wiley, New-York, **1986**.

Some quantum chemistry textbooks are critically reviewed in:
8. Strich, A. Intern. J. Quantum Chem., **1991**, 40, 719.

9. Strich, A. Intern. J. Quantum Chem., **1992**, 42, 000.

B. Specific (these are referred to in Sections I, II and V)

1. Quantum Chemistry Program Exchange, Indiana University, Bloomington, Ind. , USA.

2. Computer Physics Communications Program Library, Queen's University of Belfast, N. Ireland.

3. Pekeris, C.L. Phys. Rev., **1959** , 115, 1216.

4. Clementi, E.; Roetti, C. At. Data and Nucl. Data Tables, **1974** , 14, 177.

5. Strich, A. , SCF2EL: a personal computer SCF program for 2-electrons systems in STF basis, **1985**, unpublished.

6. Froese Fischer, C., *The Hartree Fock Method for Atoms A Numerical Approach,* Wiley, New-York, **1977**.

7. Olsen, J.; Jorgensen, P.; Simons, J. Chem. Phys. Lett., **1990** , 169, 463.

8. Walch, S.P.; Bauschlicher, C.W.; Roos, B.O.; Nelin, C.J. Chem. Phys. Lett., **1983** , 103, 175.

9. Leininger, T.; Strich, A. Proceedings of "5èmes Journées sur les Méthodes Informatiques dans l'Enseignement de la Chimie", Mulhouse (France), p. 49, September **1991.**

POSTERS

FORMATION AND PHOTOCHEMICAL DISSOCIATION OF THE 1:1 ADDUCT BETWEEN (η^5-CYCLOPENTADIENYL)(1,2-BENZENEDITHIOLATO)COBALT(III) AND QUADRICYCLANE

H. Hatano, Y. Eguchi, T. Okumachi, T. Fujita, H. Nagao, M. Kajitani, T. Akiyama, and A. Sugimori

Department of Chemistry, Faculty of Science and Technology, Sophia University, Kioi-cho 7-1, Chiyoda-ku, Tokyo 102, Japan.

FT-IR AND LASER-RAMAN SPECTROSCOPIC STUDIES OF TRANSITION METAL (II) PYRAZINE TETRACYANONICKELATE COMPLEXES

S. Akyüz[a], T. Akyüz[b], J. Eric[c], D. Davies[c]

[a] Ondokuz Mayis University, Physics Dept., 55139 Samsun, Turkey
[b] Mineral Research and Exploration Institute (M.T.A.), Ankara, Turkey
[c] Lancaster University, Enviromental Science Division , LA1 4YQ, Lancaster, U.K.

PHOTOINDUCED INTRAMOLECULAR ENERGY AND ELECTRON TRANSFER PROCESSES IN LONG CHAIN POLYNUCLEAR COMPLEXES CONTAINING RHENIUM(I) AND RUTHENIUM(II) POLYPYRIDINE MOIETIES.

C.A Bignozzi, R. Argazzi, C. Chiorboli, F. Scandola

Dipartimento di Chimica, Centro di Fotochimica CNR, Universita di Ferrara, Italy.

MEDIUM EFFECTS IN ELECTRON TRANSFERS

L. G. Arnaut

Chemistry Department, University of Coimbra, 3000 Coimbra, Portugal.

TWISTED QUATERPYRIDINE COMPLEXES OF RUTHENIUM (II). A SPECTROSCOPIC AND PHOTOPHYSICAL STUDY

E. Amouyal[a], D. Azhari[a], J. M. Lehn[b], R. Ziessel[b]

[a] Laboratoire de Physico-Chimie des Rayonnements, Bât 350, Université Paris Sud, 91105 Orsay, France
[b] Institut Le Bel, Université Louis Pasteur, 67008 Strasbourg, France.

$Ru(bipy)_2dppz^{2+}$: A MOLECULAR "LIGHT SWITCH" FOR DNA AND SDS MICELLES.

J. K. Barton[ac], J.C. Chambron[b], A. E. Friedman[ac], J. P. Sauvage[b], N. J. Turro[c]
[a] Division of Chemistry and Chemical Engineering, California Institute of Technology, Pasadena, CA 91125, USA.
[b] Université Louis-Pasteur, Institut de Chimie, 1, rue Blaise Pascal, 67008 Strasbourg, France.
[c] Department of Chemistry, Columbia University, New York, NY 10027, USA.

SYNTHESIS AND CHARACTERISATION OF NEW TRIADS WITH FERROCENE AS AN ELECTRON DONOR GROUP

C. Coudret, J.-C. Chambron, J.-P. Sauvage
Université Louis Pasteur, Institut de Chimie, 1, rue Blaise Pascal, 67008 Strasbourg, France.

SPECTROSCOPIC STUDIES OF METAL COMPLEXES WITH SULPHUR CONTAINING AMINO ACID AND PEPTIDE.

S. K. Datta, S. Chakraborty and H. G. Mukherjee
Department of Pure Chemistry, Calcutta University, 92, A.P.C. Road, Calcutta - 700 009, India.

DYNAMICAL CONTROL OF PHOTOCHEMICAL REACTIONS IN ORGANOMETALLIC COMPLEXES: SELECTIVE BOND BREAKING IN $HCo(CO)_4$

E. Kolba[a], J. Manz[a], and C. Daniel[b]
[a] University of Würzburg, 8700 Würzburg, FRG,
[b] Laboratoire de Chimie Quantique, UPR 139 du CNRS, 67008 Strasbourg, France.

PHOTOINDUCED PROCESSES IN DIADS AND TRIADS CONTAINING A Ru(II) OR Os(II) PHOTOSENSITIZER LINKED TO ELECTRON DONOR AND ACCEPTOR GROUPS

L. De Cola[a], F. Barigelletti[b], L. Flamigni[b], V. Balzani[a], J.-P. Collin[c], S. Guillerez[c], J.-P. Sauvage[c]
[a] Dip. di Chimica "G. Ciamician" University of Bologna, Italy
[b] Istituto FRAE-CNR, Bologna, Italy
[c] Laboratoire de Chimie Organo-Minérale, Institut de Chimie, Université Louis Pasteur, Strasbourg, France.

EXCIPLEXES BETWEEN POLYCYCLIC AROMATIC HYDROCARBONS AND Ag^+ - CATIONS.

H. Dreeskamp
Institut für Physikalische und Theoretische Chemie der Technischen Universität Braunschweig, Hans-Sommer-Str. 10, D-33 Braunschweig, Germany.

PHOTOADDUCT FORMATION UPON THE IRRADIATION OF Ru-TRISTETRAAZAPHENANTHRENE [$Ru(TAP)_3^{2+}$] - THE ROLE OF ELECTRON TRANSFER.

M. M. Feeney[a], J. M. Kelly[a], A. B. Tossi[a], A. Kirsch-De Mesmaeker[b] and J.-P. Lecomte[b]
[a] Chemistry Department, Trinity College, University of Dublin, Ireland.
[b] Department of Organic Chemistry, Free University of Brussels, Belgium.

THE PHOTOCHEMICAL REACTIVITY OF HIGH NUCLEARITY Os-Hg CLUSTERS

L. H. Gade[a], B. F. G. Johnson[a], J. Lewis[a], T. Kotch[b], A. J. Lees[b]
[a] University Chemical Laboratory, Lensfield Rd., Cambridge CB2 1EW, U.K.
[b] Dept. of Chemistry, State University of New York, Binghamton, N.Y. 13901, U.S.A.

PHOTOINDUCED ELECTRON TRANSFER ACROSS A RIGID GOLD (III)-ZINC(II) OBLIQUE BIS-PORPHYRIN

V. Heitz, J. P. Sauvage
Laboratoire de Chimie Organo-Minérale, UA 422, Institut de Chimie, 1 rue Blaise Pascal, 67008 Strasbourg, France.

THE CONFORMATIONAL DEPENDENCE OF THROUGH-BOND INTERACTION IN RIGID, NON-SYMMETRICAL DONOR-ACCEPTOR MOLECULES

W. Hielkema, B. Krijnen and J. W. Verhoeven
Laboratory of Organic Chemistry, The University of Amsterdam, Nieuwe Achtergracht 129, 1018 WS Amsterdam, The Netherlands.

SINGLED EXCITED STATE DYNAMICS OF OLIGOTHIENYLS: EXPERIMENTAL AND THEORETICAL STUDIES

J. L. Houben[a], R. Cimiraglia[b], R. Rossi[ac], A. Carpita[c] and M. Ciofalo[c]
[a] Istituto di Chimica Quantistica ed Energetica Molecolare,
[b] Dipartimento di Chimica e Chimica Industriale, Sezione di Chimica Fisica,
[c] Dipartimento di Chimica e Chimica Industriale, Sezione di Chimica Organica, Università di Pisa, Via Risorgimento 35, I-56126 Pisa, Italy.

ORGANOMETALLIC PHOTOCHEMICAL REACTIONS IN SUPERCRITICAL FLUIDS

M. Jobling, S. M. Howdle and M. Poliakoff
Department of Chemistry, University of Nottingham, Nottingham, NG7-2RD, England.

PROTOLYTIC REACTIONS OF RADICAL ION PAIRS: INFLUENCE ON RADICAL YIELD

A. D. Klimov and S. F. Lebedkin
Inst. of Energy Problems of Chemical Physics, USSR Academy of Sciences, 117829 Moscow, USSR.

KINETICS OF THE PHOTOCHEMICAL DISPROPORTIONATION OF $(CO)_4CoRe(CO)_3(\alpha-$diimine) COMPLEXES (α - dimine = 2,2'bipyridine, pyridine-N-i-propyl-2-carbaldehydeimine)

H. Knoll, W. G. J. de Lange, H. Hennig, D. J. Stufkens
Anorganisch Chemisch Laboratorium, J. H. van't Hoff Instituut, Universiteit van Amsterdam, Nieuwe Achtergracht 166, 1018 WV Amsterdam, The Netherlands.

ENERGY TRANSFER FROM $O_2(^1\Delta_g)$ TO CAROTENOIDS : A VARIABLE TEMPERATURE STUDY

P. F. Conn, C. R. Lambert and T. G. Truscott
Dept. of Chemistry, Keele University, Staffs. ST5 5BG, UK.

QUANTUM YIELD WAVELENGTH DEPENDENCE OF IODINE PHOTODISSOCIATION IN BENZENE AND HEXANE STUDIED BY TIME-RESOLVED THERMAL LENS METHOD

S. F. Lebedkin and A. D. Klimov
Inst. of Energy Problems of Chemical Physics, USSR Academy of Sciences, 117829 Moscow USSR.

A TOPOLOGICAL ANALYSIS OF BOND ACTIVATIONS IN PROTONATED ALCOHOLS AND FLUOROALKANES

M. Esseffar[a], M. El Mouhtadi[a], V. Lopez[b] and M. Yanez[b]
[a] Departement de Chimie, Univ. Cadi-Ayyad, Marrakech, Morocco
[b] Departamento de Quimica, C-14, Univ. Autonoma de Madrid, Cantoblanco, 28049-Madrid, Spain.

STRUCTURE, ENERGETICS AND BOND ACTIVATION OF Li^+, Na^+ and Al^+ COMPLEXES

M. Alcami, A. Luna, O. Mo and M. Yanez
Departamento de Quimica, C-14, Univ. Autonoma de Madrid, Cantoblanco, 28049-Madrid, Spain.

KINETICS OF SINGLET OXYGEN ($^1O_2$) QUENCHING: AN EPR STUDY

L.-Y. Zang, B. R. Misra and H. P. Misra
Department of Biomedical Sciences, Virginia-Maryland Regional College of Veterinary Medicine, Virginia Polytechnic Institute and State University, Blacksburg, VA 24061, USA.

PHOTOCHEMISTRY AND PHOTOPHYSICS OF SOME LOW-VALENT TRANSITION METAL α-diimine COMPLEXES

H. A. Nieuwenhuis, D. J. Stufkens, Ad Oskam
Anorganisch Chemisch Laboratorium, University of Amsterdam, J.H. van't Hoff Institute, Nieuwe Achtergracht 166, 1018 WV Amsterdam, The Netherlands.

FAST KINETICS AND SPECTRA OF THE 16 ELECTRON FRAGMENT: $Ru(dmpe)_2$

R. Osman, R. N. Perutz, D. Rooney and M. K. Whittlesey
University of York, Heslington, York, YO1 5DD, UK.

SYNTHESIS AND INTRAMOLECULAR CHARGE SEPARATION OF CONFORMATIONALLY RESTRICTED PORPHYRIN ARRAYS

A. Osuka
Department of Chemistry, Faculty of Science, Kyoto University, Kyoto 606, Japan.

PHOTOCHEMICAL CLEAVAGE OF THE METAL-HYDROGEN BOND IN ALUMINIUM PORPHYRINS: INSIGHTS FROM AF INITIO CALCULATIONS.

M. - M. Rohmer
Laboratoire de Chimie Quantique, UPR 139 CNRS, Strasbourg, France.

ENERGY TRANSFER AND PHOTOINDUCED ELECTRON TRANSFER IN MICROSTRUCTURED COPOLYMER SYSTEMS.

S. Salhi[a], C. Bied-Charreton[b], J. A. Delaire[a], J. Faure[a], R. Pansu[a], M. Sanquer-Barrié[a]
[a] Université de Paris-Sud, Centre d'Orsay, Bât. 350,
[b] Université de Paris-Sud, Centre d'Orsay, Bât. 410,
91405 Orsay, France.

A LASER-PHOTOLYSIS STUDY OF HEMOGLOBIN INTERACTION WITH THE CUPRIC IONS

R. M. Serbanescu[a], L. Frunza[b], G. Turcu[a]
[a] Department of Biophysics, Faculty of Physics, University of Bucharest, Romania
[b] Institute of Physics and Technology of Materials, Bucharest-Magurele, Romania.

COMPARISON OF FLEXIBLY AND RIGIDLY BRIDGED DONOR-ACCEPTOR SYSTEMS.

T. Scherer, R. J. Willemse and J. W. Verhoeven
Laboratory of organic Chemistry, The University of Amsterdam, Nieuwe Achtergracht 129, 1018 WS Amsterdam, The Netherlands.

ELECTRON-TRANSFER REACTIONS IN PROTEINS - A CALCULATION OF ELECTRONIC COUPLING

P. Siddarth and R.A. Marcus
Arthur Amos Noyes Laboratory of Chemical Physics, 127-72, California Institute of Technology, Pasadena, CA 91125, USA.

THE EXCITED STATE BEHAVIOUR OF SOME MIXED LIGAND POLYPYRIDYLRUTHENIUM(II) COMPLEXES

S. A. Adeyemi[abc], H. D. Burrows[b], S. J. Formosinho[b], M. G. M. Miguel[b], M. I. Silva[b], T. J. Meyer[c] and Z. Murtaza[c]
[a] Chemistry Department, Obafemi Awolowo University, Ile-Ife, Nigeria
[b] Departamento de Quimica, Universidade de Coimbra, 3049 Coimbra, Portugal
[c] Department of Chemistry, University of North Carolina, Chapel Hill, NC 27514, U.S.A.

INFLUENCE OF MOLECULAR STRUCTURE AND SOLVENT POLARITY ON THE RADIATIONLESS DECAY OF INTRAMOLECULAR EXCIPLEXES AND RADICAL ION PAIRS

P. Van Haver, N. Helsen, S. De Paemelaere, M. Van der Auweraer and F. C. De Schryver
Chemistry Department K. U. Leuven, Celestijnenlaan 200F, 3001 Leuven, Belgium.

THEORETICAL SPECTROSCOPY OF DIFLUOROMETHYLENE

R. Vetter[a], W. Reuter[b] and S. D. Peyerimhoff [b]
[a] Zentralinstitut für physikalische Chemie, Berlin, Germany.
[b] Lehrstuhl für Theoretische Chemie, Universirät Bonn, Germany.

HOW DOES ISOTOPIC SUBSTITUTION AFFECT ELECTRON-AFFINITY? PM3-CALCULATIONS ON BENZENE AND ITS RADICAL-ANION

H. Zuilhof and G. Lodder
Gorlaeus Laboratories, University of Leiden,
P.O.Box 9502, 2300 RA Leiden, The Netherlands.

INDEX

D

Q

R

S

LIST OF CONTRIBUTORS

BALZANI V.
Dipartimento di Chimica "G. Ciamician" dell'Università, 40126 Bologna, Italy

BIGNOZZI C.A.
Dipartimento di Chimica dell'Università, Centro di Fotochimica C.N.R., 44100 Ferrara, Italy

CAMPAGNA S.
Dipartimento di Chimica Inorganica e Struttura Molecolare dell'Università, 98166 Messina, Italy

CAPUANO V.L.
Department of Chemistry and Biochemistry and Center for Fast Kinetics Research, The University of Texas, Austin, Texas 78712, USA

CHIORBOLI C.
Dipartimento di Chimica dell'Università, Centro di Fotochimica C.N.R., 44100 Ferrara, Italy

DENTI G.
Laboratorio di Chimica Inorganica, Istituto di Chimica Agraria dell'Università, 56100 Pisa, Italy

Van der GRAAF T.
Anorganisch Chemisch Laboratorium, University of Amsterdam, Nieuwe Achtergracht 166, 1018 WV Amsterdam, The Netherlands

GREVELS F.W.
Max-Planck-Institut für Strahlenchemie, Stiftstrasse 34-36, D-4330 Mülheim an der Ruhr, Germany

HARRIMAN A.H.
Department of Chemistry and Biochemistry and Center for Fast Kinetics Research, The University of Texas, Austin, Texas 78712, USA

INDELLI M.T.
Dipartimento di Chimica dell'Università, Centro di Fotochimica C.N.R., 44100 Ferrara, Italy

JOHNSON M.R.
Department of Chemistry and Biochemistry and Center for Fast Kinetics Research, The University of Texas, Austin, Texas 78712, USA

KUBO Y.
Department of Chemistry and Biochemistry and Center for Fast Kinetics Research, The University of Texas, Austin, Texas 78712, USA

MAGDA D.J.
Department of Chemistry and Biochemistry and Center for Fast Kinetics Research, The University of Texas, Austin, Texas 78712, USA

MATHIS P.
Département de Biologie Cellulaire et Moléculaire, Section de Bioénergétique, Bâtiment 532, C.E. Saclay, 91191 Gif-sur-Yvette Cedex, France

MICHEL-BEYERLE M.E.
Institut für Physikalische und Theoretische Chemie, Technische Universität München, D-8046 Garching, Germany

OGRODNIK A.
Institut für Physikalische und Theoretische Chemie, Technische Universität München, D-8046 Garching, Germany

OSKAM A.
Anorganisch Chemisch Laboratorium, University of Amsterdam, Nieuwe Achtergracht 166, 1018 WV Amsterdam, The Netherlands

PADDON-ROW M.N.
Department of Chemistry, University of New south Wales, P.O. Box 1, Kensington, N.S.W. 2033, Australia

PEYERIMHOFF S.D.
Institut für Physikalische und Theoretische Chemie, Universität Bonn, Wegelerstrasse 12, D-5300 Bonn 1, Germany

RAMPI M.A.
Dipartimento di Chimica dell'Università, Centro di Fotochimica C.N.R., 44100 Ferrara, Italy

SCANDOLA F.
Dipartimento di Chimica dell'Università, Centro di Fotochimica C.N.R., 44100 Ferrara, Italy

SCHNEIDER S.
Institut für Physikalische und Theoretische Chemie der Universität Erlangen-Nürnberg, Egerlandstr. 3, D8520 Erlangen, FRG

SERRONI S.
Laboratorio di Chimica Inorganica, Istituto di Chimica Agraria dell'Università, 56100 Pisa, Italy

SESSLER J.L.
Department of Chemistry and Biochemistry and Center for Fast Kinetics Research, The University of Texas, Austin, Texas 78712, USA

STRICH A.
Laboratoire de Chimie Quantique, UPR 139 du CNRS, Université Louis Pasteur, 4 rue Blaise Pascal, F-67000 Strasbourg, France

STOR G.J.
Anorganisch Chemisch Laboratorium, University of Amsterdam, Nieuwe Achtergracht 166, 1018 WV Amsterdam, The Netherlands

STUFKENS D.J.
Anorganisch Chemisch Laboratorium, University of Amsterdam, Nieuwe Achtergracht 166, 1018 WV Amsterdam, The Netherlands

TRAMER A.
Laboratoire de Photophysique Moléculaire CNRS, Université de Paris-Sud, 91405-Orsay, France

TURNER J.J.
Department of Chemistry, University of Nottingham, Nottingham, NG7 2RD, UK

VEILLARD A.
UPR 139 du CNRS, Institut Le Bel, 4, rue Blaise Pascal, 67000 Strasbourg, France

VERHOEVEN J.W.
Laboratory of Organic Chemistry, University of Amsterdam, Nieuwe Achtergracht 129, 1018 WS Amsterdam, The Netherlands

WARMAN J.M.
Interfaculty Reactor Institute, Delft University of Technology, Mekelweg 15,2629 JB Delft, The Netherlands

LIST OF PARTICIPANTS

Takeo AKIYAMA
Department of Chemistry
Faculty of Science and Technology
Sophia University
Kioi-cho 7-1, Chiyoda-ku,
Tokyo 102
JAPAN

Pericles AKRIVOS
Aristotelian University
Chemistry Dpt.
General and Inorg. Chemistry Lab
P.O.Box 135
GR-540 06 Thessaloniki
GREECE

Sevim AKYUZ
19 Mayis University
Dept. of Physics,
Fen-Ed Fak ltesi
Samsun
TURKEY

Tanil AKYUZ
Maden Tetkik Arama Ens.,
MAT Dairesi Labratuvarlar,
Balgat
Ankara
TURKEY

Samoela ANDRIANIRINAHARIVELO
Laboratoire de Photochimie
Moléculaire et Macromoléculaire
Université Blaise Pascal
UA CNRS 433
63177 Aubière
FRANCE

Roberto ARGAZZI
Centro di Fotochimica C.N.R.
Dipart. di Chimica dell Universita
Via L. Borsari,46
44100 Ferrara
ITALY

Luis ARNAUT
University of Coimbra
Chemistry Department
3000 Coimbra
PORTUGAL

Driss AZHARI
Laboratoire de Physico-Chimie
des Rayonnements
URA 75 du CNRS
Université Paris-Sud, Bât 350
91405 Orsay
FRANCE

Vincente BALZANI
Universita degli studi di Bologna
Dipart. di Chimica "G. Ciamician"
Via Selmi 2
40126 Bologna
ITALY

Jean-Claude CHAMBRON
Laboratoire Chimie Organo-Minérale
Institut de Chimie
Université Louis Pasteur
BP 296
67008 Strasbourg
FRANCE

Jaime E. COMBARIZA
Institut für Physikalishe Chemie
Universitaet Würzburg
Marcusstrasse 9-11
8700 Würzburg
GERMANY

Christophe COUDRET
Laboratoire Chimie Organo-Minérale
Institut de Chimie
Université Louis Pasteur
BP 296
67008 Strasbourg
FRANCE

Chantal DANIEL
Laboratoire de Chimie Quantique
Institut Le Bel
Université Louis Pasteur
4, rue Blaise Pascal
67000 Strasbourg
FRANCE

Samir Kanti DATTA
B.E.S. College
5, Elgin Road
Calcutta 700020
INDIA

Luisa DE COLA
Universita degli studi di Bologna
Dipart. di chimica "G. Ciamician"
Via Selmi 2
40126 Bologna
ITALY

Maria Teresa Pitta DE LACERDA-AROSO
Universidade do Minho
Departamento de Fisica
Largo do Paço
4719 Braga Codex
PORTUGAL

Herbert DREESKAMP
Inst. f. Physikalische und
Theoretische Chemie
TU Braunschweig H. Sommer Str.10
3300 Braunschweig
GERMANY

Martin FEENEY
University of Dublin
Trinity College
Dublin 2
IRELAND

Lutz Hans GADE
Cambridge University
University Chemical Laboratory
Lensfield RD,
Cambridge CB2 IEW,
U.K.

Peter GEDECK
Institut für Physikalische Chemie I
Universitaet Erlangen-N rnberg
Egerland str. 3
W-8520 Erlangen
GERMANY

F.W. GREVELS
Max Plank Inst. für Strahlenchem.
Stiftstrasse 34-36
D-4330 Mulheim a.d. Ruhr
GERMANY

Frantisek HARTL
J. Heyrovsky Institute of
Physical Chemistry and
Electrochemistry
Dolejskova 3
182 23 Prague
TCHECOSLOVAKIA

Valérie HEITZ
Laboratoire Chimie Organo-Minérale
Institut de Chimie
Université Louis Pasteur
BP 296
67008 Strasbourg
FRANCE

Wim HIELKEMA
Laboratory of Organic Chemistry
The University of Amsterdam
Nieuwe Achtergracht 129
1018 WS Amsterdam
THE NETHERLANDS

J.L. HOUBEN
Consiglio Nationale delle Ricerche
Istituto di chimica quantistica
ed energetica molecolare
Via Risorgimento 35
56100 Pisa
ITALY

Margaret JOBLING
Nottingham University
University Park
Nottingham NG7-2RD
U.K.

Alex D. KLIMOV
Inst. for Energy Problems
of Chemical Physics
USSR Academy of Sciences
11 78 29 Moscow
USSR

Helmut KNOLL
Institut für Theoretische Chemie
de Universitaet Leipzig
Linnestr. 2
0-7010 Leipzig
GERMANY

Elise KOCHANSKI
Laboratoire de Chimie Théorique
Institut de Chimie
Université Louis Pasteur
BP 296
67008 Strasbourg
FRANCE

Christopher LAMBERT
Department of Chemistry
University of Keele
Keele Staffordsh. ST5 5BG
U.K.

Sergey F LEBEDKIN
Inst. for Energy Problems
of Chemical Physics
USSR Academy of Sciences
11 78 29 Moscow
USSR

Jean-Paul LECOMTE
Service Chimie Organique Physique
CP 165
Université Libre de Bruxelles
50 Avenue F.D. Roosevelt
B 1050 Bruxelles
BELGIUM

Visitacion LOPEZ
Departamento di Quimica C14
Universidad Autonoma de Madrid
Cantoblanco
28049 Madrid
SPAIN

Alberto LUNA FERNANDEZ
Universidad Autonoma de Madrid
Departamento de Quimica C-14
Cantoblanco
28049 Madrid
SPAIN

R.A. MARCUS
Caltech
Noyes La. of Chem. Physics, 127-72
Division of Chemistry and
Chemical Engineering
Pasadena CA 91125
USA

Maria C.A.D. MATEUS
Universidade do Algarve
Unidade de Ciencias Exactas E Humanas
Campus de Gambelas
8000 Faro
PORTUGAL

P. MATHIS
Service Biophysique
Centre d'Etudes Nucléaires
Saclay
91191 Gif-sur-Yvette
FRANCE

M.E. MICHEL-BEYERLE
Institut für Physikalische
und Theoretische Chemie
Technische Universitaet München
D-8046 Garching
GERMANY

Hara P MISRA
Department of Biomedical Sciences
College of Veterinary Medecine
VPI & SU
Virginia Tech
Blacksburg VA 24061-0442
USA

Cécile MOUCHERON
Service de Chimie Organique
Faculté des Sciences CP 160
Université Libre de Bruxelles
50, Avenue F.D. Roosevelt
1050 Bruxelles
BELGIUM

Heleen A. NIEUWENHUIS
Universiteit Van Amsterdam
Anorganisch Chemisch Lab.
Nieuwe Achtergracht 166,
1018 WV Amsterdam
THE NETHERLANDS

Zahide Ulya NURULLAHOGLU
Biology Department
University of Selçuk
Kampus / Konya
TURKEY

Alex OGRODNIK
Institut für Physikalische
und Theoretische Chemie
Technische Universitaet München
D-8046 Garching
GERMANY

Robert OSMAN
University of York
Chemistry department
Heslington
York YO15DD
U.K.

Atsuhiro OSUKA
Kyoto University
Faculty of Science
Department of Chemistry
Kyoto 606
JAPAN

S. PEYERIMHOFF
Institut für Theoretische Chemie
der Universitaet Bonn
Wegelerstrasse 12
D-5300 Bonn 1
GERMANY

A. RIERA
Departamento de Quimica C XIV
Universidad Autonoma de Madrid
Ciudad Universit. Canto Blanco
28049 Madrid
SPAIN

Maria del Carmen RIOS
University of Santiago de Compostela
Department of Physical-Chemistry
Faculty of Chemistry
E15706- Santiago de Compostela
SPAIN

Marie-Madeleine ROHMER
Laboratoire de Chimie Quantique
Institut Le Bel
Université Louis Pasteur
4, rue Blaise Pascal
67000 Strasbourg
FRANCE

B. ROOS
Department of Theoretical Chemistry
Chemical Center
P O B 124
S-22100 Lund
SWEDEN

L. SALEM
Laboratoire de Chimie Théorique
Université Paris Sud
Bât 490
91405 Orsay
FRANCE

Samira SALHI
Université Paris-Sud Centre d'Orsay
Laboratoire de Physico-chimie des
Rayonnements. Bât 350
91405 Orsay
FRANCE

J.P. SAUVAGE
Laboratoire Chimie Organo-Minérale
Institut de Chimie
Université Louis Pasteur
BP 296
67008 Strasbourg
FRANCE

F. SCANDOLA
Centro di Fotochimica C.N.R.
Dipart. di Chimica dell Universita
Via L. Borsari, 46
44100 Ferrara
ITALY

Taco SCHERER
Laboratory of Organic Chemistry
The University of Amsterdam
Nieuwe Achtergracht 129
1018 WS Amsterdam
THE NETHERLANDS

S. SCHNEIDER
Universitaet Erlangen-Nürnberg
Institut für Physikalische
und Theoretische Chemie
Egerlandstrasse 3
D 8520 Erlangen
GERMANY

Ruxandra-Mihaela SERBANESCU
c/o Dr. Alexandru Glodeanu
Theoretical Physics Group
Inst. of Phys. and Tech. Materials
PO Box MG-7
76900 Bucharest-Magurele
ROMANIA

Jonathan L. SESSLER
Department of Chemistry
University of Texas
Austin Texas 78712
USA

Prabha SIDDARTH
California Institut of Technology
Department of Chemistry
Pasadena CA 91125
USA

Maria Isilda SILVA
Departamento de Quimica
FCT.U.C.
Universidade de Coimbra
3049 Coimbra
PORTUGAL

Alain STRICH
Laboratoire de Chimie Quantique
Institut Le Bel
Université Louis Pasteur
4, rue Blaise Pascal
67000 Strasbourg
FRANCE

D. STUFKENS
University of Amsterdam
Anorganische Chemische Laboratorium
Nieuve Achtergracht 166
1018 WW Amsterdam
THE NETHERLANDS

A. TRAMER
Labo Photophysique Moléculaire
Université Paris Sud
Bât 213
91405 Orsay
FRANCE

J.J. TURNER
Department of Chemistry
University of Nottingham
Nottingham NG7 2RD
U.K.

Mark VAN DER AUWERAER
Laboratory Molecular Dynamics
And Spectroscopy,
Chemistry Department
K.U.Leuven, Celestijnenlaan 200F
B-3001 Heverlee - Leuven
BELGIUM

Saskia Ingeborg VAN DIJK
Lab. of Organic Chemistry
The University of Amsterdam
Nieuwe Achtergracht 129
1018 WS Amsterdam
THE NETHERLANDS

J.W.M. VAN OUTERSTERP
Universiteit van Amsterdam
Anorganisch Chemisch Laboratorium
Nieuwe Achtergracht 166
1018 WV Amsterdam
THE NETHERLANDS

A. VEILLARD
Laboratoire de Chimie Quantique
Institut Le Bel
4, rue B. Pascal
67100 Strasbourg
FRANCE

J.W. VERHOEVEN
Laboratory of Organic Chemistry
University of Amsterdam
Nieuwe Achtergracht 129
NL 1018 WS Amsterdam
THE NETHERLANDS

Reinhard VETTER
Zentralinstitut für
Physikalische Chemie
Rudower Chaussee 5
0-1199 Berlin
GERMANY

R. VOLTZ
Laboratoire PMOA
23 rue du Loess
BP20
67037 Strasbourg
FRANCE

Han ZUILHOF
Gorlaeus laboratories
Rijksuniversiteit Leiden
afd. Radiochemie
Einsteinweg 5, P.O. Box 9502
NL-2300RA Leiden
THE NETHERLANDS

The manufacturer's authorised representative in the EU is Springer Nature Customer Service Centre GmbH, Europaplatz 3, 69115 Heidelberg, Germany. If you have any concerns regarding our products, please contact ProductSafety@springernature.com

Printed and bound by CPI Group (UK) Ltd, Croydon, CR0 4YY
30/07/2026
02182039-0008